Pervasive Information Security and Privacy Developments:
Trends and Advancements

Hamid Nemati
University of North Carolina, Greensboro, USA

Information Science
REFERENCE

INFORMATION SCIENCE REFERENCE
Hershey · New York

Director of Editorial Content:	Kristin Klinger
Director of Book Publications:	Julia Mosemann
Acquisitions Editor:	Lindsay Johnston
Development Editor:	Julia Mosemann
Publishing Assistant:	Casey Conapitski
Typesetter:	Deanna Jo Zombro
Production Editor:	Jamie Snavely
Cover Design:	Lisa Tosheff

Published in the United States of America by
Information Science Reference (an imprint of IGI Global)
701 E. Chocolate Avenue
Hershey PA 17033
Tel: 717-533-8845
Fax: 717-533-8661
E-mail: cust@igi-global.com
Web site: http://www.igi-global.com

Library of Congress Cataloging-in-Publication Data

Pervasive information security and privacy developments : trends and advancements / Hamid Nemati, editor.
p. cm.
Includes bibliographical references and index.
ISBN 978-1-61692-000-5 (hbk.) -- ISBN 978-1-61692-001-2 (ebook) 1. Information technology--Security measures. 2. Data protection. 3. Computer security. 4. Privacy, Right of. I. Nemati, Hamid R., 1958-
QA76.9.A25P442 2011
005.8--dc22

2010030739

British Cataloguing in Publication Data
A Cataloguing in Publication record for this book is available from the British Library.

This book is dedicated to those who work tirelessly to ensure that our security and privacy are never compromised in the face of a pervasive threat stemming from our information technologies.

Editorial Advisory Board

Table of Contents

Detailed Table of Contents

Virtualization technologies are emerging as a promising solution for managing the rapidly growing complexities of modern distributed ICT infrastructures. However, a mainstream operational concern for these virtualization oriented architectures is to provide efficient security services. Establishment of in-depth security services and trust relationships are the most desirable features for the effective functioning of these systems. This chapter presents a security architecture to address the comprehensive security needs of today's virtualization oriented architectures. The concept of virtualization of security services is introduced so as to have absolute freedom to choose the underlying security mechanisms. This concept of virtualization of security services is realized through distributed virtual engines that enable unification of security service calls according to requirements and not according to the underlying technologies. A configurable mechanism for the invocation of security services is proposed to address the security needs of different kinds of users. This approach permits the evolution of efficient security infrastructure with minimal impact on the resource management functionalities. In this way, users and resource providers can configure the security services according to their requirements and satisfaction level. The resulting extensible set of security services include both core security services and contemporary security services required for the modern virtualization oriented infrastructures.

Security has been considered as an important issue in processor design. Most of the existing designs of security handling assume the chip as a single secure unit. However, such assumption is vulnerable to exposure resulted from a central failure point. This chapter proposes a secure Chip-Multiprocessor architecture (SecCMP) to handle security related problems such as key protection and core authentication in multi-core systems. Matching the nature of multi-core systems, a distributed threshold secret sharing scheme is employed to protect critical secrets. A critical secret (e.g., encryption key) is divided into multiple shares and distributed among multiple cores instead of being kept a single copy in one core that is sensitive to exposure. The proposed SecCMP can not only enhance the security and fault-tolerance in secret protection but also support core authentication. SecCMP is designed to be an efficient and secure architecture for CMPs.

Computer forensics is the discipline that deals with the acquisition, investigation, preservation and presentation of digital evidence in the court of law. Whereas anti-forensics is the terminology used to describe malicious activities deployed to delete, alter or hide digital evidence with the main objective of manipulating, destroying and preventing the creation of evidence. Various anti-forensic methodologies and tools can be used to interfere with digital evidence and computer forensic tools. However, memory-based anti-forensic techniques are of particular interest because of their effectiveness, advanced manipulation of digital evidence and attack on computer forensic tools. These techniques are mainly performed in volatile memory using advanced data alteration and hiding techniques. For these reasons memory-based anti-forensic techniques are considered to be unbeatable. This chapter aims to present some of the current anti-forensic approaches and in particular reports on memory-based anti-forensic tools and techniques.

This chapter investigates the patterns of malicious code attacks based on monthly data of the top 10 virus shares from 1998 to 2005. Three parameters were identified for study, overall pattern of the attack, the number reentries into the top 10 most prevalent attacks, and the maximum percentage share. The dataset was validated by comparing it to an independent dataset that measured the same parameters for a subset of the period of the primary dataset. The effects of malicious code that started before or disappeared outside the collection period were found to not have a significant effect. A multivariate regression analysis showed that the number of entries and the maximum share had a strong relationship with the visible life span. Multivariate cluster analysis was conducted on the reentry parameters and yielded six virus clusters classifications. The high impact viruses, 43 of the 230, are identified and further grouped.

Most access control models have formal access control rules to govern the authorization of a request from a principal. In pervasive and collaborative environments, the behaviors of a principal are uncertain due to partial information. Moreover, the attributes of a principal, requested objects, and contexts of a request are mutable during the collaboration. A variety of such uncertainty and mutability pose challenges when resources sharing must happen in the collaborative environment. In order to address the above challenges, the authors propose a framework to integrate trust management into a usage control model in order to support decision making in an ever-changing collaborative environment. First, a trust value of a principal is evaluated based on both observed behaviors and peer recommendations. Second,

In this chapter, the authors have designed and developed a framework for sharing data in an assured manner in case of emergencies. They focus especially on a need to share environment. It is often required to divulge information when an emergency is flagged and then take necessary steps to handle the consequences of divulging information. This procedure involves the application of a wide range of policies to determine how much information can be divulges in case of an emergency depending on how trustworthy the requester of the information is.

Although various European Union (EU) directives have established the proper legal framework for the provision of a trustful, legally accepted cross border transaction via electronic means, the consumers are still hesitant to use e-commerce. Lack of confidence with regard to the protection of privacy and security of electronic transactions is among the main reasons. This chapter attempts to form the appropriate confidence framework describing a set of privacy and security requirements that must be taken into account for e-ordering systems. In addition, it presents a trustful e-ordering architecture (TOES) that achieves to address these requirements based on extensible markup language (XML), XML cryptography, public key infrastructure (PKI), Web services policy language (WSPL), and Web services. TOES is an open, secure, interoperable, and affordable e-ordering system that respects the EU legislation.

The extent methods largely ignore the importance of integrating security requirements with business requirements and providing built-in steps for dealing with these requirements seamlessly. To address this problem, a new approach to secure network analysis and design is presented. The proposed method, called the SEACON method, provides an integrated approach to use existing principles of information systems analysis and design with the unique requirements of distributed secure network systems. The authors introduce several concepts including security adequacy level, process-location-security matrix, data-location- security matrix, and secure location model to provide built-in mechanisms to capture security needs and use them seamlessly throughout the steps of analyzing and designing secure networks. This method is illustrated and compared to other secure network design methods. The SEACON method is found to be a useful and effective method.

Email spam is the most effective form of online advertising. Unlike telephone marketing, email spamming does not require huge human or financial resources investment. Most existing spam filtering techniques concentrate on the emails' content. However, most spammers obfuscate their emails' content to circumvent content-based spam filters. An integrated solution for restricting spam emails is needed as content analyses alone might not provide a solution for filtering unsolicited emails. Here, the authors present a new method for isolating unsolicited emails. Though spammers obfuscate their emails' content, they do not have access to all the fields in the email header. The classification method in this chapter is based on the path an email traverses instead of content. Overall, this classifier produced fewer false positives when compared to current filters such as SpamAssassin. The authors achieved a precision of 98.65% which compares well with the precisions achieved by SPF, DNSRBL blacklists.

The wide diffusion and usage of social networking Web sites in the last years have made publicly available a huge amount of possible sensitive information, which can be used by third-parties with purposes different from the ones of the owners of such information. Currently, this issue has been addressed by enforcing into Web-based Social Networks (WBSNs) very simple protection mechanisms, or by using anonymization techniques, thanks to which it is possible to hide the identity of WBSN members while performing analysis on social network data. However, the authors believe that further solutions are needed, to allow WBSN members themselves to decide who can access their personal information and resources. To cope with this issue, this chapter illustrates a decentralized security framework for WBSNs, which provide both access control and privacy protection mechanisms. In this system, WBSN members can denote who is authorized to access the resources they publish and the relationships they participate in, in terms of the type, depth, and trust level of the relationships existing between members of a WBSN. Cryptographic techniques are then used to provide a controlled sharing of resources while preserving relationship privacy.

Preface

Let's Call it the "Security and Privacy Decade": The Consequence of Pervasive Information Technology

ABSTRACT

Pervasiveness of information technology (IT) in our lives is well underway when considering IT permeates and impacts all aspects of everything we do to the extent that we would notice its absence more intensely than we would its presence. Advances in computing and communication networks have contributed to this expanding role by making IT cheaper, faster, and more powerful resulting in capabilities that allow us to utilize information technology in ways previously unimaginable. Although this technological revolution has brought us closer and has made our lives easier and more productive, paradoxically, it has also made us more capable of harming one another and more vulnerable to be harmed by each other. Our vulnerabilities are the consequence of our capabilities. What are the security and privacy implication of this expanding role of IT? In this chapter, we discuss how the role of information technology is changing and how this change impacts information security and privacy. The evolving nature of information security and privacy brings additional challenges and opportunities that we, as a society, need to understand and prepare for in order to take full advantage of advances in information technology.

INTRODUCTION

The role of Information Technology (IT) in our lives is expanding. Pervasiveness of information technology (IT) in our lives is well underway when considering IT permeates and impacts all aspects of everything we do to the extent that we would notice its absence more intensely then we would its presence. Advances in computing and communication networks have contributed to this expanding role by making IT cheaper, faster, and more powerful resulting in capabilities that allow us to utilize information technology in ways previously unimaginable. We are the first generation of humans where the capabilities of the technologies that support our information processing activities are truly revolutionary and far exceed those of our forefathers. We speak of the age we line in as the "*information age*" and our society as "*information society*". The emergence of the society based on information signals a transition toward a new society based on the production and exchange of information as opposed to physical goods (Stephanidis et al., 1984). Information Society refers to the new socioeconomic and technological paradigms that affect human activities, individual behaviors, our collective consciousness, and our economic and social environments. The information age has important consequences for our lives as well. Essentially, it has ushered a new range of emerging computer-mediated activities that have revolutionized the way we live and interact with one another (Mesthene 1968; Nardi, 1996; Stephanidis et al., 1984). More people are employed generating, collecting, handling, processing and distributing information than any other profession and in any other time (Mason 1986). IT has made us more productive in our workplaces, has brought us closer, transformed our lives and has helped in redefining who we are as humans. We are able to communicate more freely and effortlessly with one another, make more informed decisions, and have a higher standard of living, all, resulting from advances in Information Technologies (IT). Its impacts can be felt in the ways in which we relate, interact, and communicate not just with one another but also the way we interact with the technology itself. To some extent, information technologies have become "*information appliances*". IT has also redefined our relationships with businesses we interact with and governmental agencies representing us. As a result, our world has been altered so irrevocably that we are no longer able to conduct our lives without it. Yet, many experts believe that we have only seen the tip of the iceberg. We are on the verge of the biggest societal transformation in the history of mankind traced directly to advances in the information technology. This transformation will most likely create new opportunities and challenges we have yet to fathom. But the greatest impact of all will be in the way we perceive and identify ourselves as individuals.

Information defines us. It defines the age we live in and the societies we inhibit. Information is the output of our human intellectual endeavors which inherently defines who we are as humans and how we conduct our lives. New technologies make possible what was not possible before. This alters our old value clusters whose hierarchies were determined by range of possibilities open to us at the time. By making available new options, new technologies can and will lead to a restructuring of the hierarchy of values (Mesthene, 1968). Mason argues that unique challenges facing our information society are the result of the evolving nature of information itself. Mason argues that in this age of information, a new form of social contract is needed in order to deal with the potential threats to the information which defines us. Although this technological revolution has brought us closer and has made our lives easier and more productive, paradoxically, it has also made us more capable of harming one another and more vulnerable to be harmed by each other. Our vulnerabilities are the consequence of our capabilities. Mason (1986) states "Our moral imperative is clear. We must insure that information technology, and the information it handles, are used to enhance the dignity of mankind. To achieve these goals we much formulate a new social contract, one that insures everyone the right to fulfill his or her own human potential." (Mason, 1986, p 26). In light of the Aristotelian notion of the intellect, this new social contract has a profound implication in the way our society views information and the technologies that support them. For IT to enhance the "*human dignity*", it should assist humans in exercising their intellects ethically. But is it possible to achieve this without assuring the trustworthiness of information and the integrity of the technologies we are using? Without security that guarantees the trustworthiness of information and the integrity of our technologies, appropriate uses of that information cannot be realized. This implies that securing information and privacy of that information are inherently intertwined and should be viewed synergistically. As a result, Information security and privacy have been viewed as the foremost areas of concern and interest by academic researchers and industry practitioners from a wide spectrum of different disciplines alike. We define Information security and privacy as an all encompassing term that refers to all activities needed to assure privacy of information and security of systems that support it in order to facilitate its use.

We have entered an exciting period of unparallel interest and growth in research and practice of all aspects of information security and privacy. Information security and privacy is the top IT priority facing organizations. According to the 20th Annual Top Technology Initiatives survey produced by the American Institute of Certified Public Accountants (AICPA, 2009) information security tops the list of ten most important IT priorities (http://infotech.aicpa.org/Resources/). According to the survey results, for the seventh consecutive year, Information Security is identified as the technology initiative expected to have the greatest impact in the upcoming year for organizations and is thus ranked as the top IT priority for organizations. This is first year that information privacy has shot up to second most important IT priority. This is very significant that top two issues are information security and privacy and the six out of the top ten technology initiatives discussed in this report are issues related to information security and privacy (AICPA 2009). The interest in all aspects of information security and privacy is also manifested by the recent plethora of books, journal articles, special issues, and conferences in this area. This has resulted in a number of significant advances in technologies, methodologies, theories and practices of information security and privacy. These advances, in turn, have fundamentally altered the landscape of research in a wide variety of disciplines, ranging from information systems, computer science and engineering to social and behavioral sciences and the law. This confirms what information security and privacy professionals and researchers have known for a long time that information security and privacy is not just a "technology" issue any more. It impacts and permeates almost all aspects of business and the economy.

Until recently, information security and privacy were exclusively discussed in terms of mitigating risks associated with data and the organizational and technical infrastructure that supported it. With the emergence of the new paradigm in information technology, the role of information security and ethics has evolved. As Information Technology and the Internet become more and more ubiquitous and pervasive in our daily lives, a more thorough understanding of issues and concerns over the information security and ethics is becoming one of the hottest trends in the whirlwind of research and practice of information technology. This is chiefly due to the recognition that whilst advances in information technology have made it possible for generation, collection, storage, processing and transmission of data at a staggering rate from various sources by government, organizations and other groups for a variety of purposes, concerns over security of what is collected and the potential harm from personal privacy violations resulting from their unethical uses have also skyrocketed. Therefore, understanding of pertinent issues in information security and security vis-à-vis technical, theoretical, managerial and regulatory aspects of generation, collection, storage, processing, transmission and ultimately use of information are becoming increasingly important to researchers and industry practitioners alike. Information security and privacy have been viewed as one of the foremost areas of concern and interest by academic researchers and industry practitioners from diverse fields such as engineering, computer science, information systems, and management. Recent studies of major areas of interest for IT researchers and professionals point to information security and privacy as one of the most pertinent.

Data, Data, Data Everywhere

A byproduct of pervasiveness of Information Technology is the amazingly large amount of data currently being generated. This data needs to be stored securely and privately. According to IBM (IBM 2010), worldwide data volumes are currently doubling every two years. Data experts estimate that in 2002 the world generated 5 exabytes of data. This amount of data is more than all the words ever spoken by human beings. The rate of growth is just as staggering – the amount of data produced in 2002 was up 68% from just two years earlier. The size of the typical business database has grown a hundred-fold during the past five years as a result of internet commerce, ever-expanding computer systems and mandated recordkeeping by government regulations. The rate of growth in data has not slowed. International Data Corporation (IDC) estimates that the amount of data generated in 2009 was 1.2 million Petabytes (IDC, 2010). (A Petabyte is a million gigabytes.) (IDC Report, 2010). Although this seems to be an astonishingly large amount of data, it is paled in compression to what IDC estimates that amount to be in 2020. IDC estimates that the amount of data generated in 2010 will 44 times as much as this year to an incomprehensible amount of 35 Zettabytes (A Zettabyte is 1 trillion gigabytes). IDC reports that by 2020, we will generate 35 trillion gigabytes of data. To better grasp how much data this is, consider the following: if one byte of data is the equivalent of this dot (•), the amount of data produced globally in 2009 would equal the diameter of 10,000 suns.

One of the reasons for this astonishingly large growth, according a survey by US Department of Commerce, is that an increasing number of Americans are going online and engaging in several online activities, including online purchases, conducting banking online, engaging in commerce, and interacting socially. The growth in Internet usage and e-commerce has offered businesses and governmental agencies the opportunity to collect and analyze information in ways never previously imagined. "Enormous amounts of consumer data have long been available through offline sources such as credit card transactions, phone orders, warranty cards, applications and a host of other traditional methods. What the digital revolution has done is increase the efficiency and effectiveness with which such information can be collected and put to use" (Adkinson, Eisenach, & Lenard, 2002).

Almost everything that we do in our daily lives can generate a digital footprint. Whether we are using credit cards, surfing the Internet or viewing a YouTube video, we are generating data. IDC senior vice president, John Gantz states: "About half of your digital footprint is related to your individual actions—taking pictures, sending e-mails, or making digital voice calls. The other half is what we call the 'digital shadow'—information about you—names in financial records, names on mailing lists, web surfing histories or images taken of you by security cameras in airports or urban centers. For the first time your digital shadow is larger than the digital information you actively create about yourself." Our digital shadow, the sum of all the digital information generated about us on a daily basis, now exceeds the amount of digital information we actively create ourselves (IDC, 2010). This digital footprint including our digital shadow represents us, as humans, it represents who we are, and how we conduct our lives. It needs to be secured, protected, and managed appropriately.

This proclamation about data volume growth is no longer surprising, but continues to amaze even the experts. For businesses, more data isn't always better. Organizations must assess what data they need to collect and how to best leverage it. Collecting, storing and managing business data and associated databases can be costly, and expending scarce resources to acquire and manage extraneous data fuels inefficiency and hinders optimal performance. The generation and management of business data also loses much of its potential organizational value unless important conclusions can be extracted from it quickly enough to influence decision making while the business opportunity is still present. Managers must rapidly and thoroughly understand the factors driving their business in order to sustain a competitive advantage. Organizational speed and agility supported by fact-based decision making are critical to ensure an organization remains at least one step ahead of its competitors. Several studies ((Brancheau, Janz, & Wetherbe, 1996); (Niederman, Brancheau, & Wetherbe, 1991)) have shown that data has been ranked as one of the top priorities for IT executives as an organizational resource. Similar research ((Rockart & DeLong, 1988); (Watson, Rainer Jr, & Koh, 1991)) has also revealed that data is an important part of a decision support system since it forms the basis of the information that is delivered to decision makers. The formidable challenge facing organizations involves the collection, management, and presentation of its data to enable management to make well-informed and timely decisions. With the emergence of web technologies, the collection and storage of data, both internal and external to an organization, has increased dramatically. In spite of this enormous growth in enterprise databases, research from IBM reveals that organizations use less than 1 percent of their data for analysis (Brown, 2002). This is the fundamental irony of the information age we live in: Organizations possess enormous amounts of business data, yet have so little real business information, and to magnify the problem further, a leading business intelligence firm recently surveyed executives at 450 companies and discovered that 90 percent of these organizations rely on instinct rather than hard facts for most of their decisions because they lack the necessary information when they need it (Brown, 2002). Moreover, in cases where sufficient business information is available, organizations are only able to utilize less than 7 percent of it (Economist, 2001).

Information Security

Information is a critical asset that supports the mission of an organization. Protecting this asset is critical to survivability and longevity of any organization. Maintaining and improving information security is critical to the operations, reputation, and ultimately the success and longevity of any organization. However, Information and the systems that support it are vulnerable to many threats that can inflict serious damage to organizations resulting in significant losses. The concerns over information security risks can originate from a number of different security threats. They can come from hacking and unauthorized attempts to access private information, fraud, sabotage, theft and other malicious acts or they can originate from more innocuous sources, but no less harmful, such as natural disasters or even user errors. David Mackey, IBM's Director of security intelligence estimates that IBM recorded more than 1 billion suspicious computer security events in 2005. He estimates that a higher level of malicious traffic in 2006. The damage from these "security events" can range from lose of integrity of the information to total physical destruction or corruption of entire infrastructure that support it. The damages can stem from the actions of a variety of sources, such as disgruntle employees defrauding a system or careless errors committed by trusted employees to hackers gaining access to the system from outside of the organization. Precision in estimating computer security-related losses is not possible because many losses are never discovered, and others are "swept under the carpet" to avoid unfavorable publicity. The effects of various threats vary considerably: some affect the confidentiality or integrity of data while others affect the availability of a system. Broadly speaking the main purpose of information security is to protect an organization's valuable resources, such as information, hardware, and software.

The importance of securing our information infrastructure is not lost to the government of the United States. The US Department of Homeland Security (DHS) identifies a Critical Infrastructure (CI) as "systems and assets, whether physical or virtual, so vital to the United States that the incapacity or destruction of such systems and assets would have a debilitating impact on security, national economic security, national public health or safety, or any combination of those matters." According a recent report by the DHS titled The National Strategy for Homeland Security, which identified thirteen CI's, disruption in any components of a CI can have catastrophic economic, social and national security impacts. Information Security is identified as a major area of concern for the majority of the thirteen identified CI's. For example, many government and private-sector databases contain sensitive information which can include personally identifiable data such as medical records, financial information such as credit card numbers, and other sensitive proprietary business information or classified security-related data. Securing these databases which form the back bone of a number of CI's is of paramount importance. Losses due to electronic theft of information and other forms of cybercrime against to such databases can result in tens of millions of dollars annually.

Information security is concerned with the identification of electronic information assets and the development and implementation of tools, techniques, policies, standards, procedures and guidelines to ensure the confidentiality, integrity and availability of these assets. Although Information Security can be defined in a number of ways, the most salient is set forth by the government of the United States. The National Institute of Standards and Technology (NIST) defines Information Security based on the 44 United States Code Section 3542(b)(2), which states "Information Security is protecting information and information systems from unauthorized access, use, disclosure, disruption, modification, or destruction in order to provide integrity, confidentiality, and availability." (NIST, 2003, p3). The Federal Information Security Management Act (FISMA, P.L. 107-296, Title X, 44 U.S.C. 3532) defines Information Security as "protecting information and information systems from unauthorized access, use, disclosure, disruption, modification, or destruction" and goes on to further define Information Security activities as those "carried out in order to identify and address the vulnerabilities of computer system, or computer network" (17 U.S.C. 1201(e), 1202(d)). The United States' National Information Assurance Training and Education Center (NIATEC) defines information security as "a system of administrative policies and procedures" for identifying, controlling and protecting information against unauthorized access to or modification, whether in storage, processing or transit" (NIATEC, 2006).

The overall goal of information security should be to enable an organization to meet all of its mission critical business objectives by implementing systems, policies and procedures to mitigate IT-related risks to the organization, its partners and customers (NIST, 2004). The Federal Information Processing Standards Publication 199 issued by the National Institute of Standards and Technology (NIST, 2004) defines three broad information security objectives: Confidentiality, Integrity and Availability. This trio of objectives sometimes is referred to as the "CIA Triad".

- **Confidentiality:** "Preserving authorized restrictions on information access and disclosure, including means for protecting personal privacy and proprietary information…" [44 U.S.C., Sec. 3542]. Confidentiality is the assurance that information is not disclosed to unauthorized individuals, processes, or devices (NIST, 2003 p. 15). Confidentiality protection applies to data in storage, during processing, and while in transit. Confidentiality is extremely important consideration for any organization dealing with information and is usually discussed in terms of privacy. A loss of confidentiality is the unauthorized disclosure of information.
- **Integrity:** To ensure that timely and reliable access to and use of information is possible. According to 44 United States Code Section 3542(b)(2), integrity is defined as "guarding against improper information modification or destruction, and includes ensuring information non-repudiation and

authenticity..." Therefore, integrity is interpreted to mean protection against is the unauthorized modification or destruction of information. Integrity should be viewed both from a "data" and a "system" perspective. Data integrity implies that data has not been altered in an unauthorized manner while in storage, during processing, or while in transit. System integrity requires that a system is performing as intended and is not impaired and is free from unauthorized manipulation (NIST, 2003).

- **Availability:** Timely, reliable access to data and information services for authorized users (NIST, 2003). According to 44 United States Code Section 3542(b)(2), availability is "Ensuring timely and reliable access to and use of information...". Availability is frequently viewed as an organization's foremost information security objective. Information availability is a requirement that is intended to assure that all systems work promptly and service is not denied to authorized users. This should protect against the intentional or accidental attempts to either perform unauthorized access and alteration to organizational information or otherwise cause a denial of service or attempts to use system or data for unauthorized purposes. A loss of availability is the disruption of access to or use of information or an information system.

In defining the objectives of information security, there are a number of extensions to the CIA Triad. Most prominent extensions to the CIA Triad include three additional goals of Information Security. They are: accountability, authentication, and norrepudation. One such extension appears in the National Security Agency (NSA) definition of information security as ".. measures that protect and defend information and information systems by ensuring their availability, integrity, authentication, confidentiality, and nonrepudiation. These measures include providing for restoration of information systems by incorporating protection, detection, and reaction capabilities."(CNSS, 2003) This definition is almost identical to the way "cybersecurity" was defined by the 108th US Congress. A cybersecurity bill introduced in the 108th Congress, the Department of Homeland Security Cybersecurity Enhancement Act — H.R. 5068/ Thornberry; reintroduced in the 109th Congress as H.R. 285 where cybersecurity is defined as ...the prevention of damage to, the protection of, and the restoration of computers, electronic communications systems, electronic communication services, wire communication, and electronic communication, including information contained therein, to ensure its availability, integrity, authentication, confidentiality, and nonrepudiation.

- **Accountability:** Is the cornerstone of organizational information security objective in which auditing capabilities are established to ensure that users and producers of information are accountable for their actions and to verify that organizational security policies and due diligence are established, enforced and care is taken to comply with any government guidelines or standards. Accountability serves as a deterrent to improper actions and as an investigation tool for regulatory and law enforcement agencies.
- **Authentication:** Security measure designed to establish the validity of a transmission, message, or originator, or a means of verifying an individual's authorization to receive specific categories of information (CNSS, 2003. p 5). In order for a system to achieve security, it should require that all users identify themselves before they can perform any other system actions. Once the identification is achieved the authorization should be the next step. Authorization is process of granting permission to a subject to access a particular object. Authentication is the process of establishing the validity of the user attempting to gain access, and is thus a basic component of access control,

in which unauthorized access to the resources, programs, processes, systems are controlled. Access control can be achieved by using a combination of methods for authenticating the user. The primary methods of user authentication are: access passwords, access tokens, something the user owns which can be based on a combination of software or hardware that allows authorized access to that system (e.g., smart cards and smart card readers), the use of biometrics (something the user is, such as a fingerprint, palm print or voice print), access location (such as a particular workstation), user profiling (such as expected or acceptable behavior), and data authentication, to verify that the integrity of data has not been compromised. (CNSS, 2003)
- **Nonrepudiation:** Assurance the sender of data is provided with proof of delivery and the recipient is provided with proof of the sender's identity, so neither can later deny having processed the data. (CNSS, 2003)

Any information security initiative aims to minimize risk by reducing or eliminating threats to vulnerable organizational information assets. The National Institute of Standards and Technology (NIST, 2003, p. 7) defines risk as "…a combination of: (1) the likelihood that a particular vulnerability in an agency information system will be either intentionally or unintentionally exploited by a particular threat resulting in a loss of confidentiality, integrity, or availability, and (2) the potential impact or magnitude of harm that a loss of confidentiality, integrity, or availability will have on agency operations (including mission, functions, and public confidence in the agency), an agency's assets, or individuals (including privacy) should there be a threat exploitation of information system vulnerabilities," Risks are often characterized qualitatively as high, medium, or low. (NIST, 2003, p 8). The same publication defines threat as "…any circumstance or event with the potential to intentionally or unintentionally exploit a specific vulnerability in an information system resulting in a loss of confidentiality, integrity, or availability," and vulnerability as "…a flaw or weakness in the design or implementation of an information system (including security procedures and security controls associated with the system) that could be intentionally or unintentionally exploited to adversely affect an agency's operations (including missions, functions, and public confidence in the agency), an agency's assets, or individuals (including privacy) through a loss of confidentiality, integrity, or availability" (NIST, 2003, 9). NetIQ (2004) discusses five different types of vulnerabilities that have direct impact on the governance of information security practices. They are: exposed user accounts or defaults, dangerous user behavior, configuration flaws, missing patches and dangerous or unnecessary service. An effective management of these vulnerabilities is critical for three basic reasons. First, an effective vulnerability management helps reducing the severity and growth of incidence. Second, it helps in regulatory compliance. And third and the most important reason can be summed as simply saying, it is a "good business practice" to be proactive in managing the vulnerabilities rather than be reactive by trying to control the damage from an incidence.

Information Privacy

Privacy is defined as "the state of being free from unsanctioned intrusion" (Dictionary.com, 2006). Westin (Westin, 1967) defined the right to privacy as "the right of the individuals… to determine for themselves when, how, and to what extent information about them is communicated to others." The Fourth Amendment to the US Constitution's Bill of Rights states that "The right of the people to be secure in their persons, houses, papers, and effects, against unreasonable searches and seizures, shall not be violated." This belief carries back through history in such expressions from England, at least circa 1603, "Every man's house is his castle." The Supreme Court has since ruled that "We have recognized that the principal object of the Fourth Amendment is the protection of privacy rather than property, and have increasingly discarded fictional and procedural barriers rested on property concepts." Thus, because the Amendment "protects people, not places," the requirement of actual physical trespass is dispensed with and electronic surveillance was made subject to the Amendment's requirements (Findlaw.com, 2006). Generally the definitions of privacy in regards to business are quite clear. On the Internet, however, privacy raises greater concerns as consumers realize how much information can be collected without their knowledge. Companies are facing an increasingly competitive business environment which forces them to collect vast amounts of customer data in order to customize their offerings. Eventually, as consumers become aware of these technologies new privacy concerns will arise, and these concerns will gain a higher level of importance. The security of personal data and subsequent misuse or wrongful use without prior permission of an individual raise privacy concerns and often end up in questioning the intent behind collecting private information in the first place (Dhillon & Moores, 2001). Privacy information holds the key to power over the individual. When privacy information is held by organizations that have collected the information without the knowledge or permission of the individual the rights of the individual are at risk. Individual data privacy has become a prominent issue in the United States (Dyson, 1998). In a revealing interview with MSNBC, Eric Schmidt, CEO of Google, repeated a frequently heard mantra "that if you do not want people to know what you are doing on-line, maybe you shouldn't be doing it" (Rapoza, 2010). The view of privacy concerns promoted widely is misguided at best and completely disingenuous at worst. However, for most of us, this is the prevailing view of privacy suggesting a fundamental misunderstanding of what it is all about. In his article "'I've Got Nothing to Hide' and Other Misunderstandings of Privacy," Daniel Solove provides a strong rebuttal to this line of reasoning and points out the fundamental fallacies associated with it. He argues this notion stems from a fallacious conceptualization of privacy and concedes that privacy is an "exasperatingly vague and evanescent" concept to define (Miller, 1971). It is a concept so "infected with pernicious ambiguities" (Gross, 1967) and "so complex, so entangled in competing and contradictory dimensions, so engorged with various and distinct meanings," that it may never be fully understood (Post, 2001). In defining privacy, Solvoe states that "[privacy] is not the trumpeting of the individual against society's interests, but the protection of the individual based on society's own norms and values. Privacy is not simply a way to extricate individuals from social control, as it is itself a form of social control that emerges from a society's norms. It is not an external restraint on society, but is in fact an internal dimension of society. Therefore, privacy has a social value. Even when it protects the individual, it does so for the sake of society. It thus should not be weighed as an individual right against the greater social good. Privacy issues involve balancing societal interests on both sides of the scale." (Solove, 2007, page 763)

I recently taught a graduate course in data privacy. One of the course assignments required students to analyze privacy policies of number of most popular web sites. Most students were astonished by how complex these policies were and how hard they were to understand. For example, the privacy policy of Facebook is now longer than the US Constitution with almost 50 settings and more than 170 options available to the users. Given such a large number of options and settings, how likely is for an average user to understand and to make an informed decision about which privacy settings are most appropriate to their needs. The complexities of these privacy policies make it very difficult, if not impossible, for ordinary users to comprehend the consequences of their privacy choices. Consider this; in a response to privacy critiques and in an attempt to make the Facebook the "social center of the web", in April 2010, Facebook announced new privacy policies which included the development of "*Open Graphs*", as a platform for developers to exchange ideas and information. Open Graphs is an extension of the idea of "semantic networks", which according to Tim Berners-Lee (1999), is an attempt to "bring structure to the meaningful content of Web pages thus enabling computers to understand that content and how it relates to other sites and information across the internet". Open Graphs gives Facebook the ability to integrate websites and web apps within its existing social network environment by allowing its partner sites to create categories based on users' interests and then exchange that information with one another. For example, Open Graph would make the following scenario possible. A Facebook user visits Netflex, a movie rental site, and searches for a movie to rent. Netflex, an Open Graph partner of Facebook, develops a customized review for this user based on the reviews of that movie and similar other movies uploaded by the user's Facebook friends. Once the user makes the final selection, Netflex in turn, can notify the user's Facebook friends that their movie reviews were used by the user and thereby revealing what movie the user rented. Although this is an innocuous example of what is possible, more nefarious scenarios can be envisioned. The privacy consequence of Open Graphs is far reaching and not yet well understood, not even by the experts, let alone the average user. The most significant privacy consequence of Open Graphs is the redefinition of what "*public*" means. Users need to understand that *public* no longer means public within the Facebook only (Warren, 2010). As Christine Warren states, "users need to assume that if [they] do something that is considered public, that action can potentially end up on a customized stream for everyone in [their] social graph"(Warren, 2010). Users need to know that they should be vigilant about protecting their privacy on-line and not just Facebook. The user needs to be confident that just because she has updated her Facebook profile saying that she is feeling down, she should not expect to receive e-mail solicitation for her to purchase Prozac.

Although, ultimately, the user is responsible for protecting her own privacy, she should have some measure of confidence that the protection of her privacy is a valued and measures taken to protect it. Otherwise the user may engage in privacy protecting behaviors that may be detrimental to usefulness of the data collected. One such behavior is misrepresentation of one's identity. Consider the following example. One of my graduate students excitedly called me one day to tell me about her Facebook experience. Being concerned about her privacy, she had created a new Facebook profile for herself and purposely had given an erroneous birth date in which her aged was calculated to be 63. To her amazement, she recalled, that within hours she had received an e-mail from AARP (American Association of Retired Persons) inviting her to join that organization. Her misrepresentation of her age, nullifies any value that AARP would gained from knowing her age.

This is not a criticism of Facebook's or any other company's privacy policies per-se; it is intended to be a reminder of changing landscape of privacy and its impact on our daily lives. It is a call to action. No longer should we debate whether our privacy is in danger, it is time to assume that and seek ways to protect it. Companies should remember that a good privacy policy is good business and users should never assume that their privacy is protected. They need to become a more active participant in protecting their own privacy. In practice, information privacy deals with an individual's ability to control and release personal information. The individual is in control of the release process: to whom information is released, how much is released and for what purpose the information is to be used. "If a person considers the type and amount of information known about them to be inappropriate, then their perceived privacy is at risk"(Roddick & Wahlstrom, 2001). Consumers are likely to lose confidence in the online marketplace because of these privacy concerns (Adkinson, et al, 1989; Milberg et al, 1995; Pitofsky, 2006; Smith et al, 1996). Business must understand consumers' concern about these issues and aim to build consumer trust. It is important to note that knowledge about data collection can have a negative influence on a customer's trust and confidence level online. Privacy concerns are real and have profound and undeniable implications on people's attitude and behavior (Sullivan, 2002). The importance of preserving customers' privacy becomes evident when we study the following information: In its 1998 report, the World Trade Organization projected that the worldwide Electronic Commerce would reach a staggering $220 billion. A year later, Wharton Forum on E-commerce revised that WTO projection down to $133 billion. What accounts for this unkept promise of phenomenal growth? Census Bureau, in its February 2004 report states that "Consumer privacy apprehensions continue to plague the Web and hinder its growth." In a report by Forrester Research it is stated that privacy fears will hold back roughly $15 billion in e-commerce revenue. In May 2005, Jupiter Research reported that privacy and security concerns could cost online sellers almost $25 billion by 2006. Whether justifiable or not, consumers have concerns about their privacy and these concerns have been reflected in their behavior. The chief privacy officer of Royal Bank of Canada said "Our research shows that 80% of our customers would walk away if we mishandled their personal information."

Information Security and Privacy Concerns of Medical uses of Information Technologies

Another area of concern is the growth in the use of information technology for medical purposes. Confidentiality is sacrosanct in any physician-patient relationship and rules governing this relationship going back millennia are meant to protect patient's privacy. Confidentiality, a major component of information security, is a significant mechanism by which a patient's right to privacy is maintained and respected. However, in the era of Electronic Medical Record (EMR), it is hard to achieve. Although the use of information technologies for medical purposes shows potential for substantial benefits, it is fraught with concern related to security and privacy. Since there are so many points along the EMR life cycle where security and or privacy of medical data can be compromised, wide spread use of EMR is not possible without a thorough understanding and resolution of such issues (Amit, et al, 2005; Hunt, et. al, 1998; Johnston, et. al, 1994, Kensaku, et al, 2005). For example, Clinical decision support systems (CDSSs), an EMR based system, are designed to improve clinical and medical decision making and have been shown to improve healthcare practitioners' performance and patient care (Amit, et al, 2005; Hunt, et. al, 1998; Johnston, et. al, 1994, Kensaku, et al, 2005). In a number of recent meta- analyses of several medical studies, it was reported that CDSS significantly improved clinical practice and medical decision making (Amit, et al, 2005; Hunt, et. al, 1998; Johnston, et. al, 1994, Kensaku, et al, 2005). One area of benefit that CDSS has shown considerable promise is in Comparative Effectives. Comparative Effectiveness (CE) simply means evaluating and comparing two or more possible treatments for a given medical condition in order to choose the best course of action. CE is nothing new and has been used to enhance the practice of medicine for centuries. Comparative effectiveness have successfully been applied in numerous areas of medical diagnosis and treatment, including selection of the most favorable medication among competing drugs, deciding on the use of the most efficacious medical procedures or devices, and the use of best clinical treatment management. The non-partisan Congressional Budget Office (CBO) estimates that nearly $700 billion each year goes to health-care spending that can't be shown to lead to better health outcomes. The use of comparative effectiveness can help mitigate this problem. Studies have shown that the use of comparative effectiveness improves clinical care while reducing medical costs. This can benefit insurance providers, employers, government and patients. CBO suggests that the use of comparative effectiveness of clinical care provides the best opportunity to constraint the runaway medical cost and could result in substantial reduction of the overall cost of care without sacrificing the quality of care (Please See CBO's Pub. No. 2975, 2007). Realizing the importance of CE, in 2008 legislation was introduced in Congress that established the Health Care Comparative Effectiveness Research Institute as an ambitious program to study how to achieve the best clinical outcome for patients with minimal cost. Comparative effectiveness studies make extensive use of retrospective analysis of patient data. Retrospective analysis of patient data has contributed to advancement of the art and science of medical decision making for a long time (Lavarac 1999). Retrospective analysis of patient data has contributed to advancement of the art and science of medical decision making (Lavarac 1999). Given the current pace of advancement in medicine there is a great need to develop computer assisted medical decision making systems based on retrospective analysis to enhance and support the practice of medical decision making. Such systems need to learn the decision characteristics of the diseases before they can be used to diagnose future patients with uncertain disease states (Van Bemmel, 1997). To achieve this goal, the system needs to be presented with high quality, non fragmented historic patient data as the fundamental ingredient of robust analysis. However, using patient data for retrospective analysis in support of medical decision making poses a number of significant security and privacy challenges when considering that most patient data, at the time of collection, were intended for patient care and may not have been explicitly collected for research purposes. Security and privacy concerns using retrospective analysis of medical data for comparative effectiveness are too numerous to mention involve many complex technical and non technical issues that need to be resolved before the use of such systems can become wide spread. These challenges stem from the fact that patient data sets are large, complex, heterogeneous, hierarchical, time series, nontraditional, and originate from a verity of sources with differing levels of quality and format. Further, data sources may have incomplete, inaccurate and missing elements, some may be erroneous due to human and equipment error and lastly, the data may lack canonical consistencies within and between sources (Ciosa, et al, 2002). Patient data are voluminous and are collected from various sources including medical images, patient interviews, laboratory data, and the physicians' observations and interpretations of patients' symptoms, and behavior (Ciosa, et al, 2002). Securing such diverse and voluminous type of data residing on multiple heterogeneous systems with diverse data stewardship is not a trivial task and requires a whole set of different and difficult considerations. For example, medical data lack the underlying data structures needed for mathematically based data encryption techniques. Unlike data collected using other processes, medical data consists of word descriptions by physician and nurses, with very few formal constraints on the vocabulary, medical images, hand written charts and others. Additionally, medical data also lack a canonical form that encapsulates all equivalent forms of the same concept and is the preferred notation used in most encryption algorithms. For example, all the following are medically equivalent: Colon adenocarcinoma, metastatic to liver; Colonic adenocarcinoma, metastatic to liver; Large bowel adenocarcinoma, metastatic to liver. (Ciosa, et al, 2002). Lastly, medical data are time sensitive and may have been collected at different times using different data collection methodologies. As a result, they may reside on heterogeneous systems with differing representation and stewardship. Massive quantities of patient data are generated as patients undergo different medical and health care processes and procedures. As a result, these large patient databases may contain large quantity of useful information about patients and their medical conditions, possible diagnoses, prognosis and treatments. A major challenge in using these large patient databases is the ability to properly secure and anonyomize the data. Another security and privacy issue deals with data mining of medical data. Careful and systematic mining of patient databases may reveal and lead to the discovery of useful trends, relationships and patterns that could significantly enhance the understanding of disease progression and management. This process is referred to as Data mining (DM). DM is an exciting new facet of decision support systems. Data mining derived from the disciplines of artificial intelligence and statistical analysis and covers a wide array of technologies. Using data mining, it is possible to go beyond the data explicitly stored in a database to find nontrivial relationships and information that would not have been discovered by way of standard analysis methods. Medical Data Mining (MDM) is data mining applied to patient data and has been shown to provide benefits in many areas of medical diagnosis, prognosis and treatment (Lavrac, 1999; Prather, et. al, 2009; Rayward-Smith et.al 2001; Wang et al, 2000). By identifying patterns within the large patient databases, medical data mining can be used to gain more insight into the diseases and generate knowledge that can potentially lead to development of efficacious treatments. Unfortunately, given the difficulties associated with mining patient databases, the potential of these systems are yet to be realized (Lavrac, 1999; Prather, et. al, 2009; Wang et al, 2000). Medical Data Mining is the process of discovering and interpreting previously unknown patterns in medical databases (Lavrac, 1999; Prather, et. al, 2009; Wang et al, 2000). It is a powerful technology that converts data into information and potentially actionable knowledge. However, obtaining and using new knowledge in a vacuum does not facilitate optimal decision making in a medical setting. In order to develop a successful final patient treatment management, the newly extracted useful medical knowledge from MDM that appears in form of relationships and patterns should be integrated with existing knowledge and expertise that of the physician to enhance patient care. The significance of data security and privacy has not been lost to the data mining research community as was revealed in Nemati and Barko (Nemati et al., 2001) of the major industry predictions that are expected to be key issues in the future (Nemati et al., 2001). Chiefly among them are concerns over the security of what is collected and the privacy violations of what is discovered ((Margulis, 1977; Mason, 1986; Culnan, 1993; Smith, 1993; Milberg, S. J., Smith, & Kallman, 1995), and (Smith, Milberg, & Burke, 1996). One of the most far reaching laws with privacy implication impacting medical data mining research and practitioner communities is Health Insurance Portability and Accountability Act of 1996. It provides a standard for electronic health care transactions over the Internet. As the integrity and confidentiality of patient information is critical, this requires being able to uniquely identify and authenticate an individual. Health information is subject to HIPPA. The original legislation went into effect in 2001 and the final modifications took effect in April, 2003. A core aspect of HIPAA is to appropriately secure electronic medical records. The act applies to health information created or maintained by health care providers who engage in certain electronic transactions, health plans, and health care clearinghouses. The Office for Civil Rights (OCR) is responsible for implementing and enforcing the HIPPA privacy regulation. HIPAA has strict guidelines on how healthcare organizations can manage private health information. This includes: Authentication: A unique identification for individuals using the health care system; Access control: Manage accounts and restrict access to health information; Password management: Centrally define and enforce a global password policy; Auditing: Centralize activity logs related to the access of health information. The act sets standards to protect privacy in regards to individuals' medical information. The act provides individuals access to their medical records, giving them more control over how their protected health information is used and disclosed, and providing a clear avenue of recourse if their medical privacy is compromised (Anonymous, 2006). Improper use or disclosure of protected health information has the potential for both criminal and civil sanctions. For example, fines up to $25,000 for multiple violations of a single privacy standard in a calendar year and the penalties for intentional or willful violations of the privacy rule are much more severe with fines up to $250,000 and/or imprisonment up to 10 years for knowing misuse of personal health data. There are more immediate risks of private lawsuits relying on the HIPAA standard of care.

Implications of Information Security and Privacy

A common motivation for corporations to invest in information security is to safeguard their confidential data. This motivation is based on the erroneous view of information security as a risk mitigation activity rather than a strategic business enabler. No longer should information security be viewed solely as a measure to reduce risk to organizational information and electronic assets, it should be viewed as way the business needs to be conducted. To achieve success in information security goals, it should be organization information security should support the mission of the organization. The Information Systems Security Association (ISSA) has been developing a set of Generally Accepted Information Security Principles (GAISP). GAISP include a number of information security practices including the need for involvement of top management, the need for customized information security solutions, need for periodic reassessment, the need for an evolving security strategy and the need for a privacy strategy. This implies that it should be viewed as an integral part of the organizational strategic mission and therefore, it requires a comprehensive and integrated approach. It should be viewed as an element of sound management in which the cost-effectiveness is not the only driver of the project. Management should realize that information security is a smart business practice. By investing in security measures, an organization can reduce the frequency and severity of security-related losses. Information security requires a comprehensive approach that extends throughout the entire information life cycle. The management needs to understand that without a physical security, information security would be impossible. As a result, it should take into considerations a variety of issues, both technical and managerial and from within and outside of the organization. The management needs to realize that this comprehensive approach requires that the managerial, legal, organizational policies, operational, and technical controls can work together synergistically. This requires that senior managers be actively involved in establishing information security governance.

Effective information security controls often depend upon the proper functioning of other controls but responsibilities must be assigned and carried out by appropriate functional disciplines. These interdependencies often requires new understanding of the tradeoffs that may exist, that achieving one may actually undermine another. The management must insist that information security responsibilities and accountability be made explicit and the system owners have responsibilities that may exist outside their own functional domains. An individual or work group should be designated to take the lead role in the information security as a broad organization wide process. That requires that security policies be established and documented and the awareness among all employees should be increased through employee training and other incentives. This requires that Information security priorities be communicated to all stakeholders, including, customers, and employees at all levels within the organization to ensure a successful implementation. The management should insist that information security activities be integrated into all management activities, including strategic planning, capital planning. Management should also insist that an assessment of needs and weaknesses should be initiated and security measures and policies should be monitored and evaluated continuously.

Information security and privacy professionals are charged with protecting organizations against their information security vulnerabilities and privacy threats. Given the importance and securing information to an organization, this is an important position with considerable responsibility. It is the responsibility of information security professionals and management to create an environment where the technology is used in an ethical manner. Therefore, one cannot discuss information security without discussing the ethical issues fundamental in the development and use of the technology. According to a report by the European Commission (EC, 1999, p. 7) "Information Technologies can be and are being used for perpetrating and facilitating various criminal activities. In the hands of persons acting with bad faith, malice, or grave negligence, these technologies may become tools for activities that endanger or injure the life, property or dignity of individuals or damage the public interest." Information technology operates in a dynamic environment. Considerations of dynamic factors such as advances in new technologies, the dynamic nature of the user, the information latency and value, systems' ownerships, the emergence of a new threat and new vulnerabilities, dynamics of external networks, changes in the environment, the changing regulatory landscape should be viewed as important. Therefore the management should insist on an agile, comprehensive, integrated approach to information security and privacy.

David Mackey, IBM's Director of security intelligence estimates that IBM recorded more than 1 billion suspicious computer security events in 2005. He estimates that a higher level of malicious traffic in 2006. The damage from these "security events" can range from lose of integrity of the information to total physical destruction or corruption of entire infrastructure that support it. The damages can stem from the actions of a variety of sources, such as disgruntle employees defrauding a system or careless errors committed by trusted employees to hackers gaining access to the system from outside of the organization. Precision in estimating computer security-related losses is not possible because many losses are never discovered, and others are "swept under the carpet" to avoid unfavorable publicity. The effects of various threats vary considerably: some affect the confidentiality or integrity of data while others affect the availability of a system. Broadly speaking the main purpose of information security is to protect an organization's valuable resources, such as information, hardware, and software.

The importance of securing our information infrastructure is not lost to the government of the United States. The US Department of Homeland Security (DHS) identifies a Critical Infrastructure (CI) as "systems and assets, whether physical or virtual, so vital to the United States that the incapacity or destruction of such systems and assets would have a debilitating impact on security, national economic security, national public health or safety, or any combination of those matters." According a recent report by the DHS titled The National Strategy for Homeland Security, which identified thirteen CI's, disruption in any components of a CI can have catastrophic economic, social and national security impacts. Information Security is identified as a major area of concern for the majority of the thirteen identified CI's. For example, many government and private-sector databases contain sensitive information which can include personally identifiable data such as medical records, financial information such as credit card numbers, and other sensitive proprietary business information or classified security-related data. Securing these databases which form the back bone of a number of CI's is of paramount importance. Losses due to electronic theft of information and other forms of cybercrime against to such databases can result in tens of millions of dollars annually.

In addition to specific costs incurred as the results of malicious activities such as identity theft as a result of data breaches such as theft of hardware or system break ins, or virus attacks or denial of service attacks, one of the major consequences of dealing with a security attacks are the decrease in customer and investor confidence in the company. This is an area of major concern for the management. According to an event-study analysis using market valuations done by (Huseyin Cavusoglu, Birendra Mishra, and Srinivasan Raghunathan, 2004) to assess the impact of security breaches on the market value of breached firms, announcing a security breach is negatively associated with the market value of the announcing firm. The breached firms in the sample lost, on average, 2.1 percent of their market value within two days of the announcement—an average loss in market capitalization of $1.65 billion per breach (Huseyin Cavusoglu, Birendra Mishra, and Srinivasan Raghunathan, 2004). The study suggests that the cost of poor security is very high for investors and bas for business. Financial consequences may range from fines levied by regulatory authorities to brand erosion. As a result, organizations are spending a larger portion of their IT budget in information security. A study by the Forrester Research Group estimates that in 2007 businesses across North American and Europe will spend almost 13% of their IT budgets on security related activities. The same report shows the share of security spendature was around 7% in 2006.

CONCLUSION

It is obvious that information security and privacy are top priorities for society, as they should be. Regardless of the source, the impact of a security breach or a privacy threat on an organization, either private or governmental, can be severe, ranging from interruption in delivery of services and goods, to loss of physical and other assets, to loss of customer good will and confidence in the organization. Such breaches or privacy threats to sensitive data can be very costly to a organization. Recent research shows that investing and upgrading information security and privacy infrastructure is a smart business practice. By doing so, an organization can reduce the frequency and severity of losses resulted from security breaches in computer systems and infrastructures. Information is a critical asset that supports the mission of an organization. Protecting this asset is critical to survivability and longevity of any organization. Maintaining and improving information security is critical to the operations, reputation, and ultimately the success and longevity of any organization. However, Information and the systems that support it are vulnerable to many threats that can inflict serious damage to organizations resulting in significant losses. The concerns over information security risks can originate from a number of different security threats. They can come from hacking and unauthorized attempts to access private information, fraud, sabotage, theft and other malicious acts or they can originate from more innocuous sources, but no less harmful, such as natural disasters or even user errors.

Information Security and privacy are not just technological issues alone. They encompass all aspects of business from people to processes to technology. Bruce Schneier founder and editor of Schneier.com states that "If you think technology can solve your security problems, then you don't understand the problems and you don't understand the technology." Information Security and privacy involve consideration of many interrelated fundamental issues. Among them are technological, developmental and design, and managerial considerations. The technology component of information security and privacy is perhaps the easiest to develop and to achieve. The technological component of information security and privacy is concerned with the development, acquisition, and implementation of hardware and software needed to achieve security and privacy. The developmental and design component of information security and privacy deals with issues related techniques and methodologies used to proactively development and design systems that are secure and private. The managerial and personnel component focuses on the complex issues of dealing with the human elements in information security and privacy. They deal with policies, procedures and assessments required for the management of the operation of security and privacy activities. Undoubtedly, this is the hardest part of the information security and privacy to achieve since it requires a clear commitment to security and a culture of valuing privacy by the organizational leadership, assignment of appropriate roles and responsibilities, implementation of physical and personnel security and privacy protecting measures to control and monitor collection and access to data, training that is appropriate for the level of access and responsibility, and accountability. Privacy consideration is an important antecedent to developing a customer's intension to engage with a web site for commercial, informational or entertainment purposes. As a result, privacy has become an import business driver. Studies have shown that when people (customers) feel that privacy is 'violated', they respond to it in different ways with different levels of intensity (Culnan, 1993). Still, despite this divergent and varied reaction to privacy violation, or maybe because of it, a lot of companies still do not appreciate the depth of consumer feelings and the need to revamp their privacy practices, as well as their infrastructure for dealing with privacy. Privacy is no longer about just staying within the letter of the latest law or regulation. Sweeping changes in the attitudes of people about their privacy fueled by their privacy fears will cause an intense political debate and will put once-routine business and corporate practices under the microscope, resulting in a patchwork of regulations not favoring businesses. Regulatory complexity will grow as privacy concerns surface in scattered pieces of legislation (Anonymous, 2006). Companies need to respond quickly and comprehensively. They must recognize that privacy should be a core business issue. Privacy policies and procedures that cover all operations must be enacted. Privacy Preserving Identity Management should be viewed as a business issue, not a compliance issue.

Advances in IT have allowed people to transcend the barriers of time and geography and to take advantage of opportunities not even conceivable today, opening up a new world of possibilities and opportunities. The IT revolution has transformed our lives in way unimaginable only a decade ago. Yet, we are only at the threshold of this revolution. The dizzying pace of advances in information technology promises to transform our lives even more drastically. In order for us to take full advantage of the possibilities offered by this new interconnectedness, organizations, governmental agencies, and individuals must find ways to address the associated security and privacy implications of their actions and behaviors. As we move forward, new security and privacy challenges will likely to emerge. It is essential that we are prepared for these challenges.

REFERENCES

Adkinson, W., Eisenach, J., & Lenard, T. (2002). *Privacy Online: A Report on the Information Practices and Policies of Commercial Web Sites*. Retrieved August 2009, from http://www.pff.org/publications/privacyonlinefinalael.pdf

Anonymous. (2006). *Privacy Legislation Affecting the Internet: 108th Congress*. Retrieved August 2008, from http://www.cdt.org/legislation/108th/privacy/

Anonymous. (2006). *Office for Civil Rights*. Retrieved August 2009, from http://www.hhs.gov/ocr/index.html

Brancheau, J. C., Janz, B. D., & Wetherbe, J. C. (1996). Key issues in information systems management: 1994-95 SIM Delphi Results. *MIS Quart., 20*(2), 225-242.

Brown, E. (2002, April 1). Analyze This. *Forbes*, 169, 96-98.

Businessweek. (2001), Privacy in an Age of Terror. *Businessweek*.

Ciosa, K.J., & Mooree, W. (2002). Uniqueness of medical data mining. *Artificial Intelligence in Medicine, 26*, 1–24

Classen, D. C. (1998). Clinical Decision Support Systems to Improve Clinical Practice and Quality of Care. *JAMA*, *280*(15),1360-1361.

Clifton, C., Kantarcioglu, M., Vaidya, J., Lin, X., & Zhu, M. (2002). Tools for privacy preserving distributed data mining. *ACM SIGKDD Explorations Newsletter, 4*(2), 28-34.

Culnan, M. J. (1993). How did they my name? An exploratory investigation of consumer attitudes toward secondary information use. *MIS Quart., 17*(3), 341-363.

Dhillon, G., & Moores, T. (2001). Internet privacy: Interpreting key issues. *Information Resources Management Journal, 14*(4).

Dictionary.com. (2006). *Privacy*. Retrieved July 2006, from http://dictionary.reference.com/browse/privacy

Dyson, E. (1998). *Release 2.0: A Design for Living in the Digital Age*. Bantam Doubleday Dell Pub.

Eckerson, W., & Watson, H. (2001). Harnessing Customer Information for Strategic Advantage: Technical Challenges and Business Solutions, Industry Study 2000, Executive Summary. In *The Data Warehousing Institute*.

Economist. (2001, February 17). *The slow progress of fast wires* (p. 358).

Eshmawi, A., & Sadri, F. (2009). Information Integration with Uncertainty. In *Proceedings of the 2009 International Database Engineering and Applications Conference (IDEAS'09).*

Estivill-Castro, V., Brankovic, L., & Dowe, D. L. (1999). *Privacy in Data Mining*. Retrieved August 2006, from http://www.acs.org.au/nsw/articles/1999082.htm

Evfimievski, A., Srikant, R., Agrawal, R., & Gehrke, J. (2002). Privacy preserving mining of association rules. In *Proceedings of the eighth ACM SIGKDD international conference on Knowledge discovery and data mining, July 2002, Edmonton, Alberta, Canada* (pp. 217-228).

Garg, A.X., Adhikari, N.K.J., McDonald, H. (2005). Effects of Computerized Clinical Decision Support Systems on Practitioner Performance and Patient Outcomes: A Systematic Review. , *JAMA, 293*(10), 1223-1238.

Gross, H. (1967). The Concept of Privacy, 42 New York University Law. Review. 34, 35 (1967).

Han, J., & Kamber, M. (2001). *Data Mining: Concepts and Techniques*. Morgan Kaufmann Publishers.

Hardy, Q. (2004, May 10). Data of Reckoning. *Forbes, 173*, 151-154.

Hodge, J. G., Gostin, L. O., & Jacobson, P. (1999). Legal Issues Concerning Electronic Health Information: Privacy, Quality, and Liability. *The Journal of the American Medical Association, 282*(15), 1466-1471.

Hunt, D. L., Haynes, R.B., Hanna, S.E., & Smith, K. (1998). Effects of Computer-Based Clinical Decision Support Systems on Physician Performance and Patient Outcomes: A Systematic Review. *JAMA, 280*, 1339-1346

IDC Report (2010). The Digital Universe Decade: Are You Ready? Retrieved May 2010 from http://www.emc.com/collateral/demos/microsites/idc-digital-universe/iview.htm

Iyengar, V. S. (2002). *Transforming data to satisfy privacy constraints.* Paper presented at the KDD.

Johnston, M. E., Langton, K. B., Haynes, R. B., & Mathieu, A. (1994). Effects of Computer-based Clinical Decision Support Systems on Clinician Performance and Patient Outcome: A Critical Appraisal of Research. *Ann Intern Med, 120*(2), 135-142

Kantarcioglu, M., & Clifton, C. (2004). Privacy-Preserving Distributed Mining of Association Rules on Horizontally Partitioned Data. *IEEE Trans. Knowledge Data Eng., 16*(9), 1026-1037.

Lavrac, N. (1999). Selected techniques for data mining in medicine. *Artif Intell Med,* 16, 3-23.

Lindell, Y., & Pinkas, B. (2002). Privacy Preserving Data Mining. *J. Cryptology, 15*(3), 177-206.

Liu, J. T., Marchewka, J. L., & Yu, C. S. (2004). Beyond concern: a privacy-trust-behavioral intention model of electronic commerce. *Information & Management*, 42, 127-142.

Margulis, S. T. (1977). Conceptions of privacy: current status and next steps. *J. of Social Issues*, 33, 5-10.

Mason, R. O. (1986). Four ethical issues of the information age. *MIS Quart., 10*(1), 4-12.

Miklau, G., & Suciu, D. (2004). A Formal Analysis of Information Disclosure in Data Exchange. In *SIGMOD 2004* (pp. 575-586).

Milberg, S. J., S. J., B., Smith, H. J., & Kallman, E. A. (1995). Values, personal information privacy, and regulatory approaches. *Comm. of the ACM*, 38, 65-74.

Nemati, H., Barko, R., & Christopher, D. (2001). Issues in Organizational Data Mining: A Survey of Current Practices. *Journal of Data Warehousing, 6*(1), 25-36.

Niederman, F., Brancheau, J. C., & Wetherbe, J. C. (1991). Information systems management issues for the 1990's. *MIS Quart.,* 15, 474-500.

Pan, S. L., & Lee, J.-N. (2003). Using E-CRM for a Unified View of the Customer. *Communications of the ACM, 46*(4), 95-99.

Pinkas, B. (2002). Crytographic techniques for privacy-preserving data mining. *SIGKDD Exploreations, 4*(2), 12-19.

Pitofsky, R. (2006). *Privacy Online: Fair Information Practices in the Electronic Marketplace, a Report to Congress*. Retrieved August 2006, from http://www.ftc.gov/reports/privacy2000/privacy2000.pdfFTC

Richards, G., Rayward-Smith, V.J., Sonksen, P.H., Carey, S., & Weng, C. (2001). Data mining for indicators of early mortality in a database of clinical records. *Artif Intell Med,* 22, 215-31.

Ripley, B.D. (1996). *Pattern recognition and neural networks*. Cambridge: Cambridge University Press.

Rockart, J. F., & DeLong, D. W. (1988). *Executive Support Systems: The Emergence of Top Management Computer Use.* Paper presented at the Dow Jones-Irwin, Homewood, IL.

Smith, H. J. (1993). Privacy policies and practices: Inside the organizational maze. *Comm. of the ACM*, 36, 105-122.

Smith, H. J., Milberg, S. J., & Burke, S. J. (1996). Information privacy: Measuring individuals' concerns about organizational practices. *MIS Quart.*, 167-196.

Sullivan, B. (2002). Privacy groups debate DoubleClick settlement. Retrieved August 2006, from http://www.cnn.com/2002/TECH/internet/05/24/doubleclick.settlement.idg/index.html

Vaidya, J., & Clifton, C. (2004). Privacy-Preserving Data Mining: Why, How, and When. *IEEE Security and Privacy, 2*(6), 19-27.

Van Bemmel, J., & Musen, M. A. (1997). *Handbook of Medical Informatics*. New York: Springer-Verlag.

Verykios, V. S., Bertino, E., Fovino, I. N., Provenza, L. P., Saygin, Y., & Theodoridis, Y. (2004). State-of-the-art in privacy preserving data mining. *SIGMOD Record, 33*, 50-57.

Watson, H. J., Rainer Jr, R. K., & Koh, C. E. (1991). Executive information systems: a framework for development and a survey of current practices. *MIS Quart.*, 13-30.

Westin, A. (1967). *Privacy and Freedom*. New York: Atheneum.

Chapter 1
Examining an Individual's Perceived Need for Privacy and Security:
Construct and Scale Development

Taner Pirim
Mississippi Center for Supercomputing Research, USA

Tabitha James
Virginia Polytechnic Institute, USA

Katherine Boswell
University of Louisiana – Monroe, USA

Brian Reithel
University of Mississippi, USA

Reza Barkhi
Virginia Polytechnic Institute, USA

ABSTRACT

Security and privacy issues have risen in importance as the use of technology increases. Newly developed technological devices for asset security can only be successful if people are willing to use them. Gaining an understanding of individuals' acceptance and possible use of new security technologies would be beneficial to entities that are developing, marketing, and implementing new security technologies. This study develops an instrument to determine an individual's need for security and privacy and investigates the relationship between these two constructs. The results show that the instrument developed is reliable and that a significant relationship exists between the two constructs.

DOI: 10.4018/978-1-61692-000-5.ch001

INTRODUCTION

Privacy refers to the ability of an individual to "control the terms under which personal information is acquired and used" (Henderson & Snyder, 1999; Westin, 1967). A certain expectation of an individual's right to privacy has always been present; however, there is some disagreement as to what this expectation should be (Milberg et al., 2000). Privacy concerns have garnered much attention in recent years with the rise in identity fraud and the new capabilities to collect and process information brought about by technology. In 2008 alone there were 313,982 cases of identity theft reported to the FTC, an increase of over 50,000 from 2007 (FTC, 2009). Considering that only 31,140 cases were reported in 2000, a definite upward trend can be seen in the number of identity theft cases. In fact, the identity theft category was the largest subset of fraud complaints reported to the FTC, at 26% of the total (FTC, 2009). The public concern over this threat is evident in the report by the FTC to the Ways and Means Committee of the U.S. House of Representatives in March, 2006. This report stated that the commission is contacted between 15,000 and 20,000 times per week by individuals requesting information about identity theft avoidance practices. The Consumer Sentinel Network, which is the national repository for identity theft and consumer fraud, now contains over 7.2 million complaints (FTC, 2009).

One major reason for the rise of identity fraud is that increases in Internet transactions make the authentication of persons more difficult than ever before, since there is no human contact and less opportunity for identification checks. Hence, methods for identification and verification in e-commerce environments are becoming increasingly necessary to avoid potential issues such as identity fraud. Online banking, electronic financial transactions, online data stores, and Internet commerce, for example, are becoming extremely popular. The technologies to prevent misuse of these systems continue to expand as their importance increases and the potential for financial loss grows.

With advances in technology, companies have ever-increasing abilities to collect and analyze data to make assumptions about consumer behaviors. Increasing concerns about the misuse of such data, or the use of this information in a way not intended by the individual, have pushed privacy issues to the forefront of social consciousness.

Government regulations to control the collection and use of information illustrate the growing importance of privacy to individuals. Opt-out policies for secondary use of information from credit card companies, for example, now give individuals in the United States more control over who has access to their personal information. The use of information to derive valuable insights about individuals has become an increasingly important issue due to increased capabilities in collecting, processing and joining information by corporations and government entities. This information once collected can provide companies and government agencies with data that can be used for financial gain (Mason, 1986). These increased capabilities to collect and process data create an ethical dilemma in terms of financial profitability versus an individual's right to privacy (Mason, 1986).

Security as defined by the dictionary refers to "freedom from danger" or "freedom from fear or anxiety" (http://www.m-w.com/cgi-bin/dictionary). The importance of security has been highlighted in recent years due to uncertainties in world events as well as the ever-growing threat of vulnerabilities in systems crucial to normal operation of many functions of society. Physical security has always been a concern that individuals have placed importance upon. Recent events and rises in crime have compounded this concern as individuals are now more aware of threats to their physical safety in public areas such as airports, planes, sporting events, and their places of work, in addition to their normal safety concerns for their belongings and physical well-being in their homes and vehicles. The US Department of Justice reported in the summary findings of the National Crime Victimization Survey for 2006 that the property crime rate was 159.5 per 1,000 households and the violent crime rate was 24.6 per 1,000 persons age 12 or above (Bureau of Justice, 2007). According to the FBI's Crime in the United States Report for 2007, the property crime rate was 3,263.5 per 100,000 persons and the violent crime rate was 466.9 per 100,000 persons (FBI, 2008). The National Crime Victimization Survey is based upon polling and the Crime in the United States Report is obtained by compiling data from law enforcement. While some of the statistics demonstrate decreases in crime, it is still a serious problem in the United States as the previous figures illustrate.

Computer security is now becoming as ubiquitous as physical security due to the rising use of networked computers for personal and corporate use. The protection of corporate information has risen in importance due to the possible financial implications of misuse of important data. The 2008 CSI/FBI Computer Crime and Security Survey found that approximately half of the participants of their study had between 1 and 5 cybersecurity incidents. The average estimated loss per respondent was reported to be $288,618 (Richardson, 2008). While this statistic quantifies organizational loss, on an individual level it is important for two primary reasons. First, with increases in both the reliance on personal electronic information and personal network connectivity, individuals are more frequently faced with similar vulnerabilities to computer and information security. Secondly, the increased collection and storage of personal data by organizations makes ever present the concern that malicious parties may obtain personal data, which has been released to an organization, from these stores.

Computer security was once comprised of almost only the restriction of entry into data centers and authentication onto computers. However, this has drastically changed over the last two decades. In the wired world we inhabit, not only has the focus of computer security had to change, security professionals have also had to try to keep pace with an ever-changing number and type of threats (Loch et al., 1992). Threats to computer security include: undesired physical access to data centers, hackers, viruses, natural disasters, etc. These concerns are not just important to corporations any longer, as widespread viruses and the rising use of electronic commerce have brought digital data security to the attention of every individual that uses the Internet.

It is apparent that security and privacy are both incredibly important topics in today's digital world. From an individual perspective, new threats to personal privacy and security of one's digital and physical assets are changing what these concepts mean, how they are provided, and the implications of their compromise. Privacy concerns are facing new scrutiny as technological advances often blur the cost-benefit lines of the release of information versus the usefulness of various security applications and devices. New technological devices may provide additional security but may do so at the cost of releasing information by an individual. Also, systems may provide desired services, other than security, but at the cost of allowing companies to obtain and analyze personal data. These issues provide new dimensions or vary the importance of traditional opinions on privacy debates. With the growing use of the Internet and the increasing offering of e-commerce applications, as well as an ever-increasing reliance on digital information by society, authentication of individuals and the security of access to information and transactions are extremely important. The importance of securing data, physical assets, and one's self has seen new scrutiny in recent years for a number of reasons. Technology provides new means to secure these items, and presents new vulnerabilities and considerations.

Privacy has been empirically studied in the information technology research from an organizational context. However, from a general individual perspective neither security nor privacy has been addressed in relation to IT. The research conducted in this study addresses security and privacy as very general constructs. Pulling from the social sciences literature and organizational privacy studies, the perceived need for privacy construct is developed by examining three dimensions of privacy: the degree of privacy concerns (developed personal privacy issues), information management concerns (disclosure repercussions) and interaction management concerns (trust/misuse).

The perceived need for security construct looks at an individual's perception of the importance and/or need for general security of tangible assets, physical security, and information security. The perceived need for security construct is also kept very general, yet includes different types of assets to obtain a broad picture of an individual's level of security desires. The development of a general instrument to measure perceived need for privacy and perceived need for security will provide researchers with a means of measurement of these constructs.

We argue that an individual's perception of the importance of privacy and security on a personal level may impact their behaviors towards the adoption and use of technologies in a wide variety of areas, including security technologies. As such, the measurement of these constructs may be valuable in adoption and usage studies in the IT area as well as research on the impact of corporate policies and governmental regulations. This study develops and tests an instrument to measure an individual's perceived need for security and perceived need for privacy. As security can be seen as a means to protect privacy, we also examine the relationship between these two constructs, security and privacy.

LITERATURE REVIEW

Privacy

Privacy refers to an individual's right to control the collection and use of personal information. Westin (1967) defined privacy as the individual's right to determine or control the distribution of their information, including to how it is distributed and used, to whom it is provided, and to what extent it is released. This information may be digital or not, but the heavy reliance of electronic information and communication has brought about new implications and concerns to storing and protecting personal data. This reliance on the collection and use of data in today's technology-dependent society has increased privacy concerns and led to the inclusion of privacy as one of the biggest ethical issues of the information age (Mason, 1986). Mason argues that two threats to an individual's right to privacy exist: the first is an increased ability to collect data due to the growth of information technology in areas such as surveillance, communication, computation, storage and retrieval. The second threat is an increase in the value of such information for decision making (Mason, 1986). Henderson and Snyder (1999) list three major forces driving an increased focus on personal information privacy: 1) new technological capabilities, 2) an increased value of information and 3) confusion over what is ethically right and wrong. The first two issues correspond to Mason's threats to privacy. The third issue revolves around the debate over what an individual's actual right to privacy entails. Social policies, government regulations and legal precedents have not kept pace with the ever-evolving use of technology in regard to the growing capabilities to collect, process, and use information.

In the social sciences literature, Laufer and Wolfe (1977) developed the multidimensional developmental theory of privacy. They proposed several dimensions to one's perception of privacy.

The first dimension, called the "self-ego dimension," refers to the impact of the developmental process on an individual's concept of privacy. As an individual passes through different stages of life, the influence of their experiences and their developmental stage may impact their desire for privacy. In other words, individuals have differing needs for privacy, and this is related to their developmental stage and previous experiences. The "environmental dimension" includes outside or "environmental" impacts on privacy such as barriers to intrusion that may be available due to social status or the influence of cultural norms. The third dimension, or "interpersonal dimension," deals with information management and interaction management. Information management relates to the release of information and interaction management refers to an individual's choice of who they interact with. Together these dimensions form an idea of choice and control of privacy, in which an individual develops methods to obtain a desired level of privacy by coalescing their experience, tools or status, and perceived repercussions of what information they release and to whom it is released. A study on online information fabrication proposed a combination of this theory with the theory or reasoned action, although the privacy factors were not empirically tested (Lwin & Williams, 2003).

Several studies involving the collection and use of data by corporate entities have been conducted. An instrument to measure an individual's concern over an organization's practices regarding information privacy was developed and rigorously tested (Smith et al., 1996; Stewart & Segars, 2002). This research identified four primary dimensions of an individual's concern of organizational information privacy practices: collection, unauthorized secondary use, improper access, and errors (Smith et al., 1996). Collection refers to the concern over the amount of personal information being collected and stored by organizations. The second dimension includes concerns over how the acquired data collected for one purpose may be used for a second purpose without this use being authorized by the individual. Unease over unauthorized access to data stores forms the third dimension. The last dimension refers to concern over deliberate and unintended errors in acquired data.

Culnan and Armstrong (1999) concluded that procedural fairness in the use of information has an impact on an individual's willingness to release personal information to corporations. If an individual is informed that fair practices will be employed when collecting and using information, they will be more willing to divulge personal information and relent to its use to build consumer profiles (Culnan & Armstrong, 1999).

Workplace privacy concerns have also raised debate with the increasing collection of data on employees and increased technological monitoring capabilities. Debate over an employer's right to monitor employee activity has increased due to the use of surveillance technologies being employed in corporations to monitor such areas as technology resource usage, employee movement and employee access to areas within facilities (Townsend & Bennett, 2003). The usage of surveillance and monitoring technologies raises new perceptions in light of recent world events and is no longer just a concern for the workplace. Woodman et al. (1982) found that employees have limited knowledge of how much data is actually collected and stored by their employers and that the relevancy of the data's use was important to the employees. The influence of individual concerns on organizational privacy issues is to be expected as individual consent is an important factor in the adoption of technologies or use of personal data.

This examination of the literature illustrates a personal view of privacy being formed from a combination of personal beliefs held by an individual on the importance of privacy and their control over it, the perceived repercussions of the release of information deemed private, and the trust in the entity the information is being released to or the trust in the appropriate/non-negligent usage of the information. Approaching the concept of individual privacy from this perspective, we developed an instrument to measure an individual's perceived need for privacy.

Security

Concerns about personal safety and the impacts of those concerns on individuals and society have been studied for decades. Warr, analyzing trend data on public opinion on crime, found that "fear has remained relatively constant since the early 1970's, varying no more than about 10 percent" (Warr, 1995). A heavily researched area is the cost of criminal victimization, or the threat of such victimization. Criminologists have argued that the stressful reactions due to victimization are not limited to just the individuals that are directly affected by crime; rather, individuals' behavior can be affected by this fear even if they have not been a victim of crime (Sacco, 1993). Fear of crime (FoC) measures these concerns for personal safety. There is some controversy in the literature over the definition of this measurement, however. A large number of studies define FoC as a measurement of "risk perception", or more of a "cognitive or judgment-based, general concern" (Rountree & Land, 1996). Ferraro defines fear of crime as "an emotional response of dread or anxiety to crime or symbols that a person associates with crime" (as stated in Taylor, 1995) and recognizes this measurement as a social phenomena. However, a more individualized view of the measurement has also been applied to try to identify a person's level of fear.

Regardless of the approach to this measurement, FoC has been used to study the affects, both tangible and intangible, on individuals due to their concern about crime. According to Moore and Sheppard, "It [FoC] extends to an alteration of daily living habits as well as to the negative psychological effects of living in a state of constant anxiety" (Moore & Shepard, 2006). Much of the research is very narrowly focused on issues such as FoC's impact in different environments, between ethnic groups, gender or age. However, what seems to be prevalent in much of this research is emphasis on the necessity of this type of research and the implications upon society and on individual behavior. Although the level of FoC, and thereby the impact, is unique to each individual and dependent upon a variety of factors, there is no disagreement that the level of this fear can influence an individual's behavior, by the display of negative psychological effects due to higher levels of anxiety to alteration of daily living habits, with some display differing levels of withdrawal from society. Some studies go so far as to state that FoC can affect an individual's quality of life.

Although as stated earlier, much of the literature in this area focus on specific situations that look at these types of fears with regards to possible crime and the levels of personal security associated with those fears, little to no research has been performed on these fears and the corresponding impact on behavior with regards to the impact of technology. The research in this area does illustrate that individuals may have different perceived levels of security desires due to differing fears and their corresponding influence.

Much of the research in the information technology area on security needs has been due to an increased reliance on networked computing environments and the associated risks due to this interconnectedness (Loch et al., 1992). Prior to the increased use of networking technologies and the changing landscape of corporate and personal computing, security of data in systems was primarily focused on facility security and backup mechanisms to prevent loss from natural disasters and component failure (Loch et al., 1992). Due to changes in the usage of technologies, security professionals have increasingly had to attempt to secure facilities and systems from a variety of new threats that rapidly increase in number. Loch et al. (1992) describe 13 threats to information security: accidental and intentional entry of bad data, accidental and intentional destruction of data, unauthorized access to data, inadequate control over media, poor control over manual I/O, access to data or systems by outsiders (both hackers and competitors), computer viruses/worms, weak or inadequate physical control, and natural disasters (fire, flood, power loss, etc.). The results of this study illustrate a disparity between the use of new technologies and the understanding of security risks associated with their usage. Behavioral studies in information security have focused on security awareness (Siponen, 2000) or organizational security practices (Straub, 1990). These studies illustrate the important role security plays in information technology, both in terms of securing electronic information as well as providing new means to protect both physical assets and persons.

Security as a broad concept in information technology must encompass all aspects of security, including the security of information both digital and physical, asset security and personal security of one's person. To develop a general construct with wide applicability to IT, all these areas are necessary, not only because they provide a deeper look at an individual's need for security, but also because technology now impacts all of these arenas. New security technologies to protect data, information systems, people, and facilities are constantly being developed. Securing crucial facilities and individual safety has also become increasingly important and apparent in light of recent world events. The development, implementation and adoption of these types of technologies may benefit from knowledge of the importance of security to the target individuals expected to use them. In this study, an instrument to measure an individual's perceived need from security is developed by combining all elements of asset security to examine security from a broad perspective.

DATA COLLECTION

Survey participants included faculty, staff and students from two large research universities in the southern US. A total of 429 useable responses were collected. Table 1 shows the demographic breakdown of the sample by age and sex.

The participants ranged in educational level between possessing a high school degree to a Ph.D. The educational background of the participants was also varied. Participants included individuals from a variety of academic disciplines. These academic areas included: engineering, business, and liberal arts. As can be seen from Table 1, the sample contained a relatively even distribution of males and females and range of ages (< 30 and > 30).

The method of data collection used was both web-based and traditional. Some participants electronically submitted their responses, the rest were administered a paper-based survey.

RESULTS

To gauge the perceptions of an individual's need for privacy and security at a generic level, an instrument was developed and refined. The final instrument provides a general overview of an individual's need for privacy and security outside

Table 1. Profile of Survey Participants

Total Number of Participants	492
Sex	
Male	248
Female	181
Age	
17-21	192
22-26	77
27-31	30
32-36	27
Over 36	103

of an organizational context. Figure 1 presents the conceptual model. The 18 item scale developed to test this model includes 9 items pertaining to perceived need for privacy and 9 items pertaining to perceived need for security. A 5 point scale (1=Strongly Agree; 5=Strongly Disagree) is used for all items. Descriptive statistics (means and standard deviations based on a normal distribution) for the instrument are provided in Table 2 and Table 3.

Perceived Need for Security Instrument

Table 2 illustrates that on average individuals feel a high need for security, for both their physical selves and their assets. The instrument developed includes three major areas of security: physical asset security, security of one's self, and information security. The survey was developed to give a broad gauge of an individual's perceived need for security by looking at all possible security concerns in a general context.

Figure 1. Conceptual Model

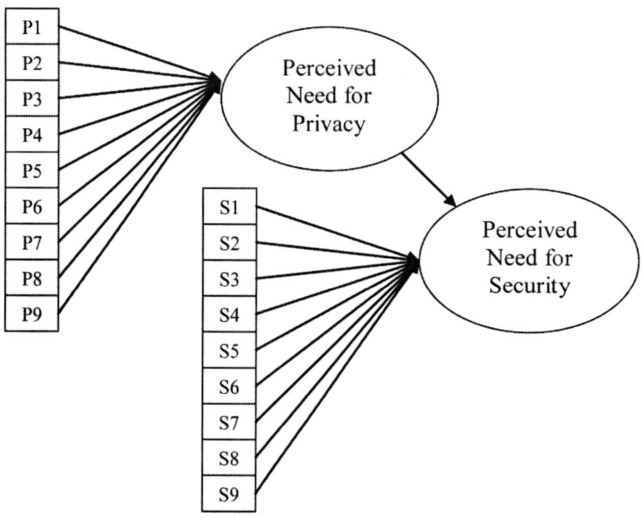

Table 2. Descriptive Statistics for Perceived Need for Security

Item NO.	Question	Mean	S.D.
S1	I feel that the safeguarding from potential external threats of my physical being is important to me.	1.56	.76
S2	I feel that my personal security at my home or in my vehicle is important to me.	1.39	.67
S3	I feel that my personal security at my place of work or other work related places is important to me	1.51	.71
S4	My security at places of public access, such as a mall or airport, or special public events, such as the Olympics or the Super Bowl, is important to me.	1.48	.64
S5	I feel that the security of my tangible assets (such as my home, vehicle, etc.) is important to me.	1.53	.70
S6	I feel that keeping my personal possessions, such as jewelry, money, electronics, etc. safe is important to me.	1.66	.74
S7	I feel that the safekeeping of my informational assets contained in digital or paper format is important to me (such as financial records, medical records, etc.)	1.53	.72
S8	I feel that the security of my personal information, such as my PC files or personal records (financial, medical, etc.)is important to me.	1.56	.72
S9	I feel that the safekeeping of information I have provided to a corporation or other entity is important to me.	1.66	.78

Table 3. Descriptive Statistics for Perceived Need for Privacy

Item No.	Question	Mean	S.D.
P1	I feel my privacy is very important to me.	1.47	.68
P2	I feel that my control over my personal information is very important to me.	1.51	.69
P3	I feel that it is important not to release sensitive information to any entity.	1.92	.97
P4	I feel it is important to avoid having personal information released that I think could be financially damaging.	1.48	.70
P5	I feel it is important to avoid having personal information released that I think could be socially damaging to me	1.65	.76
P6	I feel it is important to avoid having personal information about me released that may go against social morals and attitudes.	1.80	.86
P7	I feel that the release of personal information to individuals with whom I have a high comfort level is unacceptable.	2.62	1.19
P8	I feel that the release of personal information to entities where I feel as though I am anonymously providing the information is unacceptable.	2.27	1.11
P9	I feel that the use of personal information that has been released by me but is used in a manner not intended by me is unacceptable.	1.61	
		.86	

As shown in Table 2, on average individuals perceive a high need for security. Due to the focus on personal security and data security in current world events, this was to be expected.

Perceived Need for Privacy Instrument

The descriptive statistics for the perceived need for privacy instrument are shown in Table 3. The means indicate a high perceived need for privacy. However, while individuals may perceive their privacy as important, it is also apparent that their control over the dissemination of their private information is also important. The means for item 7 and item 8 are slightly higher, indicating that to many of the individuals the release of information to parties where there is a high level of trust or low expectation of misuse is more acceptable as was found in (Culnan & Armstrong, 1999). The standard deviation for these two items is also larger, indicating a larger range of answers.

Relationship between Perceived Need for Security and Perceived Need for Privacy

The conceptual model is presented in Figure 1. The instruments detailed above were used to gauge if a relationship existed between perceived need for privacy and perceived need for security. While a high perceived need for security existed with little variation, the perceived need for privacy tended to be more important in situations where individuals did not feel as comfortable with the possibilities of release of information to untrusted parties. These results indicate that the level of perceived security of the information released influenced their responses. The following section details the reliability of the refined instrument and illustrates the relationship between the constructs.

Table 4. Reliabilities for Model Constructs

Construct	Cronbach Alpha
Perceived Need for Security	0.90
Perceived Need for Privacy	0.85

The Cronbach's Alphas obtained for both constructs were higher than the recommended levels of 0.70 (Segars, 1997). These results are given in Table 4.

A confirmatory factor analysis was conducted to assure the measurement of the correct constructs. Table 5 presents the factor loadings. The security items all loaded very highly on factor 1. The privacy items all loaded on factor 2; however, items 1, 2, 4, and 9 also had substantial loadings on the first factor. This indicates that privacy and security may not be mutually exclusive. Privacy could be seen by individuals as relating to security in that privacy may be something that an individual wishes to secure. An individual's view on the importance of privacy could impact their opinion on the necessity of security, especially in instances where the information or asset being secured is something that the individual feels a need to keep private. This relationship was expected and is examined by running a regression analysis.

A regression was run to investigate the relationship between perceived need for privacy and perceived need for security. An individual's perceived need for privacy may impact their perceived need for security. These two constructs are related in that security measures may provide the means to ensure privacy. If an individual feels a high need for privacy, their need for security should also be high. A coefficient of .53 ($p < 0.05$) with a t-value of 15.41 was obtained which signifies a significant relationship between the variables. The model is shown in Figure 2.

Table 5. Factor Loadings

	Factor	
Item	1	2
S1	**.731**	.056
S2	**.761**	.086
S3	**.801**	.148
S4	**.753**	.119
S5	**.721**	.159
S6	**.631**	.223
S7	**.648**	.378
S8	**.652**	.384
S9	**.625**	.379
P1	.420	**.591**
P2	.444	**.616**
P3	.192	**.770**
P4	.473	**.571**
P5	.380	**.617**
P6	.305	**.619**
P7	-.077	**.725**
P8	-.034	**.688**
P9	.435	**.488**

Figure 2. Model

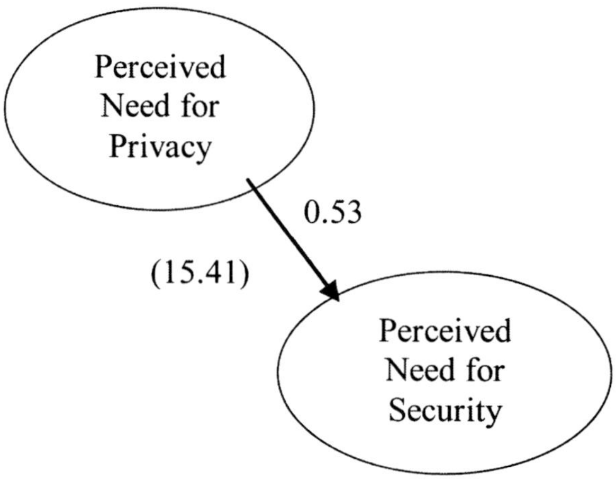

CONCLUSION AND DIRECIONS FOR FUTURE RESEARCH

The importance of security and privacy issues in today's society are increasing. The need to examine the importance of security and privacy to individuals is of utmost importance as these issues impact a variety of items such as: legislation, corporate policies, public safety concerns, and competitive advantage of corporations. The growing use of technology to both protect and provide benefits from the collection and manipulation of data makes security and privacy concerns important on several levels. Most research in the privacy area has focused on corporate or governmental contexts. Information Systems research in the security areas has focused primarily on data security practices. This study provides an instrument to generally examine an individual's need for privacy and security. Insight into an individual's perceptions of these issues can provide valuable information for a variety of situations. Corporate and governmental entities could use this information to make informed decisions on policy and technology adoption as well as aid in the smooth implementation of technologies to alleviate privacy and security concerns.

E-commerce applications and increasing reliance on networked computing environments make the security of digital information increasingly difficult as the number of threats of compromise to networked systems has been growing at an alarming rate. The introduction of new security technologies is inevitable and these technologies must gain widespread adoption to be successful in the current computing environment. In order to ease the barriers to adoption of these technologies, it is necessary to obtain an idea of the factors influencing an individual's perception of their potential benefits.

A determination of an individual's possible acceptance of a technology could help mitigate problems in deploying a new technology and provide some insight that could help lower barriers to acceptance and reduce possible financial losses. Developers of new technologies may also benefit by being able to gain insight into the acceptance of a device to be able to better market and sell the new technology as well as help deploy it in a production environment. The constructs discussed in this research are not the only factors influencing the adoption and usage of technologies. They are however applicable to a broad range of technologies in the security area and can be included with more context dependent constructs and applied in more robust models.

The instrument was found to be highly reliable and a significant relationship was found between the perceived need for privacy and perceived need for security constructs. Support for these findings was presented as well as a discussion of their implications. Future work will include the application of the constructs developed here to models for technology acceptance in specific areas.

REFERENCES

Bureau of Justice. (2007). Bureau of Justice Statistics Fact Sheet. Retrieved August, 2009, from the US Department of Justice Web site http://www.ojp.usdoj.gov/bjs/abstract/cv06.htm

Culnan, M., & Armstrong, P. (1999). Information privacy concerns, procedural fairness, and impersonal trust: an empirical investigation. *Organization Science*, *10*(1), 104–115. doi:10.1287/orsc.10.1.104

Federal Bureau of Investigations Uniform Crime reports. (2008). *FBI Releases its 2007 Crime Statistics*. Retrieved August 2009, from the FBI Web site http://www.fbi.gov/ucr/cius2007/about/crime_summary.html.

Federal Trade Commission for the Consumer. (2006). *FTC testifies on identity theft and social security numbers*. Retrieved August, 2009 from the FTC Web site: http://www.ftc.gov/opa/2006/03/ssntestimony.htm.

Federal Trade Commission for the Consumer. (2009). *Consumer Sentinel Network Data Book for January – December 2008*. Retrieved August, 2009 from the FTC Web site: http://www.ftc.gov/sentinel/reports.shtml

Henderson, S., & Snyder, C. (1999). Personal information privacy: implications for MIS managers. *Information & Management*, *36*, 213–220. doi:10.1016/S0378-7206(99)00019-1

Loch, K., Carr, H., & Warkentin, M. (1992). Threats to information systems: today's reality, yesterday's understanding . *Management Information Systems Quarterly*, *16*(2), 173–186. doi:10.2307/249574

Lwin, M., & Williams, J. (2003). A model integrating the multidimensional developmental theory of privacy and theory of planned behavior to examine fabrication of information online. *Marketing Letters*, *14*(4), 257–272. doi:10.1023/B:MARK.0000012471.31858.e5

Mason, R. (1986). Four ethical issues of the information age. *Management Information Systems Quarterly*, *10*(1), 5–12. doi:10.2307/248873

Merriam-Webster Dictionary Online. (n.d.). http://www.m-w.com/cgi-bin/dictionary

Milberg, S., Smith, H., & Burke, S. (2000). Information privacy: corporate management and national regulation. *Organization Science*, *11*(1), 35–57. doi:10.1287/orsc.11.1.35.12567

Moore, S., & Shepherd, J. P. (2006). The cost of fear: shadow pricing the intangible costs of crime. *Applied Economics*, *38*(3), 293–300. doi:10.1080/00036840500367781

Richardson, R. (2008). *2008 CSI Computer Crime & Security Survey*. Retrieved August, 2009 from the CSI Web site: http://www.gocsi.com/

Rountree, P. W., & Land, K. C. (1996). Perceived risk versus fear of crime: empirical evidence of conceptually distinct reactions in survey data. *Social Forces*, *74*(4), 1353–1376. doi:10.2307/2580354

Sacco, V. F. (1993). Social support and the fear of crime. *Canadian Journal of Criminology*, *35*(2), 187–196.

Segars, A. (1997). Assessing the unidimensionality of measurement: a paradigm and illustration within the context of information systems research . *Omega*, *25*(1), 107–121. doi:10.1016/S0305-0483(96)00051-5

Siponen, M. (2000). A conceptual foundation for organizational information security. *Information Management & Computer Security*, *8*(1), 31–44. doi:10.1108/09685220010371394

Smith, H., Milberg, S., & Burke, S. (1996). Information privacy: measuring individuals' concerns about organizational practices. *Management Information Systems Quarterly*, *20*(2), 167–196. doi:10.2307/249477

Stewart, K., & Segars, A. (2002). An empirical examination of the concern for information privacy instrument. *Information Systems Research*, *13*(1), 36–49. doi:10.1287/isre.13.1.36.97

Straub, D. (1990). Effective IS security: an empirical study. *Information Systems Research*, *1*(3), 255–276. doi:10.1287/isre.1.3.255

Taylor, I. (1995). Fear of crime: interpreting victimization risk. *Urban Studies (Edinburgh, Scotland)*, *32*(9), 1567–1568.

Townsend, A., & Bennett, J. (2003). Privacy, technology, and conflict: emerging issues and action in workplace privacy. *Journal of Labor Research*, *XXIV*(2), 195–205. doi:10.1007/BF02701789

Warr, M. (1995). The polls – poll trends: public opinion on crime and punishment. *Public Opinion Quarterly*, *59*(2), 296–310. doi:10.1086/269474

Westin, A. (1967). *Privacy and Freedom*. New York: Atheneum Publishers.

Woodman, R., Ganster, D., Adams, J., McCuddy, M., Tolchinsky, P., & Fromkin, H. (1982). A survey of employee perceptions of information privacy in organizations. *Academy of Management Journal*, *25*(3), 647–663. doi:10.2307/256087

Chapter 2
Forty Years of Federal Legislation in the Area of Data Protection and Information Security

John Cassini
Université d'Angers, France

B. Dawn Medlin
Appalachian State University, USA

Adriana Romaniello
Universidad Rey Juan Carlos, Spain

ABSTRACT

Historically, the United States has attempted to address the issues of an individual's right to information privacy and security through legislative actions. More specifically, the execution of legislation that addresses information security and privacy has been implemented in particular areas such as health care, banking, and education as well as other industries. This chapter presents an overview of the laws that have addressed the growth of security and privacy threats that havegrown over the years.

INTRODUCTION

Founded in the 1950s, the Internet was initially designed to be a military communications network. Due in part to its popularity among scientists, it was later expanded to university campuses across the United States. Next came the introduction of the Internet to the public in the early 1990s, and since that time the Internet has grown at an exponential rate. This growth has certainly led to many benefits such as banking and bill paying online as well as electronic medical records. Unfortunately, as the number of people connecting to and using the Internet increases, so does the opportunity for fraudulent and malicious activities.

While the enormous popularity of the Internet was not foreseen, neither was the need for restrictive legislative action. Therefore, there was an

DOI: 10.4018/978-1-61692-000-5.ch002

initial period of legislative non-responsiveness as policy makers could not foresee the legal issues that might surround the use of the Internet. Even if the Internet would have had laws and restrictions in place during its fast-paced evolution, chances are the Internet would have never grown to the size and popularity it enjoys today (Wisebrod, 1995).

To better understand the order that is being imposed upon that initial chaos, we will address security and privacy issues from a chronological perspective, by presenting the major laws and regulations enacted by the United States government over the last 40 years. The provisions contained in these laws and regulations show how the topic of security and privacy of information has evolved over time as our society has fully adopted a continually evolving Internet as an integral part of its daily life and activities.

A chronological view of the evolution of information security and privacy laws and regulations can provide a rich observation of the relationship between technology and the law. It can also show how developments in technology bring about new uses and misuses of that technology and call forth more specific legal controls. The chronological approach can also provide insight as to how these laws have evolved over time as the United States society has evolved in its use of the Internet and other related technologies.

BACKGROUND

Technology plays an important role in providing current and up-to-date information for consumers and organizations as well as the instantaneous sharing of information between individuals using programs such as email or instant messaging. Most consumers would agree that technology can provide the accessibility of their information, and that fact alone is advantageous, but consumers are also becoming increasingly aware of the potential harmful impact of the misuse of these same technologies.

Certainly, Internet technologies have brought with them legal challenges that are often hard to define. Because we are a society governed by laws, information technology managers and systems administrators must be aware of and address current laws, directives and regulations dealing with cybercrime issues. In addition, the growth of the Internet as a file storage and transfer medium has forced society to reexamine the notions surrounding privacy and security issues.

As the worldwide economy continues to become more dependent upon information technologies, and as economies become more interdependent, it appears that security-based threats will continue to increase, with concomitant increases in costs to individuals, organizations, and society as a whole (Conca, Medlin, Dave, 2005). Researchers continue to look at the rise of professional cybercrime as a uniquely worrying phenomenon (Antonopoulos, 2009).

No matter the organizational type, one of the greatest problems faced by any company is the myriad of ways their IT structure and services can be diminished by malware or third-party attacks. The security-based threats and vulnerabilities that currently exist, such as Trojan Horses, spyware, and other types of malicious code, have the potential to damage consumers', as well as businesses', data or other assets. When organizations determine the asset risk, they must address the vulnerabilities that exist in order to protect themselves from loss of physical and tangible assets.

In order to address these aforementioned security threats, network or security administrators often find themselves in an endless cycle of applying patch after patch in an attempt to plug all the possible holes. While patches can take care of problems in software design, they cannot account for human error, lack of security knowledge, or the ever-growing threat from malware that can be continuously introduced, as employees increasingly use the Internet for both business and personal activities.

To address the pertinent laws related to privacy and security, we provide a historical list as seen in Table 1. These laws and regulations are some of the most important privacy and security laws enacted from 1970 until the present.

Certainly, the quantity of legislation that has dealt with computer security threats has grown exponentially over the past decade in several different industries. Though we do not attempt to address each law in detail, we do concentrate on several of the key areas within the different industries of healthcare, banking and financial institutions, and personal privacy of information.

We begin our survey of affected industries with the health care area as it has been one of the most highly affected. Issues and legislation related to the privacy of patient's records are presented below.

Table 1. Major U.S. Privacy and security legislation 1970 - present

Year	Legislation
1970	U.S. Fair Credit Reporting Act
1973	U.S. Code of Fair Information Practices
1974	U.S. Privacy Act
1984	U.S. Medical Computer Crime Act
1986	U.S. Computer Fraud and Abuse Act
1986	U.S. Electronic Communications Privacy Act
1987	U.S. Computer Security Act
1991	U.S. Federal Sentencing Guidelines
1996	U.S. Economic and Protection of Proprietary Information Act
1996	Health Insurance Portability And Accountability Act Of 1996
1998	Children's Online Privacy Protection Act of 1998 (COPPA)
1998	The Digital Millennium Copyright Act
2001	Homeland Security Act: Cyber Security Enhancement Act
2003	U.S Fair and Accurate Credit Transactions Act of 2003
2003	U.S. CAN-SPAM Act of 2003
2005	The Personal Data Privacy and Security Act of 2005
2005	The Financial Data Protection Act of 2005
2009	Health Information Technology for Economic and Clinical Health Act

HEALTH CARE INDUSTRY

Enacted in 1996, the Health Insurance Portability And Accountability Act Of 1996 (HIPAA), effected every health care agency in the United States that handles protected patient health information is required to comply with the Health Insurance Portability and Accountability Act of 1996. HIPAA security standards require health care agencies to adopt policies and procedures that will show how security methods will be implemented that in a way that is "reasonable and appropriate."

HIPAA consists of three sets of standards which include: 1) transactions and codes, 2) privacy, and 3) security. The objectives of these standards are to simplify the management and administration of health insurance claims, to lower costs, and to give patients more control and access to their medical information while protecting their information from real or potential threats of disclosure or loss. HIPAA also requires that healthcare organizations take reasonable and appropriate steps to limit the disclosure of an individual's personal health information, including training employees to follow privacy procedures, designating an individual to oversee the organization's privacy initiatives, and securing access to electronic patient records. The privacy deadline for HIPAA was April 2003; the security deadline was April 2005.

There are distinct differences between privacy standards and security standards. First of all, privacy is the patient's control over the use and disclosure of his or her own personal health information (PHI), whereas security is defined as the "policies, practices, and technology that must be in place for an organization to transact business electronically via networks with a reasonable assurance of safety" (Volonino & Robinson, 2004).

Personal health information or PHI is the HIPAA term for health information in any form (i.e., paper, electronic or verbal) which personally identifies a patient. This also includes individually identifiable health information in paper records that have never been electronically transmitted (http://www.medstarsystems.com/hipaa). Privacy may also include the right to determine when, how, and to what degree PHI is shared with others. The HIPAA privacy rules grant new rights to patients to gain access to and control the use and disclosure of their personal health information.

As required by HIPAA, security and privacy administrators, as well as management, must enhance the security of their networks by setting strong password policies and procedures. Password requirements such as length and strength should be built into every health care agency's security policies. Network administrators are required to regularly change or update employee passwords. Employees should also be frequently reminded about how easy it is for hackers to get passwords through social engineering techniques. New employees should also be taught good password practices. Providing intranet resources related to network security and password security may also be helpful. Finally, the organization's password policy should be integrated into the overall security policy, and all employees should be required to read and sign the policy as a part of their employment record. Systems administrators should implement safeguards to ensure that people on their systems are using adequately strong passwords. This would require such actions as the setting of password expiration dates on all programs, keeping a password history to prevent reuse, and the lock out of accounts after 3 to 5 password attempts. Additionally, the organization should keep the number of people in the organization who have access to these passwords at a minimum. Finally, when installing new systems, they should ensure that default passwords are changed immediately.

HEALTH INFORMATION TECHNOLOGY FOR ECONOMIC AND CLINICAL HEALTH ACT OR HITECH ACT

On February 13th, 2009, The House of Representatives and Senate passed the American Recovery and Reinvestment Act to include new provisions in relation to HIPAA. Inside the new provisions are regulations that address the protection of a patient's privacy. Listed below are these provisions (American Medical Association, 2009):

- Stipulates that Federal privacy/security laws (HIPAA) be expanded to protect patient health information.
- Provides that HIPAA privacy and security laws apply directly to business associates of covered entities.
- Defines actions that constitute a breach of patient health information (including inadvertent disclosures) and requires notification to patients if their health information is breached.
- Allows patients to pay out of pocket for a health care item or service in full and to request that the claim not be submitted to the health plan.
- Requires physicians to provide patients, upon request, an accounting of disclosures of health information made through the use of an EHR.
- Prohibits the sale of a patient's health information without the patient's written authorization, except in limited circumstances involving research or public health activities.
- Prohibits covered entities from being paid to use patients' health information for marketing purposes without patient authorization, except limited communication to a patient about a drug or biologic that the patient is currently being prescribed.

- Requires personal health record (PHR) vendors to notify individuals of a breach of patient health information.
- Requires non-covered HIPAA entities such as Health Information Exchanges, Regional Health Information Organizations, e-Prescribing Gateways, and PHR vendors to have business associate agreements with covered entities for the electronic exchange of patient health information.
- Authorizes increased civil monetary penalties for HIPAA violations.
- Grants enforcement authority to state attorneys general to enforce HIPAA.

It is clear that information privacy and security issues continue to draw the attention of the healthcare industry, the public, and the Obama administration. It is expected that as more and more patient records are being sent electronically and are being converted to digital formats that more laws will be enacted.

FINANCIAL AND BANKING INDUSTRY

As the Internet matures, more and more data is constantly being uploaded to it. Thus, it should not be surprising that every type of data from personal information to financial figures is located on computers that are linked globally via the Internet. Such information linking has a dramatic affect on Internet security. During the late 1990s, a citizen named Joe Barton began a movement that demanded better security be allocated to financial data held by companies, particularly those located on the Internet. When, unknown to him, Barton's financial information was sold by his credit union to the Victoria's Secret intimate apparel merchandiser, Barton knew that legal action was necessary. Not only was Barton receiving catalogs from Victoria's Secret in the mail, but his wife and family began wondering if he had a secret partner to whom he was supplying lingerie. Tired of being harassed due to his personal information being sold, Barton took action (Hoofnagle, 2005). During Barton's legal battle, Representative Ed Markey of Massachusetts introduced an amendment that would add privacy restrictions to and protection of financial information. More specifically, the Markey Amendment was directed at the protection of people just like Joe Barton, citizens who had found that their financial data had been sold or traded to retail companies over the Internet. When Barton testified in court about his Victoria's Secret fiasco, the Markey Amendment was passed and later renamed as the Gramm-Leach Bliley Act of 1999 (GLBA).

There are three subsets defined within the scope of the GLBA: the Financial Privacy Rule, the Safeguards Rule, and a provision against pretexting, defined by the FTC as the act of obtaining financial data under false pretenses (FTC, 2005). The Financial Privacy Rule limits what financial institutions and companies who obtain financial data can do with an individual's information. Financial data can no longer be sold or traded to other companies over the Internet. Not only can the aforementioned financial information not be sold, but it cannot even be displayed to the public or to any other entity on or off the Internet (FTC, 2005). Account numbers cannot be shared with affiliated companies unless so stated in a signed waiver under the Financial Privacy Rule. Within the stated waiver, an opt-out policy must be included allowing consumers to manually remove their name from any type of third party communication. Once manually removed from the list, the consumer will have his or her financial data secured in only one firm, i.e. the initial financial agency (FTC, 2005).

The Safeguards Rule pertains to the protection of customer financial information. Securing customer data is essential to the operations of a financial firm in today's world due to the overwhelming number of Internet abusers attempting to retrieve financial data. Companies that do not directly obtain financial data (such as credit agencies) are still liable under the Safeguards Rule to protect customer financial data at all costs (FTC, 2005). Physical data, such as paperwork contained in file cabinets, are a part of the Safeguards Rule as much as are electronic data (such as databases full of customer information). The confidentiality, integrity, and overall security of financial data are covered within the Safeguards rule (FTC, 2005).

The final provision against pretexting under the GLBA prohibits the act of falsely impersonating a customer to obtain financial data. Under no circumstances can a citizen, company, or entity use deceptive means to obtain data that could compromise sensitive information (FTC, 2005).

In Joe Barton's case, the Financial Privacy Rule was broken in numerous ways. Not only did Barton's credit union sell his financial information to a credit agency using the Internet, but the credit agency in return sold his information to Victoria's Secret (Hoofnagle 2005). With such events occurring to everyday citizens, it is not unreasonable to conclude that most direct mail advertising is the result of a de facto breach in the Financial Privacy Rule of the GLBA. With financial data security covered, an in-depth look into another highly detailed Internet-related law is in order.

Another act that dealt with the financial and banking industry was the Sarbanes-Oxley Act (SOX) of July 2002. Through the early 1990s regulatory examinations were primarily focused on IT issues and the core applications (e.g. deposits, loans, general ledger, etc.) as they related to in-house computer systems. After the preparation and completion of the century date change, more attention was directed towards risk management in the regulatory approach to IT examinations, especially risk management regarding the defining of responsibility for in-house core processing as opposed to activities outsourced to a third party vendor.

Certainly, outsourcing of any type works to overcome organizational obstacles to make operations more efficient. The opportunity to outsource can also provide banks with the opportunity to transfer execution of risk to their third-party service providers. However, on February 29, 2000, the Board of Governors of the Federal Reserve System issued the Supervision and Regulation Letter SR 00-04 that provided regulatory guidance related to the issue of the outsourcing of information and transaction processing activities to either third-party providers or affiliated institutions. The Federal Reserve stated that it expected institutions to ensure that controls over outsourced information would be handled the same as if the activity were handled in-house.

The Sarbanes-Oxley Act (SOX) charged that publicly held companies must have strict corporate governance procedures that include financial disclosure, auditing and accounting. More specifically, Section 404 of SOX mandates that management assess the effectiveness of internal controls over financial reporting (http://www.sarbanes-oxley.com). Information technology plays a critical role in the financial reporting arena in most companies today and is often outsourced by community banks. Financial business processes that have been outsourced are of concern with respect to SOX requirements. When a financial process is transferred to another entity, corporate management assumes the responsibility for the accuracy of such processes and for the compliance of all parties under Section 404. This requires banks to assess whether a service provider is part of the company's internal control of financial reporting. Corporate management may not be aware that this would include any sub-contractors used by the service provider employed by the bank.

In 2005, several members of the U.S. Congress introduced the Financial Data Protection Act that would specifically address brokers of personally identifiable financial data. Essentially, H.R. 3997 is designed to stop data breaches by mandating a national standard for the protection of sensitive information concerning consumers (Moye, 2006). In order to accomplish this objective the

Act requires institutions to notify consumers if data security breaches involving their information might be used to commit financial fraud and in addition requires institutions to provide consumers with a free six-month nationwide credit monitoring service upon notification of a breach (http://www.gao.gov/new.items/d06674.pdf).

It is apparent that the United States Congress, the Federal Trade Commission as well as other government agencies are making an earnest attempt to address the issues of privacy and security concerning personally identifiable information. Without the introduction and eventual adoption of laws and regulations such as the Gramm-Leach Bliley Act which has reinforced the foundation of security concerning financial data and protected individual consumers from having their information shared over the Internet and the CAN/SPAM Act which is currently slowing down the spread of spam e-mail, the issues of privacy and security would certainly be even more of a preoccupation to consumers and regulators alike. Over time it will be seen whether laws, federal and state, can effectively protect and secure information while allowing the Internet to continue to fulfill its potential as a legitimate information sharing medium.

PERSONAL PRIVACY PROTECTION

The Children's Online Privacy Protection Act of 1998 was created to protect children under the age of 13 and to provide to parents control over what information is collected from their children online. In April 2009, the U.S. Federal Trade Commission issued a staff report that addressed the issue of privacy issues of children and smart phones. Their report stated the following:

"The increasing use of smart phones to access the mobile Web presents unique privacy challenges, especially regarding children. The FTC will expedite the regulatory review of the Children's Online Privacy Protection Rule to determine whether the rule should be modified to address changes in the mobile marketplace. This review, originally set for 2015, instead will begin in 2010 and provide opportunity for extensive public comment." (http://www.ftc.gov/opa/2009/04/mobilerpt.shtm)

In addition, the act also provides that if parents do decide to give third parties the right to their child's information that they ultimately have control over how such information may be used. Another central theme of the act is the educating of consumers related to the issue surrounding the privacy of children.

PATRIOT ACT

In October of 2001, President George Bush signed into public law the Uniting and Strengthening America by Providing Appropriate Tools Required to Intercept and Obstruct Terrorism Act of 2001, more commonly known as the Patriot Act (http://www.security-port.com/patriot-act.htm). Among the provisions of the Act that have specific relations with information security and privacy are the following sections (Martins, 2005).

- Section 204 - allows stored voice mail communications to be obtained by a search warrant rather than by having to meet the more stringent wiretap requirements. However, messages on an answering machine tape are not accessible through this provision.
- Section 210 - expands the type of information that an electronic communications provider must disclose. This could include records of session times and duration, temporarily assigned network addresses, and means of payment, and it is not limited to investigations of suspected terrorist activity.
- Section 211 - makes cable companies that provide telephone or Internet services sub-

ject to existing laws that cover telecommunications providers and Internet service providers (ISPs).

- Section 215 - allows the government to seek a court order to obtain personal records such as library, financial, phone, travel, and medical records. This is done by amending the Foreign Intelligence Surveillance Act and is based on a much lower probable-cause standard than that for a regular warrant.
- Section 216 - applies telephone monitoring laws to Internet traffic, including e-mail, Web page, and Internet protocol addresses.
- Section 314- provides for information sharing among financial institutions and between the government and financial institutions.
- Section 319(b) - amends Section 5318 of Title 31 of the U.S. Code to include a "120-hour rule." This provision requires that a financial institution must produce records relating to "any account opened, maintained, administered, or managed in the United States" upon request from an appropriate federal banking agency. This section also provides instruction on maintaining foreign bank records.
- Section 326 - requires that financial institutions verify a person's identity when that person seeks to open an account and to maintain records of the information used in such identification, amending Section 5318 of Title 31, U.S. Code.
- Section 505 - allows the government to seek personal records with no judicial approval through the use of an administrative subpoena. This provision does not expire at the end of 2005 and has been used many times since 2001. It was, however, struck down as unconstitutional by a New York Federal District Court in September 2004. The case is currently awaiting appeal.

Many of the act's provisions were set to sunset on December 31, 2005, and in July 2005, the U.S. Senate passed a reauthorization bill (S.1389) with substantial changes to several sections of the act, while the House reauthorization bill (H.R. 3199) kept most of the act's original language (http://www.cnn.com/2005/POLITICS/12/21/patriot.act/index.html|) The two bills were then reconciled and a "compromise bill," was created which removed most of the changes from the Senate version. The new bill was passed in March 2006 and was signed into law by President Bush that same month.

The Controlling the Assault of Non-Solicited Pornography and Marketing Act of 2003 (or CAN-SPAM) was enacted in late 2003 due to the rampant growth of Internet e-mail spamming. Spamming is the sending out of massive waves of advertising e-mails in hopes of "hooking" a certain amount of sales with each wave. These e-mails, or spam, typically contain messages that may easily fool the average user and infect computers with malicious applications. The most recent wave of e-mail spam contains pornography and other adult content that should not be accessible to younger users.

In a staunch defense of consumer rights, the Federal Trade Commission (FTC) passed the CAN-SPAM Act of 2003 to combat spamming and protect individuals from its effect. Within the CAN-SPAM Act of 2003, various issues ranging from misleading e-mails to sender identification are now legally binding on the e-mail creator. The header of every e-mail sent must now contain legitimate "To and From" information including the originating domain name (FTC, 2004). Further promoting the new legitimacy rules of the CAN-SPAM Act, e-mail subject lines can no longer be deceptive or misleading about the contents of the e-mail. In line with the Gramm-Leach Bliley Act, the CAN-SPAM Act now requires advertising or promotion e-mails to offer an opt-out choice allowing each recipient to remove his or her e-mail address from the e-mail list (FTC, 2004). Arguably the most demanding aspect of the CAN-SPAM Act relates to a physical address requirement. Not only do e-mails containing advertising have to be labeled as an official advertisement, but each e-mail sent out (every piece of spam) must contain the physical address of the spam sender (FTC, 2004).

The United States Senate introduced the Personal Data Privacy and Security Act of 2005 which attempted to define security breaches and the handling of sensitive information as did HR 4127. However, the Senate bill included additional amendments. First, an individual who knowingly hides a security breach causing damage to one or more persons can be fined and/or imprisoned for up to five years. Second, those individuals who steal and use consumer information in a fraudulent manner can have up to two years tacked onto their mandatory prison sentence that is already in effect. Third, the bill would allocate $25 million annually nationwide for enforcement. This allocation of funds, in conjunction with other nuances of the law makes for a solid first try at requiring accountability from companies.

This bill is also significant in that it addresses the question of Government databases that are handled by private contractors. Requirements include an evaluation of potential contractors involved in handling personal data; an audit of commercial data brokers hired for projects involving personal data; and a privacy impact assessment on the use of commercial databases by federal departments.

CONCLUSION

Internet technologies have brought technical, social, ethical, political, and policy challenges to the arena of information security and privacy laws. Advanced technologies such as data mining, Web 2.0, robust Internet infrastructures and expert systems have further fueled many interested parties in the exploitation of personal and supposedly private information. In addition, with the ever increasing number of security risks as evidenced by the rise of identity theft, social engineering, and cyber crime, the influx of these activities has weakened the ability of individuals to secure their own information privacy.

As our society has become ubiquitously interconnected, an increasing number of online users continue to become vulnerable to security breaches and privacy intrusion. Governing bodies from states to national and to international entities have passed laws and regulations to protect an individual's privacy and the security of an individual's information. However, most of these laws have not been able to keep pace with the new challenges accelerated by the development of new technologies. Lack of communication and coordination among countries poses still another challenge to creating a secure global society.

In this chapter, we examined the progression of major information security and privacy laws and regulations. A chronological and holistic view of causes for the continuous development of new laws and regulations can provide a basis for future global collaboration in effectively combating the existing and future challenges in the areas of information security and privacy.

REFERENCES

Antonopoulas, A., (2009, February 17). ATM hack: Organized crime or market forces? *Network World Newsletters*.

California State Senate. (2007). Retrieved on April 2, 2007 from http://info.sen.ca.gov

Conca, C., Medlin, D., & Dave, D. (2005). Technology-based security threats: Taxonomy of sources, targets and a process model of alleviation. *International Journal of Technology Management*, *4*(2), 166–177. doi:10.1504/IJITM.2005.006765

Federal Trade Commission Protecting America's Consumers. (2004). Retrieved on March 22, 2007 from http://www.theiia.org/chapters/index.cfm/view.resources/cid/90.

Federal Trade Commission Protecting America's Consumers. (2005). Retrieved on March 22, 2007 from http://www.ftc.gov/privacy/privacyinitiatives/promises.htm

Hoofnagle, C. J. (2005). *Privacy Self Regulation: A Decade of Disappointment*. Retrieved on March 13, 2007 from http://www.epic.org/reports/decadedisappoint.html.

http://www.medstarsystems.com/hipaa. Retrieved on February 16, 2007.

http://www.security-port.com/patriot-act.htm. Retrieved on February 15, 2007.

Kita, C. I. (2003).. . *IEEE Annals of the History of Computing*, *25*(3), 62–67. doi:10.1109/MAHC.2003.1226656

Library of Congress. (2007). Retrieved on March 19th from http://Thomas.loc.gov.

Martins, C. S., & Martins, S. J. (2005). The Impact of the USA PATRIOT Act on Records Management. *Information Management Journal*, *39*(3), 52–58.

Moye, S. (2006). Congress Assesses Data Security Proposals. *Information Management Journal. 40*(1), 20-23.

Politics: Senate Gives Patriot Act Six More Months. (2005). Retrieved on March 13, 2007 from http://www.cnn.com/2005/POLITICS/12/21/patriot.act/index.html.

Sarbanes-Oxley Financial and Accounting Disclosure Information. (2007). Retrieved on March 13, 2007 from http://www.sarbanes-oxley.com/.

United States Government Accountability Office. Personal Information. (2006). Retrieved on March 25, 2007 from http://www.gao.gov/new.items/d06674.pdf.

United States Government Department of Health and Human Services Fact Sheet. (2003). *Protecting the Privacy of Patients' Health Information.* Retrieved On April 2, 2007 from http://www.hhs.gov/news/facts/privacy.html.

Volonino, L., & Robinson, S. R. (2004). *Principles and Practice of Information Security: Protecting Computers From Hackers and Lawyers*. Upper Saddle River, NJ: Pearson Prentice Hall.

Wisebrod, D. (1995). *Visions of the Future: Governments Wrestle with Information Technology Directed Research Project, University of Toronto.* Retrieved on March 12, 2007 from http://www.catalaw.com/dov/docs/dw-govts.htm.

Chapter 3
The Impact of Privacy Legislation on Patient Care: A Canadian Perspective

Jeff Barnett
University of Victoria BC, Canada

ABSTRACT

This chapter looks at how privacy legislation in Canada may interfere with patient care and the use of electronic health records. A review of the literature shows that legislation across Canada is not uniform and varies to the degree in which it addresses issues of privacy and confidentiality. What is not clear is what impact legislation has on the movement towards electronic health records. A case study was undertaken to look at a specific project involving the design of an electronic health record as a means of sharing information between physicians and prostate cancer patients. Through interviews with those associated with the project, it was clear that legislation itself was not a barrier. The problem was that the legislation was open to interpretation. The author hopes that an understanding of the issues raised by this chapter will lead to further discussion and research on this topic.

INTRODUCTION

Patient privacy is a topic of great interest both in Canada and internationally. Many health conferences and academic journals often have several papers devoted to this topic. This issue has gained more attention in the last five years as there is a continuing trend in healthcare towards more electronic capture and sharing of information between providers and patients. It has been recognized that healthcare has typically been behind other industries like the airlines and banking when it comes to embracing its users in the sharing of electronic information. Patients and their healthcare providers have expressed an interest in sharing of information to facilitate optimal patient care. Unfortunately as more patient information is made available electronically there is an increased potential for loss of privacy and confidentiality.

In Canada most provinces have legislation to ensure that there are safeguards in place to protect confidentiality and security of sensitive personal

DOI: 10.4018/978-1-61692-000-5.ch003

data. However it may be that this legislation is too restrictive and interferes with patient care activities. Many initiatives such as those led by federal and provincial governments are looking at novel ways of providing access to and sharing of electronic patient information. Often these initiatives involve the use of the Internet or some form of information and communication technology (ICT). These initiatives may run into issues with current legislation to the extent that they never get off the ground or, at best, are significantly delayed in implementation. In general, the topics presented usually deal with how to be compliant with legislation or how to go about addressing issues related to privacy. Little has been discussed that addresses how legislation may have an impact on patient care. Indeed there is scarce information in the literature that looks at how the legislation in any particular jurisdiction may have an impact on novel approaches to the delivery of care.

The aim of this chapter is to look at a new and innovative project for the sharing of information for prostate cancer patients in the Canadian province of British Columbia (BC). A case study was undertaken in order to examine whether current legislation has had an impact on this project, known as the Provider Project.

BACKGROUND

The Provider Project

The British Columbia Cancer Agency (BCCA) provides cancer care programs for the people of BC, including prevention screening, early detection, diagnosis and treatment services, support programs, community programs, research and education. BCCA, along with seven other provincially mandated healthcare agencies is a member of the Provincial Health Services Authority (PHSA), which is responsible for governing, managing, and funding this group of healthcare agencies and services.

BCCA is committed to providing all patients with access to a full range of quality cancer services, regardless of where they live. To serve the people of BC, BCCA has a complete network of facilities, such as regional cancer centres, community cancer centres, research centres, and foundations. There are five regional cancer centres located in Victoria, Vancouver, Kelowna, Surrey and Abbotsford. Through these five cancer centres located around the province, various cancer care programs are provided to patients, including assessment and diagnosis, chemotherapy and radiation therapy, counselling services and follow-up care.

One of the many types of cancers treated at the BCCA is prostate cancer. Prostate cancer patients are often exposed to a variety of treatment options such as surgery, radiotherapy, hormonal therapy and, rarely, chemotherapy. Prostate cancer is often referred to as a chronic disease. This is because the tumour is slow growing and with currently available treatments, patients remain very active and alive for many years after diagnosis. Caring for these patients during treatment is complicated and often involves a multidisciplinary approach to care. The members of the care team are usually radiation oncologists (specialized physicians), radiation therapists, nurses, pharmacists, social workers and, most importantly, their primary care physician.

More recently with the increasing involvement of patients in shared decision making, patients have become very active members of their own care team. This is especially true for prostate cancer (PC) patients. Several studies have indicated that these patients are keenly interested in their care and often are very active in seeking out information. However with the complexity of their care, gaps in information between PC patients and their care team can exist. It has been demonstrated that a growing number of patients now refer to the Internet as a source of information in order to fill those gaps. However the web sites dealing with PC are often generalized and do not provide patient specific information. PC patients, just like other patients with a chronic disease, need information pertinent to their condition. More recently attempts have been made to develop personalized web-based electronic health information as an innovative method of delivering patient specific information. The Provider Project is an initiative led by a physician at the BCCA'S Vancouver Island Cancer Centre. The Provider Project has been designed to provide a prototype web-based patient electronic medical record system geared specifically for PC patients. This project has been ongoing for the last five years. The issues around privacy and confidentiality have been a major obstacle for getting this project launched. The vision of Provider Project is to give a group of PC patients access to a subset of data in a customized version of their current paper-based and electronic medical record held at BCCA. However the main challenge has been in dealing with issues of privacy and confidentiality.

In BC in particular there are two pieces of legislation that have direct authority as far as privacy and patient care in BC are concerned. One is the *BC Freedom of Information and Protection of Privacy Act* or FOIPPA. This Act of the Province of BC was introduced in 1997. The main purpose of this act is to "make public bodies more accountable to the public and protect personal privacy by... **preventing the unauthorized collection, use or disclosure of personal information by public bodies...**" Emphasis is added here because this is where there has been controversy about what constitutes "unauthorized collection or disclosure."

A second important piece of legislation in BC is Bill 73 which is an amendment of the BC Freedom of Information and Protection of Privacy Act. It was drafted in response to the U.S. Patriot Act, a Federal Act of the United States that "... allows access to personal information located in Canada that is involved in outsourcing of public body functions to a US linked contractor." Bill 73 came into law in October 2004. Bill 73 in effect eliminates access to any personal information outside Canada. It states "A public body must ensure that personal information in its custody or under its control is stored only in Canada and accessed only in Canada." Again the interpretation of this act has invoked controversy as to what constitutes access and storage. All health authorities and public bodies in BC and many of the research bodies are grappling with the implications of both pieces of legislation. Any clinician or program leader creating new initiatives involving the collection of patient data may be required to fill out a Privacy Impact Assessment (PIA) document and quite often a "Bill 73" Impact statement.

Bill 73 had no effect on the Provider Project and is out of scope of this chapter.

METHODOLOGY

Research Design

The research design used in this chapter consisted of a case study of the Provider Project that was broken out into three tasks:

- The first task involved a literature review to determine the major issues surrounding privacy legislation and how they may impact patient care. This was limited to North America with a focus on Canadian Law as there are many differences in legislation between Canada and the United States.
- The second task was to interview two experts. These individuals were selected for their advice and knowledge in the area of privacy legislation and their knowledge of the Provider Project. One of these experts is a legal specialist from the University of Victoria. The other is a senior privacy officer with the PHSA. The intent of the discussions with the experts was to gain a clear understanding of the relevant legislation in BC and any other legislation that may have an impact on the Provider Project.

- The third phase of this project employed several instruments to collect information about the privacy and confidentiality issues faced by the Provider project during this phase. One of the main techniques used was structured interviews with the project sponsor and other stakeholders associated with the Provider Project.

Data Collection

The structured interviews were centred primarily on the understanding and interpretation of the main issues related to the privacy and confidentiality of patient data. The main focus of the interviews was to look at the challenges and potential roadblocks the stakeholders had experienced in this project. Here the intent was to obtain key factors and "lessons learned" in putting together a project of this nature. In-depth interviews with two experts in privacy were undertaken. The lead physician and members of the Provider Project were interviewed in depth. A record of each interview was recorded using a recording device. Consent to use the device was obtained prior to its use. Once the interviews had taken place the recordings were reviewed and a transcript of each interview was created using Transana Software (Transana version 2.10 open source software from the University of Wisconsin). After the transcripts were written up, themes and keywords were identified using the software.

Secondary Data Sources

In addition to the interviews the case study also involved direct observation, and documentation review. Direct observation included attending formal team meetings of the Provider Project to understand how issues of privacy impact on the project. As well, with the permission of the investigator passive observations also took place whereby the researcher sat in on discussions with Patient Information Management and Information Technology Staff at the BC Cancer Agency.

Several key documents were also reviewed:

- The Provider Project Plan
- An Executive Summary of the Provider Project prepared for the BCCA Executive
- The Privacy Impact Assessment (PIA) for Provider
- Provider team meeting minutes
- Other Provider Project documentation
- Email communications between the various stakeholders.

RESULTS

Review of Privacy Legislation in Canada

Most of the Canadian provinces have legislation covering the protection of personal information, in the form of a *Freedom of Information and Protection of Privacy Act* (FOIPPA). Some provinces like Saskatchewan and Alberta have legislation that is specific to personal health information in the form of Health Information Acts. As provincial governments develop these acts and implement policies addressing the protection of health information, it remains unclear as to what effect legislation has on patient care and an already complex healthcare system.

The world of research has been struggling with these issues for some time and there are many papers in the literature that deal with health research and patient privacy. The use of health information for research purposes is largely governed by the federal *Personal Information Protection and Electronic Documents Act* (PIPEDA), enacted in 2000. This is a federal act that sets out 10 principles of privacy provisions based on the Canadian Standards Association's Model Code for the Protection of Personal Information, recognized as a national standard in 1996.

In the research arena, several authors both in Canada, North America and Europe have identified how legislation has an impact on research.

McCarthy suggests that efforts to protect patient privacy may come into conflict with the ability to produce timely and valid research to safeguard and improve health (McCarthy et al., 1999). The use of health records for research is well known but Upshur suggest that more strict rules and strengthening of consent legislation may have adverse effects on health services research. He argues that it may also lead to an "authorization bias" in health services research (Upshur et al., 2001). Earlier work by Gordis and Gold emphasizes the value of using medical records in research particularly for diseases like cancer, cardiovascular disease and pediatrics (Gordis & Gold, 1980). Yet there are very few examples in the literature where the use of personal data for research purposes has caused harm to Canadians (Upshur, 2001). A further discussion of legislation and privacy issues related to healthcare research is beyond the scope of this review.

When it comes to healthcare delivery and healthcare information we are only now just exploring the issues that are being brought by the movement to electronic health records (EHRs) and the exchange of health information by electronic means. In Canada, federal bodies such as Canada Health Infoway are pushing towards the goal of EHRs for all Canadians. All across Canada there are many new and innovative projects. The main aim of these projects is to improve the access, quality and productivity of Canada's healthcare system. Many projects involve new and innovative approaches to patient care that involves the interchange of patient information in an electronic format. E-health is becoming an accepted part of healthcare. Back in 2002, Richard Alvarez, the current head of Canada Health Infoway who was at that time the President & CEO of the Canadian Institute for Health Information (CIHI), suggested that e-health now focused on the consumer as we have evolved to a new consumer centric model for healthcare. He maintained that Canadians are demanding more from their healthcare system and wish to become more involved, yet at the same time they are increasingly concerned about privacy issues (Alvarez, 2002).

This poses an essential paradox in that the needs of patient and their care must be balanced with their concern for privacy. A commentary in the Canadian Medical Association Journal suggests that "A paradox looms: Canadians demands high-quality, accessible and efficient healthcare *and* privacy for their personal health information" (Upshur et al., 2001). Yet has this contradiction always been here regardless of whether health information is kept in an electronic or paper format? Prior to the advent of e-health, initiative privacy breaches did occur and usually involved individuals finding medical records in a back alley. Whenever a breach like this has occurred it always would make front page news but invariably not last long in the public eye. After the initial shock and dismay the issue would die and be regarded as an isolated occurrence. With the advent of ICT in healthcare and the need for healthcare reform there is a demand to look more closely at privacy issues. Some authors suggest that the issues of privacy and confidentiality of health information have always been with us. The only thing that has changed is the move from paper to electronic data capture. Berger (citation)states that "having one's electronic medical records accessible to any interested employee in a hospital or clinic (or to members of their family when the hospital allows telecommuting) is not that much different than having the various papers, charts and notes stuck in a manila folder in a filing cabinet that anyone in a white coat can access." In his book Database Nation, Garfinkel documents several major cases in the United States where personal health information was abused (Garfinkel, 2001). All the cases he documents have nothing to do with electronic records but are results of unethical actions or mistakes made by healthcare providers. He goes on to suggest that the introduction of computerization into healthcare poses several threats that could severely impact patient privacy. These are based on the assumption that electronic records can be more readily accessed than paper records and are less secure. However, several authors have argued that this is not necessarily the case. In fact there are many articles that suggest moving to electronic records will improve security. Morrissey suggests that handling patient data electronically may actually be more secure than paper-based records (Morrissey, 2006). In his paper entitled, "A day in the life of a Medical Record", he demonstrated how relatively unsecured a paper chart can really be. He goes on to say that short of putting it in a locked filing cabinet, it will never be as secure as an electronic record can be made.

Healthcare in Canada and around the world continues to evolve into a multidisciplinary and multiple-agency networking approach (Weitz et al., 2003). There is a greater degree of service coordination in order to meet the complex needs of individuals. The move toward electronic health records and electronic patient records coupled with the advancement in technologies, especially the Internet, are fuelling this. The irony is that as regionalization continues across Canada and small healthcare groups are collapsed into larger healthcare networks the demand for information sharing intensifies. The so-called healthcare networks are being created in order to gain efficiencies and thereby save scarce healthcare dollars. Yet this is not enough. The healthcare system is also under increasing pressure to do more with less and ICT is seen as one of the ways to help solve these problems (Weitz et al., 2003).

Patients as consumers of healthcare are also demanding more and have become more informed. Physicians and other healthcare providers are expected to keep their patients well-informed while at the same time keeping patient information confidential. Keeping patients informed is one of the main expectations in the consent process. The Canadian Medical Association has developed a privacy code that defines both confidentiality and consent within the scope of medical practice (CMA Health Information Privacy Code, 1998). It states:

Confidentiality, confidential mean that health information that is confided by a patient is to be kept secret and not disclosed or made accessible to others unless authorized by patient consent. A breach of confidentiality occurs whenever a health professional discloses or makes health information available to others without or inconsistent with the patient's consent.

Consent means a patient's informed and voluntary agreement to confide or permit access to or the collection, use or disclosure of his or her health information for specific purposes. Express consent is given explicitly, either orally or in writing. Express consent is unequivocal and does not require any inference on the part of the provider seeking consent. Implied consent arises where agreement may reasonably be inferred from the action or inaction of the individual and there is good reason to believe that the patient has knowledge relevant to this agreement and would give express consent were it sought.

This is where discussion about consent and confidentiality of health information gets mired in debate. Weitz (2003) suggest that the argument comes back to the balance between protecting the individual's rights to privacy and autonomy while at same time allowing providers access to information they need to care for their patients.

In Canada, legislation has been brought in as an attempt to provide a framework of how health information can be collected, shared and accessed. However there is a problem in Canada in that there is a patchwork of federal statutes, provincial legislation and codes of conducts of the various healthcare professionals (Bennet, 1996). The type and extent of legislation varies from province to province and has a different impact on users of healthcare depending on the province they live in. As stated previously, most Canadian provinces have legislation covering the protection of personal information in the public sector, in the form of FOIPPA.

The approach across Canada is far from uniform. The different rules, practices and customs within the various provinces, health districts, hospitals and clinics do not allow for consistent, effective practices for sharing of data or protecting patient privacy (Berger, 2002). Clearly this is a problem for healthcare delivery. Saskatchewan for example has the *Health Information Protection Act* that actually allows individuals to refuse to have their information stored in an electronic format in the Saskatchewan Health Information Network (SHIN). Patients also have the right to restrict access to other providers of health. Weitz suggests that Saskatchewan's Act may be one of the most limiting in Canada as individual rights may place unreasonable limits on what type of health information providers can access.

In Canada the current state of privacy suggests that aside from legislation there is already some infrastructure in place. A survey undertaken by Infoway in 2002 indicated that 9 out of 10 organizations require staff to sign confidentiality agreements and 4 out of 5 regulate access of employees and physicians to clinical records. On the other hand over 50% have policies regulating remote access to clinical records and less than 50% have policies governing patient consent to information sharing (Canada Health Infoway, 2002). The concept of an electronic record is well supported by the Canadian public. An Ekos survey by Infoway suggested that 85% of Canadians do support electronic health records. The same survey showed that Canadians say the following will make them feel more comfortable about electronic health records (Infoway EKOS Survey, 2002):

- Find out when and who accessed their record 71%
- Serious criminal offense for unauthorized access 64%
- Clear and accessible privacy policy 61%
- Ability to access, verify and correct record 57%
- Supported by their doctor 57%

When it comes to information sharing and patient perception there is very little data available in Canada. Weitz suggests that there is a lack of public input into the discussion (Weitz, 2003). He suggests that little is known about how patients feel about this issue. In 1997, Saskatchewan's Minister of Health sent out consultation packages to the public, health providers, district health boards, consumer groups, special interest health groups, researchers, MLAs and the media. This package was intended as part of a public consultation process on the protection of health information (Saskatchewan Health, 1997). When the respondents were asked specifically about information sharing, over 95% agreed that health information necessary for treatment should be available to service providers where and when they need it. In addition 89% agreed that hospitals, health boards, physician's offices and others should be able to exchange health information when it is important for health services. Ironically, Saskatchewan has legislation in the form of the *Health Information Act* that is more restrictive than some of the other provinces, like BC. Sometimes the public may be consulted on issues such as this, but do not ultimately have some bearing on the text of legislation. The consultation process may be more of a public relations exercise.

A similar survey done in the Netherlands on inpatients suggested that patients expect that members of their healthcare should be able to share vital personal health information without their consent for that communication (Klop et al., 1991). In the provinces like Alberta and Saskatchewan that have legislation governing health information, the legislators have tried to take this into account. The legislation that sets out the rules regarding disclosure of information particularly between healthcare providers was created with intent to clarify how disclosure may come about and with less legal or regulatory obstacles. The use of ICT in healthcare has grown rapidly over the past decade but legislation is only now trying to keep pace. It is ironic that health information legislation addressing disclosure has come about in response to advances in technology. Yet as previously mentioned the issues it addresses have always been with us and are not entirely new. A paper chart in a hospital can be accessed by any staff member – for that matter anyone walking into the hospital.

Initially some provincial legislation was considered too restrictive and not practical to implement. For example, when the *Health Information Act* (HIA) was first introduced in Alberta, section 59 required consent from individuals before information could be disclosed. This section stated that for consent to be valid it must include:

a. An authorization for the custodian to disclose the health information specified in the consent;
b. The purpose to which the information may be disclosed;
c. The identity of the person to whom the health information may be disclosed;
d. An acknowledgement that the individual providing the consent has been made aware of the reasons why the health information is needed and the risks and benefits to the individual of consenting or refusing consent;
e. The date the consent is effective and the date if any on which the consent expires; and
f. A statement that the consent may be revoked at any time by the individual providing it.

Section 59 was repealed in 2003 because it was felt that it was too costly to comply with this legislation. Physicians and other healthcare providers felt that they had to spend extra time obtaining patient consents. The Alberta Privacy Commissioner in fact suggested that getting consent to provide information to a provincial EHR would be too costly and impractical. Cornwall in *Connecting Health* looked at Alberta's Pharmaceutical Network and found that physicians were spending an inordinate amount of time explaining the system to patients (Cornwall, 2002). This was largely driven by the current legislation and the physicians' fears around legal liability. Shaw suggested that the desire to provide provincial health networks where information can be shared is bogged down by the lack of privacy legislation in some provinces (Shaw, 2000). He suggests that some projects are slow to move ahead and are suffering because policy is not keeping up with technological change. He raises the suggestion that some projects may be suffering from policy paralysis.

As the EHR is gradually being developed in each province, policy makers both in government and in health authorities will continue wrestling with the issues of privacy and confidentiality. However it is important to point out that in order to make the roll out go smoothly several authors suggest that the public must be consulted. The success of Newfoundland's provincial patient registry is in part due to the involvement of stakeholders and getting a good understanding of the privacy issues way up front. Rather than waiting for legislation to be created by the provincial government the project leaders created privacy guidelines which they shared with their stakeholders that were then presented to the government.

Reis writes that legislation specifically for health information is still relatively fresh as is the projects associated with information sharing. A balance must be struck between the need to protect the privacy and confidentially and the ability to administer the new systems coming into place (Reis, 2005). However she notes that in some provinces such as Ontario, current provisions of legislation places the burden of implementing appropriate security measures on the custodians of information. This is where problems may occur as it places the onus on the custodians to create reasonable safeguards. Failure to do so could be subject to investigations by provincial privacy commissioners (Reis, 2005).

Interviews

Over the summer of 2006 interviews took place with the two privacy experts and members of the Provider team. Each of the interviews lasted over one hour. The two privacy experts were given the same interview instrument. This was slightly different than the one administered to the members of the Provider team. A set of 10 questions was given to the privacy experts and a different set of 10 questions were administered to the team members. As the questions were open-ended there was a great deal of discussion between the interviewer and the interviewee. Once the interviews were recorded a transcript of the interviews was created and this was analyzed using the Transana software. From the interviews with both the team members and the experts the following themes were apparent:

- Privacy Legislation in BC is highly subject to individual interpretation;
- Challenges around staff and resources had an effect on Provider;
- There is a need for better regulatory interoperability between jurisdictions;
- Protection of patient information is not well defined;
- Patient consent is an overriding principle; and
- BC lacks a Health Information Act similar to Alberta.

Privacy Legislation can be Highly Subject to Individual Interpretation

All of the interviewees stated that the current privacy legislation can be open to interpretation depending on the individual and their own biases and background. In the case of the BCCA the legislation that was always referred to in regards to Provider was FOIPPA. Although this legislation has been around since 1996, there appears to be a difference of opinion as to how it affects use of patient information in healthcare. What is important to note is that this piece of legislation does not mention keywords like patient, health record or electronic record. The Act does not in fact specifically address health information. So it is not surprising that it is subject to interpretation. In the case of Provider, FOIPPA was used initially as a reason not to allow for the release of information in the early stages. The stakeholders all echoed the comment that the initial resistance to allow for the transfer of patient information came from individuals within BCCA who said it could not be done because of privacy concerns and FOIPPA. As the Provider project evolved it became clear that there was a willingness on the part of senior IT staff at the BCCA to move ahead. The issues around release of information became more focused on the release of information by electronic means. BCCA had policy that allows for the release of patient information directly to the patient. Like most healthcare institutions patients were able to go to the health record department and request their paper chart. They could sign a release of information form. This would allow health records staff to give out a copy of all or part of their health record. The Provider proponents were well aware of this but felt that if the patient signed a release they could also have that information released in an electronic format. Here in theory if the patient was consenting to the release of information, the format should be irrelevant. Initially Provider team members were told that this was not possible due to privacy legislation. Subsequent requests for the release of the information in an electronic form were denied not due to legislation but by the simple fact that health records staff were not set up to provide information in an electronic format. It became clear that no one at BCCA or PHSA were interested in looking at Provider as a new way of delivering information to patients. To be successful it had to become a research project. As with any research project, it had to have an ethics review and approval by the BCCA, which it obtained successfully. In addition, the project team was required to fill out a Privacy Impact Assessment (PIA) document as part of the process. The intent of a PIA was to ensure that a new program or initiative avoided any violation of the relevant legislation and could mitigate any security risks. As Provider moved into a design phase issues of privacy were a still a prime concern. The members of the project team were initially given limited advice on how to best deal with privacy and confidentiality issues. In the end the team obtained legal help from an expert in the area of health law who helped draft the PIA. This PIA addressed many potential issues that the Provider Project could encounter. Doing the full PIA was time-consuming and required several iterations before it was acceptable to senior management at BCCA. Even then it did not help solve the issue of electronic transfer of information.

Challenges around Staff and Resources had an Effect on the Provider Project

One of the issues that came up during the interviews was that the Provider Project started to evolve at a time when BC was undergoing health restructuring. The project was initially discussed with the Chief Information Officer (CIO) at BCCA. At that time, BCCA was a semi-autonomous body that reported directly to the Ministry of Health. Under restructuring it became an agency under PHSA. The departments of Information Management and Information Technology at BCCA were merged into centralized departments of PHSA. The CIO of BCCA became the CIO of PHSA. One of the managers in information management in BCCA became the Corporate Director for Information Management at PHSA. Both individuals were familiar with the Provider Project but once they assumed new roles they took on corporate projects and priorities. The Provider Project was not seen as a high profile project and it was felt that there were limited or no IT resources that could be devoted to the technical tasks that were required in the project. At one point the main investigator for Provider offered to fund PHSA IT staff to work on the project but this request was not acted upon.

The main issue for the project that required assistance from IT was getting some specific information out of the Cancer Agency Information System (CAIS). CAIS is BCCA's electronic patient record that contains an electronic version of most information contained in the paper chart. The main pieces of data required for Provider residing in CAIS were specialized lab values such as Prostate Specific Antigen (PSA), appointment data, radiation therapy treatment and records of drug therapy. The intent was to have this data transferred electronically into Provider for specific patients. Help was requested from PHSA IT to accomplish this. In the end because IT support from PHSA was not forthcoming due to issues related to workload and concerns about privacy, the Provider team abandoned this approach. In its place they came up with an approach by which patients who had consented to be part of the Provider Project would obtain a paper copy of their information through the release of information process. This would be manually entered into Provider by a research assistant. This process was outlined in the PIA and agreed to by PHSA.

There is a Need for Better Regulatory Interoperability between Jurisdictions

As mentioned in the literature review, privacy legislation has been enacted at the federal and provincial levels. Different provinces have different statutes and regulations concerning the exchange of patient information. This varies from province to province. The problem can be compounded within a province where different health authorities or regional health districts have jurisdictions over an organization. These Health Authorities may have different rules and procedures. Provider was initially planned when the BCCA was an agency of the Ministry of Health and had its own policies and procedures around patient information. When BCCA became part of PHSA, it was governed by the rules and regulations of PHSA. Overlying both organizations was the provincial FOIPPA legislation. In addition, Provider eventually moved into a project involving researchers from the University of Victoria (UVIC). UVIC had policies and rules that had to be taken into consideration as well. Of particular concern was the proposed electronic transfer of information from the BCCA to the University. The computers that would house the Provider Project were located in a secure computing facility at UVIC. The intent was to transfer pertinent data from BCCA to the University. A collaborative agreement was reached between BCCA and UVIC to allow for the handling and sharing of data. This was all part of the PIA. UVIC was willing to allow for the information exchange to take place electronically but PHSA would not, for reasons already discussed. One of the difficulties for the Provider team was that there was a great deal of confusion about the need for data sharing agreements versus a collaborative research agreement between the University of Victoria research office and PHSA. Several discussions took place which created a delay for the project. In the end, no agreement was reached between PHSA, UVIC and the Provider team. An alternate possibility was discussed with regard to getting lab data into Provider. It was thought that because most of the lab data in CAIS originated from labs outside of BCCA, this data could be obtained from the original source. Patients seen at the Vancouver Island Cancer Centre as part of the Provider Project routinely would be getting their lab test done at labs in the city of Victoria. The lab test results were generally available within the Vancouver Island Health Authority (VIHA). One possibility considered was lab data originating in VIHA could be obtained directly from VIHA rather than going through BCCA/PHSA. However it was felt that the security and privacy issues that would arise in dealing with another organization would further complicate and delay the project.

While discussions between the Provider team and PHSA were taking place, there were no other data sharing agreements in existence between organizations that would allow for innovative projects like Provider to take place. Health Authorities in BC were relatively new and were starting to develop policies and procedures around health information. Only within the last few years has there been a move for the privacy officers in each of the health authorities to get together to share information. In the past privacy officers were operating in an ad hoc fashion addressing only what they found out about as opposed to being proactive.

Protection of Patient Information is Not Well Defined

A good deal of the discussion around the Provider Project centre on the protection of patient information. As mentioned the Provider proponents did complete a PIA. In general, the point of doing a PIA is to perform an appraisal of the possible effects that a particular activity or proposal may have on privacy. The Provider team used a PIA template designed by PHSA. This template states that a project would require that PIA be completed if "Transition from paper-based system to electronic system (even if it involves no new uses, collection, disclosure or flow of data)". It also states that this should be completed for information sharing to/from other systems, public bodies, or agencies.

The PIA written for the Provider Project did take this into account. In spite of this the transfer of information from BCCA to the Provider website could not take place electronically. The Provider team settled on having information transferred from the paper-based chart. One of the points that came up in the interviews with Provider team members was the comment that it was difficult to believe that a paper-based process was any more protective of patient information. FOIPPA is vague in this regard as section 30 of the Act states: "A public body must protect personal information in its custody or under its control by making reasonable security arrangements against such risks as unauthorized access, collection, use, disclosure or disposal." It does not specifically address the issue of paper versus electronic. All interviewees appreciated that having patient information in an electronic format raises a different set of issues, but the legislation focuses on the fact that a public body must protect the information. One of the experts suggested that electronic transfer can be made more secure than a paper process. Yet the perception existed that the opposite was true.

FOIPPA also mentions that disclosure can take place provided that the patient has consented in writing. Again it does not say how that disclosure can or cannot take place. This was a central problem for the Provider team. Consent for the release of information was to be obtained from the patient but the interpretation of the legislation was that it could not be released electronically.

Patient Consent is an Overriding Principle

Ironically, one of the reasons that Provider came about was to address the information needs of prostate cancer patients. A literature review and survey by Provider's proponents suggested that this group of patients was highly information-seeking. They contended that prostate cancer patients are motivated to be involved in their care and obtain as much information as possible about their condition. The patients surveyed by the proponents of Provider all were keen to participate in the project and all desired access to their health information. Both experts interviewed agreed that patient consent was an important issue. Both stated that if patients consent, legislation should not prevent transfer of the information. They felt that current legislation in BC does not provide any barriers. Both also stated that there may be other issues such as workload or lack of procedures that would be more of a barrier. But more often than not, the legislation is used as a "scapegoat".

The legal expert interviewed suggested that once consent has been given, there was no legislative or regulatory reason that data from BCCA/PHSA could not be used to populate Provider. Both experts agreed that informed consent should be the overriding principle but more often than not it is overlooked. Both experts also stated that consent is not the piece of paper that the patient signs. It is an ongoing process involving the patient and from which the patient can withdraw at any time. This sometimes gets forgotten and issues irrelevant to the consent process evolve to become barriers. In the case of Provider, all interviewees agreed that patients were very supportive of the project and were keen to have information about their disease available in a timely fashion. This echoes what is reported in the literature about patient access to health information (Denton, 2001; Hassol et al.; 2004; & Pai and Lau, 2005).

The Province of BC Lacks Health Information Similar to the Province of Alberta

An issue that came up with both experts is that BC lacks legislation similar to other provinces like Alberta. Both FOIPPA and the private sector legislation called the *Personal Information Protection Act* (PIPA) make little reference to healthcare and electronic records. On the other hand, Alberta has legislation called the *Health Information Act* (HIA or the Act) which became law on April 25, 2001. This legislation protects the privacy of Albertans and the confidentiality of their health information. It attempts to balance the protection of privacy and confidentiality with the need to enable health information to be shared and accessed where appropriate, to provide health services and to manage the health system. The legislation also provides individuals with the right to access their own information and to have that information protected from inappropriate collection, use and disclosure. At the same time, it defines who are the custodians of the health information and provides a legislative framework for how those custodians can collect, use and disclose health information. BC does not have similar legislation. The experts agreed that this is one reason why the existing legislation can be subject to interpretation and is often referred to as a "moving target". Although the Act may be lacking in some situations or considered overkill, it does specifically cover electronic records and electronic databases. Most importantly one of the main reasons for its existence is "... to enable health information to be shared and accessed, where appropriate, to provide health services and to manage the health system." This Act, as pointed out by the legal expert sits at the provincial level and sets out a legislative framework for all health regions in Alberta. For the Provider, similar legislation in BC may not have assisted in getting it off the ground more quickly but the perceived legislative barriers around privacy would not have occurred. It would also provide a framework for the health authorities so that issues of privacy and sharing of health information would not be subject to interpretation by individuals within each authority.

DISCUSSION

The concept of providing patients access to all or part of their health records is not new. Several papers (Denton, 2001; Hassol et al.; 2004; Matthews & Johnson, 2002; Smith & Protti, 2001) have appeared in recent years that discuss the concept of the Personal Health Record (PHR). Indeed there are many websites in the US where patients can interact with their care provider and track their own healthcare. The spectrum is quite wide all the way from reading disease information to actually monitoring treatment outcomes such as lab results. One example is Epic System's MyChart which is used by many hospitals and clinics throughout the US. At Fairview Health Services in Minneapolis patients use MyChart (http://www.fairview.org/mychart/) to perform many functions such as read lab results, schedule appointments, access family records and communicate with providers just to name a few. There are many examples of this type of PHR in the US but in Canada it is very limited. One site however stands out and this is the Grand River Hospital in the Waterloo region. They have created a patient portal called My Care Source (https://mycaresource.grandriverhospital.on.ca/portal/). Like Epic's, My Care Source patients are enabled to access information and tools to help manage their care. There is an argument that this type of information should be available in Canada. Perhaps the personal health record is the ultimate patient centric approach and should be driving the process. But clearly it is not as yet. There are papers in the literature that support the notion that patients are interested. An early study in 2001 by Denton (Denton, 2001) suggested that patients are interested in keeping some sort of personal record. The author surveyed 330 patients who had been provided a commercially available electronic personal health record (EPHR). Over 90% of the respondents felt that they should keep some sort of EPHR. One important aspect of this study was that 63 respondents were happy to have an EPHR. But they did not want the computer that housed the EPHR connected to the Internet due to Internet fraud. By comparison only 4 respondents were comfortable with their information being on a computer connected to the Internet. A more recent study by Hassol of the MyChart system (Hassol, 2004) used in the Gesinger Health system suggested that patients have a positive attitude to online access to their EHR and only a minority had concerns about privacy and confidentiality.

On a national level the creation of Canada Health Infoway came in 2001 and the move to electronic health records was boosted by the release of the Romanow Report in 2002. Canada began to embark on a journey towards developing an electronic health record. At the same time some clinicians and researchers were also looking at the concept of a patient centred care. Several authors recommended that patients must be at the centre of the health system and must be treated as partners by health professionals – "as equals with different expertise" (Coulter, 2002). Provider was an attempt at creating a form of a patient health record or PHR for a group of prostate cancer patients. At the time the concept of a PHR was very new and so that IT departments and managers did not really know how to deal with it. The project raised several issues around privacy and security that were highly subject to interpretation as current legislation on BC was vague.

Through discussions with the experts and various stakeholders, it was evident that the Provider project was started in at a very challenging time. It was important for the project team to be aware of the issues around privacy but there was confusion as to what the issues and more importantly how to resolve them. As already stated, privacy legislation does not specifically address issues around patients, electronics record or the manner in which information can be disclosed. What FOIPPA does state under section 33 is *"A public body must ensure that personal information in its custody or under its control is disclosed only as permitted under section 33.1 or 33.2."*

Section 33.1 states that:

"A public body may disclose personal information referred to in section 33 inside or outside Canada as follows:

(a) in accordance with Part 2;

(b) if the individual the information is about has identified the information and consented, in the prescribed manner, to its disclosure inside or outside Canada, as applicable."

However the initial advice given was that the transfer of information from the electronic record in CAIS to Provider could not take place due to privacy concerns. An important question in this regard was that no one clearly had an idea of what the privacy concerns were and how the legislation prevented the transfer of data from the electronic record required for Provider. At first the issue was whether this data could ever be obtained and then the issue changed to whether it could be obtained in an electronic format. The advice that the proponents eventually got was that transfer of this information could take place but not by electronic means. Throughout the years when Provider was discussed and planned, records of Provider meetings and emails indicated that managers and agencies voiced concerns about privacy legislation and this prevented the project from moving forward. Yet the existing legislation, FOIPPA, had no direction mention of electronic health records or patient privacy. The Act is written in general terms and was subject to interpretation. This indicates that the legislation was actually not a barrier to Provider as was suggested by senior managers in IT. This lack of clarity might have been reduced if BC had legislation similar to Alberta, as has been mentioned. Some may argue that legislation is not perfect but at least it has set out the legislative groundwork that would have been of great assistance to the Provider team. Others may also suggest that BC health information legislation would come with its own interpretations by people and biases.

One could also argue that the desires and wishes of the patients were not being addressed and this may be considered another issue of balance. Here patients had been consulted and were interested in a new approach to their care. They were willing to take part in a pilot study and would consent to the transfer of information. In spite of this expressed desire on the part of the patients the Provider Project has been slow to evolve.

It is important to state that once a legal expert was consulted and brought into assist with the writing of a PIA, the barriers began to fall. The advice from the legal expert ensured that the PIA addressed any of the issues that could possibly be questioned by individuals with regard to FOIPPA legislation. More importantly her legal expertise was used to clarify what the pertinent issues were and how to proceed once the PIA was written. Her help was invaluable because it was based on knowledge of the relevant legislation and not subject to bias or personal opinion. However the fact that it took a great deal of time to do this was an indication that there was and still is a "legislative void" in BC. Today most of the health authorities in BC in fact do have Privacy Officers who are charged with the task of being familiar with the current legislation and policies of the health authority. These individuals belong to the Information Privacy & Security Working Group (IPSWG) where information is shared and current projects within each health authority are discussed.

Yet without legislation similar to the other provinces there is still the potential for interpretation and bias when dealing with these issues. Gostin suggests in reference to a similar situation in the U.S. "The lack of a uniform policy on interstate dissemination of healthcare information imposes hardships on almost everyone" (Gostin, 1997). This is very similar to Canada in that different provinces have different legislation. Should this be dealt with at a federal level? Gostin thinks so as he argues that "National safeguards that protect the privacy of healthcare information should be based on fair information practices. Federal legislation should establish uniform and comprehensive privacy protection of healthcare information. Privacy protection should cover all healthcare information regardless of its form (paper, microfilm, or electronic), location (in storage, transit, or archives), or user or holder (government, provider, or private organization)."

In Canada this is a challenge in that most Canadian healthcare issues fall under the jurisdiction of the provincial government. Could more health information acts in Canada help solve some of the problems highlighted in Provider? Plater et al (1998) suggest that the provincial "...FOIP-PAs make protecting privacy one of their stated purposes, privacy issues are secondary to enhancing the flow of health information in HIAs". As information technology becomes more complex and the demands by consumers to address complex health concerns the issue of access to and sharing of information will become more complex.

The move on the part of patients to gain more access to information in their electronic chart is happening. The Provider Project was ahead of its time as the concepts expressed in the project were not as well discussed or understood as they are now. However at the heart of this is the thought that patients should be able to have readily available information about their healthcare in an electronic format. As mentioned above several organizations in the US have made this happen. However, again advances in technology have started to push the envelope even more. Google Health and Microsoft's Health Vault are just two examples of new technology. As Health Care Organizations (HCO's) in Canada grapple with the desire for patients to have access they are constrained by the need for privacy protection. The use of electronic health records (EHRs) is considered to be the Holy Grail of Medicine, (Alvarez, 2005). Yet battles around access to the EHR by care providers is still a major issue in Canada that has not been fully resolved. One of the dangers expressed around privacy protection is that if patients believe that their privacy is not being maintained, it may lead to patients forgoing tests or treatment resulting in a worsening of the condition and ultimately compromising the quality of care. It can also lead to an undermining of research, public health, and quality projects due to an imprecise medical record (Goldman, 2009). In spite of these issues both Microsoft and Google are forging ahead to offer consumers their own version of EHRs. Yes these consumer based so-called personal health records (PHRs) are not without problems of their own controversy. Proponents in the US like Ozdemir and Bandyopadhyay (2009) suggest that "the experience may unfold in a manner akin to that of online banking, a somewhat controversial idea when it was first introduced, but now is accepted as matter-of-fact. To continue the banking analogy further, PHRs can be to health records what personal software Quicken or Microsoft Money (which have recently moved online as well) is to financial records".

On the other hand, there is concern that these PHRs may create more problems than they will solve. The control of access to and sharing of information is in the hands of patients. This in itself could create more problems. In particular all access to necessary data may be blocked unless there is consent on the part of the patient. Microsoft's privacy policy projects an ideology of total personal consent for any e-data access (Manos, 2007). In addition both Google Health and Health Vault are not subject to government legislation. Researchers from North Carolina State University question in fact whether these types of PHRs are in fact protecting the individual's privacy. Anton raises serious questions around the ability of individuals to realize what protection may be in place for their personal health information. Her organization, http://theprivacyplace.org, seriously questioned whether patients actually understand the terms expressed in the privacy statement of HealthVault (Anton, 2007). HealthVault has only recently come to Canada so it is too early to tell whether some of the issues raised across the border with regard to Google Health and HealthVault will be similar here in Canada.

In Canada, Dr. Ann Cavoukian current Information and Privacy Commissioner for the Canadian province of Ontario argues that there is a place for citizen access and co-management of their record is possible thanks to the so-called Web 2.0 technologies. In her opinion, this must be facilitated by government. She argues that "government to protect personal information within their care against privacy breaches and ensure control by minimizing personal data at every stage of the information life cycle, from collection to retention and including secure destruction and protecting the data against unauthorized access, tampering or inappropriate disclosure" (Cavoukian, 2009). The argument in context of Canadian healthcare comes back to government, again the crucial player in the evolution of this phenomenon. What has clearly happened is that there is a disconnect between policy and technological growth. The speed with which technical innovations have occurred over the last twenty years is amazing. The introduction of the Apple III in 1983 coincided with the introduction of the Federal Privacy Act in Canada. Apple and other vendors have not stood still and introduced new technologies like the iPod and the Blackberry. Yet the introduction of newer legislation to cover privacy protection has occurred at a snail's pace.

Recently the Province of British Columbia introduced legislation that was created to deal with some of the issues raised with regard to access of information in an electronic format. Bill C24 or the *e-Health (Personal Health Information Access and Protection of Privacy) Act* was introduced in 2008. This legislation deals with some of the key issues that had previously not been addressed by legislation with respect to rights of individuals to access their records. Although several privacy experts such as the Privacy Commissioner of the Province of BC have welcomed this legislation, it still falls short in several areas. This Act does provide a specific legislative framework governing access and privacy for electronic health information databases in the public domain but it does not cover private industry such as software companies like Google and Microsoft. It also does not deal with some of the issues faced by the Provider team. The legislation being sufficiently new is subject to interpretation and still does not address the format and manner for how release of information can take place.

CONCLUSION AND RECOMMENDATIONS

The Provider Project encountered several challenges as it grew from a concept in response to a patient care need. Throughout its development issues around protection of personal privacy were very much on the minds of those involved. As an isolated innovative concept that grew into a research project it encountered several hurdles from inception to proposal. The main point of this case study was to look at the impact that provincial legislation had on the project. In doing this research it was clear that several issues need to be addressed:

- Health organizations will need to look at the logistics for the release of patient data electronically when it is requested by patients or researchers. In the case of Provider, information requested for the project website is available as part of an electronic chart. Patients taking part in the project are expected to obtain access to their own information, which they could get through a release of information request. However the only way those patients could get their own information was in a paper format. No mechanism exists to provide this information electronically. Although Provider is a small scale project with a small number of select patients, this is worrisome for projects on a larger scale;
- Legislation itself did not prevent Provider from moving forward. Rather it was how the legislation was interpreted or how individuals felt it should be applied that impacted the project. The apparent lack of specific healthcare legislation had an impact on this project. Although the health information acts of provinces like Alberta may not be the perfect solution, they could clarify the process of information exchange and disclosure between different organizations. This would help to minimize the amount of interpretation, opinion and bias with regard to how and when information can be released;
- Further research into the privacy legislation and its impact on patient care is needed as there are many projects moving forward that have the potential to encounter some of the issues that the Provider Project ran into. The recent introduction in British Columbia's *e-Health Act* is sufficiently new so it remains to be seen whether its introduction will have a positive impact on the ability of pilot studies like Provider to take place; and
- As software companies like Google and Microsoft continue to develop consumer solutions for healthcare, it will be interesting to examine what new issues around privacy and security for health information will be raised.

REFERENCES

Alvarez, R. C. (2002). The promise of e-Health - a Canadian perspective. *Ehealth International*, *1*(1), 4. PubMeddoi:10.1186/1476-3591-1-4doi:10.1186/1476-3591-1-4

Alvarez, R. C. (2005). *In Search of the Holy Grail. One Year Later.* Retrieved on July 24, 2009 from http://www.longwoods.com/website/events/docs/BWTCTAlvarezMar082005.pdf

Anton, A. (2007). *Is That Vault Really Protecting Your Privacy?* Retrieved on July 24, 2009 from The PrivacePlace.org http://theprivacyplace.org/2007/10/is-that-vault-really-protecting-your-privacy/

Anzen Consulting Inc. and Sextant for Canada Health Infoway (2007). *A 'Conceptual' Privacy Impact Assessment (PIA) on Canada's Electronic Health Record Solution (EHRS)* Blueprint Version 2.

Bandyopadhyay, S., & Ozdemir, Z. (2009, May 8). *On the Potential Role of Independent Personal Health Record Platforms in the Digitization and Sharing of Health Record.* Retrieved on July 24, 2009 from SSRN: http://ssrn.com/abstract=1402983.

Bennett, C. (1996). Regulating Privacy in Canada. An Analysis of Oversight and Enforcement in the Private Sector. *Advisory Council on Health Infostructure, Paths to Better Health: Final Report.* Ottawa: Health Canada Publications.

Berger, E. (2002). Attitudes to privacy, health records and interconnection: Implications for healthcare organizations. *Hospital Quarterly, 5*(4), 40-5, 2.

Buckovich, S. A., Rippen, H. E., & Rozen, M. J. (1999). Driving toward guiding principles: A goal for privacy, confidentiality, and security of health information. *Journal of the American Medical Informatics Association, 6*(2), 122–133. PubMed

Canada Health Infoway. (2002). *Pan-Canadian EHR Survey Phase I: Results and Analysis.* Retrieved April 1, 2008 from http://www.longwoods.com/view.php?aid=17335. Ottawa: Canada.

Canadian Medical Association. (1998). Health information privacy code. *Canadian Medical Association Journal, 159*, 997–1006. PubMed

Carter, M. (2000). Integrated electronic health records and patient privacy: Possible benefits but real dangers. *The Medical Journal of Australia, 172*(1), 28–30. PubMed

Cavoukian, A. (2009). *Privacy and Government 2.0: The Implications of an Open World Information and Privacy Commisioner of Ontario.* Retrieved on July 3, 2009 from http://www.ipc.on.ca/images/Resources/priv-gov-2.0.pdf.

Cornwall, A. (2003). Connecting health: a review of electronic health record projects in Australia, Europe and Canada. Retrieved on September 1, 2006 from http://www.piac.asn.au/publications/pubs/churchill_20030121.html. Sydney, Public Interest Advocacy Centre.

Coulter, A. (2002). After Bristol: Putting patients at the centre. *BMJ (Clinical Research Ed.), 324*(7338), 648–651. PubMeddoi:10.1136/bmj.324.7338.648doi:10.1136/bmj.324.7338.648

Denton, I. C. (2001). Will patients use electronic personal health records? responses from a real-life experience. *Journal of Healthcare Information Management, 15*(3), 251–259. PubMed

Freedom of Information and Protection of Privacy Act (FOIPPA), (1997). Retrieved on March 1, 2008 from http://www.qp.gov.bc.ca/statreg/stat/F/96165_00.htm. BC

Garfinkel, S. (2001). To know your future. In *Database nation: the death of privacy in the 21st century.* Sebastopol, CA: O'Reilly & Associates, Inc.

Giokas, D. (2004). *Privacy Implication for the Interoperable EHR.* Retrieved on September 1, 2007 from http://www.rileyis.com/seminars/sept04/GiokasSept04.pdf

Goldman, J., & Hudson, Z. (1999). *Promoting Health/Protecting Privacy: A Primer.* Retrieved on July 24, 2009 from http://www.chcf.org/topics/view.cfm?itemID=12502.

Goldman, J., & Hudson, Z. (2000). Virtually exposed: Privacy and e-health. *Health Affairs (Project Hope), 19*(6), 140–148. PubMeddoi:10.1377/hlthaff.19.6.140doi:10.1377/hlthaff.19.6.140

Gordis, L., & Gold, E. (1980). Privacy, confidentiality, and the use of medical records in research. *Science, 207*(4427), 153–156. PubMeddoi:10.1126/science.7350648doi:10.1126/science.7350648

Gostin, L. (1997). Health care information and the protection of personal privacy: Ethical and legal considerations. *Annals of Internal Medicine, 127*(8 Pt 2), 683–690. PubMed

Hassol, A., Walker, J. M., et al. (2004). Patient Experiences and Attitudes about Access to a Patient Electronic Health Care Record and Linked Web Messaging. *Journal of the American Medical Informatics Association, 11*, 505–513. PubMed-doi:10.1197/jamia.M1593doi:10.1197/jamia.M1593

Health Information Act, Alberta (1999). Retrieved on September 1, 2007 from http://www.oipc.ab.ca/hia/

Hollaman, L. (2002). Practical realities: Potential impact of the draft privacy of personal information act on healthcare providers. *Hospital Quarterly, 5*(4), 46–48. PubMed

Klop, R., van Wijmen, F. C., & Philipsen, H. (1991). Patients' rights and the admission and discharge process. *Journal of Advanced Nursing, 16*(4), 408–412. PubMeddoi:10.1111/j.1365-2648.1991.tb03430.xdoi:10.1111/j.1365-2648.1991.tb03430.x

Kosseim, P. (2003). The landscape of rules governing access to personal information for health research: A view from afar. *Health Law Journal, 11*, 199–215. PubMed

Lombardi, R. (2008). *Consumer control over personal medical data is coming to Canada*, Retrieved on July 24, 2009 from InterGovWorld.com, 08 September 2008, http://www.itworldcanada.com/a/search/42adc562-c0a8-0006-001d-88f64bbf545d.html

Manos, D. (2007) *Microsoft joins lawmakers, activists to demand patient privacy rights*. (Healthcare IT News, 18 October 2007).

Matthew, I. K., & Johnson, K. B. (2002). Personal Health Records: evaluation of functionality and utility. *Journal of the American Medical Informatics Association, 9*(2), 171–180. PubMed-doi:10.1197/jamia.M0978doi:10.1197/jamia.M0978

McCarthy, D. B., Shatin, D., Drinkard, C. R., Kleinman, J. H., & Gardner, J. S. (1999). Medical records and privacy: Empirical effects of legislation. *Health Services Research, 34*(1 Pt 2), 417–425. PubMed

McGaw, D. (2008). Comprehensive Privacy and Security: Critical for Health Information Technology. *Centre for Democracy and Technology Version 1.0.* Washington, DC: CDT Health Privacy Project.

Morrissey, J. (2006). *A Day in the Life of a Medical Record - Lifting the veil on the security of today's paper-based environment.* The National Alliance for Health Information Technology. Retrieved on September 1, 2007 from http://www.nahit.org/dl/A_Day_in_the_Life.pdf

Myers, J., Frieden, T., Bherwanj, K., & Henning, K. (2008). Ethics in public health research: privacy and public health at risk: public health confidentiality in the digital age. *American Journal of Public Health, 98*(5), 793–801. PubMed-doi:10.2105/AJPH.2006.107706doi:10.2105/AJPH.2006.107706

Pai, H., & Lau, F. (2005). Web-based electronic health information systems for prostate cancer patients. *The Canadian Journal of Urology, 12*(3), 2582–2591. PubMed

Plater, S., Seeley, E., & Dixon, L. A. (1998). Two routes to privacy protection: A comparison of health information legislation in Canada and the United States. *Journal of Women's Health / the Official Publication of the Society for the Advancement of Women's Health Research, 7*(6), 665-672.

Policy and Planning Division. Saskatchewan Health. *Consultation paper on Protection of Personal Health Information*. Retrieved on September 1, 2007 from http://www.health.gov.sk.ca/adx/aspx/adxGetMedia.aspx?DocID=286,94,88-,Documents&MediaID=126&Filename=protection-of-person-health-information.pdf

Reis, N., & Moysa, G. (2005). Legal Protections of ElectronicHealth Records: Issues of Consent and Security. *Health Law Review*, *14*(1), 18–25. PubMed

Robinson, D. M. (1994). Health information privacy: Without confidentiality. *International Journal of Bio-Medical Computing*, *35*(Suppl), 97–104. PubMed

Rozovsky, L. E., & Rozovsky, F. A. (1985). The computer challenge: More data, less privacy. *Canadian Doctor*, *51*(3), 46–48. PubMed

Shaw, A. (2000). *Provincial Health Networks Slowed Down by the Lack of New Privacy Legislation.* Canadian Health Care Technology (March 2000).

Smith, J. R., & Protti, D. J. (2001). *Personal Health Records: part II: a potential model for Canada?* Healthcare Information Management and Communications.

Tapp, A. (2004). CNPS: Respecting your privacy rights. *The Canadian Nurse*, *100*(4), 36–37. PubMed

Upshur, R. E., Morin, B., & Goel, V. (2001). The privacy paradox: Laying Orwell's ghost to rest. *Canadian Medical Association Journal*, *165*(3), 307–309. PubMed

Weitz, M., Drummond, N., Pringle, D., Ferris, L. E., Globerman, J., Hebert, P., et al. (2003). In whose interest? current issues in communicating personal health information: A Canadian perspective. *The Journal of Law, Medicine & Ethics*, *31*(2), 292–301. PubMeddoi:10.1111/j.1748-720X.2003.tb00090.xdoi:10.1111/j.1748-720X.2003.tb00090.x

Willison, D. J. (1998). Health services research and personal health information: Privacy concerns, new legislation and beyond. *Canadian Medical Association Journal*, *159*(11), 1378–1380. PubMed

Yung, A., & Li, Y. (2009). *A Literature Review of Developments in Health Information Banks,* Retrieved on July 24, 2009 from *ElectronicHealthcare, 7*(4), e1-e5.

Chapter 4
A Privacy Protection Model for Patient Data with Multiple Sensitive Attributes[1]

Tamas S. Gal
University of Maryland Baltimore County (UMBC), USA

Zhiyuan Chen
University of Maryland Baltimore County (UMBC), USA

Aryya Gangopadhyay
University of Maryland Baltimore County (UMBC), USA

ABSTRACT

The identity of patients must be protected when patient data is shared. The two most commonly used models to protect identity of patients are L-diversity and K-anonymity. However, existing work mainly considers data sets with a single sensitive attribute, while patient data often contain multiple sensitive attributes (e.g., diagnosis and treatment). This chapter shows that although the K-anonymity model can be trivially extended to multiple sensitive attributes, L-diversity model cannot. The reason is that achieving L-diversity for each individual sensitive attribute does not guarantee L-diversity over all sensitive attributes. The authors propose a new model that extends L-diversity and K-anonymity to multiple sensitive attributes and propose a practical method to implement this model. Experimental results demonstrate the effectiveness of this approach.

INTRODUCTION

Patient data is often shared for research and disease control purposes. For example, the Center for Disease Control and Prevention has a National Program of Cancer Registries which collects data on cancer patients. Such data is made available to public health professionals and researchers to understand and address the cancer burden more effectively.

Privacy is one of the biggest concerns in sharing patient data because without appropriate protection, personal information is vulnerable to misuse. For example, identity theft remains the top concern among customers contacting the Federal Trade Commission (Federal Trade Commission,

DOI: 10.4018/978-1-61692-000-5.ch004

2007). According to a Gartner study (Gartner Inc., 2007), there were 15 million victims of identity theft in 2006. Another study showed that identity theft cost U.S. businesses and customers $56.6 billion in 2005 (MacVittie, 2007). Therefore, legislation such as the Health Insurance Portability and Accountability Act (HIPAA) requires that health care agencies protect the privacy of patient data. This chapter focuses on models that protect identity of patients and at the same time still allow analysis to be conducted on the sanitized data.

K-Anonymity and L-diversity privacy protection model: The two most commonly used privacy protection models for identity protection are K-anonymity (Sweeney, 2002b) and L-diversity (Machanavajjhala et al., April 2006). K-anonymity prevents *linking attack*, which recovers private information by linking attributes such as race, birth date, gender, and ZIP code with publicly available data sets such as voter's records. Such attributes that appear in both public and private data sets are called *quasi-identifiers*. The K-anonymity model divides records into groups with sizes ≥ K such that each group has identical value or range on quasi-identifier attributes.

Example 1. Figure 1 shows some patient records, where age is the quasi-identifier and disease type and treatment are sensitive attributes (i.e., attributes with privacy sensitive information). Figure 2 shows the anonymized data where the first four rows belong to the same group and have the same range of age. Linking attack cannot discover the identity of a patient using the age attribute because there are at least K (K = 4) patients with the same age range.

L-diversity further enhances K-anonymity by preventing another type of privacy attack called *elimination attack* (which was used by Sherlock Holmes to solve mysteries by excluding the impossible). We use an example to illustrate elimination attack. In Figure 2, if K=3, then the first 3 patients satisfy 3-anonymity. However, they have only 2 different disease type values: heart disease and flu. If someone knows that the patient with ID 3 is unlikely to have heart disease, then he can infer that the patient most likely has flu.

L-diversity prevents elimination attack by requiring that the values of privacy sensitive attributes (e.g., the attribute disease type) in a group have enough degree of diversity. Several

Figure 1. Original patient data

Patient ID	Age	Disease Type	Treatment
1	42	Heart disease	Medicine
2	41	Heart disease	Surgery
3	49	Flu	Intravenous therapy
4	43	Stomach disease	Intravenous therapy
…	…	…	…

Figure 2. Anonymized patient data with K=4

Patient ID	Age	Disease Type	Treatment
1	41-50	Heart disease	Medicine
2	41-50	Heart disease	Surgery
3	41-50	Flu	Intravenous therapy
4	41-50	Stomach disease	Intravenous therapy
…	…	…	…

different definitions of diversity were proposed in (Machanavajjhala et al., April 2006). In this chapter, we use the definition of L-diversity where a sensitive attribute in a group contains at least *L* different values. For example, in Figure 2, the group of first four patients has 3 different disease types and is thus 3-diverse. If the attacker knows that a patient does not have heart disease, he cannot decide which type of disease the patient has because he may have either stomach disease or flu. In general, L-diversity can protect privacy against attackers who are able to eliminate up to *L-2* values.

Problem of data with multiple sensitive attributes: There has been a rich body of work on implementation and application of K-anonymity and L-diversity (Bayardo et al., 2005; LeFevre et al., 2005; Menon et al., 2006; Samarati, 2001; Xiao et al., 2006). However, existing research considers data sets with only one sensitive attribute, while many data sets such as patient data have multiple sensitive attributes (e.g., disease and treatment). Since K-anonymity does not have any condition on sensitive attributes, it can be applied to data sets with multiple sensitive attributes. However, we will next show that L-diversity cannot be directly applied to data sets with multiple sensitive attributes.

Suppose in Figure 2 both disease type and treatment are sensitive attributes. The group of the first four patients has 3 distinct values on both attributes. However, if the attacker knows that a patient does not have heart disease, he can decide that this patient has IV (Intravenous therapy) as treatment because only patients with heart diseases received the other two types of treatment. Thus this group is not 3-diverse. The root cause of this problem is that the elimination of rows containing one sensitive attribute value may eliminate multiple values of other sensitive attributes. In this case, the elimination of rows containing the value heart disease (i.e., the first two rows) also eliminates values medicine and surgery. Therefore, preserving L-diversity on each individual sensitive attribute will not preserve L-diversity for multiple sensitive attributes.

Our contributions: This chapter has made the following contributions.

- We propose a privacy model that extends K-anonymity and L-diversity to data with multiple sensitive attributes.
- We provide a method to implement our model. Experimental results show that our method also introduces little distortion to data, which will help subsequent data analyses.

The rest of the chapter is organized as follows. The next section describes related work. We then describe the proposed privacy model for multiple sensitive attributes and describe how to implement the model. Finally we present the experimental results and conclude the chapter.

RELATED WORK

The existing work on privacy protection techniques can be divided into two categories: those that protect personal identity (called data anonymization) and those that protect sensitive attribute values or sensitive patterns. We first review literature in both categories and then describe the relationship of this chapter to the literature.

Data Anonymization: The research in this field is based on two privacy protection models: K-anonymity and L-diversity. The K-anonymity model was first proposed by Sweeney in (Sweeney, 2002b) and it protects the data against linking attacks. The L-diversity model was proposed as a complement to K-anonymity in (Machanavajjhala et al., April 2006). L-diversity further protects the data against elimination attack. It is a general consensus in the field that both models need to be implemented to protect privacy.

There are two approaches to implement K-anonymity and L-diversity. The first approach is generalization, i.e., replacing values of quasi-identifiers with more general values (Samarati, 2001; Sweeney, 2002a). The second approach is called anatomy and it divides data into multiple tables, one storing the quasi-identifier attributes and group ID of each record, and the others storing the count of sensitive attribute values in each group (Xiao et al., 2006). This approach introduces less distortion to data (Xiao et al., 2006). LeFevre et al. proposed a method based on full domain generalization (LeFevre et al., 2005), which maps the domain (or range) of attribute values to a more general domain. For example, all five-digit zip codes can be generalized to a domain that contains only the first four digits (e.g., 21250 will become 2125*). A method that uses k-d tree to divide data into groups was proposed in (LeFevre et al., 2006). K-d tree puts data with similar values into the same group, leading to less data distortions.

Hiding sensitive values and patterns: Research in this field is also called privacy-preserving data mining because the goal is to preserve privacy and at the same time allow data mining on the modified data. A survey can be found in (Verykios et al., March 2004). The most well known method in this field is random perturbation, which adds some random noise to sensitive attribute values (Agrawal et al., 2001; Agrawal et al., May 2000; Zhu et al., 2004). However, Kargupta et al. showed that random perturbation method is subjected to attacks using correlations of data (Kargupta et al., 2003). A similar method for association rule mining was proposed in (Evfimevski et al., June 2003; Evfimevski et al., July 2002). A tree-based approach (Li et al., 2006) was also proposed. The basic idea is to divide data into groups using k-d tree, and replace values of sensitive attributes with the average of their values within a group.

There has also been work to hide sensitive patterns such as association rules in the data (Hintoglu et al., 2005; Menon et al., 2006; Menon et al., September 2005; Oliveira et al., 2002; Saygin et al., 2001; Verykios et al., April 2004).

Relationship of this chapter to existing research: This chapter focuses on data anonymization (i.e., the first category) for two reasons. First, it is important to hide the identity of patients. Second, many privacy protection methods proposed in the second category of research only work for numerical attribute values, while patient data contain many categorical attributes. For example, for the tree-based approach (Li et al., 2006), it is unclear how to compute the average of a categorical attribute such as disease type.

As mentioned in the Introduction, existing work on data anonymization mainly considers the data set with only one sensitive attribute, while patient data often contains multiple sensitive attributes. We have also shown in the Introduction, that the L-diversity model cannot be directly extended to multiple sensitive attributes. This chapter proposes a privacy protection model that works for multiple sensitive attributes.

A more recent work (Li et al., 2008) implements the L-diversity model proposed by (Machanavajjhala et al., April 2006)), which requires that each attribute have at least *L* different values for all records with the same values on the other attributes. However, such requirement will lead to very large group sizes because the group size is at least L^m, where m is the number of attributes. Large group sizes will introduce too much data distortion, making the data not useful for subsequent analysis. The methods proposed in this chapter do not have this problem.

PRIVACY MODEL OVER MULTIPLE SENSITIVE ATTRIBUTES

This section proposes our privacy model. The first subsection reviews the existing K-anonymity and L-diversity models over a single sensitive attribute. The second subsection proposes a novel privacy model over multiple sensitive attributes. The third subsection proposes a variant of our model that deals with data sets with very few distinct values on some sensitive attributes.

Existing Model

Let *T* be the data to be shared. *T* contains *d* quasi-identifiers attributes $A_{Q1}, ..., A_{Qd}$, and *m* sensitive attributes $A_{S1}, ..., A_{Sm}$. Based on the literature (Sweeney, 2002b), we assume quasi-identifier attributes and sensitive attributes have no overlap because sensitive attributes normally do not appear in public available data sets. Next we give a few definitions.

Definition 1. (Partition/Group) A **partition** consists of several subsets of *T*, such that each record in *T* belongs to exactly one subset. We refer to these subsets as **groups** and denote them as $G_1, ..., G_p$.

Next we describe two approaches to implement K-anonymity and L-diversity: generalization (Samarati, 2001; Sweeney, 2002a) and anatomy (Xiao et al., 2006).

Definition 2. (Generalization) Given a partition of *T*, generalization of T makes all records in group G_i to have the same values on quasi-identifier attributes $A_{Q1}, ..., A_{Qd}$.

For example, Figure 2 in the Introduction shows a generalization of T, where the first four records belong to one group. Numerical values such as ages can be generalized to ranges, and categorical values can be generalized according to a predefined hierarchy.

Definition 3. (Anatomy) Anatomy of *T* produces a quasi-identifier table (QIT) and *m* sensitive tables (ST) as follows. The QIT stores all quasi-identifier attributes $A_{Q1}, ..., A_{Qd}$ and the group ID of each record. Each ST stores the group ID, distinct values of a sensitive attribute A_{Sj} $(1 \leq j \leq m)$, and the count of the value of A_{Sj} in each group.

For example, Figure 3 shows anatomy of T where the table on the left is the QIT table, and the other two tables are ST for sensitive attribute "disease type" and "treatment".

It has been shown in (Xiao et al., 2006) that anatomy and generalization are equivalent in terms of privacy protection. For example, suppose an attacker knows that a patient's age is 49 (the third patient). All four patients in the generalized table (Figure 2) have age in the range of 41-50. Thus the attacker cannot decide which record belongs to this patient. He can do a random guess and the probability of getting the correct disease type (flu) is 25% because only one of four patients has flu. Similarly, when the attacker looks at the anatomized table in Figure 3, the attacker can learn that the patient with age 49 belongs to the first group using the QIT table. However, he cannot figure out the exact disease type of the patient because the ST table only stores counts of disease types for the first group. Thus the attacker can only do a random guess of the patient's disease type and the probability of a correct guess is again 25%. This chapter uses the anatomy approach to implement our privacy model because as shown in the work of (Xiao et al., 2006)), anatomy often leads to a smaller degree of data distortion and benefits subsequent analysis of anonymized data.

Definition 4. (K-anonymity) A data set T satisfies K-anonymity if it is divided into a partition and

Figure 3. Anatomized patient data

Patient ID	Age	Group ID
1	42	1
2	41	1
3	49	1
4	43	1
...	...	...

Group ID	Disease Type	Count
1	Heart disease	2
1	Flu	1
1	Stomach disease	1
...	...	...

Group ID	Treatment	Count
1	Medicine	1
1	Surgery	1
1	IV	2
...	...	...

each group G_i $(1 \leq i \leq p)$ in the partition contains at least K records, and T is either generalized or anatomized.

Definition 5. (L-diversity for a single sensitive attribute) A data set T satisfies L-diversity if it is divided into a partition and the sensitive attribute A_{S1} contains at least L different values in each group G_i. T is also generalized or anatomized.

The data in Figure 2 satisfies 4-anonymity because each group contains at least 4 records. If disease type is the only sensitive attribute, this data set is also 3-diverse.

It is clear that K-anonymity has no condition on sensitive attributes, thus K-anonymity model applies to data sets with any number of sensitive attributes. However, as shown in the Introduction, L-diversity cannot be directly extended to multiple sensitive attributes because making each sensitive attribute L-diverse does not guarantee that all sensitive attributes are L-diverse.

Our Model

This section presents our privacy model. We assume that when a distinct sensitive attribute value is deleted from a group, all rows containing that value will be deleted. For example, if the value "heart disease" is deleted in Figure 1, the first two rows are both deleted. **Definition 6. (K-anonymity and L-diversity for multiple sensitive attributes)** T satisfies both K-anonymity and L-diversity if T is divided into a partition and each group G_i $(1 \leq i \leq p)$ in the partition satisfies that: (1) the group contains at least K records, and (2) at least L distinct values (possibly from multiple sensitive attributes) need to be deleted to delete all rows in the group. T is also anatomized or generalized.

The first 4 patients in Figure 2 satisfy 4-anonymity. They also satisfy 2-diversity because we need to delete at least two values (e.g., "heart disease" and "IV") to delete all the rows in the group. The new model survives linking attack because it ensures K-anonymity. The next theorem shows that it also survives elimination attacks.

Theorem 1. Our privacy model survives elimination attacks with up to L-2 values, the same as the L-diversity model over a single sensitive attribute.

Proof: If the attacker eliminates L-2 sensitive attribute values, the remaining rows must have at least 2 distinct values for each sensitive attribute. This is because if the remaining rows have only one distinct value, all rows in the group will get deleted after deleting L-1 values, and this violates the second requirement of L-diversity.

Machanavajjhala et al. (Machanavajjhala et al., April 2006) proposed a definition of L-diversity over multiple sensitive attributes where each sensitive attribute must have at least L distinct values for records that have the same values on all other attributes. For example, if there are two sensitive attributes, disease type and treatment, then each disease type must have L corresponding treatment values and vice versa. For a data set with m sensitive attributes, the group must have at least L^m rows. Note that rows in the same group will be generalized or anatomized. Thus having such large group sizes will introduce too much data distortion, making the data not useful for subsequent analysis. We will compare our model with this model experimentally in Experimental Evaluation.

Variant of Our Model with Column Wise-Constraints

The model proposed so far treats all sensitive attributes uniformly. However, in practice some sensitive attributes may have very few distinct values while other attributes may have many. Thus sometimes it makes sense to allow a different degree of diversity on different attributes. We propose a variant of our model that adds this flexibility.

Definition 7. (Privacy model with column-wise *constraints*)T satisfies both K-anonymity and L-diversity with column-wise constraints if T is divided into a partition and each group G_j $(1 \leq$

$j \leq p$) in the partition satisfies that: (1) the group contains at least K records, and (2) to delete all rows in the group, at least L distinct values need to be deleted, and *at most* L_i ($0 \leq L_i \leq L$) of them are from sensitive attribute A_{Si}. T is also anatomized or generalized.

The definition is the same as the general model in Definition 6 except that at most L_i values will be deleted from attribute A_{Si}. For attributes with few distinct values, the user can set a $L_i < L$ such that fewer values can be deleted from this attribute. For attributes with many distinct values, the user can set $L_i = L$ (i.e., still allow L values to be deleted from that attribute).

Consider the data set in Figure 4 (quasi-identifier attributes are not shown). Let L_1 and L_2 be the constraints on disease type and treatment, respectively. Disease type attribute has 6 distinct values and treatment attribute only has 3. Thus we can set $L_2 = 2$ for treatment such that at most 2 treatment values can be deleted. After the deletion of 2 treatment values (e.g., T1 and T2), there are always 2 disease types left (e.g., D5 and D6). Hence at least 4 deletions (e.g., deleting T1, T2, and D5, D6) are needed to delete all rows and $L = 4$. We set $L_1 = 4$ because disease type has many values. Without column-wise constraints, L can be 3 at most because there are only 3 distinct treatment values. The next theorem shows the relationship of these two models.

Theorem 2. If a group G satisfies L-diversity without column-wise constraints (the general model), then G also satisfies L-diversity with constraints where $L_i \leq L$.

Figure 4. Example For Column-Wise Constraints

Disease Type	Treatment
D1	T1
D2	T1
D3	T2
D4	T2
D5	T3
D6	T3

Proof: Suppose G satisfies the L-diversity model without column-wise constraints. If G is not L-diverse with constraints, then there exists $L' < L$ such that G satisfies L'-diversity with constraints. Thus all rows in G will be deleted after deleting L' values (with at most L_i of them from attribute A_{Si}). This conflicts with the condition that G satisfies L-diversity without constraints (i.e., at least L ($L > L'$) values need to be deleted to delete all rows in G). Thus G satisfies L-diversity with constraints. The reverse of Theorem 2 is not true. For example, the data in Figure 4 satisfies the column-wise model with $L=L_1=4$, $L_2=2$, but not the general model with $L=4$.

METHOD TO IMPLEMENT THE MODEL

This section describes how to implement our privacy model. The first subsection shows an overview. The second subsection proposes a method to check L-diversity. The third subsection shows how to check L-diversity with column-wise constraints.

Overview

Figure 5 shows the algorithm to anonymize the data. It has three input parameters: the data set T, and the parameter K and L in our privacy model. It contains two steps. In the first step, data is divided into a partition such that each partition contains at least K records and satisfies L-diversity. In the second step data is anatomized. Next we describe these two steps.

Figure 5 also shows the algorithm (Split) to partition the data. The algorithm follows the K-d tree approach to generate the partition (LeFevre et al., 2006). The benefits of K-d tree is that records with similar values will be put in the same group, thus there will be less data distortion. The algorithm works top down, i.e., starting with the whole data set as a single group G and then split-

Figure 5. Anonymize algorithm

```
Anonymize(data set T, K, L)

    1) P = empty set
    2) Split(T, K, L, P)
    3) Anatomize(T, P)


Split(Current group G, K, L, partition P)

    1) A_Qj = choose_dimension()
    2) splitVal = find_median(G, A_Qj)
    3) G_L = {records in G and with value on A_Qj ≤ splitVal}
    4) G_R = {records in G and with value on A_Qj > splitVal}
    5) if Satisfy-Model(G_L, K, L)
    6)    Split(G_L, K, L, P)
    7) if Satisfy-Model(G_R, K, L)
    8)    Split(G_R, K, L, P)
    9) if neither G_L nor G_R satisfies our model
    10) Add T to P
Satisfy-Model(G, K, L)
```

ting the existing groups into smaller groups. The algorithm stops when further splits will violate K-anonymity or L-diversity conditions.

At line 1 the algorithm selects a quasi-identifier attribute to split. Following the literature (LeFevre et al., 2006), we select the splitting attribute as follows. We first normalize each quasi-identifier attribute by subtracting the mean and then dividing the difference by the standard deviation of that attribute (i.e., computing the Z-score). We then select the attribute with the largest range. For categorical attributes, we first represent them using integer values such as 1, 2, 3 (this is often already done in many patient data sets because categorical attributes are often represented by integer code), and then use the above method. Note that we only need to apply this conversion to quasi-identifier attributes. Sensitive attributes are not touched.

At line 2 to line 4, the algorithm splits the current data into two groups G_L and G_R by the median of the selected attribute. At line 5 to 8, the algorithm checks whether G_L and G_R satisfy our privacy model. The check for K-anonymity is straightforward: the group is K-anonymous if the group contains at least K records. The check for L-diversity is more complicated and will be discussed in the following subsections.

If G_L or G_R satisfies our model, it will be split further by recursively calling the algorithm. If neither of them satisfies our model, no further split is possible. Thus at line 10 the algorithm adds the current group G to the partition.

Once the partition is generated, the data set T will be anatomized. Based on (Xiao et al., 2006), we use the QIT table to store the quasi-identifier attributes and group ID for each record. The count of values of each sensitive attribute is stored in a separate ST table.

Complexity of Our Algorithm: let n be the number of records, m be the number of sensitive attributes, d be the number of quasi-identifier attributes, and $|G|$ be number of rows in G. The cost of building the k-d tree is $O(d\ n \log n)$ (LeFevre et al., 2006), excluding the cost of checking L-diversity. The following subsection will show that the cost of checking L-diversity for a group G_i is $O(|G_i|\ m + |G_i| \log |G_i|)$. Note that in each level of the k-d tree, the union of all groups equals the complete data set T. Thus the cost of checking L-diversity for one level of k-d tree is $O(n\ m + n \log n)$. There can be at most $O(\log n)$ levels of the tree. Thus the total cost of checking L-diversity is $O(m\ n\ log\ n + n\ (log\ n)^2)$. The total cost of partitioning is thus $O((m+d)\ n\ log\ n + n\ (log\ n)^2)$. The cost of generating the anatomized the data is $O((m+d)\ n)$ because the data only needs to be scanned once to generate the ST and QIT tables. Therefore, the total cost of the algorithm is $O((m + d)\ n\ log\ n + n\ (log\ n)^2)$. The last subsection will show that the cost for checking column-wise constraints is $O(|G_i|\ m\ log\ m + |G_i|\ log\ |G_i|)$ and the total cost is $O((m\ log\ m + d)\ n\ log\ n + n\ (log\ n)^2)$. Since $log\ m$ and $log\ n$ are quite small, both costs are almost linear with the data size $(m+d)\ n$.

Checking L-Diversity

This section describes how to check L-diversity for a group.

Theorem 3. Checking for L-diversity for multiple sensitive attribute is NP hard.

We can prove this theorem by reducing the minimal set cover problem to the problem of checking L-diversity. The detail of the proof is omitted due to space constraints. Here we just give some intuition about how we link these two problems. For each sensitive attribute value *v*, we create a set *RID(v)* that records the IDs of rows that contain that value. For example, consider the group of first four records in Figure 1, we have RID(heart disease)={1,2}, RID(flu) = {3}, RID(stomach disease)={4}, RID(medicine)={1}, RID(surgery)={2}, and RID(IV) = {3,4}.

The definition of L-diversity means that at least *L* values need to be deleted to delete all rows in the group. This is equivalent to state that at least *L* RID sets are needed to cover all rows in the group. In the above example, at least 2 RID sets are needed to cover the group of 4 records. For example, we can choose RID(heart disease)={1,2} and RID(IV)={3,4}.

However, finding the minimal number of RID sets to cover all row IDs is the minimal set cover problem which is NP hard. Thus we use a heuristic algorithm to check L-diversity in polynomial time. The next two theorems give the basis of our algorithm.

Theorem 4. if there exists a set *C* of at least *L* rows, and no two rows have the same value on any sensitive attribute, then *C* is *L*-diverse.

Proof: If no two rows have the same value on any sensitive attribute, then deleting one distinct sensitive value can delete at most one row in *C*. Thus at least *L* such deletions are needed and *C* is *L*-diverse.

Theorem 5. If a subset *C* of a group *G* is L-diverse, *G* is at least L-diverse.

Proof: If at least *L* distinct sensitive attribute values need to be deleted to delete a subset of a group, at least that many deletions are needed to delete the whole group.

The heuristic algorithm tries to find a subset *C* that has no common values on sensitive attributes. Figure 6 shows the algorithm. It starts with an empty set *C*, and then repeatedly adds

Figure 6. Algorithm to check L-diversity

```
L-diverse(G)

 1) compute for each sensitive attribute value v the number of rows in G contains that
    value, call it f(v)
 2) for each row r_i, compute a total frequency f(r_i) = Σ f(v) for v in r_i
 3) sort rows in ascending order of total frequency, store them in G'
 4) C = empty set
 5) while G' is not empty
 6)   pick the row r_i with the minimal f(r_i), delete it from G'
 7)   add r_i to C if it does not share any common value with existing rows in C
 8)   return yes if C contains L rows
 9) end while
```

rows to this set when no two rows in the set have common values on any sensitive attribute. Line 1 to 3 also compute the total frequencies of sensitive attribute values in a row and sort rows in ascending order of the frequency. This allows us to add rows that have values that are less frequent in the group first. The intuition is that such rows have smaller chances of sharing common values on sensitive attributes.

For the group of the first four records in Figure 1, suppose $L = 2$. Line 1 computes the frequencies of all sensitive attribute values. Thus, we have f(heart disease) = f(IV) = 2, and f(flu) = f(stomach disease) = f(medicine) = f(surgery)=1. Line 2 computes the total frequency of each row. The total frequencies of all rows are 3. Suppose row 1 is added to *C*. Next we try row 2, but it shares the disease type value with rows 1. Thus we try row 3 and it is added to *C*. Now *C* contains 2 rows and the algorithm returns yes because $L=2$.

Let *m* be the number of sensitive attributes and $|G|$ be the group size. The frequency of values and rows can be computed in $O(m\ |G|)$ time. The sort at line 3 takes O(|G| log |G|) time. We can use a hash table to keep track of the values of sensitive attributes in *C*. Thus checking whether a row contains values already in *C* (line 7) takes O(m) time. Since at most |G| rows can be added to *C*, the total complexity of the algorithm is thus O(m |G| + |G| log |G|).

The algorithm is sound in the sense that for any group that the algorithm returns yes, the group is indeed L-diverse. Thus using this algorithm will not affect privacy protection. On the other hand, the algorithm does not check all possible subsets of the group (doing so require exponential time). Thus some of the groups may be L-diverse but the algorithm may return no. In consequence, the algorithm may generate groups larger than the optimal case because these groups may be split further. This is the price we pay for not spending exponential time.

Checking L-Diversity with Column-Wise Constraints

This section presents the algorithm to check L-diversity with column wise constraints. Figure 7 shows the algorithm. Figure 8 shows an example of how the algorithm works. The algorithm consists of two steps. In the first step (line 1), it finds a subset *C* without common sensitive attribute values as in the algorithm in Figure 6. For example, for the data set in Figure 4, *C* will contain 3 records as shown in Step 1 in Figure 8.

Now there are two possible cases. In the first case, *C* contains at least *L* rows. By Theorem 4, *C* is L-diverse without column-wise constraints. By Theorem 2, *C* is also L-diverse with column-wise constraints. Using Theorem 5, the group *G* is also L-diverse and the algorithm returns yes.

In the second case, *C* contains less than *L* rows. The example in Figure 8 is in this case because *C* contains 3 rows and L=4. The algorithm repeatedly adds rows to *C* and check whether *C* satisfies L-diversity. Each round the algorithm adds a remaining row with the minimal total frequency to *C* (line 6) and checks *C*'s diversity. A variable *x* is used to keep track of *C*'s diversity and $x = |C|$ initially.

The algorithm uses the following lemma and theorem to check *C*'s diversity.

Lemma 1. Let $v_1, v_2, \ldots, v_x$ be the values of sensitive attributes in a *C*, and $f_C(v_i)$ be the frequency of v_i in *C*. Let *V* be a set of values to be deleted. Let $SumF(V) = \Sigma f_C(v_i)$, where v_i is in *V*. Then the maximal number of rows to be deleted is $SumF(V)$.

Proof: Let $RID(v_i)$ be the IDs of rows containing value v_i. Size of $RID(v_i) = f_C(v_i)$. When all values in *V* are deleted, the set of rows that gets deleted is the union of all $RID(v_i)$ for v_i in *V*. The size of union is at most the sum of sizes of each RID sets, which equals $SumF(V)$.

Theorem 6. Suppose frequencies of values are sorted in descending order, i.e., $f_C(v_1) \geq f_C(v_2) \geq \ldots \geq f_C(v_x)$. Suppose set *V* contains the *L* most frequent

Figure 7. Algorithm to check L-diversity with column-wise constraints

```
L-diverse-Column(G)

1)  find a subset C with no common sensitive attribute values using the
    algorithm in Figure 6.
2)  If |C| ≥ L, return yes.
3)  sort rows not in C in ascending order of total frequency and store them
    in G'
4)  x=|C|
5)  while G' is not empty
6)    pick row r_j in G' with the minimal f(r_j), delete it from G', add r_j to C
7)    select values v_1, ..., v_{x+1} in C with the highest frequencies in C and
    with at most L_i values from sensitive attribute i.
    Let v_i's frequency in C be f_C(v_i)
8)    if Σ f_C(v_i)> |C|, 1 ≤ i ≤ x
9)      delete r_j from C // C is not x-diverse
10)  else
11)    if Σ f_C(v_i)≤|C|, 1 ≤ i ≤ x+1
12)      x = x+1 // C is x+1-diverse
13)      if x = L return yes
14)  end while
```

Figure 8. Example of check L-diversity with column-wise constraints

Step 1: initial C

D1	T1
D3	T2
D5	T3

Step 2: add row (D2, T1) to C

D1	T1
D3	T2
D5	T3
D2	T1

3 most frequent values: T1(2), T2(1), D1(1)

Step 3: add row (D4, T2) to C

D1	T1
D3	T2
D5	T3
D2	T1
D4	T2

3 most frequent values: T1(2), T2(2), D1(1)

Step 4: add row (D6, T3) to C

D1	T1
D3	T2
D5	T3
D2	T1
D4	T2
D6	T3

4 most frequent values: T1(2), T2(2), D1(1), D2(1)

values, with at most L_i values from attribute A_{Si} (if A_{Si} has more than L_i values, we select the L_i most frequent ones). If $SumF(V) \leq |C|$, then C is L diverse with column-wise constraints.

The proof is straightforward. Consider any set U containing L values. By Lemma 1, deleting U will delete at most $SumF(U)$ rows. Since set V contains the most frequent values, $SumF(U) \leq SumF(V)$. Since $SumF(V) \leq |C|$, $SumF(U) \leq |C|$. Thus at least L deletions (with at most L_i from attribute A_{Si}) are needed to delete all rows in C and C is L-diverse.

Based on Theorem 6, the algorithm checks C's diversity as follows. At line 7, the algorithm selects the x most frequent values from C, with at most L_i values selected from attribute i. At line 8, the algorithm checks whether the total frequency of these values is greater than size of C. If so, the newly added row is rejected because we cannot prove that C is x-diverse using Theorem 6. Otherwise, we can prove that C is x-diverse and the algorithm keeps the newly added row in C. Next, the algorithm checks whether C is $x+1$ diverse by computing the total frequency for the $x+1$ most frequent values. If the total frequency is less or equal to the size of *C*, *C* is x+1 diverse and *x* is increased by 1 at line 12. Finally, if the value of *x* reaches *L*, *C* is L-diverse. By Theorem 5, *G* is also L-diverse and the algorithm returns yes.

For example, consider the data in Figure 4. Figure 8 shows the process of the algorithm where the numbers in parenthesis are frequencies. At step 2, the row with value *(D2, T1)* is added to *C*, and the total frequency of the 3 most frequent values *(T1, T2, D1)* equals the size of *C*. Thus *C* remains 3-diverse. At step 3, the row with value *(D4, T2)* is added and C is still 3-diverse. At step 4, the row with values *(D6, T3)* is added. Now the 4 most frequent values are *T1, T2, D1*, and *D2*. Note that T3 is not counted because at most 2 treatment values can be selected according to the column-wise constraint $L_2=2$. The total frequency is 2+2+1+1=6, which equals size of *C*. Thus *C* is 4-diverse. Since $L=4$, the algorithm returns yes.

Complexity: This algorithm calls the algorithm in Figure 6 first. Thus it takes at least $O(m\,|G| + |G|\,log\,|G|)$ time. Line 7 is the most expensive step because it needs to find the *x* most frequent values in *C*. Each time only the *m* values in the new row r_j will get their frequency increased by one (the frequency of other values stay the same). Thus the *x* most frequent values must come from the list of *m* values in the new row and the list of previously *x* most frequent values. If the previously most frequent values are already sorted, we just need to sort the *m* new values on their frequency and merge them with the previous list. Thus line 7 can be done in $O(m\,log\,m + x)$ time. Since x $\le$ L, the time is $O(m\,log\,m + L)$. Line 7 can be executed at most |G| times, thus the complexity of the algorithm is $O(|G|\,log\,|G| + |G|\,(m\,log\,m+L))$. The value of L is typically quite small. If *L* and *m* are in the same order, the complexity becomes $O(|G|\,log\,|G| + |G|\,m\,log\,m)$.

EXPERIMENTAL EVALUATION

The first subsection describes the setup of experiments. The second subsection reports the results of our privacy model. The third subsection reports the results of our model with column-wise constraints.

Setup

Machine: Experiments were run on a Dell PowerEdge Server with 3 GHz CPU and 2 GB memory, running Windows Server 2003.

Data: We used a patient data set obtained from the Kentucky Cancer Registry. It contains information about 72,194 patients. We used 4 quasi-identifier attributes: birth date, gender, race, and zip. We used 7 sensitive attributes: tumor topography, histology, survival years, diagnose date, age at diagnosis, tumor size, and tumor site. This is a real data set and the data distribution is skewed for many attributes. Our method was implemented in Perl. **Metrics:** A successful privacy protection method protects privacy and introduces little data distortion. *K* and *L* indicate the degree of privacy protection. We use two metrics to measure distortion. The first is *discernability* (Bayardo et al., 2005; LeFevre et al., 2006), which is the average group size of all records. Since each group needs to have at least *K* rows, the best possible value of discernability is *K*. Smaller discernability means less distortion.

The second metric for distortion is the average relative error for a large number of randomly generated SQL queries. Ideally, if we know the details of the subsequent data analysis, we can measure data distortion by its impact on all types of data analysis that can be performed. However, it is difficult to know beforehand all types of data analysis, thus we use random SQL queries. These queries return the number of patients satisfying several randomly generated conditions. The number of conditions was randomly selected from 1 to 4. At least one condition was on a randomly selected sensitive attribute. The other conditions were on randomly selected quasi-identifier attributes. The conditions were randomly generated equality or range conditions. 8800 different queries were tested. The error is computed to compare the counts over the sanitized data to the counts over the original data.

Algorithms: We compare the following three algorithms:

- Anonymize: this is our algorithm proposed in Figure 6.
- Anonymize-Column: this is our algorithm implementing column-wise constraints in Figure 8.
- *Exponential-L*: this algorithm implements the L-diversity model proposed by (Machanavajjhala et al., April 2006)), which requires that each attribute have at least *L* different values for all records with the same values on the other attributes. This algorithm is the same as our method except the way it checks L-diversity. A similar method has been implemented in (Li et al., 2008).

Results for Our Model

Discernability results: There are three important parameters: *K*, *L*, and the number of sensitive attributes (*m*) included in the data set. We fixed two of them and varied the third. Figure 9 reports the discernability for various *L* values when *K*=50 and *m*=3 (the first three sensitive attributes are used). Figure 10 reports the discernability for various *K* values when *L*=10 and *m*=3. Figure 11 reports the discernability when various number of sensitive attributes are included and *K*=50 and *L*=10.

The discernability of Exponential-L is very high (meaning very high degree of data distortion). Exponential-L generates a single group that contains the whole data set for all cases except when there is only one sensitive attribute (in Figure 11). Exponential-L requires each attribute have at least *L* different values for all records with the same values on the other attributes. This leads to very large group sizes and large distortions to data.

The discernability of our method (Anonymize) is lower than that of Exponential-L by several orders of magnitude (note that the scale for y-axis is logarithmic), and is not much higher than the

Figure 9. Varying L, K=50, m=3

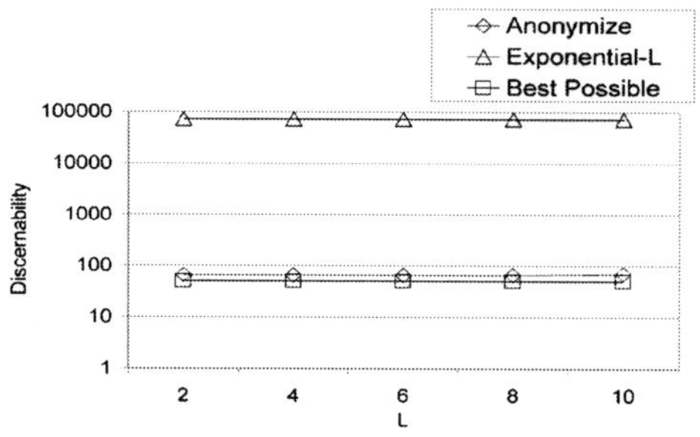

Figure 10. Varying K, L=10, m=3

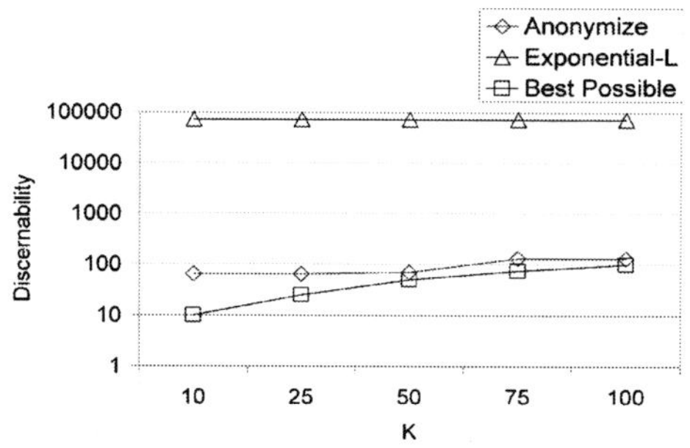

Figure 11. Vary number of sensitive attributes, K=50, L=10

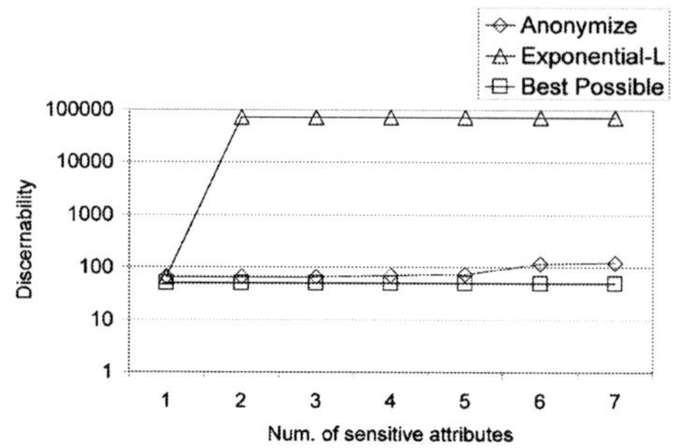

best possible discernability (which is *K)*. The discernability increases with *K* and *L*, because more data distortion is introduced by providing a higher degree of privacy. The increase of discernability is quite small when *L* increases because most groups cannot be split further due to violation of K-anonymity. Thus, the increase of *L* has a smaller impact on group size. The discernability of Anonymize also increases with the number of sensitive attributes. As the number of sensitive attributes increases, two rows are more likely to have the same values on a sensitive attribute, and thus larger groups are needed to achieve L-diversity. However, the discernability of our approach is still not much higher than the best possible case (less than a factor of two in most cases).

Error of random queries: Table 1 shows the average relative error for random queries when K=50, L=10, and m=3. The error of Anonymize algorithm for all queries is 11.3%. The error for queries returning less than 1% of the total number of patients is 14.6% and the error for queries returning 1% or more patients (there are 1117 such queries) is only 1.6%. The error for the first subset is higher because data distortion has a larger impact on these queries. For example, if a query returns just one patient and the distortion makes it return two, the error is 100%. The results show that our method introduces small error, especially when medium to large number of records are returned. This property is suitable for data analysis because it is not very meaningful to study a very small fraction (e.g., less than 1%) of a data set.

Table 1. Average relative error of random queries when K=50, L=10, and m=3

Algorithm	All queries	Queries returning < 1% of patients	Queries returning ≥ 1% of patients
Anonymize	11.3%	14.6%	1.6%
Exponential-L	23.2%	28.6%	8.5%

The relative error for Exponential-L is about twice of the error for Anonymize on all queries, and is about 5 times of the error for Anonymize on queries returning more than 1% of patients. This is expected because Exponential-L generates very large groups and introduces larger data distortion. Since Exponential-L has very bad performance, we only show the results of Anonymize and Anonymize-Column in the remaining sections.

Results for L-Diversity with Column-Wise Constraints

Figure 12 shows the results for L-diversity with column-wise constraints. We use three sensitive attributes: tumor topography, histology, and recur status. The attribute "recur status" has very few distinct values (31). Let L_1, L_2, and L_3 represent the column wise constraints for each attribute, we set $L = L_1 = L_2 = 10$, and varies L_3 from 0 to 10. Note that L_3=0 means none of the values of "recur status" attribute will be deleted and thus there is no L-diversity requirement on that attribute. L_3 = 10 means it has the same requirement as other attributes. The results show that the discernability increases for larger L_3 values, and becomes the same as the discernability of the case without column-wise constraints as L_3 = 8. This is expected because a smaller L_3 means fewer values are deleted from "recur status" attribute, making it easier to achieve L-diversity because this attribute has very few distinct values. The average relative error of random queries for L_3 = 2 is 10.4%. The average relative error for L_3 = 10 is 11.7%. Thus using column-wise constraints also leads to lower average relative error. The difference is not as big as the difference for discernability, because in both settings many groups are generated and thus the estimation of query results is quite accurate.

Figure 12. Vary L_3, $L=L_1=L_2=10$, $K=50$, $m = 3$

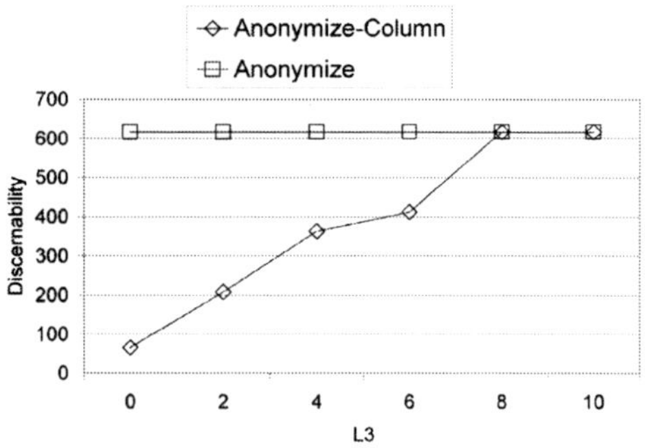

Execution Time

Figure 13 and 14 report the execution time of our algorithms when the number of rows and number of attributes are varied. We vary the number of sensitive attributes in the same way as in the previous subsection and use all rows in the data set. We vary the number of rows by selecting a fraction of rows in the data set, and use all 7 sensitive attributes. $K=50$ and $L=10$ in all cases, and L_i of the last attribute is set to 2 for Anonymize-Column. The results show that the execution time of both Anonymize and Anonymize-Column scale almost linearly with the number of rows and number of attributes. The execution time also increases at a slower pace when the number of sensitive attributes increases, because it is more difficult to satisfy L-diversity for more sensitive attributes, and thus fewer groups are generated. Anonymize-Column also takes slightly more time than Anonymize, because it calls the Anonymize algorithm first. However, the difference is not big because many groups also satisfy L-diversity without constraints, and Anonymize-Column does not need to do extra work for these groups.

CONCLUSION

This chapter proposes a privacy model that protects identity of patients for data with multiple sensitive attributes. A variant of this model is also proposed, which allows the user to specify lower degree of diversity for attributes with very few distinct values. This chapter also proposes efficient algorithms to implement the model. Experiments show that the proposed approach introduces distortion orders of lower magnitude than the distortions introduced by the existing approach in the literature, and introduces small relative error for random SQL queries. As future work, we will study how to extend other formats of L-diversity to multiple sensitive attributes.

Figure 13. Execution time when varying number of sensitive attributes

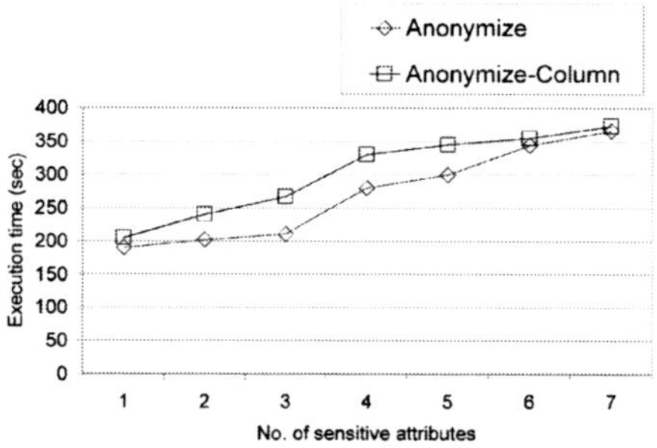

Figure 14. Execution time when varying number of rows (as fraction as total data set)

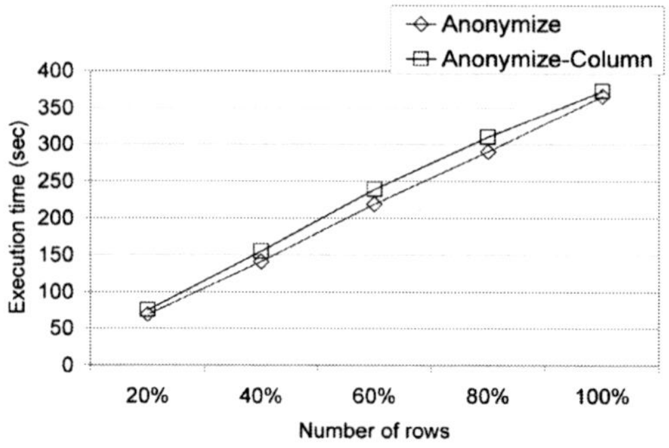

REFERENCES

Agrawal, D., & Aggarwal, C. C. (2001). *On the design and quantification of privacy preserving data mining algorithms.* Paper presented at the 20th ACM SIGMOD SIGACT-SIGART Symposium on Principles of Database Systems, Santa Barbara, CA.

Agrawal, R., & Srikant, R. (May 2000). *Privacy preserving data mining.* Paper presented at the 2000 ACM SIGMOD Conference on Management of Data, Dallas, TX.

Bayardo, R. J., & Agrawal, R. (2005). *Data Privacy through Optimal k-Anonymization.* Paper presented at the ICDE.

Evfimevski, A., Gehrke, J., & Srikant, R. (June 2003). *Limiting privacy breaches in privacy preserving data mining.* Paper presented at the 22nd ACM SIGMOD-SIGACT-SIGART Symposium on Principles of Database Systems, San Diego, CA.

Evfimevski, A., Srikant, R., Agrawal, R., & Gehrke, J. (July 2002). *Privacy preserving mining of association rules.* Paper presented at the 8th ACM SIGKDD International Conference on Knowledge Discovery and Data Mining (KDD'02), Edmonton, Alberta, Canada.

Federal Trade Commission. (2007). Identity Theft Resource Center: Facts and Statistics: Find out more about the nation's fastest growing crime.

Gartner Inc. (2007). Gartner Says Number of Identity Theft Victims Has Increased More Than 50 Percent Since 2003.

Hintoglu, A. A., Inan, A., Saygin, Y., & Keskinoz, M. (2005). *Suppressing Data Sets to Prevent Discovery of Association Rules.* Paper presented at the ICDM.

Kargupta, H., Datta, S., Wang, Q., & Sivakumar, K. (2003). *On the Privacy Preserving Properties of Random Data Perturbation Techniques.* Paper presented at the ICDM.

LeFevre, K., DeWitt, D. J., & Ramakrishnan, R. (2005). *Incognito: Efficient Full-Domain K-Anonymity.* Paper presented at the SIGMOD.

LeFevre, K., DeWitt, D. J., & Ramakrishnan, R. (2006). *Mondrian Multidimensional K-Anonymity.* Paper presented at the ICDE.

Li, X.-B., & Sarkar, S. (2006). A Tree-Based Data Perturbation Approach for Privacy-Preserving Data Mining. *IEEE Transactions on Knowledge and Data Engineering, 18*(9), 1278–1283. doi:10.1109/TKDE.2006.136

Li, Z., & Ye, X. (2008). Privacy Protection on Multiple Sensitive Attributes (LNCS 4861, pp. 141-152).

Machanavajjhala, A., Gehrke, J., Kifer, D., & Venkitasubramaniam, M. (April 2006). *L-Diversity: Privacy Beyond k-Anonymity.* Paper presented at the 22nd IEEE International Conference on Data Engineering (ICDE 2006), Atlanta, Georgia.

MacVittie, D. (2007, Aug. 31). Javelin 2006 Identity Fraud Report. *Network Computing.*

Menon, S., & Sarkar, S. (2006). *Exploiting Problem Structure to Efficiently Sanitize Very Large Transactional Databases.* Paper presented at the WITS.

Menon, S., Sarkar, S., & Mukherjee, S. (2005, September). Maximizing Accuracy of Shared Databases when Concealing Sensitive Patterns. *Information Systems Research, 16*(3), 256–270. doi:10.1287/isre.1050.0056

Oliveira, S., & Zaiane, O. R. (2002). *Privacy Preserving Frequent Itemset Mining.* Paper presented at the IEEE International Conference on Privacy, Security and Data Mining, Maebashi City, Japan.

Samarati, P. (2001). Protecting Respondents' Identities in Microdata Release. *TKDE, 13*(6), 1010–1027.

Saygin, Y., Verykios, V. S., & Clifton, C. (2001). Using unknowns to prevent discovery of association rules. *SIGMOD Record*, *30*(4), 45–54. doi:10.1145/604264.604271

Sweeney, L. (2002a). Achieving k-anonymity privacy protection using generalization and suppression. *International Journal on Uncertainty, Fuzziness and Knowledge-based Systems*, *10*(5), 571–588. doi:10.1142/S021848850200165X

Sweeney, L. (2002b). K-anonymity: a model for protecting privacy. *International Journal on Uncertainty, Fuzziness and Knowledge-based Systems*, *10*(5), 557–570. doi:10.1142/S0218488502001648

Verykios, V. S., Bertino, E., Fovino, I. N., Provenza, L. P., Saygin, Y., & Theodoridis, Y. (2004, March). State-of-the-art in Privacy Preserving Data Mining. *SIGMOD Record*, *33*(1), 50–57. doi:10.1145/974121.974131

Verykios, V. S., Elmagarmid, A. K., Elisa, B., Saygin, Y., & Elena, D. (2004, April). Association Rule Hiding. *IEEE Transactions on Knowledge and Data Engineering*, *16*(4), 434–447. doi:10.1109/TKDE.2004.1269668

Xiao, X., & Tao, Y. (2006). *Anatomy: Simple and Effective Privacy Preservation.* Paper presented at the VLDB.

Zhu, Y., & Liu, L. (2004). *Optimal randomization for privacy preserving data mining.* Paper presented at the KDD.

ENDNOTE

[1] This work is in part supported by National Science Foundation under Grant Numbers IIS-0713345.

Chapter 5
Protecting Patient Information in Outsourced Telehealth Services:
Bolting on Security When it Cannot be Baked in

Patricia Y. Logan
Marshall University Graduate College, USA

Debra Noles
Marshall University Graduate College, USA

ABSTRACT

Hospitals have increasingly employed outsourcing to lower the cost of healthcare delivery and improve efficiency and quality, thereby, enabling more focus on core competencies of patient care, teaching, and research. Outsourcing presents a challenge for protecting patient information when new services are implemented or integrated into an existing healthcare information system. Enabling new outsourced telehealth services often requires "bolting on" security to legacy systems rather than "baking" it into the system. This chapter addresses security practices necessary for healthcare organizations implementing new telehealth services as part of an outsourced relationship. While a number of recommendations are available for security readiness assessments pursuant to HIPAA compliance, none directly addresses the challenge of implementing security for outsourced clinical services. A case study is presented for a recent implementation of teleradiology services within a large regional hospital. Using the case, system vulnerabilities are demonstrated and relevant best practices to mitigate exposing patient information are discussed.

INTRODUCTION

Multiple pressures exerted within the healthcare industry have driven the move toward outsourcing of clinical services. They include:

- rising costs of healthcare
- market competition
- need to focus on revenue generating core competencies (i.e., patient care, teaching, and research)
- lack of available specialists in key specialties (e.g., radiology)
- demand for quality healthcare services by the customers (Bedi, Sarma, & Arya, 2002)

The Hospital Outsourcing Trends in Clinical Services Survey (2006), questioned 266 executives from hospitals of all sizes, and reported 78% of the hospitals outsource at least one patient service, and 83% expect the level of outsourcing at their facilities to stay the same or increase over the next two to three years (Hill, Bartrum, & Guy, 2006). While hospitals have outsourced non-core activities, the clinical and diagnostics services have been slower to be adopted as outsource candidates (Outsourcing in Most Hospitals, 2007). Many clinical services have moved to offshore locations to realize greater cost savings and efficiencies (Goelman, 2007). In particular, a recent trend is to send patient radiological studies offshore to be interpreted by United States certified physicians. This process assists hospitals finding it increasingly difficult to attract and maintain on-site radiologists during evening hours. The need for such specialists has grown due to the increase in emergency room patients given scans to help diagnose their condition (Goelman, 2006; Levy & Goelman, 2005). Increasingly more common, the first interpretation of a radiology study is performed offshore in Europe or Australia with a second read done the following day within the hospital (Pollack, 2003).

The electronic delivery of sensitive data to offshore locations has sparked interesting security debates on how healthcare institutions should protect a patient's right to privacy. There are security implications and regulatory requirements at state, national, and international levels for hospitals entering into outsourced relationships. Equally important, hospitals must also determine if the security practices of the outsourced companies are acceptable. When the outsourced companies do not have adequate security for patient data, the hospital will be held liable, even if a breach of patient information happens at the third-party company. If hospitals do not protect the sensitive medical data traveling to third-party entities, the risk of exposing sensitive personal information to cybercriminals increases dramatically. The Identity Theft Resource Center (ITRC) reports that in 2007, almost four million patient records were exposed, representing 65 incidents and 14.9% of all breaches reported (Data Breach Report, 2007). While HIPAA represents the primary security guidelines for the United States, it does not mandate reporting of a security breach to patients. With no requirement for reporting, it is likely that the data from the ITRC significantly understates the actual occurrence of data loss from healthcare organizations.

The complexity of hospital data security management is a result of balancing the requirement to provide proper access to the data vs. the requirement to sufficiently protect the data. Clinical users expect the data to be available when it is needed, and view security as a secondary concern, especially in emergency scenarios. Presently, there is no guidance for hospitals involved in outsourcing clinical services in how best to comply with the security expectations of patients and regulatory agencies.

The goal of building any clinical information system is to improve the quality of patient care. The methods for protecting the data should not interfere. Hospitals must acknowledge patient data security is about risk management, and they must balance implementation costs with potential costs for data breach or compromise.

Security Implications of Outsourcing Clinical Services

A popular mantra in IT circles is the notion that security should be "baked" into an application during the system development life cycle. Hospitals, as do many other organizatons, largely buy rather than build applications and then integrate them into a legacy environment (Schwartz, 2007). Outsourced services (with back-end vendor applications/systems) are frequently introduced into this legacy environment. In an outsourced environment, where applications are integrated across multiple hardware platforms, networks, operating systems, and standards, following the best practice of building security into a new system during development is seldom a possibility. The process of integrating systems and providing security more closely resembles a "bolting on" of processes and technology. It could be argued that in *any* legacy environment, security is seldom "baked in" due to the heterogeneity of these systems that make retrofitting functionality very complex. Given the difficulty of adding security to newly integrated systems, is an outsource vendor any more likely to provide secure systems? Vendors have a monetary incentive to provide security, as their reputation and profitability depend upon providing systems that meet current HIPAA requirements. Outsource vendors are also aware that healthcare administrators believe security to be less important than the quality and pricing of a product or service (Booz-Allen, 2006). If decisions are made on the quality of the products and services and on competitive pricing, then outsourcing may represent a vulnerability with both sides of the transaction less likely to be concerned about the security of a system. The challenges in outsourced telehealth relationships occur in two areas: (1) the absence of practice standards for implementing security within outsourced relationships; and (2) over-reliance on the assurances of vendors that security has been built into the new system. At present, HIPAA expects the responsibility for security to rest on the primary (i.e., outsourcing) healthcare institution, and much less on the outsourced vendor. In this chapter, we will provide a case study of a recent implementation of teleradiology services outsourced to a new business partner which includes the integration of services. We will use the experience from this implementation to inform practitioners in the IT healthcare field of the practice points necessary to achieve an implementation that meets the security requirements of HIPAA.

The Case Study: Outsourcing Teleradiology at FHC

The Family Hospital Corporation (FHC[1]) is a regional hospital in the mid-Atlantic region of the United States. It is a nonprofit, 893-bed, regional referral center with more than 5,000 employees. It is the state's largest teaching hospital, and prepares a variety of healthcare professionals in a number of clinical specialties. It has a Level I trauma center that serviced 2,900 patients in 2006. The emergency room provides care for 98,633 patients annually. FHC has three primary divisions: Memorial, South, and Children's. In addition, there are several ambulatory care centers and also various physician practices affiliated with the FHC organization.

FHC has implemented new technologies which help manage the increasing number of studies and images: Picture Archiving and Communication Systems (PACS) and Radiology Information Systems (RIS). A PACS system gives hospitals the ability to digitally store medical images so that these images can be viewed from any location or workstation capable of retrieving the digital image. This expands the possibilities of conventional systems and allows medical professionals to view images from their homes and other off-site locations over broadband Internet connections.

It also removes any geographical boundaries to specialists capable of reading the images (Beck & Neufeld, 2007). Removing geographical distances as a barrier to providing care is a primary benefit of telemedicine. PACS technology is maturing and platforms are flexible and fully featured, allowing the storage of images from multiple imaging devices (called modalities) including x-ray, CT, MRI, ultrasound, and nuclear medicine (Jackman, n.d.). RIS systems offer hospitals customizable workflow management of images from patient registration through the production of a report. RIS can integrate with PACS for a complete system that also includes storage and archive features. The PACS and RIS systems at FHC were installed approximately four years ago.

A staff of over 100 IT analysts is responsible for a variety of healthcare information systems. FHC maintains an outsourced relationship with a group of radiologists that have a contract through 2008 to read and interpret all scans and images at FHC facilities. Radiologists, Inc. (RI)[1] was formed in the 1960s and employs 17 board certified radiologists. Recruiting radiologists to provide 24/7 coverage required by their contract with FHC has been difficult for RI due to the remote location of the regional medical center. Over 40% of the emergency radiology scans occur at night with the most critical cases often presenting during the evening hours from 10 p.m. to 7 a.m (NightHawk, n.d.). Due to the volume of patients seen in the emergency rooms during these evening hours requiring scans, RI began investigating the potential for outsourcing the evening services to another vendor.

RI focused on a vendor with a reputation for quality and 15% of the outsource market for teleradiology services: NightHawk Radiology Services (Kelly, 2006). NightHawk Radiology Services is headquartered in Couer d'Alene, Idaho, and provides services to over 1,000 hospitals in the United States (see www.nighthawk.net). It employs United States board certified, state-licensed, and hospital-privileged radiologists who provide radiological interpretations to customers 24 hours a day, seven days a week, from centralized reading facilities located in the United States, Australia, and Switzerland (NightHawk, 2007). NightHawk establishes a secured virtual private network (VPN) connection with hospital customers. Images are sent compressed and encrypted across the Internet to radiologists in Sydney, Australia or to other NightHawk destinations. Reports are sent back in approximately 20 minutes and travel via encrypted fax to a hospital's radiology department. Other centralized reading offices in Zurich, Switzerland, and Coeur D'Alene, Idaho, provide redundancy and "follow-the sun" availability. NightHawk has reported 99.9% availability with only one incident of catastrophic failure (a failure in MCI's Internet network between the United States and Australia resulted in a 30 minute downtime for the company). In September 2006, NightHawk earned a three year renewal of its JCAHO Gold Seal of Approval (NightHawk, 2006). The JCAHO accredits healthcare organizations from the perspective of confidentiality and data security.

In June 2006, a NightHawk representative presented NightHawk's teleradiology services to the FHC Board of Executives and to radiologists from RI. The major concern for FHC was the quality of the interpretations. FHC approved a pilot and began testing the new service on a limited basis with evening studies being sent to Australia, for preliminary reads under the supervision of a radiologist. By December 2006, FHC was sending all of its late evening radiological studies to NightHawk for preliminary reads. The volume increased from approximately 600 studies sent in Fall 2006 to over 3500 studies by March 2007. Because of the Medicare/Medicaid restrictions for radiology studies being interpreted on foreign soil, the final read is always done at FHC by RI in order for reimbursement to be processed. Figure 1 shows the workflow of the proposed system using the legacy applications with PACS and RIS. The new outsourced component of the digital workflow is found in the shaded area.

Figure 1. Proposed workflow

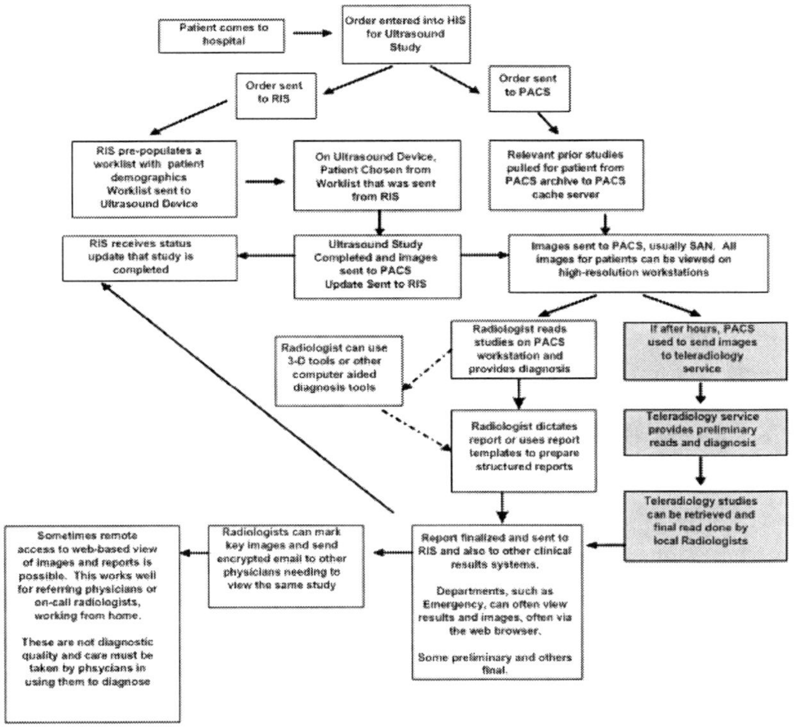

FHC executives worked with both RI and NightHawk on the paperwork related to HIPAA's business associate designation to provide adequate coverage for each entity in the disclosure of protected health information (PHI). Because there was a three-way agreement between RI, NightHawk, and FHC, the business associate agreements were executed in multiple counterparts, but all of them summed up to one agreement. In addition, the FHC Privacy Officer drafted the Patient Consent form for sending studies overseas and ensured the new NightHawk workflow in the FHC Radiology Department followed both HIPAA and JCAHO requirements. The forms and the new workflow ensured (1) patient consent, (2) patient privacy, and (3) protection of patient data. The security of the patient data is provided through encryption, a secure VPN, password-protected access, and physical isolation of the system to a secure region of the hospital. There are two dedicated virtual private network connections to NightHawk destinations from FHC: Chicago, Illinois, and Coeur d'Alene, Idaho. Figure 2 demonstrates the VPN connections that exist. Studies are sent out from the FHC connection to Chicago for quality assurance purposes only. Quality technicians at NightHawk in Chicago verify the number of images and patient identifying information and then forward the studies via VPN to reading centers in the NightHawk network. For FHC, this is the reading center in Australia. The second VPN connection is to the reading center in Idaho for backup purposes. FHC network analysts worked with the NightHawk implementation team to provide two secured and encrypted VPN connections to specific host (Nighthawk) IP addresses into the NightHawk network, from the secured VPN router on the FHC network. On the FHC side, the VPN connection points to two hosts (FHC) that are part of the PACS storage archive. The VPN logs are monitored by FHC staff to detect unauthorized traffic (i.e., not in the trusted IP address range).

Figure 2. VPN connections

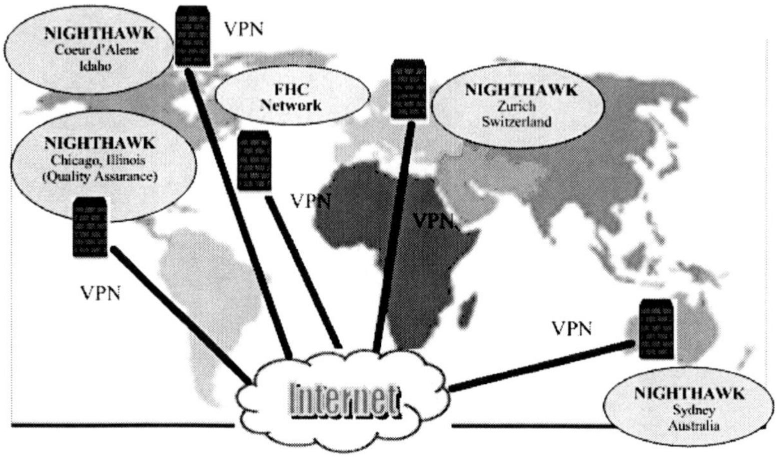

Figure 3 visualizes the workflow for the new outsourced relationship. All FHC divisions follow this NightHawk workflow starting at 11 p.m. each day until 7 a.m. the next morning. When a study is sent to NightHawk, the radiology technologists complete an online requisition form with patient information including patient demographics, type of exam performed, clinical history, evidence of trauma (if any), and also the number of images to be sent to NightHawk. Once the form and images are received by NightHawk in Illinois, the Quality Assurance Center will determine if all the images made it across the VPN. This center also determines which radiologist can read the study and routes it to a specific individual in Australia. The Web-based tool used by FHC to submit studies to NightHawk is then updated with a status that the study has been assigned to a radiologist. The Web tool flashes an update every 20 seconds and within 5 minutes, the tool will flash an update that the radiologist has started reading the study. Within 20 minutes, the Web tool flashes that the radiologist has finished reading the study and that a preliminary report is being faxed back to the Radiology Department at FHC. This communication method allows the technicians waiting at FHC to know the status of the studies sent to NightHawk and be able to give important updates to physicians waiting on the results, most often from the Emergency Department.

The VPN router logs determine the level of traffic sent across these two connections. Any failure in the public Internet would impact the availability of NightHawk services. In the event of a complete Internet outage, FHC will direct any emergency studies to an "on-call" radiologist from RI. The radiologist could either come to the hospital Radiology Department or a FHC facility to read studies during an outage.

Although the NightHawk implementation at FHC was generally successful, there have been several items for which perceived outcomes differed from the actual outcomes. Due to the requirement that all images had to be received at the remote end before an interpretation could begin, the NightHawk application appeared very sensitive to data interruption. Initially, the size of the images was reported as a problem. Compression tools were used and the packet loss and delay were alleviated. The advertised reliability of NightHawk service has also been uncertain. An outage in March 2007 was due to another NightHawk customer sending studies that exceeded maximum study size limits, but yet still passed into the NightHawk network. This impacted the inability of other hospitals to send studies during

Figure 3. Outsourced workflow

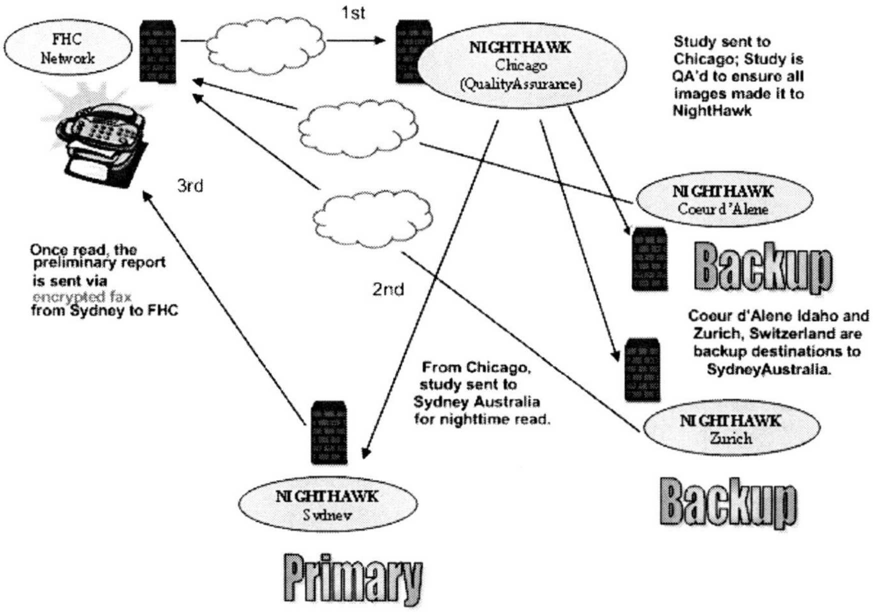

a 45-minute outage period. Initially, NightHawk was unable to determine the root cause of the outage, but later, on the evening of the outage, NightHawk reported a single customer had inadvertently flooded the NightHawk interface with larger than anticipated (larger than normal) images. There are questions that remain about the effectiveness of the compression software used by Nighthawk and the level of monitoring of network loads. The team from FHC requested NightHawk improve the method of alerting customers to outages, perhaps by using an automated call-down or pager notification with more details. Overall, the NightHawk implementation at FHC was successful from the executive perspective, despite the service interruption.

While the FHC case does not involve a security breach, it does highlight areas where vulnerabilities exist and could increase the likelihood that PID will be compromised. There are a number of lessons-learned from this implementation. Specific vulnerabilities exist in the following areas: (1) Security was not a major factor in the selection of the outsource partners; (2) There was no security readiness assessment on any of the systems that were bolted to the outsourced application; (3) The performance of the proposed technology was not tested and verified; (4) Mobile devices were not restricted; (5) No preparations were made for incident response; and (6) No pilot study was performed or benchmarks for performance from the security perspective were used. The following discussion provides practice points for practitioners in how to achieve success in outsourced projects and for guidance in bolting on security to legacy systems that have integration to outside systems. The practice points offered can be used for any telehealth initiative and are not restricted to those involving teleradiology applications.

Practice Pointer 1: Security from the Top

The FHC case highlights the absence of selection criteria and proper "vetting" for outsource vendors. The sole focus of interest was on a single criterion: quality of the image interpretations. In the case study, the radiologists chose the outsource vendor, Nighthawk, based solely on their market share and reputation. The due diligence effort was left to FHC IT staff and consisted largely of a short pilot to verify the quality of the interpretations. There were no discussions of security beyond the verification that the vendor had received certification for their practice from the JCAHO. Interestingly, the JCAHO only accredits for quality and safety, not security. The Gold Seal Certification also applies only to U.S.-based inspections rather than offshore locations such as the reading center in Sydney, Australia. An over-reliance on the presence of certification or vendor assurances that security is consistent with industry best practices can lead to compromises of patient data.

Booz-Allen Hamilton conducted a survey in 2006 to assess the critical concerns of executives in selecting an outsource vendor and found that quality of service ranked as the first concern, with pricing and security, in second and third places respectively. Often in a rush to implement new services and realize cost savings, C-level executives ignore the usual level of due diligence required to insure the protection of PID as required by HIPAA. Due diligence requires that quality from a technology performance perspective and security be verified and that pricing should be a secondary consideration. Healthcare organizations should begin to think of quality as encompassing security as a benchmark. Such due diligence does not just include the outsourced vendor, but also the existing systems that will be integrated into the new workflow. Many of the service providers delivering outsource services are new to the healthcare arena and may not yet have policies in place to address protection of sensitive data or the technical expertise to deliver the level of service required by the covered entity. Those in IT are often assured that the vendor meets the standard required for the responsible party but it is difficult to police service providers geographically removed from the healthcare organization. Hospitals should work to develop enterprise standards for the outsource vendor selection process. These standards must include a very thorough process of investigating the security practices of all possible vendors. Security documentation from each vendor should become a requirement and failure to provide as much should be a serious consideration for elimination of the vendor. Hospitals are beginning to rely on vendors' assertions of benchmark certifications similar to those used in other industries.

Modernization of healthcare systems would certainly be easier if a unified certification body existed. In 2006, the Certification Commission for Healthcare Information Technology (CCHIT) was officially designated by the United States Department of Health and Human Services as a Recognized Certification Body because it has developed thorough certification criteria for vendors of Electronic Health Record (EHR) software. The mission of CCHIT is to create a sustainable certification program for healthcare information systems. According to CCHIT Chairman Mark Leavitt, MD, "vendors see significant market impact from certification" (Monegain, 2008). CCHIT has received endorsements from professional medical organizations, including the American College of Physicians.

When a new telehealth partnership is formed, clearly identifying and actively engaging key stakeholders is an important practice point because the message that security is an enterprise concern, not just a technical concern, must be kept alive. The potential for new vulnerabilities to be introduced due to increased exposure, improper access by new stakeholders, or interaction with poorly secured systems will require a new understanding of the potential risks.

Hospital executives must view security as a business concern and a policy effort. Enterprise security should be questioned regularly in board meetings whenever a new relationship and system is established. Executives must aggressively and visibly sponsor information privacy and security campaigns that align with the policies and procedures for employee conduct (Collman & Grimes, 2005). The C-level officers must begin to ask questions about both clinical impact and protection of patient data. Physician "champions" and organizational staff must carry the same message and execute the work to protect data. Targeted training and continuous awareness of the need to protect patient information will be required whenever outsourcing of a clinical service occurs. From the beginning, investment in security and privacy should be considered as any other strategic initiative: adequate resources must be provided, and proper education and training related to privacy and security must be made available.

By identifying stakeholders and finding champions to carry the message early in the formation of the partnership, the enterprise and the outsourced entity will be setting up the strongest foundation possible for acceptance of the additional security needed to support the partnership.

The following practice points include:

- Outsource vendors should have a performance standard that encompasses clinical quality and technology performance.
- Offshore sites used in outsourcing should be accredited or inspected for adherence to the security practices of the hospital.
- Board room discussions should include questions about security.
- Due diligence must encompass investigation of the outsource provider's experience with privacy and data security, including examining any privacy complaints from other clients, how the service provider protects its clients' PID, and the service provider's privacy and security policies.
- Policies and procedures for tracing, reporting, and managing security incidents should be documented.
- The data protection laws in force in the host country should be reviewed.
- Outsourced service providers should be restricted from subcontracting without the healthcare organization's prior written consent in each instance.
- Outsource service providers must remain contractually liable for the functions that are subcontracted.

Practice Pointer 2: Due Diligence is Not Just about Business Risk

The FHC case highlights the absence of a security readiness assessment in preparing for the integration of systems. The readiness assessment should be performed on all systems that will be impacted by the introduction of the outsourced application. It can also identify required upgrades, changes in process, and training needs that can ultimately impact security. The PACS and RIS systems had been in place for four years without an audit or assessment with respect to security. Bolting on an application without a prior assessment of potential security issues that could be amplified by the vendor application can lead to exposure of patient data. Used to correctly identify patients for services and for billing, significant amounts of PID are revealed in these systems and would be of strong interest to cybercriminals.

The Healthcare Information and Management Systems Society (HIMSS) Security Work Group created an Application Security Questionnaire that can be used as a guide in assessing security readiness in advance of implementations (http://www.himss.org/content/files/ApplicationSecurityv2.3.pdf). Within the information security context, due diligence involves both a review of policy documents and an examination of the infrastructure, in order to determine where necessary "bolts" of security must be applied. Hospitals must also determine whether the newly proposed architecture will sufficiently protect PID.

The following practice pointers should be applied to effectively "bolt on" security:

- As part of a mandatory readiness assessment, an internal privacy and data audit should be performed (1) to understand PID processing activities for the targeted systems and (2) to identify the specific PID to be transferred to the service provider.
- Open architectures should be identified and partners selected based on having common platforms.
- Standard application protocols should be required and be consistent with the legacy environment.

Practice Pointer 3: Monitor the Technology

In the FHC case, the VPN connections were not adequately monitored which could lead to intrusion and interception of transmissions. VPN's typically encapsulate the data into a special VPN data frame that provides information to the receiving site so that it can reassemble the data at the destination (Dennis, 2002). Authentication is the primary means used to verify that transmissions originate from the authorized parties using the VPN tunnel. The least secure method of authentication is the user identification and password combination. Best practices suggest that a two- or three-factor method be used that involves a passcode in addition to a code generated by a key device (RSA, 2005). FHC used only the user ID and password combination. The critical layers of security that protected the data transmission at FHC include:

- **Encryption**: The data are converted into a format unreadable by unauthorized individuals.
- **Authentication:** Secure HTTPS transfer of basic user name and password.

FHC had not performed an audit on the exisiting PACS and RIS systems and did not audit the new teleradiology system after implementation. On an annual basis, the hospital CIO should hire an independent security consultant to challenge the various security systems and provide comprehensive reports on the vulnerabilities of the hospital's system, particularly, the remote access portion of the network. Also, particular attention should be given to the internal portion of the network. Employees have proven to be some of the worst offenders for security breaches of sensitive patient information. Overall, security should be addressed at various levels: transaction, application, and client.

- **Transaction**: each patient record that is sent/read should be encrypted and the protected health information should not be discernable in transmission;
- **Application**: the specific clinical software should require individual log-in and should keep audit trails of the specific use of the features; and
- **Client**: the PC workstation should be hardened with VPN software, the operating systems patches should be current and the software certificates aligned with the outside systems to which it has access.

The type of threats that are possible if the end devices are not adequately protected are reconnaissance attacks, where hackers will scan for answering IP addresses and port numbers, and IP spoofing, as a first step to attempted unauthorized access to the network devices and client machines. The device doing the routing/firewalling can be hacked, opening up the hospital network to an intruder. If a hacker manages to log into the router/firewall, then it can be reconfigured to allow free access over the Internet to the client machine, bypassing any firewall rules or access lists. If hospital uses a hardware VPN terminator, then there are not any firewall safeguards to break, and it is an easy matter to get to the network port of the client. From there an intruder may be able to access the client itself, giving them an unfettered view of confidential patient information that is not yet encrypted. Potentially, an intruder can hijack a client to gain access to the hospital network, send and receive e-mails, and log in to internal servers. Even if an intruder cannot get into the client, once on the router, they can use it to access any network service. Determined intruders can simulate a client's connection (it is an easy matter to determine IP addresses from DHCP scopes or arp caches) and gain access to a network. With outsourced companies such as NightHawk, the hospital should discuss the partner's present security practices and request documentation to prove privacy and security provisions are adequate to meet the expectations of hospital privacy officers, executives, and the security group in the Information Services Department. The main tenants should cover the confidentiality, availability, and integrity of the data, particularly protected health information that flows between the two sites. The contractual agreement between the two parties should spell out in enough detail how the actual data flow works and should include a visual representation that is easy to understand by all parties involved in the agreement. In this drawing, all points of access to the data should be identified. This includes identifying the level of protection that will be provided by the carrier networks, including bandwidth vendors. Additional information to address the protection measures for both the security and privacy at any of these individual access points is necessary. The hospital should not finalize the agreement with the third party until all protection is in place.

In setting up security to protect the transmission of data to the third party entity, the hospital must retain the control of the connection and should be able to disconnect or fix the transmission should a security breach or a problem arise. If the hospital is the "sender" of the protected information, then it should retain the rights to control the data getting to the "destination." Often hospitals with smaller information services departments may be encouraged by the third party entity to hand over the maintenance and repair of the connection. These institutions should consider hiring additional information security resources to support the third party connections or consider hiring independent security managers who do not have any relationship with the third party entity receiving the data. Often, independent security consultants can find improvements to be made and may suggest additional measures to protect the data.

FHC, Nighthawk, and RI entered into a three-way business associate agreement. FHC maintained control of the VPN connectivity between its network and the entry points into the NightHawk network. Only users from the Radiology Department were given access via username and password to the NightHawk Web site to track studies sent to be read. At present, the NightHawk system cannot be accessed from anywhere except hospital workstations and it is not available for remote use at home by physicians. VPN connections also involve client stations that require review and management.

As the amount of PID being transmitted increases, and other avenues of interception are hardened, it can be expected that cybercriminals will turn to the weaknesses in hospital networks as their next source of identity theft and intrusion.

The important practice points include:

- A standard for auditing the VPN connection at all sites that transmit and receive patient data should be established.
- One-time passwords, using token mechanisms should be required for authentication.
- Host Intrusion Detection should be installed on the end stations that can identify anyone trying to hack into them or install back-door software.
- Split tunneling should not be used as it can open a network to the outside.
- Remote routers and firewalls should be installed with strong protection for administration access.

- Daily monitoring of VPN connections for IP addresses, traffic, and load should be done by both the outsource vendor and the healthcare organization.
- All attempted access and abnormal behavior on firewalls, routers, and end stations should be logged and investigated.
- Client log-ins should be routinely audited.
- Remote access VPNs should provide sufficient encryption to prevent man-in-the-middle attacks.
- Patches should be applied on clients used in outsourced systems.
- Multiple VPN clients running on the same PCs should be discouraged.

Practice Pointer 4: Secure the Perimeter

Hospital CIO's should anticipate requests from clinicians to use personal devices to access data on the hospital network, including PDAs, cell phones, or small handheld devices. Firm decisions must be made to either provide protected access to patient data from these devices or to not allow any personal devices to access the hospital network. The reports of data loss at healthcare organizations by the Privacy Rights Clearinghouse show that in 2007 there were 30 incidents involving hospitals or medical centers that involved exposure or theft of PID (for the complete list, see http://www.privacyrights.org/ar/ChronDataBreaches.htm#Total). Of the data thefts, there were 19 that involved the loss of a mobile device that contained identifying and unencrypted data from hospital systems. The convenience of a mobile device offered to physicians must be weighed against the substantial risk of loss or theft both inside and outside the healthcare organization. Some concerns with respect to a nonhospital asset or nonthird party asset are the following scenarios:

1. A physician is at a conference, hotel, or business center and loads the VPN client on a machine, but forgets to remove it completely before they leave. VPN clients that have expiring certificates and clear out browser cache when exiting Internet Explorer will prevent potentially leaving patient information on a workstation that can be viewed by anyone else using the client.
2. A radiologist uses the VPN client on their own laptop, but other people, including family members, also have access to that laptop and may use it at other times. If the VPN software uses a soft-token (already existing on the PC in the software), the VPN connectivity may not require additional authentication and the family members may be able to connect to the hospital network. If the physician is not using strong passwords, and is typical of some, keeping a list of passwords stored in a wallet, desk drawer, and so forth, this presents an easier access point for unauthorized use of the VPN connection into the hospital network and potential use of the radiology system, whereby sensitive patient data could be accessed.
3. Web pages that allow access to the PACS systems can be easily bookmarked. Often settings on the PC with respect to the operating system can offer to "remember" passwords for users. If physicians have these pages bookmarked and have told the operating systems to assist in remembering passwords, there is the potential to allow access to hospital systems by unauthorized users, especially if they are allowed unfettered or unmonitored access to the physician's laptop.
4. The security features offered through local service providers (cable or DSL) may not have adequate levels of protection for data passing through these hubs.
5. Home or small office wireless networks shared by multiple PCs may not be

completely secure and are often easy to eavesdrop on with simple sniffing tools. The physician's son or daughter could be watching him/her work and be in a different room of the house.

For today's physicians, the remote access to patient data is important for consultation and following the progress of a patient's care while away from the hospital facility. But the risks of exposure are great if simple protections and enforced access policies are not instituted. Even more critical, physicians accessing patient data outside of secure facilities in offshore locations must use the same mitigations against theft of data and intrusion as are followed inside the hospital.

The following practice points include:

- VPN clients with expiring certificates or that solidly clear out browser cache when exiting Internet Explorer should be used.
- Violations or security breaches resulting from negligence on the part of physicians should result in penalties for the person violating the rules, up to and including suspension of privileges at the hospital.
- The physicians and other outside parties accessing information or patient data should read, sign, and abide by a written policy concerning access to patient data. The level of access should be designed on a "need-to-know" scale, rather than a full house-wide access to the hospital data systems.
- Outsourced entity users should be given access to only the hospital systems necessary. The levels of access should be designed so that their requests match up to only the specific systems to which they need access.
- Systems should prevent saving data to workstations that exist outside of the hospital infrastructure.
- The system should prevent the download and installation of the VPN client to unauthorized workstations (such as home PCs).

Practice Pointer 5: Practice Incident Response

Hospitals should develop specific incident response plans that can be followed should confidentiality or security be violated. A governance process should be established to handle escalation of security issues and to enforce security policies and procedures. These plans and procedures should be routinely updated and tested. Employees should know that these documents exist and know how to use them. Such a document is often helpful in determining the root cause of the information breach.

Some medical centers have full-time executive level chief security officers (CSO) who actively engage vendors, clinicians, and patients in the protection of digital medical records. Also, the privacy officer can be given the responsibility for drafting security policies and procedures and the necessary forms that are considered part of the information security policies and procedures. The CSO should also be given responsibility for incident response including how breaches will be publicized to patients. Forms that were developed at FHC for the new teleradiology application and represent best practices include:

- **Remote Access Policy**: hospital employees, healthcare providers who access patient information and other outside entities must complete a remote access request form and read and sign a remote access policy statement. This document remains on file in the privacy officer's location. The statements contained in the document represent that the individual has read, understood, and acknowledged that patient health information is private and must remain secure.
- **Patient Consent Form**: Whenever patient information is to be released to an outside entity beyond the infrastructure of the hospital, the patient must sign a consent form to authorize the release of specific information.

On the form, the patient can specifically deny certain types of information from these outside entities and also can specify the destinations to receive or not receive patient information.

- **Business Associate Agreements**: The chief security officer keeps the business associate agreements that have been signed by vendors and third party entities with the designation as business associates with the hospital under the provisions of the HIPAA regulation.
- **Incident Response Plan**: A security incident response plan is followed whenever there is an apparent breach in the confidentiality or security of patient data. This plan is activated whenever a security incident is reported at the individual, departmental, or corporate level.

Practice Pointer 6: Pilot Studies and Benchmarks

The outsource relationship is no different than any other technology product or service that is offered to provide cost savings, efficiencies, or competitive advantage: it requires a proof of concept that the advertised virtues can be realized once installed. To maximize security and ensure perceived benefits, hospitals must plan a specific pilot study that carefully determines the system's behavior and matches the design to the actual implementation. The pilot should be designed and implemented by all parties to the outsource partnership. Failure to implement a pilot and rushing instead into production based on the outsourced vendor's implementation can create problems both for the later operation of the system and the effective establishment of security protocols. Early in the project planning phase, healthcare organizations should determine what the benchmarks for success will be, including the appropriate security metrics. For FHC, no benchmarks were identified except the readability of the transmitted image, and response time for the report to be faxed. Without any technology or security metrics in place, FHC is vulnerable to performance degradation and compromise of the data.

The following practice points apply:

- A pilot system should be designed and implemented before moving into production.
- The pilot system should be evaluated according to the preproject established benchmarks and deficiencies corrected prior to the move to production.
- A formal review with the outsource partners of any potential risks for security should be performed.

CONCLUSION

Once the telehealth system is established, hospitals cannot rest on their laurels for implementing an outsourced system. Continued vigilance is required both from the technical team supporting the connected systems as well as the C-level executives who must regard security as a strategic initiative. New threats continue to develop and network analysts supporting the hospital network must move risk mitigation forward on a daily basis. Department managers within the hospital, including the lab and radiology managers, must carry a strong position that breaches the security of patient data will not be tolerated. Any weaknesses in the systems used within the hospital must be reported immediately. Even patients must acknowledge that sometimes treatment requires sharing data among clinicians, but that it should not be at the expense of the integrity of the data. All the involved parties must work together to ensure the highest level of security is continuously obtainable.

HIPAA and, to a certain extent, the JCAHO have laid out the expectations for the protection of data in healthcare settings. The FHC case suggests that hospital management may be placing too much emphasis on the cost and quality aspects represented in outsource opportunities and less emphasis on the security aspects of the proposed system. Shortcuts to system implementation are an invitation to a compromise of patient data. Based on the FHC case, the practice points offered in this paper can provide mitigation of threats and lower the risk inherent in outsourced systems, particularly those systems wherein security must be "bolted" on because it was not built-in originally.

REFERENCES

Beck, G., & Neufeld, J. (2007). 2007 Indiana Critical Access Hospital Clinical and Technology Needs Survey. Retrieved June 12, 2008, from http://www.in.gov/isdh/publications/partner_rel/pdfs/2007IndianaCriticalAccessHospitalSurvey-p1to17.pdf

Bedi, S., Sarma, R. K., & Arya, S. (2002). Outsourcing CT scan: A boon for patients, physicians and hospital administrators in a general hospital. *Journal of the Academy of Hospital Administration, 14*(2), 7–12.

Collman, J., & Grimes, S. (2005). What healthcare executives should know and do about information security. *HIMSS Privacy and Security Toolkit.* Retrieved June 12, 2008, from http://www.himss.org/content/files/CEOWhitePaperFinal.pdf

Data Breach Report (2007). Identity Theft Resource Center. Retrieved June 12, 2008, from http://idtheftmostwanted.org/ITRC%20Breach%20Stats%20Report%202007.pdf

Dennis, A. (2002). *Networking in the Internet Age* (p. 208). New York: John Wiley and Sons, Inc.

Goelman, A. (2006). *Work without boundaries: Information technology and the extension of radiology work in time and space.* Paper presented at Work Beyond Boundaries, Vancouver, BC.

Goelman, A.(2007, April 26). Teleworks that work: Teleradiology and the emergence of Nighthawk Radiology firms. Presented at Sloan Foundation Industry Studies Annual Conference. Retrieved June 12, 2008, from http://web.mit.edu/sis07/www/goelman.pdf

Hill, R., Bartrum, T., & Guy, B. (2006). *Hospital outsourcing trends in clinical services.* Retrieved June 12, 2008, from http://www.wallerlaw.com.sitemason.com/files/WallerSurvey.pdf

HIMSS. Retrieved June 12, 2008, from http://www.himss.org/ASP/topics_privacy.asp

HIPAA Security Guidance. U.S. Department of Health and Human Services, Centers for Medicare and Medicaid Services. Retrieved June 12, 2008, from http://www.cms.hhs.gov/SecurityStandard/Downloads/SecurityGuidanceforRemoteUseFinal122806.pdf

Information Security Risk a Top Concern Among Outsourcing Executives. (2006, March 23). Booz Allen Hamilton, Inc. McLean, VA. Retrieved June 12, 2008, from http://www.boozallen.com/capabilities/services/services_article/1876648?lpid=827466

Jackman, M. W. A digital imaging transformation in radiology departments. *Healthcare Technology: The Digital Healthcare System.* Retrieved June 12, 2008, from http://www.hctproject.com/content/PDF/HCT2_wp_jackman.pdf

Joint Commission on Accreditation of Healthcare Organizations. (2000). *2000 Hospital Accreditation Standards.* Retrieved June 12, 2008, from http://www.jcaho.org

Kelley, M. (2006, April 9). Web handling radiology night work. *The Charleston Gazette.* Retrieved June 12, 2008, from http://redorbit.com/news/health/463659/web_handling_radiology_night_work/index.html

Levy, F., & Goelman, A. (2005). Offshoring and radiology. In L. Brainard & S. M. Collins (Eds.), *Offshoring white-collar work: The issues and implications*. The Brookings Trade Forum.

Monegain, B. CCHIT receives slew of applications just under deadline. Retrieved June 12, 2008, from http://www.healthcareitnews.com/story.cms?id=9060

NightHawk Radiology Holdings, Inc. releases record third quarter 2007 results. (2007, October 30). Cour D'Alene, ID. Retrieved June 12, 2008, from http://ir.nighthawkradiologyservices.net/releasedetail.cfm?ReleaseID=272186

NightHawk Radiology Services, Inc. (n.d.). Retrieved June 12, 2008, from http://www.nighthawkradiologyservices.net/

NightHawk Radiology Services awarded accreditation from joint commission. (2006, September 29). Retrieved June 12, 2008, from http://www.nighthawkrad.net/admin/editor/uploads/files/JCAHONewsRelease.pdf

Outsourcing in most hospitals. Retrieved June 12, 2008, from http://www.offshoringtimes.com/Pages/2007/BPO_news1584.html

Pollack, A. (2003, November 16). Who's reading your x-ray? *The New York Times*. Retrieved June 12, 2008, from http://query.nytimes.com/gst/fullpage.html?sec=health&res=9503E5D91238F935A25752C1A9659C8B63&fta=y

Privacy Rights Clearinghouse. Retrieved June 12, 2008, from http://www.privacyrights.org/identity.htm

RSA, best practices in authentication (2005). Retrieved June 12, 2008, from http://www.accusys.com/Best_Practices%20_in_Authentication_WP_0905.pdf

Schwartz, M. (2007, December 6). Application developlment: Buy, build or cobble? Retrieved June 12, 2008, from http://www.bmighty.com/hardware_software/showArticle.jhtml?articleID=204701872

ENDNOTE

[1] The name and location of the medical center have been changed. Some distinguishing details have been omitted to protect the people and partnerships.

This work was previously published in the International Journal of Information Security and Privacy, Volume 2, Issue 3, edited by H. Nemati, pp. 55-70, copyright 2008 by IGI Publishing (an imprint of IGI Global).

Chapter 6
Obtaining Patient's Information from Hospital Employees through Social Engineering Techniques:
An Investigative Study

B. Dawn Medlin
Appalachian State University, USA

Joseph Cazier
Appalachian State University, USA

ABSTRACT

Social engineering can be briefly defined as the obtaining of information through deceptive methods. The intention of the action is to acquire information that will be of use in order to gain access to a system or use of information obtained from the system. There are benefits gained by allowing health care workers access to patient data, but the ability to maintain security of that information may be compromised due to the accessibility. Using methods such as social engineering, health care workers may innocently provide sensitive information without realizing that they have participated in the process of deception. This chapter addresses the issue of social engineering used to obtain health care worker's passwords, as well as the laws that govern health care workers in relation to the privacy and security of confidential patient information.

INTRODUCTION

For most organizations, an employee uses a self-selected username and password as a form of authentication. Even when an organization chooses to enhance security by assigning information used by employees for authentication, social engineers are able to obtain information by using a variety of tricks and techniques (Ciampa, 2005). Social engineering is defined as the "act of manipulating a person or persons into performing some action" (McQuade, 2006). That action may be the "master key" to the health care agency's vault containing patient information.

DOI: 10.4018/978-1-61692-000-5.ch006

Most hackers rely on employees to unknowingly help them attack company networks and systems by simply answering a series of simple questions. Today, most health care agencies have intrusion detection/prevention systems such as firewalls that can be used to alert organizations in the event of a security breach, but these systems cannot prevent employees from inadvertently sharing information with others. Therefore, the question still remains, "how much information might an employee provide to a stranger or to a co-worker?"

The social engineer can, and often does, utilize an arsenal of methods that allow him or her to involve the emotions of a victim to aid in an attack. According to Mitnick & Simon (2002), the social engineer may flirt with the employee in an attempt to trick the individual into releasing information or another approach sometimes taken is to convince the employee that their job depends on supplying the attacker with the requested information. No matter the technique employed, if relevant and meaningful information is supplied, the entire network and all of the information it contains has been placed at risk.

Managers must be vigilant in their efforts to protect patient information as required by several laws. Most recently, on February 17th, 2009, President Obama signed into law the Health Information Technology and Clinical Health Act (HITECH) as part of the American Recovery and Reinvestment Act. The HITECH Act enhances the security and privacy provisions as well as the penalties contained in the Health Insurance Portability and Accountability Act of 1996 (http://www.nixonpeabody.com/publications_detail3.asp?ID=2621). This new law also requires patients be notified in the event of a security breach.

In this study, we simulated how a social engineer might obtain personal information from unsuspecting hospital employees. As previously mentioned, health care agencies and their employees must be especially vigilant in their effort to guard against the sharing of patients personal and/or private information.

BACKGROUND

Social engineers have traditionally used the telephone as the mechanism to obtain information. But today's social engineer is just as likely to approach an employee of an organization and act as though they need to obtain information in order to complete their job. Another method used by social engineers is to present themselves as an employee and act as though they are assisting others. Of course, depending upon the shrewdness and professionalism of the social engineer, not all attempts are successful.

If the social engineer is attempting to find out about one particular patient, they may target that person's medical health record. A patient's medical record may include gender, race, family history, sexual history including types of birth control, sexual activity and treatment, any history or diagnosis of substance abuse, and diagnosis of mental illness. Other medical information, such as HIV status, may also be included. The accessibility of this confidential information may open the door to various forms of discrimination. For instance, chronic diseases such as HIV and AIDS may result in an increase in insurance rates or even denial of coverage, due to the extensive medical treatment usually needed by these patients. Individuals may even be ostracized or stigmatized because of their disease type. Patients expect the information contained in their records to remain secure and private, to be seen only by those individuals whose access is medically or administratively necessary.

Unfortunately, patient's medical records are being illegally accessed and often when a breach occurs, the incident is seen in the news. Table 1 represents recent security breaches of patient information ranging from occurrences that affected individual patients, to an occurrence that wreaked havoc on thousands of patients.

As noted earlier, the HITECH Act includes a series of privacy and security provisions that expand the current requirements under the Health Insurance Portability and Accountability Act

Table 1. List of recent hospital security breaches

Date	Organization	Event	Records Affected
2009	Memorial Medical's Lee Campus, Johnstown, PA.	Police found a red notebook in the purse of a female employee of the hospital with passwords from the hospital. She admitted using the passwords to gain access to personal information.	200 patients
2009	Cedars-Sinai Medical Center, Los Angeles, California	A male previously employed in the billing department was sentenced to prison after he pleaded guilty to stealing patient records and defrauding insurance companies of hundreds of thousands of dollars.	1005 patients
2009	Kaiser Permanente Hospital, Bellflower, California	Hospital workers improperly accessed the medical records of Nadya Suleman, the mother who gave birth to octuplets. The hospital has since fired 15 employees. California health regulators fined Kaiser Permanente's Bellflower hospital $250,000 on May 14 for failing to keep employees from snooping in the medical records of Nadya Suleman	Individual
2008	NY Presbyterian Hospital/Weill Cornell Medical Center, New York, New York	A man who worked in the admissions department at a Manhattan hospital was charged with stealing and selling information on nearly 50,000 patients. Prosecutors said the man exploited his access to the hospital's computer system to acquire lists of patient names, phone numbers and Social Security numbers over a two-year period.	50, 000 patients
2008	Sparrow Hospital, Lansing, Michigan	Employees of Sparrow Hospital were fired or disciplined in July after it was discovered they attempted to access computerized medical information about Gov. Jennifer Granholm when she was admitted there for abdominal surgery April 29	Individual

(HIPAA). Among other things, the legislation strengthens enforcement of the HIPAA privacy rule and includes a mandate requiring patients be notified in the event of a breach of identifiable health information.

More specifically, the Federal Trade Commission (FTC) and the Department of Health and Human Services (HSS) have issued the first set of HIPAA privacy/security guidance under the new HITECH Act requirements. The new guidance relates to the security breach notification requirement, which is expected to go into effect September 2009 (http://compliance.utimaco.com/na/tag/hitech-act/). "Under this requirement, health plans and personal health record (PHR) vendors must provide individual notification if there has been a security breach of protected health information (PHI). Notification must be provided to individuals in writing within 60 days of discovery of the breach. If the breach involves more than 500 individuals, notice also must be made in prominent media outlets and to the Secretary of Housing and Health Services or to the FTC for PHR vendors (http://compliance.utimaco.com/na/tag/hitech-act/).

Both HIPAA and HITECH, simply put, make it illegal for unauthorized personnel to access or release information from someone's medical records. More specifically, HIPAA addresses security and privacy measures in relationship to passwords, either directly or indirectly, in the following standards (http://www.hhs.gov/news/facts/privacy.html). These standards, as listed in Table 2, include management processes, user education and training, and access control.

Within a health care system governed in part by regulations such as HIPAA and HITECH, both employees and agencies must do everything humanly possible to protect patient's information. Passwords are just one of the methods by which employees can assist in that protection. They act as barriers between the "outside" world and the

Table 2.

Security Management Process [161.308(a)(1)] Healthcare organizations must show that they have a consistent set of internal processes, with implementation that is widespread and institutionalized. Processes range from establishing criteria for who has access to what, and who can request certain resources; to ensuring that access rights are revoked immediately upon employee termination.
Security Awareness and Training [161.308(a)(5)] HIPAA requires that staff members be trained and educated concerning the proper handling of PHI. This basic-level security training should include measures such as password management.
Access Control [161.312(a)] HIPAA security regulations require a definition of who has access to PHI within the organization, as well as the rules determining an individual's right of access, and the reasons for denying access to some individuals.

world of networked systems at large. If a social engineer is able to obtain a password and other identifiable information, they can likely impersonate the employee and gain admittance to the system (Medlin & Cazier, 2007); the intrusion will most likely go unnoticed.

There are many examples in the literature that emphasize the importance of good password practices, describe problems related to passwords and offer guidelines on how to address the shortcomings. A few examples include a study by Vu, Proctor, Bhargav-Spantzel, Tai, Cook and Schultz, who performed experiments on how to improve password security and enhance password memorability (Vu *et al*, 2007); Ives, Walsh and Schneider, who gave recommendations on how to improve authentication systems (Ives *et al*, 2007); and Furnell, who in a study on website password practices suggests that more should be done to advise, support and protect users and to ensure efficient password use (Furnell, 2007).

The use of passwords in computer systems was adopted after the invention of the computer terminal. At that time passwords were short, generally about six characters in length, making them easier for individuals to remember, resulting in self-created passwords that are personally identifiable and reflect the personalities and interests of those individuals creating them. With a few taps on the keyboard, employees can self-describe themselves with passwords such as redsoxfan, lovedaisy, or boonenc. As psychologists have discovered, people are fairly predictable in their patterns and are very predictable in the aggregate, resulting in predictable passwords (Andrews, 2004).

For health care agencies, the password works like the key to a lock, anyone who has it can get in. Toward that end, many sites offer advice on how to construct a password. One of the first guidelines in creating good passwords was published in 1985 by the Department of Defense and is still relevant today (http://www.alw.nih.gov/Security/FIRST/papers/password/dodpwman.txt). Their guidelines recommended the following: 1) passwords must be memorized; 2) passwords must be at least six characters long, 3) passwords must be replaced periodically, and 4) passwords must contain a mixture of letters (both uppercase and lowercase), numbers, and punctuation characters.

Unfortunately, enforcement of most of the above recommended password actions are not always put into practice. For most employees it is much easier to create easy to remember passwords. A person's ability to remember a sequence of items is limited to around seven items, plus or minus two (Kanaley, R., 2001). This sequence of items cannot be drawn from an arbitrary range, necessitating the use of familiar chunks of words or numbers. Human memory also thrives on redundancy and repetition. Studies have shown that an individual's short term memory will only retain a password for about 30 seconds, thereby requiring individuals to attempt to memorize their passwords immediately (http://brain.webus.com/memory/human_memory.htm). This attempt at memorizing passwords is often ineffective and inefficient and can lead to such poor practices as using short or easily guessable passwords.

To combat the issue of having to remember so many different passwords, some users have resorted to selecting familiar terms such as a pet or family name, their own name, their phone number, or other common terms that could be found in a dictionary. British psychologist Helen Petrie, Ph.D., a professor of human/computer interaction at City University in London, analyzed the passwords of 1,200 British office workers who participated in a survey funded by CentralNic, an Internet domain-name company in 2001. She found that most individuals' passwords fell into one of four distinct password categories (Table 3).

The first category, labeled by Dr. Petrie as "family," comprised nearly half of the respondents. These individuals selected their own name, the name of a child, partner or pet, birth date, or significant number such as a social security number. Dr. Petrie further found that individuals also chose passwords that symbolized people or events with emotional value or ties.

One third of the survey participants were identified as "fans," using names of athletes, singers, movie stars, fictional characters, or sports teams. Fantasists made up eleven percent of survey responses. Dr. Petrie found their passwords were comprised of sexual terms or topics. The final ten percent of participants were identified as "cryptics." These users were seemingly the most security-conscious, but it should also be noted that they were the smallest of the four identified categories. These individuals selected unintelligible passwords that included a random string of letters, numerals, and symbols such as Jxa+157.

Dr. Petrie's four categories (Table 4) were enlarged in 2005 after a study of health care passwords was conducted by Medlin and Cazier (2005). The newly expanded list was found to be more specific in relation to the previously shown list of password categories. It could be argued that the following categories as seen in Table 4 of "Other," "Faith," "Place," and "Numbers," should be included in the category of "Family," as it may represent words that relate to an individual's favorite food, religious choices, favorite places with emotional ties, and numbers that may be personally meaningful to them. But each of the passwords collected were reviewed by four individuals, and collectively determined to fall into one of the eight categories.

If an employee created a password that was easy to guess because of identifiable information or

Table 3. Petrie's category definition

Category	*Definition*
Family	Name or nickname, name of a child, partner or pet, birthday
Fan	Names of athletes, singers, movie stars, fictional characters or sports teams
Fantasists	Interest in sex is evident in passwords such as "sexy" "stud" and "goddess"
Cryptic	Unintelligible passwords or a random string of letters, numbers and symbols

Table 4. Additional categories defined

Category	*Definition*
Other	Common English dictionary terms that did not include religious terms or places
Faith	Terms associated with religion or religious activities
Place	Names associated with towns or cities
Numbers	A string of all numbers

weakness in relation to good passwords, it created the opportunity for a social engineer to "guess" that password and to possibly have access to the system. Social engineers who obtain information from the employee by pretending to innocently ask questions about hobbies, family members and pets, or the employee's birth location can then assume the legitimate employee's identity and gain access to all data the employee is authorized to view. It is therefore imperative that employees be taught about social engineering tactics and the use of good, strong passwords.

RESEARCH METHODOLOGY

Instrument

Since a large number of attacks are carried out by insiders and often present the most danger to health care agencies due to their knowledge of the system, we developed a survey instrument to test how much sensitive information they would supply. To simulate a real attack and to obtain a fair statistical representation of the security weaknesses, a two-page survey was administered to employees of five hospitals of varying sizes and in different regions of the state. Hospital administration approval was obtained before administering the instrument, but the administration did not endorse the survey to respondents and did not specifically ask them to participate.

Data was gathered to not only determine how many employees would disclose their passwords and other personal information such as their address, phone number and email, but also simulated the types of information individuals were willing to share with co-workers, colleagues, or friends of colleagues. The information that employees were willing to share, including their passwords and other personal information, would certainly make it easier to hack into a system instead of having to "guess" at the necessary authentication information

In the first section of the survey, employees were asked to supply their place of birth, their gender and their job title. It is important to note that employees may innocently and in casual conversation share with co-workers and others, these aforementioned types of information which could be used to build a profile of an individual that would assist a social engineer in hacking into an organization's computerized system where authentication is required.

The next section requested non-personal information. Questions included the employee's favorite movie and their favorite television show. They were then asked question like their pet's name. This question sounds innocuous, but as earlier noted may be one way to guess a portion or their entire password. The aforementioned categories of gender, place of birth, favorite movie and television show, as well as pet's name were selected and based on previous research studies conducted by Botelho and Cazier (2008) and password creation research performed by Medlin, Cazier (2006) and Petrie (2001).

The next section of the survey was designed to determine the existence of any security training programs offered by the health care agency. Information included whether or not the organization offered security training programs, if the employee had attended security training programs and, if attended, the date of their last class.

The last section contained questions related to their password practices and, as noted earlier, was based on previous research. Subsequent questions were developed to determine their password length and composition. In addition, employees were asked if they selected their own password or had one assigned to them by the organization. It has been shown that individuals will generally create passwords that are easy to remember, making it easier for hackers and others to guess.

Data Collection

The number of respondents and the types of information collected are of particular interest, as they show current hospital workers' willingness to share personal, as well as password, information that may increase the accessibility of information to those that should not be privy to patient's health care records.

The data set was comprised of 118 responses. Eighty-eight percent (88%) of the respondents were female and twelve percent (12%) were males. Respondents were from various locations, ranging from small rural areas with approximately 5,000 people, to larger, more urban populations of 500,000. Fifty-three of the respondents filled out entry forms for a drawing, and thus provided the researchers with additional personal and identifiable information.

Analysis and Results

Interestingly, the findings noted in Table 5 indicate that most respondents were often required to use a password to access systems, but rarely changed their passwords. As further indicated, most of the respondents used the same password on multiple accounts. This type of behavior should certainly be a concern for all health care agencies. The practice of rarely changing passwords and/or using the same password for multiple accounts would assist social engineers and other hackers, allowing them to easily attain access to one system and possibly more.

Analyzing the results related to employees' other password practices found that eighty-nine percent (89%) were allowed to choose their own passwords, with the average password being about seven characters in length. In addition, only sixteen percent (16%) of the employees included special characters, adding to the problem of less than secure passwords.

As seen in Table 5, half of the respondents created passwords consisting of family names, including their own name or nickname, the name of a child, or significant other. It is obvious that a very small percentage of employees are using most of the best practices recommended by governmental, educational, and private organizations.

The next largest category was that of "other," with forty-five percent (45%) of the respondents indicating that their passwords included a number. The choice to integrate a number is important, but just as important is the placement of that number and whether or not the number relates to meaningful and informative information such as a phone number or birth date.

Fifteen percent (15%) of the respondents self-reported the inclusion of "fan-based" words, which could include names of athletes, singers, movie stars, and fictional characters or sports teams. "Place" was the next highest category, with fourteen percent (14%), using another identifiable piece of information such as the city where the employee works or lives.

The smallest of all of the self-identified password categories was "fantasy," followed closely by the categories of school and faith. Five percent (5%) of the employees selected the "cryptic" category, suggesting that these employees are security-conscious since that category includes passwords that are unintelligible or include a random string of letters, numbers, and symbols. Unfortunately, as noted earlier, is it also the smallest of all of the eight self-reported categories.

Most interesting, of the 118 respondents, seventy-three percent (73%) of the employees shared their passwords with a co-worker or the friend of a co-worker through this survey instrument. It should be noted that one of the largest threats is that of the internal employee and again, the confidentiality of the password. Internal employees can also act as social engineers to gain access to additional resources.

In the area of password and security training, most of the respondents, fifty-four percent (54%) indicated that their employer had offered password security training, with fifty-eight percent (58%) of

Table 5. Descriptive statistics

Variable Name	Question	Answers	N	Mean	Std Dev
Gender	Please mark your gender	1 = Male, 0 = Female	118	0.12	0.32
Pass_Freq	How often do you use a password to access systems?	1 = Very Often / 5 = Never	118	1.23	0.59
Pass_Change	How often do you change your passwords?	1 = Very Often / 5 = Never	117	2.85	1.13
Reuse	Most people use the same password on multiple accounts. How often do you do this?	1 = Very Often / 5 = Never	118	2.47	1.32
Pass_Train	Does your employer offer password security training?	1 = Yes, 0 = No	115	0.54	0.50
Awar_Train	Does your employer offer any other security awareness training?	1 = Yes, 0 = No	113	0.58	0.50
Current_Train	When was the last time you participated in either a password or another security awareness training program?	1= Last week / 5 = Never	115	4.09	1.08
Choose_Pass	On average, do you choose your own password or have one assigned?	1 = Choose Own, 0 = Assigned	117	0.89	0.32
Family	Does your password fit into this category?	1 = Yes, 0 = No	118	0.50	0.50
Cryptic	Does your password fit into this category?	1 = Yes, 0 = No	118	0.05	0.22
Number	Does your password fit into this category?	1 = Yes, 0 = No	118	0.45	0.50
Fan	Does your password fit into this category?	1 = Yes, 0 = No	118	0.15	0.95
Faith	Does your password fit into this category?	1 = Yes, 0 = No	118	0.03	0.18
School	Does your password fit into this category?	1 = Yes, 0 = No	118	0.02	0.13
Fantasy	Does your password fit into this category?	1 = Yes, 0 = No	118	0.00	0.00
Place	Does your password fit into this category?	1 = Yes, 0 = No	118	0.14	0.34
Other	Does your password fit into this category?	1 = Yes, 0 = No	118	0.51	0.50
Characters	How many characters are in your most commonly used password?	1 = 1-3, 2 = 4, 3 = 5, 4 = 6, 5 = 7, 6 = 8, 7 = 9, 8 = 10+	116	5.03	1.71
Numbers	Do your passwords contain any numbers?	1 = Yes, 0 = No	117	0.87	0.34
Special_Char	Do your passwords have any special characters in them (@, #, %, &, etc)	1 = Yes, 0 = No	118	0.16	0.37
Password	Please tell us your password	1 = Shared 0 = Did Not Share	118	0.73	0.45

the hospitals offering some other type of security awareness training (Table 5). Attendance by the employee in either a current password or security awareness training program was measured on a likert scale of 1 being last week and 5 being never. The employees indicated that currently, they almost never attended the security awareness programs.

A T-test was conducted to see if there were significant differences between those that shared their password versus those that did not share their password in relation to the categories established by Petri and others. Those that used family as a part of their password were also more willing to share their password (see Table 6). Regarding the frequency of changing passwords, a significant difference was found between those who shared their passwords with others in comparison to those who did not.

Several findings were significant. Sixty-three percent (63%) of those who included family as part of their password were willing to share the password with others. Even more surprising, those individuals who included numbers in their

Table 6. Employees willingness to share or not share passwords

	Shared Password			Did Not Share Password			Difference	
	N	*Mean*	*Std Dev*	*N*	*Mean*	*Std Dev*	*Mean Dif*	*Sig.*
Gender	86	0.14	0.35	32	0.06	0.25	0.08	0.25
Pass_Freq	86	1.14	0.41	32	1.47	0.88	-0.33	0.01
Pass_Change	85	2.69	1.09	32	3.28	1.14	-0.59	0.01
Reuse	86	2.48	1.26	32	2.44	1.50	0.04	0.89
Pass_Train	84	0.51	0.50	31	0.61	0.50	0.10	0.34
Awar_Train	83	0.53	0.50	30	0.70	0.47	0.17	0.11
Current_Train	85	4.13	1.09	30	3.97	1.07	0.16	0.48
Choose_Pass	85	0.93	0.26	32	0.78	0.42	-0.15	0.02
Family	86	0.58	0.50	32	0.28	0.46	0.30	0.00
Cryptic	86	0.05	0.21	32	0.06	0.25	-0.02	0.73
Number	86	0.50	0.50	32	0.31	0.47	0.19	0.07
Fan	86	0.20	1.10	32	0.03	0.18	0.17	0.40
Faith	86	0.02	0.15	32	0.06	0.25	-0.04	0.30
School	86	0.01	0.11	32	0.03	0.18	-0.02	0.47
Fantasy	86	0.00	0.00	32	0.00	0.00	0.00	NA
Place	86	0.14	0.35	32	0.13	0.34	0.01	0.84
Other	86	0.55	0.50	32	0.41	0.50	0.14	0.18
Characters	86	5.26	1.62	30	4.37	1.81	0.89	0.01
Numbers	86	0.91	0.29	31	0.77	0.43	0.13	0.06
Special_Char	86	0.20	0.40	32	0.06	0.25	0.14	0.08

passwords were willing to share their passwords at a rate of fifty percent (50%); this seems counter intuitive, as one would assume that employees who have created stronger passwords by including numbers would be less likely to share their passwords. As expected by most security experts, those who were more security conscious and included a special character in their password were not as willing to share it.

In addition to the survey, respondents were asked to provide several pieces of identifiable information such as their name, phone number, address, and an email address for a drawing of a gift card. Out of the 118 respondents, forty-five (45%) signed up for the drawing. Again, this was on a voluntary basis.

FUTURE TRENDS

Certainly as more electronic medical records are being created and shared, both patients and health care providers will be interested in the tools that are available to them in order to protect their information. The question therefore becomes "What tools are out there?"

For health care administrators, security is enhanced by using systems tools that are already available, such as Active Directory and LDAP (Lightweight Directory Access Protocol). Most likely, one or the other, or a combination of both is already in use to help in the securing of information. Even when other front-end access management products, like IBM Tivoli, Citrix or Sun Microsystems' Java System Identity Manager are in use, the directory server on the back end is likely to be Active Directory, LDAP or both.

With LDAP, it's possible to set a minimum password length, minimum number of alphabetic and numeric characters, number of repeat characters and the number of characters which must be different from a user's previous passwords. Group Policy Objects (GPO) in Active Directory also allows for those settings, in addition to settings that prevent an employee from reusing up to 24 of his or her last passwords, force password resets after a set time frame and require passwords to be complex with a combination of numbers, uppercase and lowercase letters. Since both Active Directory and LDAP integrate with third-party access management provisioning tools, password compliance would no longer be a concern.

Other solutions include thin clients, which are low-cost, centrally managed computers with no CD-ROM players, disk drives, or expansion slots. These devices use a central system to store data, providing high levels of availability, reliability, and security. The idea is to limit the capabilities of these computers to only essential applications, which eliminates the need to store data on a desktop device. There is no local storage, no local processing, and no local opportunity to gain access to sensitive data other than in its displayed form.

Additionally, more health care agencies may consider adopting biometrics. Biometrics is the science of identifying people through physical characteristics. Usually not one technology but a cluster of several, biometrics uses fingerprints, handprints, retina scans, voice recognition, facial structure, and even hand motions while writing a signature-to identify individuals (Simpson, 2002).

HIPAA calls for a tiered approach to data access in which staff members only have access to the information they require to perform their jobs. Biometrics makes possible such a tiered approach, while eliminating the security breaches that result from shared passwords or lost badges. Biometric applications are extremely limited even though they have been around for nearly two decades; however, that is changing, due to decreasing cost, increasing accuracy, emerging technology, public acceptance, and stricter compliance regulations.

Smart cards may also be used as these operate with a chip that includes stored memory, and an operating system. A patient's entire clinical history is stored on the smart card which can only be accessed via reading devices in a physician's office, primary care center, hospital, or other medical institution. Through the use of this device, exposed paper records will not be a concern. An added benefit of smart cards is the ability for users to electronically forward patient information to other health care authorities and insurers. Specifically, Java-based card technology emerges as a leading platform because of its ability to support multiple health care applications securely, while incorporating biometrics for positive identification and authentication.

As existing technological trends advance and new technology enters the marketplace, it is important to remember that both the employee and the patient must always be vigilant in protecting the information within the patient's record and on health care agency networks. Computerized systems and security methods cannot prevent individuals from talking or providing information to the shrewd, cunning and calculating social engineer.

CONCLUSION

Findings of the present study indicate that employees are willing to share personal information with co-workers and friends of co-workers. Seventy-three percent (73%) of the employees shared information that a social engineer could use to create a profile of an employee and gain access to the employer's network and other confidential patient information. It is imperative that employees understand the consequences of sharing information, as well as the importance of creating and maintaining strong passwords.

The simulation that was carried out during this study demonstrated that many employees may currently be in violation of HIPAA and HITECH regulations due to their willingness to share their information and their practice of creating weak passwords, thus allowing for easy access into a system. Hospitals and other health care agencies must identify ways to educate employees regarding HIPAA and HITECH regulations to protect patients and prevent penalties for sharing or misusing information.

REFERENCES

Analytics, I. D. (n.d.). *Data Breach Harm Analysis from ID Analytics Uncovers New Patterns of Misuse Arising from Breaches of Identity Data*. Retrieved November 12, 2007, from http://www.idanalytics.com/news_and_events/20071107.html.

Andrews, L. W. (2004). *Passwords reveal your personality*. Retrieved March 13, 2007, from http://cms.psychologytoday.com/articles/pto-20020101-000006.html.

Botelho, C. M., & Cazier, J. A. (2008). Guarding Corporate Data from Social Engineering Attacks . In *Handbook of Research on Information Security and Assurance*. Hershey, PA: IGI Global.

Burnett, M., & Kleiman, D. (2006). *Perfect Passwords. Selection, Protection, Authentication*. Syngress.

Ciampa, M. (2005). *Security Guide to Network Fundamentals* (2nd ed.). Boston, MA: Course Technology, Cengage Learning.

Damie, P. (2002). *Social Engineering: A Tip of the Iceberg*. Retrieved January 10, 2008 from http://www.isaca.org/Content/ContentGroups/Journal1/20023/Social_Engineering_A_Tip_of_the_Iceberg.htm.

Diagnosis Identity Theft. (2007). Retrieved on November 30, 2007, from http://articles.moneycentral.msn.com/Banking/FinancialPrivacy/DiagnosisIdentityTheft.aspx

Freeman, L. (2009). *Health Law Alert*. Retrieved June 30, 2009, from http://www.nixonpeabody.com/publications_detail3.asp?ID=2621.

Furnell, S. (2007). An assessment of website password practices. *Computers & Security*, *26*, 445–451.

Gragg, D. (2007). A Multi-Level Defense Against Social Engineering. SANS. Retrieved July 1, 2007, from http://www.sans.org/reading_room/whitepapers/engineering/920.php.

Human Memory. (n.d.). Intelegen, Inc. Retrieved December 12, 2007, from http://brain.web-us.com/memory/human_memory.htm.

Hupp, M. (2007). Protecting patient medical records from the nosy. Retrieved November 30, 2007, from http://www.bizjournals.com/milwaukee/stories/2007/11/12/focus3.html?t=printable.

Insider Threat Research. (n.d.). Retrieved December 1, 2007, from http://www.cert.org/insider_threat.

Ives, B., Walsh, K. R., & Schneider, H. (2007). The domino effect of password reuse. *Communications of the ACM*, *47*(4), 75–78. doi:10.1145/975817.975820

Kanaley, R. (2001). Login error trouble keeping track of all your sign-ons? Here's a place to keep your electronic keys, but you better remember the password. *San Jose Mercury News*, 3G.

Laughlin Memorial Hospital Discovers a Treatment for Password Problems. (n.d.). Retrieved December 12, 2007, from http://www.imprivataonesign.com/documents/casestudies/memorial_healthcare.pdf

McQuade, S. (2006). *Understanding and Managing Cybercrime*. Upper Saddle River, New Jersey: Prentice Hall.

Medical identity theft on the rise as health care desperation leads to crime. (n.d.). Retrieved December 1, 2007, from http://www.naturalnews.com/020528.html.

Medlin, B. D., & Cazier, J. A. (2005). An Investigative Study: Consumer Password Choices on an E-Commerce Site. *Journal of Information Privacy and Security*, *1*(4), 33–52.

Medlin, B. D., & Cazier, J. A. (2007). An Empirical Investigation: Health Care Employee Passwords and Their Crack Times in Relationship to HIPAA Security Standards. *International Journal of Healthcare Information Systems and Informatics*, *2*(3), 39–48.

Melville, N. (2009). Today's interconnectivity can also put patient data in jeopardy. Retrieved June 1, 2009, from http://compliance.utimaco.com/na/tag/hitech-act/

Mitnick, K., & Simon, W. L. (2002). *The Art of Deception*. Hoboken, New Jersey: John Wiley & Sons.

O'Connor Hospital Privacy Policy. (n.d.). Retrieved December 3, 2007, from http://www.oconnorhospital.org/contact/Pages/HospitalPrivacyPolicy.aspx.

Pfleeger, C. P., & Pfleeger, S. L. (2007). *Security in Computing* (4th ed.). Prentice Hall Protecting the Privacy of Patients' Health Information. Retrieved November 13, 2009, from http://www.hhs.gov/news/facts/privacy.html

Simpson, R. L. (2002).. . *Nursing Management. Chicago*, *33*(12), 46–48.

Thompson, S.T. Helping the Hacker? (2006). Library Information, Security, and Social Engineering. Information Technology and Libraries. *Chicago, 25*(4), 222-226.

Vu, K. L., Proctor, R. W., Bhargav-Spantzel, A., Tai, B. L., Cook, J., & Schultz, E. (2007). Improving password security and memorability to protect personal and organizational information. *International Journal of Human-Computer Studies*, *65*, 744–757. doi:10.1016/j.ijhcs.2007.03.007

APPENDIX A

Condensed Survey

Table 7.

Gender: [ ] Male [ ] Female Place of Birth: ______________________________
Favorite TV Show: ________________ Favorite Movie: _______________________
Favorite Sport and Team: ____
Favorite Singer/Band: ____________ Favorite Song: _________________________
Favorite Type of Pet: __________________ Pet Name(s): ____________________
Do you have kids: [ ] Yes [ ] No
Please list the first names of the ***one (or two)*** most important person (or people) in your life:
____
To help with our research study, please tell us a little about your organizations security
How often do you use a password to access systems?
[1] Very Often [2] Often [3] Occasionally [4] Not Often [5] Never
How often do you change your passwords?
[1] Very Often [2] Often [3] Occasionally [4] Not Often [5] Never
Most people use the same password on multiple accounts. How often do you do this?
[1] Very Often [2] Often [3] Occasionally [4] Not Often [5] Never
Does your employer offer password security training? [1] Yes [0] No
Does your employer offer any other security awareness training? [1] Yes [0] No
When was the last time you participated in either a password or another security awareness training program?
[1] Last week [2] Last month [3] Last 6 months [4] Last Year [5] Never
On average, do you choose your own password or have one assigned?
[1] Choose Own [0] Assigned
Most passwords fall into the following categories, please mark if yours fits in any of these (select all that apply) Family, Fan, Fantasy, Cryptic, Faith, Place, Numbers, School, Other
How many characters are in your most commonly used password?
[1] 1-3 [2] 4 [3] 5 [4] 6 [5] 7 [6] 8 [7] 9 [8] 10+
Do your passwords contain any numbers? [1] Yes [0] No
Do your passwords have any special characters in them (@, #, %, &, etc) [1] Yes [0] No
In order to facilitate creation of solid data, we would appreciate if you would provide one of your passwords. This information will be held in the strictest of confidence, and will be used only to generate a number that describes the characteristics of your password and then destroyed.
My Home/Work/Both password is: ____________________
Comments:

Chapter 7
Effects of Individual Trust in Broadcast Media and the Internet on Privacy–Risking Uses of E–Health: An Expanded Analysis

E. Vance Wilson
Arizona State University, USA

David D. Dobrzykowski
Eastern Michigan University, USA

Joseph A. Cazier
Appalachian State University, USA

ABSTRACT

People claim to be concerned about information privacy on the Internet, yet they frequently give out personal information to online vendors and correspondents with whom they have little, if any, prior experience. This behavior is known as the privacy paradox and is particularly relevant to the context of e-health, due to the special risks of health information exposure. Using data from the 2005 Health Information National Trends Survey (HINTS), this study addresses a key question regarding online health information privacy: Do individuals self-police risky e-health activities (i.e., uses where personal information is given out) or are they indifferent to risk based upon generalized trust in broadcast media and the Internet as sources for health information? In exploring this question, our study contrasts several alternative models of health trust, and recommends a new partial-mediation model for future studies. The results show that Internet trust mediates most effects of broadcast media trust on Internet use and that Internet trust plays a much smaller role in motivating Internet uses that are risky than is the case for low-risk uses. These results have important implications for researchers, policymakers, and healthcare administrators in determining the level of privacy protection individuals need in their use of e-health applications.

DOI: 10.4018/978-1-61692-000-5.ch007

INTRODUCTION

Early e-health offerings were primarily informational, but an increasing variety of online healthcare services are now being developed. These services include online access to health records (Masys, Baker, Butros, & Cowles, 2002), electronic billing and payment services (Altinkemer, De, & Ozdemir, 2006), and public health reporting (Järvinen, 2009). Patients can interact with peers and mentors in online support groups (Zrebiec & Jacobson, 2001) and engage in computer-mediated communication with physicians and clinical staff (Wilson, 2003). And even though many of the early e-health vendors failed (Itagaki, Berli, & Schatz, 2002), leading e-business organizations, including Microsoft and Google, are now turning their attention toward e-health development (Lohr, 2007). The new e-health services are prized by the public (Homan, 2003), yet they entail important information privacy risks that are inherent to communication and personalization.

Exchange of information during communication creates opportunities for personal information to be exposed, either through an accident, such as inadvertently overhearing a conversation, or by design, as is the case with surreptitious "phishing" strategies (Hesse, Nelson, Kreps, Croyle, Arora, Rimer & Viswanath, 2005). Information privacy risks also emerge in personalization, defined as "the ability to proactively tailor products and product purchasing experience to tastes of individual consumers based upon their personal and preference information" (Chellappa & Sin, 2005, p. 181). In the case of e-health, personalization can be applied to acquire and organize health information according to the patient's preferences, to automatically generate health forms and records, or to provide monitoring capabilities to help manage chronic disease. However, information used for personalization can be exposed due to events beyond an individual's control, as illustrated by the recent exposure of records relating to 1.8 million patients and physicians that occurred when a laptop computer belonging to the U.S. Veterans Administration was stolen (Gaudin, 2007). Although individuals clearly benefit from having access to personalized e-health that can do more than simply provide health information, personalization does tend to increase the risk that privacy will be compromised.

Population sample surveys show that individuals strongly desire privacy in their use of the Internet (Fox, Rainie, Horrigan, Lenhart, Spooner, & Carter, 2000), yet they are increasingly pragmatic about providing personal information online (Taylor, 2003). This *privacy paradox* of individuals seeking privacy while giving out their information is especially relevant to the context of e-health due to the sensitivity of health information. Harm from the exposure of health information can have unique financial and emotional effects, such as obstructing insurance coverage, limiting job prospects, damaging personal relationships, and inviting social ostracism. If individuals perceive the potential for privacy risk harm to be especially high in the context of health information, this will reduce their motivation to use e-health (Cazier, Wilson, & Medlin, 2007).

Healthcare policymakers and regulators have responded to the public's privacy concerns by creating specialized professional privacy standards (Mason, McCall, & Smith, 1999) and stringent privacy regulations, such as the Health Insurance Portability and Accountability Act of 1996 (HIPAA) which instates severe penalties for violations. These standards and regulations are intended to protect individuals from having their health information exposed, but they have proved to be cumbersome and relatively ineffective (Choy, Hudson, Pritts, & Goldman, 2001). Thus, current standards and regulations that apply to health information may actually reduce benefits and increase costs of e-health use.

We propose that it is important to learn more about the factors that drive use of e-health services, the nature of the relationships, and the impacts of the relationships on privacy risk. This knowledge will allow us to better understand and model the privacy trade-offs individuals make in their decisions involving personal health information. In addition, assessing the ability of individuals to self-police e-health privacy risks can help policymakers and regulators create standards and laws that are more attuned to individual needs and preferences. In this study, we use data from the 2005 Health Information National Trends Survey (HINTS) to address a key question regarding online health information privacy: *Do individuals self-police risky e-health activities (i.e., where personal information is given out) or are they indifferent to risk based upon generalized trust in broadcast media and the Internet as sources for health information?*

BACKGROUND

Our research approaches information privacy risks from the perspective of trust—the willingness to assume risks of disclosure (Mayer, Davis, & Schoorman, 1995). Trust mitigates the effects of individuals' perceptions of privacy risk on subsequent behaviors (Dinev & Hart, 2006) and increases the willingness of individuals to share personal information in online settings (Cazier, Shao & St. Louis, 2007). When trust is present, it is an important stimulator of online transactions (Quelch & Klein, 1996), and lack of trust is a crucial reason why people drop out of online business interactions (Luo, 2002).

For online organizations it is essential to build and maintain trust because the primary recipient of an individual's trust in online contexts is the organization itself; in online transactions there often are no salespeople or other human agents in whom to develop trust separately from the organization (Chow & Holden, 1997). Zucker (1986) identified three key bases from which trust in organizations is developed. *Process*-based trust arises from a process of social exchange and shared experiences between organizations and customers. Successful prior and current experiences build trust for future exchanges. *Characteristic*-based trust derives from a sense of commonality in characteristics that are shared with the other party, for example, similar values, background, ethnicity or experiences. *Institution*-based trust develops through the influence of a third party, such as a government agency, a financial institution or some other organization that assures the trustworthiness of the target organization. An example of institution-based trust is the WebTrust Assurance Program developed by the American Institute of Certified Public Accountants (AICPA) (Srivastava & Mock, 1999).

Trust acts to reduce the fundamental tension underlying what we have described previously as a privacy paradox. Although individuals desire complete security and control over personal information, they simultaneously seek benefits that can be obtained only by relinquishing some of this security and control. In the case of e-health, communication and personalization services can be obtained only if the user assumes additional information privacy risks. One example is on-line health risk assessment (HRA) services. HRAs ask patients to provide information about their health status, family history, and health risk behaviors in exchange for personalized resources such as access to tailored chronic disease management information and online health coaching programs. Providing this information has the potential to benefit the patient, yet this involves inherent privacy risks should security be compromised.

It is theorized that individuals apply a "privacy calculus" to determine whether benefits of disclosing personal information exceed the costs of increased risks (Culnan & Armstrong, 1999, p. 108), and this calculus is based to a large degree upon trust. A number of recent empirical studies in online settings have further confirmed the relationship between trust and various aspects of privacy, including privacy risk perception (Dinev & Hart, 2006; Jarvenpaa, Tractinsky, & Vitale, 2000), privacy risk concerns (Eastlick, Lotz, &

Warrington, 2006; Malhotra, Kim, & Agarwal, 2004) and privacy protection characteristics (Liu, Marchewka, Lu, & Yu, 2005). These findings suggest that it can be useful to apply trust measures to study how individuals decide whether to accept privacy risks.

Trust is important to many forms of e-health use. E-health is an increasingly common source of health information (Krane, 2005), and many people have come to trust e-health for this purpose. In a recent population sample survey, nearly two-thirds of respondents indicated a trust level of "some" or "a lot" toward Internet health information (Hesse et al., 2005, p. 2620). Searching online for health information does not represent a significant privacy risk, as typical e-health sites offer anonymous public access to informational content. However, individuals are also adopting e-health services that emphasize communication and personalization (Sillence, Briggs, Harris, & Fishwick, 2007), including e-prescribing, remote disease monitoring, and patient-physician email (Taylor & Leitman, 2002). As discussed earlier in the chapter, communication and personalization create opportunities for information exposure and thereby increase information privacy risks.

Previous research has described health-related Internet trust factors (Hesse et al., 2005) and modeled development of trust in e-health services (Sillence, Briggs, Harris, & Fishwick, 2006). However, we are not aware of any studies that address how trust influences individual decisions to use e-health services representing differential privacy risks. This gap in the literature motivates our research design as described in the following sections.

RESEARCH DESIGN

Our research design is based on a mediated trust model of Internet behavior originating from studies of direct-to-consumer (DTC) advertising. Menon, Deshpande, Perri, and Zinkhan (2002) examined the relationship between individuals' trust in various sources of prescription drug information and their information search behaviors. They observe:

The findings reveal that trust in drug information from traditional media sources such as television and newspapers transfers to the domain of the Internet. Furthermore, a greater trust in on-line prescription drug information stimulates utilization of the Internet for information search after exposure to prescription drug advertising. (Menon et al., 2002, pp. 17-18)

Subsequent research corroborates the contention by Menon et al. (2002) that trust in Internet information sources mediates effects of trust in traditional media sources on information-seeking via the Internet (Huh, DeLorme, & Reid, 2005). In addition, Internet trust is found to influence other behavioral factors, including intention to revisit a website (Hong, 2006) and follow-up communication by individuals with physicians and others (Huh et al., 2005). These findings underpin the full-mediation health trust model presented in Figure 1.

The full-mediation health trust model posits that individuals' trust in broadcast media health information and their use of e-health services is fully mediated by trust in Internet health information (Huh et al., 2005; Menon et al., 2002). This

Figure 1. Full-Mediation Health Trust Model

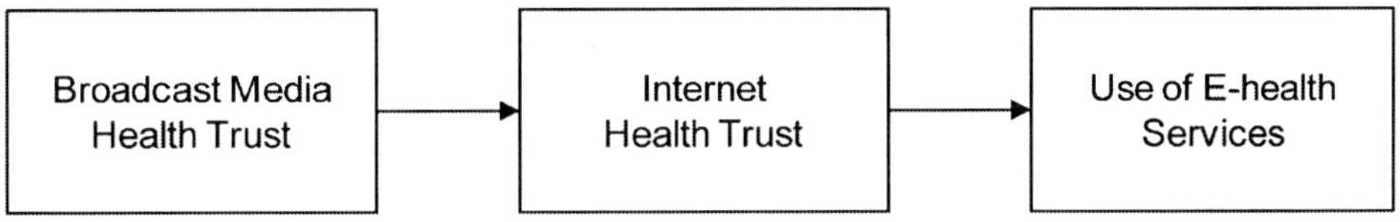

relationship provides a structure for understanding how generalized trust factors (i.e., trust in broadcast media and the Internet as sources for generalized health information) influence individuals' health-related behaviors. In the present study, we apply the model to understand how individuals approach differential privacy risks in the use of e-health services. We propose to do this by contrasting the relationships between Internet health trust and uses of e-health services at two different levels of risk. We conceptualize low-risk e-health services as those in which no personal information is exchanged, e.g., looking on the Internet for information about exercise and physical activities. High-risk e-health services require exchange of personal information, e.g., using the Internet to buy medicine or vitamins. If individuals self-police high-risk uses of e-health, then we would anticipate finding significantly lower association between Internet trust and use of high-risk vs. low-risk e-health services. If individuals are indifferent to risk in the use of e-health services, however, we would anticipate finding similar levels of association for the two levels of risk.

The full-mediation health trust model described above has performed well in prior empirical tests. To the best of our knowledge, however, the relationships in this model have not been tested rigorously against alternative structures. In addition to testing the model presented in Figure 1, we expand our analysis in this chapter to include tests of three alternative health trust models: a partial-mediation model, a joint effects model, and a single-factor model (see Figure 2). The partial-mediation model posits that broadcast

Figure 2. Alternative Health Trust Models

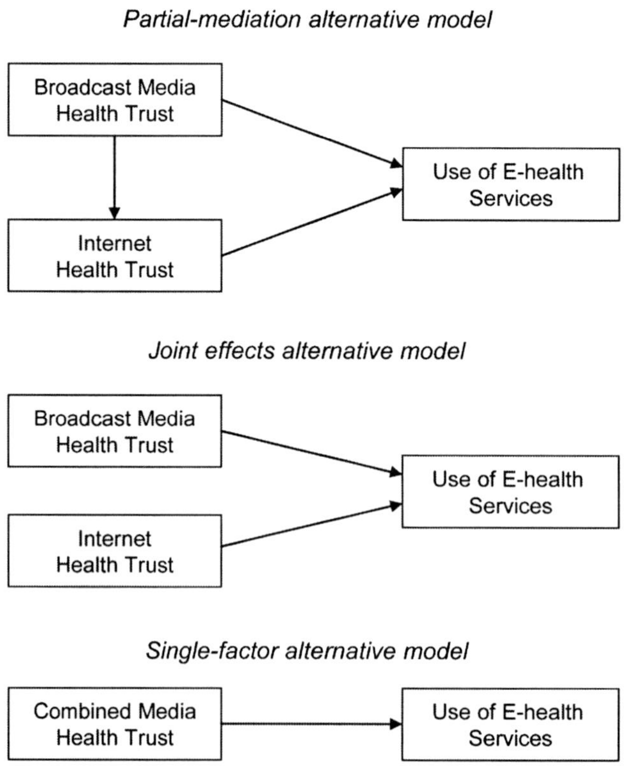

media health trust has direct effects on e-health use in addition to effects mediated by Internet health trust. The joint effects model posits that broadcast media health trust and Internet health trust provide distinct predictions of use of e-health services. The single-factor model posits that items comprising broadcast media health trust and Internet health trust are best modeled as components of a single health trust factor. Each model represents a plausible alternative structure of the relationship between trust beliefs and e-health use behaviors, thus, it is important to conduct explicit testing of their predictions and fit relative to the full-mediation model.

RESEARCH METHOD

Data for this study were obtained from the 2005 Health Information National Trends Survey (HINTS), conducted by the National Cancer Institute. HINTS utilizes a national probability sampling methodology to assess U.S. residents' knowledge and perceptions regarding cancer information and other issues surrounding healthcare (Davis, Park, Covell, Rizzo, & Cantor, 2005). HINTS is directed toward documenting changing patterns in use of health information (especially information relating to cancer), identifying health communication trends, assessing how cancer risks are perceived, and testing theories relating to health communication (HINTS, 2008). The portion of the survey used in the present study relates to perceptions of trust, use of the Internet for healthcare purposes, and demographic data.

HINTS data were collected by telephone interviews and online questionnaires between February and August of 2005. List-assisted, random digit samples of all telephone exchanges in the U.S. were used to develop a nationally representative sample of households. Residents of age 18 and above were recruited to participate in the survey, with intentional oversampling of residents who were self-identified as Black or Hispanic. The overall response rate of contacted households was approximately 21%, producing a total dataset representing 5587 individuals who completed the survey.

Sample Characteristics

Our research design focused on the subset of HINTS participants who (1) had previously used the Internet to access health information or other health services and (2) provided codable responses to all health trust and Internet health use items. This reduced dataset contained 2121 participants. Due to the requirement of Internet use, our participants are younger and better educated, have higher household incomes, and are slightly more likely to be males than is the case in the full HINTS dataset (see Table 1). These differences necessarily reflect the digital divide which characterizes Internet use at present (Norris, 2001), thus we propose our use of the subset is appropriate given the focus of the present study.

Measures

In addition to demographic measures, the subset of HINTS items used in the present study assess health trust in broadcast media and the Internet as information sources and several health uses of the Internet. All survey items used in the present study are documented in the Appendix.

Broadcast media health trust is measured through responses on four-point Likert scales to items which ask "How much do you trust information about health or medical topics" in newspapers, magazines, radio, and television. These factors have been shown to produce a unidimensional measure in previous research (Menon et al., 2002), suggesting that broadcast media health trust should be modeled as a reflective construct.

Internet health trust is assessed through responses on a four-point Likert scale to the single item, "How much would you trust information about health or medical topics on the Internet?"

Internet Health Use is developed as a measure of behavioral variety that is calculated as the total number of different services in a predefined set that participants report having used. The rationale for counting the number of different behaviors is based upon the premise that higher counts represent more extensive utilization of the Internet across the population for the represented activities (health uses). Although behavioral variety measures are not as prevalent in information technology (IT) research as behavioral frequency measures, such measures have proved to be effective in a wide range of field research including studies of antisocial behavior (Morizot & Le Blanc, 2005), effects of childhood play on social development (Pellegrini, 1988), and management education (Lengnick-Hall & Sanders, 1997).

Table 1. Demographic data for the HINTS dataset and the subset used in this research

Factor	Category	% in HINTS	% in Subset
Age	18-34	18.6	24
	35-39	8.3	10.3
	40-44	9.2	13.5
	45 +	63.9	52.1
Gender	Male	34.5	36.8
	Female	65.5	63.2
Education	Less than High School	12.3	3.3
	High School Graduate	25.9	18.2
	Some College	27.7	33.2
	Bachelor's Degree	18.0	26.7
	Post-Baccalaureate Degree	12.4	18.6
Annual Household Income	< $10,000	4.9	2.6
	$10,000 to < $15,000	6.0	2.5
	$15,000 to < $20,000	5.2	3.0
	$20,000 to < $25,000	5.4	4.6
	$25,000 to < $35,000	10.1	8.2
	$35,000 to < $50,000	11.7	15.2
	$50,000 to < $75,000	16.5	26.1
	$75,000 to < $100,000	8.7	15.2
	$100,000 to < $200,000	9.5	18.4
	>= $200,000	2.4	4.1

The HINTS dataset contains several yes-no measures that assess Internet health uses during the preceding 12-month period. These measures range from looking for health or medical information to buying medicines and vitamins online. We divide these measures of Internet health use into two levels of privacy risk. *Low-risk Internet health uses* do not offer the opportunity for personal information to be exchanged. These uses focus on looking for health information. *High-risk Internet health uses* promote the exchange of personal information through activities including buying a health product online, participating in an online support group, and communicating with one's doctor or clinic via email. Three items are used to measure each Internet health use factor, as shown in the Appendix.

RESULTS

Our study applies four alternative research models to assess the degree to which individuals self-police risky e-health activities. In order to conduct this assessment, we developed separate path models to predict low-risk and high-risk Internet health use for each alternative research model.

Model Development

Our path models were developed using AMOS 4.0 structural equation modeling (SEM) software. In the full-mediation health trust model, broadcast media health trust is modeled as a reflective latent factor and Internet health trust and Internet health use are modeled as single-indicator factors. This model predicts that any effects of broadcast media health trust on Internet health use will be fully mediated by Internet health trust. In the partial-mediation model, a relationship is added between broadcast media health trust and e-health use. In the joint effects model, broadcast media health trust and Internet health trust are modeled as having only direct relationships with e-health use. In the single-factor model, a combined media health trust reflective latent factor is created from all trust items and is modeled as having a direct relationship with e-health use.

Model Fit

Fit of the two measurement models was tested using absolute fit, incremental fit, and chi-square methods, as recommended by Hair, Black, Babin, Anderson, & Tatham (2006). Absolute fit indices directly measure how well the specified model reproduces the observed data. We applied Goodness-of-fit index (GFI) and RMSEA and as indices of absolute fit. Incremental fit indices examine fit of a specified model relative to an alternative baseline model, commonly referred to as a null model. We applied the Comparative Fit Index (CFI) as an incremental fit index. Chi-square results examine the extent that a perfect fit exists between observed and estimated covariance matrices in a specified model. We report Chi-square per degree of freedom.

Path Model Results

Results of model testing showing the explained variance and fit of each model are presented in Table 2. The best overall predictions and fit characteristics are found in the partial-mediation and joint effects models. Explained variance is identical within these models as they incorporate the same direct relationships between both trust belief factors and e-health use. Fit indices are slightly better overall for the partial-mediation model. Fit of the full-mediation health trust model is similar to that of the partial-mediation and joint effects models but explained variance is reduced, especially in the case of high-risk e-health uses (R^2 = .013 vs. .026). The single-factor model demonstrates poorer predictions and fit characteristics than any of the other three alternatives.

These results suggest it is important to model direct effects of broadcast media health trust on behavior, but there is only slight quantitative basis for choosing between the partial-mediation and joint effects models. We recommend the partial-mediation health trust model and base our subsequent discussion on this model primarily because we believe it presents a superior conceptual perspective of the relationship between trust beliefs and e-health use.

In the partial-mediation health trust model (see Figure 3) broadcast media health trust is strongly associated with Internet health trust (path coefficient = .56). The path coefficient between Internet health trust and low-risk use is .31 compared to .15 for the relationship between Internet health

Table 2. Explained variance and fit of research models

Model	E-health Use Measure	Explained Variance in Use (R^2)	Fit Measures			
			GFI	RMSEA	CFI	χ^2/df
Full-mediation health trust model	Low-risk uses	.088	.991	.049	.980	6.02
	High-risk uses	.013	.992	.051	.980	6.17
Partial-mediation model	Low-risk use	.095	.991	.050	.962	6.23
	High-risk uses	.026	.990	.045	.965	5.35
Joint effects model	Low-risk use	.095	.991	.056	.980	7.64
	High-risk uses	.026	.993	.051	.982	6.55
Single-factor model	Low-risk use	.045	.975	.091	.931	18.49
	High-risk uses	.002	.99	.054	.974	7.17

trust and high-risk use ($Z = 6.68$, $p < .0001$). A significant association was found in the direct relationship between broadcast media health trust and high-risk use, however, the strength of this relationship did not vary significantly between the low-risk and high-risk models. The path model results generally corroborate prior findings (Hong, 2006; Huh et al., 2005; Menon et al., 2002) but predictive effects of Internet health trust were significantly more prominent in the model predicting low-risk Internet health use.

DISCUSSION

The findings present several important implications for privacy researchers, policymakers, and healthcare administrators. First, we find the mediated trust model of Internet behavior on which we based our research can explain how generalized media and Internet health trust influence use of the Internet for health activities reasonably well where personal information is not at risk. In such circumstances, most trust in information provided by traditional media (such as pharmaceutical advertising) must be transferred to trust in Internet information sources before it can motivate individuals to perform low-risk Internet behaviors, such as searching for information on exercise and physical activities. This finding compares closely with prior studies of the mediated trust model and its encompassed relationships (Hong, 2006; Huh et al., 2005; Menon et al., 2002).

Second, our test of alternative model structures suggests that explanatory power is increased by including a direct relationship between broadcast media health trust and e-health use as was done in the partial-mediation health trust model. Although this is not a large effect, it is important in highlighting the distinct role that broadcast media continues to play in guiding online behaviors.

Third, we find that generalized trust has relatively little consequence in motivating high-risk Internet health uses, explaining less than one-quarter the variance in this factor of that explained in low-risk uses. This finding indicates that individuals do not rely heavily on generalized trust of traditional media and the Internet

Figure 3. Partial-Mediation Model Results (Paths Significant Unless Otherwise Indicated)

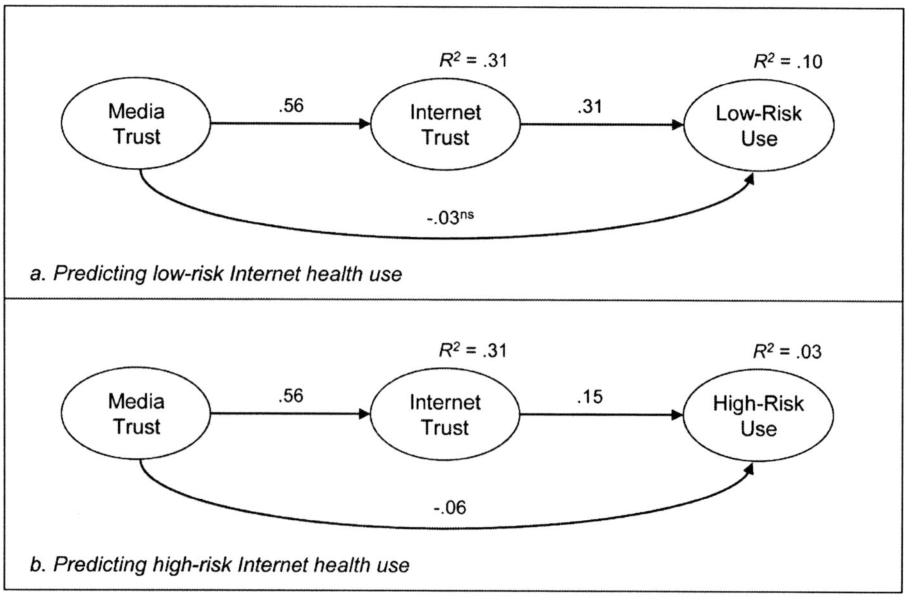

in deciding whether to risk exposure of personal health information and implies that self-policing is prevalent in such health-related Internet activities as buying, medicine, participating in a support group, or emailing doctors. It must be recognized that this is a tentative interpretation, as it is possible that aspects of trust which are not measured by our research design—such as trust in a vendor's online security and privacy protection measures—could be more predictive of high-risk Internet health uses.

The findings suggest that additional research is needed to identify other antecedents of risky online behaviors, as generalized trust does not appear to be an overriding cause of peoples' willingness to give out personal information while desiring privacy—a phenomenon referred to earlier in the chapter as the privacy paradox. For policymakers and administrators, the results caution against assuming that individuals routinely misapply trust and therefore need a high level of overall protection in their online dealings. The finding that generalized trust is not strongly predictive of high-risk online behaviors suggests that regulations and policy interventions should be implemented primarily to avoid *designed* health information privacy losses, such as health information phishing schemes, rather than addressing every aspect of e-health security.

Third, we find an important distinction between trust in traditional media and trust in the Internet as information sources. Although these factors are correlated, Internet health trust is a unique factor that largely mediates the influence of broadcast media health trust on the online behaviors we studied. Several explanations for this distinction are plausible, and these deserve additional study:

- Unlike traditional media, the Internet provides the activity domain as well as the informational domain, making it imperative for individuals to achieve some level of Internet-specific trust before they will initiate Internet activities. This interpretation would suggest that health promotion and advertising for online services should be moved away from traditional media and onto the Internet, thereby avoiding the need for customers to transfer trust to the Internet.
- Where traditional media emphasize broadcast (one-way) communication, the Internet provides a mix of broadcast and interpersonal (two-way) communication (Hoffman & Novak, 1996). Interpersonal interaction is especially important in health domains as people expect speedy and relevant responses to their health questions and concerns. For this reason, interactive elements of the Internet may be key to development of Internet health trust, implying that promotional resources should be directed toward increasing opportunities for online interpersonal communication.
- Alternatively, social presence of informational communication on the Internet, e.g., in health-related chat groups, may be higher than with traditional media. If increased social presence leads to greater Internet health trust, this would imply that promotional resources should be directed toward increasing the social presence of online informational resources.

Identifying the source(s) of the distinction between broadcast media health trust and Internet health trust that we have found in the present study will be an important area for future research. In addition, we expect this issue to have ramifications for healthcare administrators in allocating resources and directing the design of future e-health services.

Finally, the findings suggest that finding ways to increase Internet health trust among potential e-health uses can be effective in increasing use of e-health services, especially where information privacy risk is low. This is an important consideration for healthcare administrators, as seemingly important e-health services are often underutilized by intended users (Payton & Brennan, 1999). Our research shows that Internet health trust largely mediates effects of broadcast media health trust, presumably through transfer to the Internet of institution-based trust in broadcast media as a health information source (Zucker, 1986). It will be important to explore other sources for developing institution-based Internet health trust, such as endorsements by professional medical organizations, certification by security audit services, and partnering e-health development with Microsoft, Google, or other leading e-business organizations that are already well-trusted Internet information sources.

LIMITATIONS

The key limitations of the present study arise from use of the HINTS dataset. From our perspective, HINTS provided an outstanding opportunity to conduct a study using data from a large, current, and nationally-representative population sample. However, this also required us to make several trade-offs. Our choice of measures for Internet health trust and Internet health uses was highly constrained, requiring us to use some single-indicator constructs which can attenuate strength of relationships. Some dimensions we would have liked to assess, such as Internet health use frequency, were not available in the dataset. In addition, it was not possible for us to develop certain types of manipulation checks or to direct follow-up questions to clarify participants' responses to the survey. We propose that these trade-offs are reasonable, given the exploratory nature of the present research. However, it will be important for future researchers to follow up the issues that were limited in our research design.

CONCLUSION

This study applied measures of trust to assess whether individuals self-police privacy-risking e-health activities or are indifferent to them. Addressing this question is important to researchers, who benefit both from empirical validation of the underlying mediated trust model of Internet behavior as well as illumination of the relationships between generalized trust and Internet health uses at two levels of privacy risk. Policymakers and healthcare administrators benefit from improved understanding of the role trust plays in mitigating the privacy paradox that faces e-health users, which can be applied to guide development of better e-health services and more efficient resource allocation processes.

REFERENCES

Altinkemer, K., De, P., & Ozdemir, Z. D. (2006). Toward a consumer-to-healthcare provider (C2H) electronic marketplace. *Communications of the Association for Information Systems*, *18*(19), 413–430.

Cazier, J. A., Shao, B. B. M., & St. Louis, R. D. (2007). Sharing information and building trust through value congruence. *Information Systems Frontiers*, *9*, 515–529. doi:10.1007/s10796-007-9051-6

Cazier, J. A., Wilson, E. V., & Medlin, B. D. (2007). The role of privacy risk in IT acceptance: An empirical study. *International Journal of Information Security and Privacy*, *1*(2), 61–73.

Chellappa, R. K., & Sin, R. G. (2005). Personalization versus privacy: An empirical examination of the online consumer's dilemma. *Information Technology Management*, *6*, 181–202. doi:10.1007/s10799-005-5879-y

Chow, S., & Holden, R. (1997). Toward an understanding of loyalty: The moderating role of trust. *Journal of Managerial Issues*, *9*(3), 275–299.

Choy, A., Hudson, Z., Pritts, J., & Goldman, J. (2001). Exposed online: Why the new federal health privacy regulation doesn't offer much protection to Internet users. *Pew Internet & American Life Project.* Retrieved from http://www.pewinternet.org/pdfs/PIP_HPP_HealthPriv_report.pdf

Culnan, M. F., & Armstrong, P. K. (1999). Information privacy concerns, procedural fairness, and impersonal trust: An empirical investigation. *Organization Science*, *10*(1), 104–115. doi:10.1287/orsc.10.1.104

Davis, T., Park, I., Covell, J., Rizzo, L., & Cantor, D. (2006) *Health information national trends survey (HINTS): Final report.* National Cancer Institute. Retrieved from http://cancercontrol.cancer.gov/hints/docs/HINTS2005FinalReport-0523.pdf

Dinev, T., & Hart, P. (2006). An extended privacy calculus model for e-commerce transactions. *Information Systems Research*, *17*(1), 61–80. doi:10.1287/isre.1060.0080

Eastlick, M. A., Lotz, S. L., & Warrington, P. (2006). Understanding online B-to-C relationships: An integrated model of privacy concerns, trust, and commitment. *Journal of Business Research*, *59*, 877–886. doi:10.1016/j.jbusres.2006.02.006

Fox, S., Rainie, L., Horrigan, J., Lenhart, A., Spooner, T., & Carter, C. (2000). Trust and privacy online: Why Americans want to rewrite the rules. *Pew Internet & American Life Project.* Retrieved from http://www.pewinternet.org/reports/toc.asp?Report=19

Gaudin, S. (2007, July 3). VA investigator blames IT specialist, lax security for major data loss. *InformationWeek.* Retrieved from http://www.informationweek.com/news/showArticle.jhtml?articleID=200900263&cid=iwkPrintURL

Hair, J. E., Black, W. C., Babin, B. J., Anderson, R. E., & Tatham, R. L. (2006). *Multivariate Data Analysis* (6th ed.). Upper Saddle River, N.J: Pearson, Prentice Hall.

Hesse, B. W., Nelson, D. E., Kreps, G. L., Croyle, R. T., Arora, N. K., Rimer, B. K., & Viswanath, K. (2005). Trust and sources of health information. *Archives of Internal Medicine*, *165*, 2618–2624. doi:10.1001/archinte.165.22.2618

HINTS. (2008) *How Americans find and use cancer information: Health Information National Trends Survey.* National Cancer Institute. Retrieved from http://hints.cancer.gov/about.jsp

Hoffman, D. L., & Novak, T. P. (1996). Marketing in hypermedia computer-mediated environments: Conceptual foundations. *Journal of Marketing*, *60*(3), 50–68. doi:10.2307/1251841

Homan, Q. (2003) *Healthcare satisfaction study final report.* Harris Interactive/ARiA Marketing. Retrieved from http://www.harrisinteractive.com/news/downloads/harrisariahcsatrpt.pdf

Hong, T. (2006). Contributing factors to the use of health-related websites. *Journal of Health Communication*, *11*(2), 149–165. doi:10.1080/10810730500526679

Huh, J., DeLorme, D., & Reid, L. N. (2005). Factors affecting trust in online prescription drug information and impact of trust on behavior following exposure to DTC advertising. *Journal of Health Communication*, *10*(8), 711–731. doi:10.1080/10810730500326716

Itagaki, M. W., Berlin, R. B., & Schatz, B. R. (2002) The rise and fall of e-health: Lessons from the first generation of Internet healthcare. *Medscape General Medicine, 4*(2). Retrieved from http://www.medscape.com/viewarticle/431144_Print

Jarvenpaa, S. L., & Tractinsky, N. (1999). Consumer trust in an Internet store: A cross-cultural validation. *Journal of Computer-Mediated Communication, 5*(2). Retrieved from http://www.ascusc.org/jcmc/vol5/issue2/jarvenpaa.html.

Jarvenpaa, S. L., Tractinsky, N., & Vitale, M. (2000). Consumer trust in an Internet store. *Information Technology Management, 1*, 45–71. doi:10.1023/A:1019104520776

Järvinen, O. P. (2009). Privacy management of patient-centered e-health . In Wilson, E. V. (Ed.), *Patient-Centered E-Health*. Hershey, PA: IGI Global.

Krane, D. (2005) Number of 'cyberchondriacs'—U.S. adults who go online for health information—increases to estimated 117 million. *Healthcare News, 8*(5). Retrieved from http://www.harrisinteractive.com/news/newsletters_healthcare.asp

Lengnick-Hall, C. A., & Sanders, M. M. (1997). Designing effective learning systems for management education: Student roles, requisite variety, and practicing what we teach. *Academy of Management Journal, 40*(6), 1334–1368. doi:10.2307/257036

Liu, C., Marchewka, J. T., Lu, J., & Yu, C. (2005). Beyond concern–a privacy-trust-behavioral intention model of electronic commerce . *Information & Management*, (42): 289–304. doi:10.1016/j.im.2004.01.003

Lohr, S. (2007, August 14) Google and Microsoft look to change health care. *New York Times*. Retrieved from http://www.nytimes.com/2007/08/14/technology/14healthnet.html

Luo, X. (2002). Trust production and privacy concerns on the Internet: A framework based on relationship marketing and social exchange theory. *Industrial Marketing Management, 31*(2), 111–118. doi:10.1016/S0019-8501(01)00182-1

Malhotra, N. K., Kim, S. S., & Agarwal, J. (2004). Internet users' privacy concerns (IUIPC): The construct, the scale, and a causal model. *Information Systems Research, 15*(4), 336–355. doi:10.1287/isre.1040.0032

Mason, J. K. McCall, & Smith, R.A. (1999) *Law and Medical Ethics* (5th Ed.). London: Butterworths.

Masys, D., Baker, D., Butros, A., & Cowles, K. E. (2002). Giving patients access to their medical records via the Internet: The PCASSO experience. *Journal of the American Medical Informatics Association, 9*(2), 181. doi:10.1197/jamia.M1005

Mayer, R. C., Davis, J. H., & Schoorman, F. D. (1995). An integrative model of organizational trust. *Academy of Management Review, 20*(3), 709–734. doi:10.2307/258792

Menon, A. M., Deshpande, A. D., Perri, M., & Zinkhan, G. M. (2002). Trust in online prescription drug information among Internet users: The impact on information search behavior after exposure to direct-to-consumer advertising. *Health Marketing Quarterly, 20*(1), 17–35. doi:10.1300/J026v20n01_03

Morizot, J., & Le Blanc, M. (2005). Searching for a developmental typology of personality and its relations to antisocial behavior: A longitudinal study of a representative sample of men. *Journal of Personality, 73*(1), 139–182. doi:10.1111/j.1467-6494.2004.00307.x

Norris, P. (2001). *Digital divide: Civic engagement, information poverty, and the Internet worldwide*. Cambridge, UK: Cambridge University Press.

Payton, F. C., & Brennan, P. F. (1999). How a community health information network is really used. *Communications of the ACM, 42*(12), 85–89. doi:10.1145/322796.322814

Pellegrini, A. (1988). Rough-and-tumble play and social competence. *Developmental Psychology*, *24*, 802–806. doi:10.1037/0012-1649.24.6.802

Quelch, J. A., & Klein, L. R. (1996). The Internet and international marketing. *Sloan Management Review*, *37*(3), 60–76.

Sillence, E., Briggs, P., Harris, P., & Fishwick, L. (2006). A Framework for understanding trust factors in Web-based health advice. *International Journal of Human-Computer Studies*, *64*, 697–713. doi:10.1016/j.ijhcs.2006.02.007

Sillence, E., Briggs, P., Harris, P., & Fishwick, L. (2007). Going online for health advice: Changes in usage and trust practices over the last five years. *Interacting with Computers*, *19*, 397–406. doi:10.1016/j.intcom.2006.10.002

Srivastava, R. P., & Mock, T. J. (1999). Evidential reasoning for web trust assurance services. *Journal of Management Information Systems*, *16*(3), 11–32.

Taylor, H. (2003). Most people are 'privacy pragmatists' who, while concerned about privacy, will sometimes trade it off for other benefits. *Harris Interactive*. Retrieved from http://www.harrisinteractive.com/harris_poll/printerfriend/index.asp?PID=365

Taylor, H., & Leitman, R. (2002). The future use of the Internet in four countries in relation to prescriptions, physician communication and health information. *Health Care News, 2*(13). Retrieved from http://www.harrisinteractive.com/news/newsletters_healthcare.asp

Wilson, E. V. (2003). Asynchronous health care communication. *Communications of the ACM*, *46*(6), 79–84. doi:10.1145/777313.777316

Zrebiec, J. F., & Jacobson, A. M. (2001). What attracts patients with diabetes to an Internet support group? A 21-month longitudinal website study. *Diabetic Medicine*, *18*(2), 154–158. doi:10.1046/j.1464-5491.2001.00443.x

Zucker, L. G. (1986). Production of trust: Institutional sources of economic structure, 1840-1920. *Research in Organizational Behavior*, *8*, 53–111.

APPENDIX

The following survey items, labeled below with identifiers from the original HINTS dataset, were used in the present analysis.

Table 3.

Broadcast Media Health Trust
Response Measures: 1 = "A Lot", 2 = "Some", 3 = "A Little", 4 = "Not At All"
HC-13c How much would you trust information about health or medical topics in newspapers?
HC-13d How much would you trust information about health or medical topics in magazines?
HC-13e How much would you trust information about health or medical topics on the radio?
HC-13g How much would you trust information about health or medical topics on television?

Table 4.

Internet Health Trust
Response Measures: 1 = "A Lot", 2 = "Some", 3 = "A Little", 4 = "Not At All"
HC-13f How much would you trust information about health or medical topics on the Internet?

Table 5.

Use of Low-Risk E-Health Services (no personal information is exchanged)
Response Measures: Recoded to 1 = "Yes" and 0 = "No" for analysis
HC-14a In the past 12 months, have you done the following things while using the Internet: Looked for health or medical information for yourself?
HC-14f In the past 12 months, have you done the following things while using the Internet: Looked for information about physical activity or exercise?
HC-14i In the past 12 months, have you done the following things while using the Internet: Looked for information about quitting smoking?

Table 6.

Use of High-Risk E-Health Services (personal information exchange is required)
Response Measures: Recoded to 1 = "Yes" and 0 = "No" for analysis
HC-14c In the past 12 months, have you done the following things while using the Internet: Bought medicine or vitamins on-line?
HC-14d In the past 12 months, have you done the following things while using the Internet: Participated in an on-line support group for people with a similar health or medical issue?
HC-14e In the past 12 months, have you done the following things while using the Internet: Used e-mail or the Internet to communicate with a doctor or a doctor's office?

Table 7.

Demographic Items
GA-0A May I please have your age? *Open-Ended Response Measure*
GA-0C Are you male or female? *Response Measure: 1 = "Male", 2 = "Female"*
DM-03 What is the highest level of school you completed? *Response Measure: 1 = "Never Attended", 2 = "Grades 1 Through 5 (Elementary)", 3 = "Grades 6 Through 8 (Middle)", 4 = "Grades 9 Through 12 (Some High School But No Diploma)", 5 = "High School Graduate (High School Diploma Or Equivalent", 6 = "Vocational Or Trade School Graduate", 7 = "Some College, But No Degree", 8 = "Associate Degree In College", 9 = "Master's Degree", 10 = "Professional School Or Doctorate Degree"*
HHIncC What is your {combined} annual household income? *Response Measure: 1 = "< $10,000", 2 = "$10,000 - <$15,000", 3 = "$15,000 - <$20,000", 4 = "$20,000 - <$25,000", 5 = "$25,000 - <$35,000", 6 = "$35,000 - <$50,000", 7 = "$50,000 - <$75,000", 8 = "$75,000 - <$100,000", 9 = "$1000,000 - <$200,000", 10 = ">$200,000"*

Chapter 8
Safeguarding the Privacy of Electronic Medical Records

Jingquan Li
Texas A&M University-San Antonio, USA

Michael J. Shaw
University of Illinois at Urbana-Champaign, USA

ABSTRACT

The continued growth of healthcare information systems (HCIS) promises to improve quality of care, reduce harmful medical errors, and streamline the entire healthcare system. But the resulting dependence on electronic medical records (EMRs) has kindled patient concern about who has access to sensitive medical records. Healthcare organizations are obliged to protect patient medical records under the Health Insurance Portability and Accountability Act (HIPAA) of 1996 and the economic stimulus bill of 2009. The purpose of this study is to develop a formal privacy policy for safeguarding the privacy of EMRs. This study describes the impact of EMRs and HIPAA on patient privacy. It proposes access control and audit logs policies to protect patient privacy. To illustrate the best practices in the healthcare industry, this chapter presents the case of the University of Texas M. D. Anderson Cancer Center. The case demonstrates that it is critical for a healthcare organization to have a formal privacy policy in place.

INTRODUCTION

The strategic utilization of information systems/information technologies (IS/IT) has played a central role in enabling organizations across many industry segments to address many business challenges and achieve a level of sustainable competitive advantage (Croasdell, 2001; Hammond, 2001; Holt *et al.*, 2000). Healthcare is noted for embracing new scientific discoveries and using leading edge technologies to enable better cures for diseases and better means to enable early detection of most life threatening diseases. Ironically, the healthcare industry in the US, which has a greater need for more accurate and timely information, has experienced less development of IS/IT than have other industries such as banks or airlines. The Health Insurance Portability and Accountability Act of 1996 (HIPAA) is the largest governmental law in healthcare since Medicare.

DOI: 10.4018/978-1-61692-000-5.ch008

HIPAA mandates new federal standards for electronic transactions, such as payment processing, patient's medical information privacy, and security procedures that secure the privacy protections. Currently healthcare organizations are contending with relentless pressures not only to implement IS/IT technologies but also to become compliant with the HIPAA regulatory.

The growing use of healthcare information systems (HCIS) has provided healthcare organizations with tremendous benefits, including significantly reduced costs, reduced harmful medical errors, and improved quality of care. But the resulting dependence on electronic medical records (EMRs) has also kindled patient concern about the privacy and security of patient medical records. EMRs often contain some of the most sensitive information who and what we are, such as mental and physical illness. Perhaps more than any other type of data, the confidentiality of EMRs is absolutely essential. When doctors' file cabinets held the bulk of medical records, the employees working in those practices had access to them. As hospitals and clinics switch to electronic record keeping, however, access to private medical records will soon be very easy for anyone with a computer and information systems access. Under HIPAA, new healthcare privacy provisions designed to protect data transmitted and stored electronically went into effect April 14, 2003. The requirements of HIPAA and compliance issues are getting the attention of top executives in the healthcare industry.

Having a formal privacy policy is a key step in implementing any HIPAA compliance program. It should expressly cover how a health organization is protecting EMRs; the rules and limits on who can use and need access to EMRs; and the capability to track who has disclosed sensitive data and the circumstances of disclosure. A positive, formal, and continually practiced privacy policy by all employees can establish rules and limits on who can gain access to EMRs and thus minimize the possibilities of privacy breaches. On the other hand, a poorly defined and improperly implemented and managed privacy policy can make EMRs ripe for privacy abuse. The HIPAA privacy rule puts an emphasis on access control and audit trails to protect patient data. This study investigates the use of access control and audit policies for protecting the privacy of EMRs.

The objective of this study is to develop formal access control and audit policies for safeguarding the privacy of EMRs. The development of HIPAA compliance program and security policies has been addressed by several studies (Li and Shaw, 2004; Kieke, 2003; Messmer, 2003; Anonymous, 2001; DeMuro, 2001). There are also several papers addressing the issue of the privacy of EMRs (Miller and Tucker, 2009; Runy, 2008; Swartz, 2004; Ateniese and Medeiros, 2002). The closely related works to this study include the following. Zunkel (2005) studied how to use biometric technology to protect personal information and found that biometric technology does not endanger personal information; it protects it. Borrowing the principles of reporting and auditing from the accounting sector, Stevens (2002) found that through comprehensive reports of network activity logs and regular auditing of security measures and devices, healthcare organizations could generate the proof of HIPAA compliance. While these studies are devoted to technical aspects and particular access control and audit technologies, this study takes a management-oriented approach to develop access control and audit logs policies for protecting the privacy of EMRs strategically.

The rest of the chapter is organized as follows. In the second section, we discuss the issue of patient privacy in the healthcare industry. In the third section, we describe the HIPAA privacy rule and its privacy implications. In the fourth section, we investigate access control and audit logs policies for protecting patient privacy. To illustrate the impact of EMRs on patient privacy and the importance of having a privacy policy in the healthcare system, we present a case example of the University of Texas M. D. Anderson Cancer Center in the fifth section. We conclude with a summary in the final section.

PATIENT PRIVACY IN HEALTHCARE

Traditionally, our medical records were kept in paper documents in different doctor's offices. Cumbersome paper records not only contain too many errors and inefficiencies but also hinder communication between healthcare providers. According to the *Journal of the American Medical Association*, each year, as many as 98,000 patients die in U.S. hospitals from preventable medical errors, such as receiving the wrong medication; nearly half of all patients do not get all the treatments or tests that they should have received (Weise 2005). These problems persist because of industry-wide failures connected to the lack of accurate and reliable health data.

As many hospitals and health systems begin to implement healthcare information systems (HCIS), electronic medical records (EMRs) have become the primary means for storing health data, and electronic communications have been the preferred way of doing business. Storing and sharing patient health records electronically benefits both patients and doctors by speeding up access to patient information and ensuring that information is accurately and reliably available, in the right place at the right time. Healthcare information systems enable doctors, nurses, and lab technicians to gain access to medical records from a computer, instead of having to track down paper files. HCIS can also use networked computer databases to inform doctors about best practices for treating patients. With HCIS, for instance, a click of a mouse pulls up a patient's full medical record. The electronic systems can alert doctors to everything from a patient's current medications to drug allergies. According to research firm Harris Interactive (Taylor and Leitman, 2001), the major causes of medical errors include multiple physicians treating the same patient without all having access to all the patient's medical records and with each storing different, incomplete medical records in different places. There is near consensus among healthcare industry experts that the widespread use of EMRs, accessible to all those seeing and treating a patient as well as to the patient, would substantially improve the coordination and quality of health care, in addition, electronic prescribing would further reduce errors that result from handwritten, hard to decipher prescriptions. Converting paper files into electronic medical records can significantly improve the quality of health care, reduce harmful medical errors, lower costs, streamline the entire healthcare system. As a consequence, most healthcare entities are now fully dependent upon their EMRs and HCIS.

The implementation of healthcare information systems has provided healthcare organizations with tremendous opportunities to improve quality of care. But the resulting dependence on EMRs and electronic communications also creates significant potential problems that can result in major harm to healthcare organizations. One of the problems will be the invasion of patient privacy. EMRs contain some of the most sensitive information who and what we are, such as mental illness, sexual behaviors, HIV/AIDS, communicable diseases, cancer, and other stigmatized conditions. Perhaps more than any other type of data, it is important for healthcare organizations to keep medical records confidential and ensure that only personnel authorized to see the data get to view it.

When a patient's health records were kept in paper files in doctor's offices, these records have been protected through the doctor-patient code of confidentiality, and third-party access to records was at the doctor's discretion. As hospitals and clinics switch to electronic record keeping, however, access to private medical records will soon be very easy for anyone with an information systems access. The Internet age, with free and easy exchange of data over open networks, exacerbates the situation. It is realized that the code of confidentiality is inadequate to protect the patient's interests because too many people will see your medical records. Through networked electronic medical records systems, doctors today interact with other health practitioners, diagnostic services and hospitals, all of whom contribute to the medical records. Others who have access to patient data are: support staff, receptionists, clerks, administrators, transcriptionists, laboratory technicians, insurers, and healthcare providers. Once your electronic medical records are exposed intentionally or accidentally, they will live forever on the Internet. But what happens to healthcare when the privacy of patient data is not protected and insurers and employers have access to an unprecedented amount of information? Could they discriminate against people – or their children – based on their future potential for acquiring certain stigmatized disease? What happens to healthcare when the privacy of EMRs is not protected and criminals get their hands on EMRs (EMRs also contain demographic and financial data)? Will it result in massive identity theft and massive fraud in the healthcare industry (billing for people who are not real patients or for cure that did not happen)? There will be (and there has been) potential for real harm to patients. For instance, some patients are likely to withhold health information from their physicians or reluctant to seek care if they do not trust that their personal data will be kept confidential. For that reason, implementing systems that protect patient privacy, yet are able to communicate needed information is the key to whether healthcare information systems move forward or back. Furthermore, patient privacy and security concerns are driven by the federal medical privacy rule.

FEDERAL MEDICAL PRIVACY RULE

Healthcare organizations have long been under scrutiny when it comes to implementing patient privacy policies and procedures but the Health Insurance Portability and Accountability Act of 1996 (HIPAA), passed by the U.S. Congress and signed into law in 2000, has elevated its importance and placed comprehensive new privacy requirements on the U.S. healthcare industry. HIPAA calls for standardization of electronic patient health, administrative and financial data, and unique health identifiers for individuals, employees, health plans and healthcare providers. The greatest challenges among these standards for health organizations will be those regarding patient privacy (Li and Shaw, 2004). HIPAA mandates that healthcare organizations follow certain guidelines of privacy protection and that patients are given some specific control over how personal medical information is used and transmitted electronically. The U.S. Department of Health and Human Services (HHS) issued those patient-data privacy guidelines, as ordered under the HIPAA regulations. Subject to interpretation, the HIPAA privacy rule demands that any organization providing healthcare services – and any of their business associates handling protected patient data—apply "administrative, physical and technical safeguards" to ensure confidentiality.

HIPAA defines three main groups as "covered entities:" healthcare providers, health plans, and healthcare clearinghouses (DeMuro and Grant, 2001). "Healthcare providers" refer to organizations that provide medical or health services and transmit any health data in electronic form. "Health plan" is defined broadly to include any individual or group plan that provides or pays the cost of medical care. HMOs, health insurers, and group health plans, including employee benefit plans, are considered to be health plans. "Healthcare clearinghouse" is defined as a public or private entity that processes or facilitates the processing of nonstandard data elements of health information into standard data elements. Billing companies are considered to be healthcare clearinghouses. Working closely with these groups are "Business Associates." Business associates are defined by HIPAA as: persons who perform functions or activities involving the use or disclosure of protected health information (PHI) on behalf of "covered entities." This includes claims processing or administration, data analysis, utilization review, quality assurance, billing, benefit management, practice management, legal, actuarial, management, accreditation or financial services.

Under the HIPAA privacy rule, healthcare organizations are obliged to protect patient records. A central feature of the HIPAA privacy rule is a set of limitations on the use and disclosure of protected health information. According to the regulation, PHI is individually identifiable health information that is a subset of health information, including demographic information collected from an individual, medical data related to the past, present, or future physical or mental health or condition of an individual, and financial information related to the past, present, or future payment for the provision of healthcare to an individual. Examples of PHI include personal information, such as name, address, phone, social security number, credit card number, e-mail, identities of doctor and patient, and so forth. Since PHI includes the most sensitive information in medical records, the HIPAA regulation prohibits PHI from being released to third parties, or used by the healthcare industry, except as directly by the patient in signed consent or authorization forms.

The HIPAA privacy rule outlines severe penalties to individuals or organizations misusing patient information, intentionally or otherwise. Compliance is required by April 14, 2003; noncompliance will carry stiff civil and criminal penalties, including fines up to $25,000 for multiple violations of the same standard in a calendar year and fines up to $250,000 and/or imprisonment up to 10 years for knowing misuse. Because of HIPAA, healthcare organizations are now open to a range of potentially damaging consequences including fines, liability claims, criminal charges, and even prison terms. Healthcare organizations should become knowledgeable about HIPAA and consider the requirements of HIPAA in their planning, management, and implementation efforts. A formal privacy policy must be in place to protect patient privacy.

The recent economic stimulus bill, passed by the US Congress and signed into law on February 17, 2009, mandates the utilization of an electronic health record for every American by 2014. "Our recovery plan will invest in electronic health records and new technology that will reduce errors, bring down costs, ensure privacy, and save lives," President Barack Obama told members of Congress on February 24, 2009. One of the government's purposes with electronic health records is to make doctors easier to share medical records and use electronic medical records in ways that are likely to benefit patients, not only within a hospital but potentially nationwide as well. However, the pervasive use of EMRs without strengthening the federal medical privacy rule potentially threaten individual privacy because EMRs can be easily duplicated and transmitted to unauthorized people. Drafters of the stimulus bill took steps to expand federal patient-privacy regulations, such as requiring healthcare organizations to notify individuals if there is a breach of unsecured protected health information and to keep audit trails of those with whom they have shared patient information. The stimulus bill also calls for increased penalties for those who break the federal medical privacy rule.

PRIVACY POLICIES FOR SAFEGUARDING THE PRIVACY OF EMRS

Under the HIPAA privacy rule, healthcare organizations must not only ensure the privacy of EMRs, but also ensure that business associates with which they do business maintain the privacy. Today as medical databases become networked in more complex healthcare information systems, healthcare organizations are facing a rising number of privacy attacks. In fact, many medical organizations are unprepared to safeguard the confidentiality of patient medical records as the healthcare industry prepares to move from archaic paper files to EMRs. Loss or compromise of sensitive EMRs can occur from hackers and identity thieves, or even from internal sources in the form of disgruntled employees or corrupt executives. It is fundamental for any healthcare organization to have a formal privacy policy because technical mechanisms must be accompanied by formal organizational policies for controlling the access to EMRs. Without formal privacy policies and practices in place to protect the privacy of EMRs, patients' right to privacy is at risk. This is the reason why it is so important to take the strategic, management-oriented approach starting with privacy management policies and practices.

Privacy policies are the management directives that give guidance to personnel on how their organization will handle EMRs and patients' right to privacy. These policies identify classification of medical data based on their sensitivity. The policies include requirements to implement appropriate policies and practices to prevent intentional and accidental disclosures of protected information; establish training and awareness programs for all employees of the covered entity who may have access to protected health information; designate a privacy officer with primary responsibility for ensuring compliance with the federal and state privacy regulations; establish a system for receiving and responding to complaints regarding the covered entity; and implement appropriate sanctions for violations of the privacy guidelines (DeMuro and Grant, 2001). Some weak and ignored policies include: (1) access controls are not being applied; (2) inadequate monitoring and auditing of information systems; and (3) lack of a security incident and disaster recovery plan (Schneier, 2000). The HIPAA privacy rule puts an emphasis on access control and audit logs policies for protecting the privacy of EMRs. This study investigates the importance of access control and audit logs around patient privacy.

Access Control

If access is more limited and tightly controlled, EMRs are more privacy sensitive than paper files. Two most common access control approaches are credentialed data access and authentication. In the approach of credentialed data access, access control policies should grant access to patient data on a need-to-know basis only, and they should be the driving force in the implementation of HCIS. An organization should identify the persons or classes of persons, as appropriate, in its workforce who need access to the sensitive data to carry out its duties. For each such person or class of persons, the organization should identify the category(s) of potential sensitive health data to which access is needed and any conditions appropriate to such access. The organization should make best efforts to limit the access of such persons or classes of persons identified above to the types and amount of sensitive data required to do their jobs. Access to confidential medical data is limited to persons with a legitimate "need to know" to perform their jobs within the organization. Credentialed data access thereby reduces the exposure of sensitive patient data.

Another approach is authentication. Authentication usually involves a user name and a password, but more companies are implementing multifactor authentication that includes other methods of demonstrating identity, such as access cards, hardware tokens, fingerprints, or voice recognition. The acceptable forms by which humans can authenticate themselves are something the person is which can be biometrically identified such as voice or retinal patterns, fingerprint, vein, iris, hand geometry, face and signature; something the person has such as an access card or hardware token; and something the person knows such as a password or personal identification number.

Information systems managers have devised clever and complex password systems requiring frequent changes, rules for alphanumeric and symbolic characters, and prohibitions against repetition of character strings. Password management and allocation often becomes so complex that a user resorts to typing the password into a text file and leaving it on his or her electronic desktop, or worse, writing it in a Post-it note and sticking it to the monitor. When IT managers also require a "what you have" factor—such as an access card that transmits the password to the terminal—it may end up in a desk drawer or hanging on a cord suspended from a push-pin, readily accessible to others.

The use of biometrics provides a means by which healthcare records can be kept confidential (Zunkel 2005). Biometric technology provides a "who you are" solution to these issues, ensuring the person gaining access to data is, in fact, the person authorized. As with any credential-based access system, the biometric technology employed must be reliable, user friendly and, most of all, manageable, and technology developers have pushed to meet those requirements.

Audit Logs

While many healthcare organizations have wisely deployed technical safeguards, such as firewalls and encryption technologies, to keep patients' data confidential, proving that those safeguards are successful is harder. Audit logs provide the capability to track all accesses to EMRs and audit tools that allow more frequent examination of audit trails to detect inappropriate accesses to EMRs. In the field of databases, database engine records every transaction in a transaction log to maintain database consistency and to aid in recovery. The log is a storage area that automatically tracks changes to a database. If the system fails, the automatic recovery process uses the transaction log to roll forward all committed transactions and roll back any incomplete transactions. The concept of transaction log in database systems can be extended to defend the privacy of EMRs in the healthcare information systems. Every time someone accesses the patient record, the healthcare information systems have got to know what she or he has accessed, when it was accessed and what they did.

Audit logs provide several benefits. First, through comprehensive reports of activity logs in the healthcare information systems, healthcare organizations can generate the proof of compliance that HIPAA demands. In the case of litigation, it is interesting to note that in general, courts have been lenient when dealing with organizations that have privacy policies and procedures in place, even if these policies are not perfect. They are generally less lenient with organizations that can demonstrate nothing in regards to managing patient privacy. In this sense, through comprehensive reports of HCIS activity logs and regular auditing of established privacy policies and safeguards, healthcare organizations can produce the proof of HIPAA compliance, which is extremely important.

Second, audit logs help track criminals when a privacy breach occurs. For those companies that suffered privacy breaches, they found that it was difficult to figure out where the data was initially compromised and to find the trail back to where that the data was compromised. Therefore, it is very important for organizations to know exactly where the data lies and to be able to provide an auditable trail of their activities.

Finally, audit logs demonstrates a commitment to the highest level of legal and ethical conduct in the manner in which an organization gathers, stores, uses and protects patient medical information. Knowing that the organization takes seriously its obligation to follow the federal medical privacy rule makes patients and their loved ones feel safer that their personal details are not floating around in the public domain. Privacy management is a continuous process. The organization needs to continue to reassess and change its technical safeguards over time, as business environment is dynamic. Therefore, it is very important to collect and analyze information from the audit logs to make improvements.

THE M. D. ANDERSON CASE OVERVIEW AND FINDINGS

To illustrate the impact of EMRs on patient privacy and the importance of having a formal privacy policy in the health-care industry, we now consider the case of the University of Texas M. D. Anderson Cancer Center. Electronic medical records promise to promote better medical outcomes, lower costs and streamline the entire healthcare system. The development of the electronic medical records system with a major focus on the integration of research and clinical data has been identified as a key strategic initiative at M. D. Anderson. Meanwhile, M. D. Anderson is strongly committed to maintaining the security and privacy of patient medical records under the HIPAA privacy rule. To this end, M. D. Anderson has developed data access and audit trails policies for protecting the privacy and confidentiality of EMRs. In this section, we illustrate how a best practices standard, such as the EMR initiative, would work in conjunction with the enhanced privacy policy that M. D. Anderson proposes. We focus on how M. D. Anderson applies data access control and audit trails to ensure the privacy and security of EMRs.

M. D. Anderson Cancer Center in Houston is a premier cancer hospital in the United States. With more than 16,000 faculty and staff and 1,600 volunteers working in more than 25 buildings around Houston and in central Texas, M. D. Anderson is one of the largest cancer centers in the world. The center, which treats more than 70,000 patients a year, includes an in-patient pavilion with 512 beds, an out-patient clinic, two research buildings, a faculty building and a patient-family hotel. M. D. Anderson is also a top research center; the center invests more than $400 million on cancer research each year; about 11,000 patients participated in therapeutic clinical trials each year. M. D. Anderson has identified improvement in electronic processing and communications of medical records has been identified as a key to enhancing the safety and efficiency of its patient care. The center has developed an electronic medical records system that would integrate patient records with clinical research data. In a healthcare institution like M. D. Anderson, everyone wants to keep all the medical data forever and has it accessed immediately. The electronic medical records system has now between 600 and 800 terabytes of data that is in a high-availability environment. Probably the best example of that is the center's imaging environment. Technologies such as computed tomography (CT) and magnetic resonance imaging (MRI) has been an integral part of clinical cancer care and imaging techniques are continually improving and finding more and more applications in clinical and research situations. The center has more than 200 million images online available for its clinicians. That environment is completely replicated, so the center actually have 400 million images online – a copy of each one – and the center has a third backup it maintains off-site that is an extract of the images. The creation of the electronic medical records system promotes better medical outcomes by speeding up access to cancer patient information and ensuring that information is more complete and readily available for diagnosis, treatment and claims-payment purposes. The electronic system also promises to improve medical research and innovative treatment at the center by integrating clinical trials data with clinical data. The increased availability of EMRs carries significant benefits for M. D. Anderson. But this has put increasing demands on the underlying privacy protection mechanisms. Like other medical providers, M. D. Anderson is bound by the HIPAA privacy rule to provide adequate protection for the personal medical data.

M. D. Anderson has always been a leader in standing up for patients' rights to privacy since it was established in the 1940s. The center is committed to safeguard patient confidences and privacy with the constraints of the HIPAA privacy rule. It has developed a formal privacy policy that explicitly articulates how EMRs to be protected. The center believes that patients "own" the data about themselves (as defined by HIPAA's PHI definition) and they only allow its use by others under certain circumstances. Since access is one of the areas most vulnerable to privacy breaches, the center needs the capability to authorize access to EMRs and keep audit logs over complex heterogeneous information systems and networks. The truth is that the more access and availability there is, the more risk there is.

The center stresses on formal role-based information access policies and practices that clearly define who in the organization have privilege and need to access to particular data, the rules determining a person's right of access, the reasons for denying individual access, including how access is granted, modified and denied. If you are a physician, you can only see certain things; if you are a nurse, you can only see certain things; if you are a pharmacist, you can only see certain things; if you are a cancer researcher, you can only see certain things; if you are an administrator in the Department of Cancer Medicine, you can only see certain things. For example, clinical EMRs may be used and/or disclosed for certain research purposes. All research projects at M. D. are subject to review and approval by an Institutional Review Board (IRB). An IRB committee is responsible for protecting individual research subjects and insuring that research is conducted ethically. You will not be enrolled in a research project that is not reviewed and approved by an IRB. The goal is to restrict the access and/or release of protected health information to authorized individuals and prevent unreasonable possibility of unnecessary, inadvertent, wrongful, or malicious access to protected health information.

M. D. Anderson has established a formal auditing policy based on HIPAA regulatory requirements and security best practices. The center has a substantial department of information security that watches over access to the electronic medical records system very vigilantly. The system maintains records of all accesses to EMRs. A required accounting must include: (1) the date of the disclosure; (2) the name and address of the entity or person who received the medical information; (3) a brief description of the medical information disclosed; and (4) a brief statement of the purpose of the disclosure. Audit logs for systems containing EMRs should be retained and reviewed regularly according to the M. D. Anderson privacy policy. No security information system is foolproof. The center has audit trail tools that allow more frequent examination of audit logs to detect inappropriate accesses to EMRs. Once a privacy breach happens to medical data, IT professional will be indispensable in locking down the systems, then determining the extent of the breach, data compromised, and source of attack.

The case of M. D. Anderson demonstrates that it is critical for a healthcare organization to have a formal privacy policy. Electronic record keeping presents a massive threat to patient privacy – PHI in particular. Keeping patient records private and secure is a huge concern for any healthcare organization. By establishing privacy policies and implementing them with state-of-the-art security technologies, a healthcare organization can actually turn the privacy issue into a benefit when it uses EMRs. Traditional paper medical records are not immune to privacy risks. There is no way to restrict who sees them, or to detect inappropriate accesses to them. In contrast, EMRs provide the potential for greater protections with access control and audit trails. The access control policy expressly defines who in the organization have the privilege and need to access to particular data and ensures that they gain access to only a small part of EMRs to carry out their job duties. The formal auditing policy maintains audit logs of all accesses to EMRs and allows frequent examination of audit logs in order to detect privacy breaches.

CONCLUSION

Patient privacy is a major concern for most health-care organizations today due to the continued growth of healthcare information systems and the resulting dependence on EMRs and electronic communications. Healthcare organizations in the United States face governmental and regulatory pressures to become compliant with the HIPAA privacy rule. In this study, we described the importance of patient privacy in the context of EMRs and HIPAA compliance. We found that it is crucial for a healthcare organization to have a formal privacy policy. We developed and analyzed access control and audit logs policies to protect the privacy of EMRs. More importantly, to illustrate the best practices in the healthcare industry, we presented the case of the University of Texas M. D. Anderson Cancer Center and developed insights into the impact of the EMRs on patient privacy and the importance of having a privacy policy. Further research about privacy protection in healthcare should try on a systematic view of the security and privacy of EMRs. We also notice that the explosive progress of IS/IT, including web-based systems and mobile computing devices, causes healthcare information systems more porous. In this case, new privacy management policies and practices need to be developed to safeguard patient privacy.

REFERENCES

Ateniese, G., & Medeiros, B. (2002). Anonymous E-prescriptions. *WPES'02*, Washington, DC (pp. 19-28).

Croasdell, D. C. (2001). It's role in organizational memory and learning. *Information Systems Management*, *18*(1), 8–11. doi:10.1201/1078/43194.18.1.20010101/31260.2

DeMuro, P. R., & Grant, A. H. III. (2001). Health privacy standards raise complex implementation issues. *Healthcare Financial Management*, *55*(1), 42–47.

Hammond, C. (2001). The intelligent enterprise. *InfoWorld*, *23*(6), 45–46.

Holt, G. D., Love, P. E. D., & Li, H. (2000). The learning organization: toward a paradigm for mutually beneficial strategic construction alliances. *International Journal of Project Management*, *18*(6), 415–421. doi:10.1016/S0263-7863(99)00066-6

Kieke, R. L. (2003). What's next, now that the April deadline has passed? *Journal of Health Care Compliance*, *5*(4), 41–43.

Li, J., & Shaw, J. (2004). Protection of health information in data mining. *International Journal of Healthcare Technology and Management*, *6*(2), 210–222. doi:10.1504/IJHTM.2004.004977

Matt, S. (2002). Auditing and reporting for HIPAA compliance. *Health Management Technology*, *23*(12), 12–13.

Messmer, E. (2003). HIPAA deadline ups health-care anxiety. *New World (New Orleans, La.)*, *20*(10), 30.

Miller, A., & Tucker, C. (2009). Privacy protection and technology diffusion: the case of electronic medical records. *Management Science*, *55*(7), 1077–1093. doi:10.1287/mnsc.1090.1014

Runy, L. A. (2008). The best defense. *Hospitals & Health Networks*, *7*(3), 22–26.

Schneier, B. (2000). *Secrets & lies: digital security in a networked world.* New York: John Wiley & Sons, Inc.

Swartz, N. (2004). A prescription for electronic health records. *Information Management Journal*, *38*(4), 20–26.

Taylor, H., & Leitman, R. (2001). U.S. trails other English speaking countries in use of electronic medical records and electronic prescribing. *Harris Interactive*, *1*(28), 1–3.

Weise, E. (2005). Medical errors still claiming many lives. *USA Today*. Retrieved May 17, 2005 from http://www.usatoday.com/news/health/2005-05-17-medical-errors_x.htm.

Zunkel, D. (2005). The other side of privacy: protecting information with biometrics. *Security Technology Design*, *15*(6), 46–53.

Chapter 9
Using Statistical Texture Analysis for Medical Image Tamper Proofing

Samia Boucherkha
Mentouri University, Algeria

Mohamed Benmohamed
Mentouri University, Algeria

ABSTRACT

This chapter discusses an approach for both authentication of medical images and confidentiality for the related textual data in an online medical application paradigm. The image authentication is achieved in a soft manner through a feature-based digital signature while the confidentiality of the related patient information is achieved through reversible data hiding. The selected features are robust towards geometric transformations, while fragile towards texture alterations that are characteristic of medical images. The processing scheme is done in a block by block basis to permit the localization of tampered image's regions. The effectiveness of the scheme, proven through experiments on a sample of medical images, enables us to argue that implementing mechanisms lying on this approach will help to maintain personal patient privacy and medical image integrity.

INTRODUCTION

Advanced computer based medical equipment is nowadays widely used in management of disease. Image processing technology is often used in computer based medical applications; the images are processed, stored, or transferred to a distant location over Internet or local networks. The resulting processed images can then be made available with patient information to medical researchers, second opinion physicians, students, and social security institutions. In such medical applications, the protection of the integrity and confidentiality of healthcare records is a critical issue.

Indeed, manipulation of medical images might lead to wrong interpretation and false diagnosis while the secrecy of the related patient information is imposed by the legal and ethic considerations, since patient data are very personal, and therefore, sensitive. The risks of unauthorized manipulation and misappropriation are increased when dealing with an open environment. So, before patient records can be distributed online, it is necessary to integrate in the medical application, mechanisms that maintain personal privacy and data integrity.

Encryption is a widely used technology to ensure several security services such as confidentiality, integrity, source authentication, and non-repudiation. The issues concerning medical applications are source authentication, non-repudiation, and integrity (Coatrieux, Maitre, Sankur, Rolland, & Collorec, 2001).

Source authentication is required to verify if the doctor who is claiming to have sent the record has actually sent it; Non-repudiation ensures that the doctor who sent the record cannot claim that he did not, while integrity will verify that the received record is exactly the same as the one sent and has not been modified by unauthorized people. Besides this, confidentiality of patient information is highly mandatory.

The traditional methods used for data authentication are digital signature (DS) (Stallings, 2003). To obtain the DS of an image, the digest of this latter is computed using a one-way hash function, then encrypted with the private key of a public-key cryptosystem. At the receiver's side, the DS is decrypted with the sender's public key, and again the image's digest is computed with the same hash function, then compared to the received one. Thanks to public-key cryptography, the resulting scheme ensures source authentication, non-repudiation, and integrity services.

Given the one-way characteristics of the hash function, it is unlikely that two different images will have the same digest, and even if a single bit changes, the DS may be totally different. This property may be convenient for text messages, but it does not offer the flexibility often asked with digital images. In fact, many image applications may tolerate more or less manipulations, such as format conversion, moderate level of filtering, and contrast adjusting without suffering from content changing. Thus, the image will be improperly categorized as inauthentic by the previous hash-based mechanism. In this case, an authentication procedure that uses some special-purpose functions to extract essential image invariant features and aims at authenticating the semantic content of images, in place of the hash digest is more convenient. As a result, the process can distinguish content-preserving manipulations such as filtering, compression, and scaling from content-changing manipulations such as cropping, objects addition, deletion, and modification. This way to proceed is generally referred to as features-based digital signature and leads to a more "soft" authentication. The current literature includes many of such authentication techniques which features are based on color, intensities, edges, or shapes (Bas, Chassery, & Macq, 2002; Li, Lou, & Chen, 2000; Thiemert, Sahbi, & Steinebach, 2005). The obtained DS may be appended to the transmitted image file, or alternatively embedded in the image itself via data hiding.

Data hiding is an emerging technology that enables the insertion of secret information invisibly throughout the image without degrading its visual quality. This technique takes advantage of the fact that digital images contain a lot of redundant information due to large spatial correlations, so some of this redundancy could be replaced by the secret information without perceptually changing the appearance of the image. A useful data hiding technique should be robust against malicious attacks to remove the embedded information and should not greatly affect the quality of the original image.

Data hiding has been adapted to medical imagery to connect visual and related textual data by several authors (Acharya, Anand, Bhat, & Niranjan, 2001; Boucherkha & Benmohamed,

2004; Kong & Feng, 2001; Macq & Dewey, 1999). Related data types may be patient data, context of image acquisition, parameters of image acquisition, primary evaluation of image, parameters of visualization, administrative data, and so forth. There are two potential benefits of embedding textual information in the related image: They remove any need to store separate metadata, and patient personal information is preserved from innocuous eyes.

In general, data hiding schemes proceed by modifying a predefined number of pixels in the original image in an irreversible way in order to embed the secret information. This is not acceptable in medical imagery where the least modification might have severe consequences. In fact, any degradation that will affect the reading of images can lead to an erroneous diagnosis, and consequently may have clinical, ethical, and legal incidences. Invertible data hiding is a new paradigm which enables the exact recovery of the original image upon extraction of the embedded information (Fridrich, Goljan, & Du, 2002; Tian, 2002).

In this work, we propose a system that can be used in medical image archiving or transfer, providing image authentication and confidentiality of the patient info in the same time, to achieve the aforementioned requirements. Soft authentication is achieved through features based digital signature, and confidentiality through data hiding. To this end, we adapt a recent data hiding scheme of large embedding capacity and low computational cost, and propose a content based digital signature (CBDS) which is especially sensitive to texture alterations that are characteristic of medical images.

The next section presents the basic technology of content based digital signatures for image authentication. Next, we describe the chosen data hiding technique followed by our own approach. Selected results on a sample of medical images are reported then, and finally we give some concluding remarks.

Features-Based Digital Signature

Features extraction refers to representing various image characteristics in terms of a collection of quantitative descriptors, often referred to as a feature vector, in a form that can be easily compared with the corresponding qualities in the received image. The choice of appropriate descriptive parameters will radically influence the reliability and effectiveness of subsequent feature qualification through the target application. In a semifragile signature context, the goal is to find binary representatives for the source image such that the hash values satisfy two requirements: they remain invariant under a constrained set of attacks, and they are different for perceptually different inputs with high probability. It is important to carefully select which of the measures discriminate the type of manipulations that are allowed.

There are many different ways to extract content dependent features from an image (Li, Lou & Chen, 2000; Lu & Liao, 2003). One approach considers image positions of edges, or contours in the spatial domain whose existence is proved during the verification stage. Another approach considers the relationships between pairs of coefficients in the transform domain (DCT, DWT, or DFT). The two approaches have in common that the gathered features will be quantized to allow some amount of predefined nonmalicious distortion. Figure 1 shows the block diagram of a typical features based image authentication scheme.

The proposed features detection scheme is especially sensitive to texture alterations of the host image. This provides significant advantage for authentication of medical images, whose analysis is strongly texture based. Furthermore, the scheme is invariant towards geometric transformations that are generally considered nonmalicious manipulations in most medical applications. It is based on first order image statistics collected from local pixels blocks, which represent basic texture descriptors of the considered image. The choice of such features was highly inspired from those

Figure 1. Features based image authentication

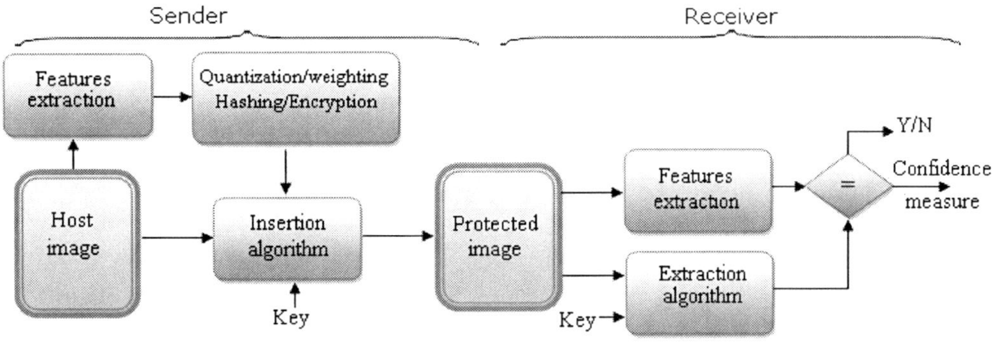

commonly used in image segmentation, pattern recognition, and classification.

Texture is an important characteristic that is generally used to identify and recognize objects in an image. In this case, the extracted features represent discriminant properties of the surface appearance. A variety of techniques for analysing image texture have been proposed by several authors (Haralick, 1979; Zhang & Tan, 2002). These techniques have been classified into three categories as statistical, structural, and spectral. Structural methods determine the primitives which compose the texture and their placement rules while spectral methods correspond to filtering the image with a set of spatial masks in the frequency domain. Statistical methods analyze the spatial distribution of grey values by computing local grey-level statistics which are constant or slowly varying over a textured region. Different textures can be discriminated by comparing the statistics computed over different subregions. Hence, statistical methods allow relatively simple and reliable textural analysis; they also have been used very successfully to characterize images in many computer vision domains such as classification, pattern recognition, and content based image retrieval (CBIR) by several authors (Clausi, 2002; Fauzi & Lewis, 2006; Howarth & Rüger, 2004; Manjunath, Wu, Newsam, & Shin, 2000).

The specific statistics considered in this work are the mean value of the histogram block, the standard deviation, the skewness, and the kurtosis. The first three statistics are proven to be scale and rotation invariant, while the last is translation invariant (Haralick, 1979). Standard deviation and mean are the two mostly used statistic estimates. Mean is the average of a random distribution, while standard deviation measures the "spreadness" of the distribution. Skewness measures the degree of asymmetry of a distribution. For example, the skewness of a normal distribution is zero since a normal distribution is symmetric. Positive skewness indicates a distribution with an asymmetric tail extending toward more positive values, whereas negative skewness extends toward more negative values. Kurtosis characterizes the relative peakedness or flatness of a distribution compared to the normal distribution. Positive kurtosis indicates a relatively peaked distribution while negative kurtosis indicates a relatively flat distribution. The general definition of these attributes is given by:

$$\mu = \frac{1}{N}\sum_{i,j} g(i,j) \tag{1}$$

$$\sigma^2 = \frac{1}{N}\sum_{i,j}(g(i,j) - \mu)^2 \tag{2}$$

$$SKEW = \frac{1}{N}\sum_{i,j}(g(i,j) - \mu)^3 \tag{3}$$

$$KURT = \frac{1}{N}\sum_{i,j}(g(i,j) - \mu)^4 \tag{4}$$

where g(i,j) is the grey level value.

Embedding Strategy

The chosen method for embedding the extracted features is a modified version of Tian's data hiding technique (Tian, 2002). In addition of being a high capacity low computational cost technique, it is fully reversible, that permits to remove completely the distortion due to embedding after authentication has taken place. This data hiding is based on the computation of a difference value between adjacent pixels, then this difference is expanded to embed secret information. The processes of Tian's scheme can be described as follows:

First, an integer transformation is defined as:

$$l = \left\lfloor \frac{(x+y)}{2} \right\rfloor, \; h = x - y \qquad (5)$$

where x and y are two adjacent pixels.

The inverse integer transformation can be represented as:

$$x' = l + \left\lfloor \frac{h+1}{2} \right\rfloor, \; y' = l - \left\lfloor \frac{h}{2} \right\rfloor \qquad (6)$$

The embedding process consists of the following stages:

1. Compute l_i and h_i for the ith pair of pixels, denoted as $\{x_i, y_i\}$, by using Equation 5.
2. Compute $h'_i = 2 \text{ x } h + w_i$ (expansion), where w_i is the corresponding bit to be embedded.
3. Get x'_i et y'_i by replacing the parameters l_i and h'_i in Equation 6.
4. Repeat the former steps until all pixel pairs are processed.
5. The embedded image can be constructed by replacing each pixel pair (x_i, y_i) by (x'_i, y'_i).

In the retrieval phase, the corresponding pixel pairs are collected again and substituted in Equation 5:

$$l_o = \left\lfloor \frac{(x'_0 + y'_0)}{2} \right\rfloor, \; h'_0 = x'_0 - y'_0 \qquad (7)$$

Extracting the least significant bit of h'_0, we get the embedded bit w_0 and h_0.

Finally the original pixels are calculated from Equation 6 as:

$$x_0 = l_0 + \left\lfloor \frac{h_0+1}{2} \right\rfloor, \; y_0 = l_0 - \left\lfloor \frac{h_0}{2} \right\rfloor \qquad (8)$$

The overall flowchart of Tian's scheme is given in Figure 2.

To prevent overflow and underflow problems, the difference value h (after data embedding) must satisfy:

$$0 \le l + \left\lfloor \frac{h+1}{2} \right\rfloor \le 255$$

and

$$0 \le l - \left\lfloor \frac{h}{2} \right\rfloor \le 255 \qquad (9)$$

Figure 2. Flowchart of Tian's data hiding scheme

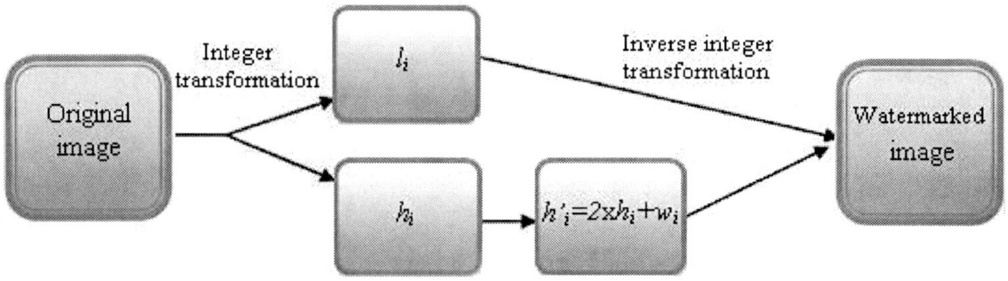

Tian (2002) solved this problem by embedding secret information only into the difference of pixel pairs from those pairs that are not expected to cause an overflow or underflow. The location map that indicates the modified pairs is compressed and included in the payload. This will greatly reduce the capacity as well as increase the processing time due to compression.

In this work we propose to solve the overflow/underflow problem in a simple but efficient manner. In the difference expansion stage, we represent the difference value h in the binary reflected gray code (BRGC). This latter is a code assigning to each of a contiguous set of integers, a word of bits such that each two adjacent code-words differ by only one bit, yielding to a Hamming distance of 1 between adjacent codes. A fundamental characteristic of this code is that the LSB values of contiguous codes are alternatively 11 00 11 00 11 00… This "cyclic" property is exploited here to expand the difference value without inducing an overflow/underflow to the modified pixels. The difference expansion stage becomes:

- After computing h, obtain the BRGC of h, say h_G,
- If w_i = LSB of h_G then h' := h_G;
else if h_G is even then h' := h_G+1 else (h_G is odd) h':= h_G -1;

Hence, the LSB of h' is forced to contain w_i while still belonging to the interval [0-255].

The rest of the procedure is almost the same.

We can note that the capacity of the scheme is increased to 1 bit per pixel pair.

Signature Generation

The system first divides the image into non-overlapping blocks of size 16-by-16 pixels. A feature vector is then extracted from each block. The block size is chosen to compromise between texture effectiveness and computation time. Smaller block sizes may preserve more texture details but increase the computation time. Conversely, increasing the block size can reduce the computation time but lose texture information and increase the features coarseness. Each features vector consists of the four aforementioned features which are normalized to fit an 8-byte width. The encoder scans the blocks one at a time, evaluates the texture features in the block, and embeds the previously computed features vector. The blocks are accessed in a pseudorandom order with a secret key as a seed. Hence, the features vector output of a block is embedded into another distant block. In this way, an attack modifying a single block causes authenticity problems along the whole chain. The block chaining nature of the signature calculation process renders the scheme resistant to content cutting and pasting attacks.

In the other hand, all the vectors are concatenated into one bit-string, which is then hashed using a one way hashing function to generate a global signature which is embedded in the first visited block. In this way, the algorithm can be customized differently for different needs: one may use only the local signature to authenticate individual blocks with or without attack localization or use the global cryptographic digital signature contained in the first visited block within a public key context, if available.The overall flowchart of the signature generation process is given in Figure 3.

Let us consider an example of processing a 512-by-512 image. The image is partitioned into 1024 non-overlapping 16-by-16 blocks, each one containing 128 pairs of pixels. Four double-byte coefficients are computed per block; thus, 64 bits are extracted per block which are interleaved with 64 bits of patient information according to a secret key, then embedded into another distant block, randomly selected. About 1024 bytes of patient information may be embedded in a given image.

Signature Verification

When a suspect image is received, the authenticity is verified in the following way: First, the image is again divided into non-overlapping 16-by-16

Figure 3. Block diagram of the signature generation process

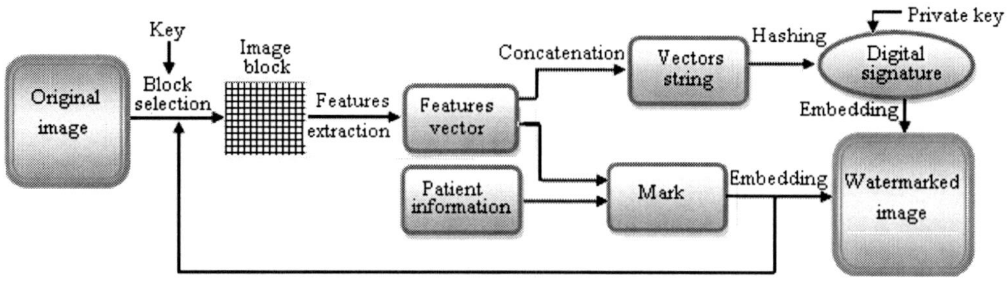

blocks which are randomly selected using the same key as in the embedding stage. The signature is picked from the first block then an iterative process is engaged as shown in Figure 4:

- extract the mark from a block and separate it in patient information and authentication information
- generate texture features from the current block and compare them with the authentication information; if the match fails, the block will be marked as attacked
- obtain the next block

Experimental Results

Several tests were conducted to attest the scheme efficiency using MATLAB, due to its useful Image Processing Toolbox. The experiments are carried out on a collection of 512-by-512 medical images from different modalities, collected mainly from the World Wide Web. For illustration, we included some results on two representative images. Figure 5 and Figure 6 show the impact of data embedding on these images. From Figure 5, no perceptual change can be noticed between original images and embedded ones, while Figure 6 shows the intensity histograms of the images (a) and (b) respectively, before and after embedding. We can notice that the grey levels distribution histograms are nearly the same. To evaluate the scheme over attack localization, we modified slightly the first image feigning a false fracture. The attack was successfully detected as shown in Figure 7.

To attest to the robustness of the extracted features against geometrical attacks, we used the Hamming distance defined by:

$$d_1(v,v') = \frac{1}{k}\sum_{i=0}^{k}\left|(v(i) - v'(i))\right| \qquad (10)$$

Where v and v' are respectively the features extracted from the mark and the features vector of

Figure 4. Block diagram of the signature verification and attack localization processes

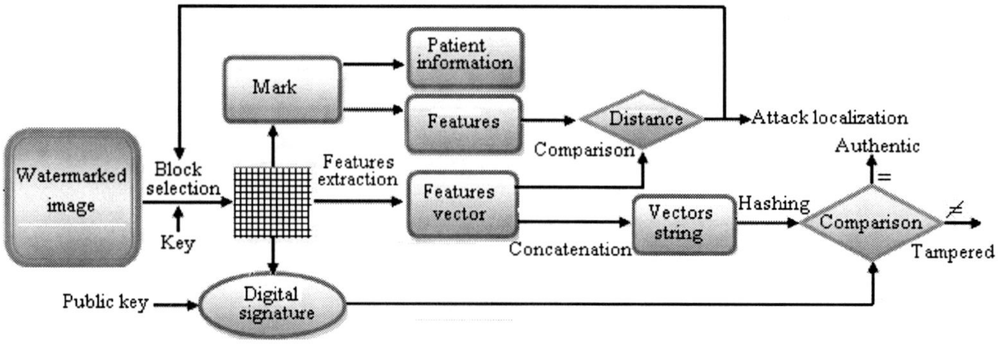

Figure 5. Original images (a) and (b) and corresponding embedded images

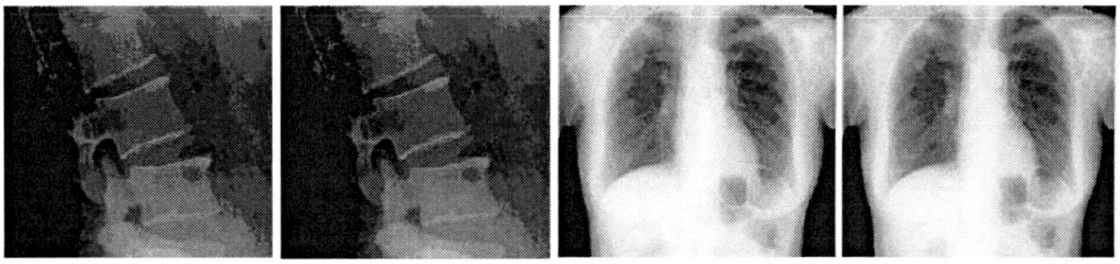

Figure 6. Intensity histograms of image (a) and image (b) before and after embedding

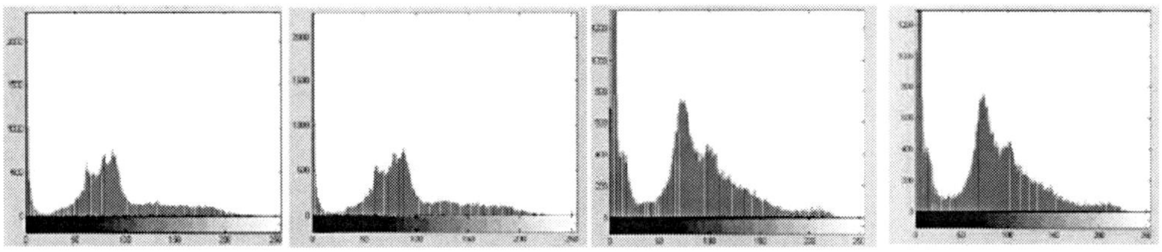

Figure 7. Original, attacked, and verified image (a)

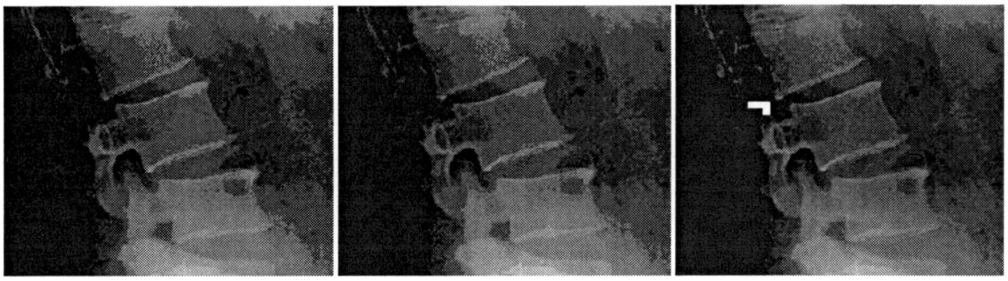

the current block. This measure gives a value close to zero for similar vectors and close to 0.5 for very dissimilar ones. The more a block is modified, the more this value is expected to increase. The binary representation of the features vector of an original image block was compared to those obtained from the corresponding manipulated block using xor and shift operations. Table 1 shows three sample features vector gathered from different blocks of the same image, while Table 2 summarizes the difference between features of original and corresponding geometrically transformed samples in terms of Hamming distance. As expected, the rotation was the most penalized transformation.

Table 1. Normalised extracted features

Features	Sample 1	Sample 2
Mean	59.011	62.755
Standard deviation	16.296	16.988
Skewness	65.19	18.46
Kurtosis	3.816	4.165

Table 2. Hamming distance between features of original and attacked samples

Attacks	Sample 1	Sample 2
Scaling by 10%	0.0013	0.0016
Scaling by 20%	0.0035	0.0280
Scaling by 30%	0.0139	0.0617
Rotation by 10°	0.1145	0.1220
Rotation by 20°	0.1410	0.1591
Rotation by 30°	0.2731	0.2918
Cropping by 10%	0.0111	0.0451
Cropping by 20%	0.0723	0.0883
Cropping by 30%	0.0918	0.1103

CONCLUSION

This chapter proposes a features-based authentication method that is robust to typical medical image manipulations but fragile to malicious ones. The features are generated based on the texture characterization of the image, then embedded in the difference expansion domain, after being interleaved with patient information for protection. Data hiding embedding is performed on a block-by-block basis to permit the localization of altered regions and can be extracted without the original image. The experimental results indicate that the proposed scheme is feasible. Given its relative simplicity, it can be applied to the image at the time of acquisition to serve in many medical applications concerned with privacy protection, safety, and management.

REFERENCES

Acharya, R., Anand, D., Bhat, S., & Niranjan, U. C. (2001). Compact storage of medical images with patient information. *IEEE Transactions on Information Technology in Biomedicine, 5*, 320–323.

Bas, P., Chassery, J. M. & Macq, B. (2002). Geometrically invariant watermarking using features points. *IEEE Transactions on Image Processing, 11*, 1014–1028.

Boucherkha, S., & Benmohamed, M. (2004). A lossless watermarking based authentication system for medical images. *Transactions on Engineering, Computing and Technology, 1*(1), 240–243.

Clausi, D.A. (2002). An analysis of co-occurrence texture statistics as a function of grey level quantization. *Canadian Journal of Remote Sensing, 28*(1), 45–62.

Coatrieux, G., Maitre, H., & Sankur, B. (2001). Strict integrity control of biomedical images. *Proceedings of Security and Watermarking of Multimedia Contents III, SPIE, 4314*, 229–240.

Coatrieux, G., Maitre, H., Sankur, B., Rolland, Y., & Collorec, R. (2001). Relevance of watermarking in medical imaging. In *Proceedings of the IEEE CIT: Applications in Biomedicine* (pp. 250–252).

Fauzi, M. F. A., & Lewis, P. H. (2006). Automatic texture segmentation for content-based image retrieval application. *Pattern Analysis & Applications, 9*(4), 307–323.

Fridrich, J., Goljan, M., & Du, R. (2002). Lossless data embedding: New paradigm in digital watermarking. *EURASIP Journal on Applied Signal Processing, 2*(2), 185–196.

Haralick, R. (1979). Statistical and structural approaches to textures. *Proceedings of the IEEE, 67*(5), 786–804.

Howarth, P., & Rüger, S. (2004). Evaluation of texture features for content-based image retrieval (Book chapter: Image and video retrieval). *Lecture Notes in Computer Science, 3115*, 978–1039. Berlin/Heidelberg: Springer.

Kong, X., & Feng, R. (2001). Watermarking medical signals for telemedicine. *IEEE Transactions on Information Technology in Biomedicine, 5*(3), 195–201.

Li, C.-T., Lou, D.-C., & Chen, T.-H. (2000). Image authentication via content-based watermarks and a public key cryptosystem. In *Proceedings of IEEE International Conference on Image Processing, III*, 694–697.

Lu, C.-S., & Liao, H.-Y. (2003). Structural digital signature for image authentication: An incidental distortion resistant scheme. *IEEE Transactions on Multimedia, 5*, 161–173.

Macq, B., & Dewey, F. (1999). Trusted headers for medical images. In *Proceedings of the DFG VIII-D II Watermarking Workshop.*

Manjunath, B., Wu, P., Newsam, S., & Shin, H. (2000). A texture descriptor for browsing and similarity retrieval. *Journal of Signal Processing: Image Communication, 16*, 33–43.

Qibin, S., & Chang, S.-F. (2005). A secure and robust digital signature scheme for JPEG2000 image authentication. IEEE Transactions on Multimedia, 7(3), 480–494.

Stallings, W. (2003). *Cryptography and network security: Principles and practices* (3rd ed.). Prentice Hall.

Thiemert, S., Sahbi, H., & Steinebach, M. (2005). Applying interest operators in semifragile video watermarking. IS&T SPIE International Symposium on Electronic Imaging *Security, Steganography and Watermarking of Multimedia Contents VII, 5681*, 353–362.

Tian, J. (2002). Reversible watermarking by difference expansion. In *Proceedings of the Workshop on Multimedia and Security* (pp. 19–22).

Zhang, J., & Tan, T. (2002). Brief review of invariant texture analysis methods. *Pattern Recognition, 35*, 735–747.

This work was previously published in the International Journal of Information Security and Privacy, Volume 2, Issue 3, edited by H. Nemati, pp. 18-27, copyright 2008 by IGI Publishing (an imprint of IGI Global).

Chapter 10
Privacy Inference Disclosure Control with Access-Unrestricted Data Anonymity[1]

Zude Li
The University of Western Ontario, Canada

ABSTRACT

This chapter introduces a formal study on access-unrestricted data anonymity. It includes four aspects: (1) it analyzes the impacts of anonymity on data usability; (2) it quantitatively measures privacy disclosure risks in practical environment; (3) it discusses the potential factors leading to privacy disclosure; and (4) it proposes the improved anonymity solutions within typical k-anonymity models, which can effectively prevent privacy disclosure that is related with the published data properties, anonymity principles, and anonymization rules. The experiments have found these potential privacy inference violations and shown the enhanced privacy-preserving effect of the new anti-inference policies to access-unrestricted data publication.

INTRODUCTION

The term "*privacy*" is generally used as "the right to select what personal information about me is known to what people" (Westin, 1976). In a technical view, data privacy protection is to manage individual data in a privacy aware way, including determination of *when*, *how* and *to what extent* such data can be communicated to others with *what special features*. Privacy is becoming a hot concern for both individuals and organizations.

DOI: 10.4018/978-1-61692-000-5.ch010

In the past thirty years, a lot of research studies have been contributed to data privacy. Roughly, these studies can be classified into two types: *access-restricted* and *access-unrestricted*. The *access-restricted* type focuses on how to protect privacy through restricting user access to the private data: A user can access the data only if he/she satisfies the access constraints predefined on the data. The *access-unrestricted* type concerns on data publication, making them oriented to most users and at the same time avoiding possibly compromised privacy disclosure. In the access-restricted type studies, researchers usually extend access-based control models for privacy enhancing (Antón et al., 2003; Crook et al., 2005). For instances, *Purpose-Based Access Control* (PBAC) (Byun et al., 2005) and *Hippocratic Database* (HDB) (Agrawal et al., 2002) are two *access control* based privacy techniques. In these models, *direct privacy disclosure* (i.e., access some unauthorized data or with illegal purposes) can be efficiently prevented, while *indirect privacy disclosure* (i.e., infer unauthorized or purpose-illegal private data based on accessed data aggregation) is still a challenging problem. Some *inference control* techniques (Staddon, 2003) are available for handling this issue.

Access-unrestricted data privacy techniques are widely required in practice, such as individual data dissemination, voting data proclamation, health-care data publication, etc. Data anonymity is generally the privacy solution used for this type of applications. The key issue for enhanced access-unrestricted privacy protection is to quantitatively measure the risk of privacy disclosure and information loss which are coupled by data anonymity. One of recently proposed privacy techniques for handling access-unrestricted data privacy is *k-anonymity* (Sweeney, 1997). In nature, it is a data processing model towards maintaining data usability as well as avoiding privacy disclosure during unrestricted data access. No matter of its implementation complexity in practice, privacy disclosure on k-anonymized access-unrestricted dataset is still existed (Machanavajjhala et al., 2006; Li et al., 2006a).

This chapter aims to a formal research on access-unrestricted privacy protection, mainly focusing on the data anonymization process and the privacy inference risk analysis with anonymized data. As the best we know, there are no literatures related to privacy inference control study for access-unrestricted privacy applications. Most anonymity risk detection techniques focus on the anonymization process on only the resulting dataset but not any other external information, which, subsequently, cannot guarantee the survivability of real application systems. This chapter, based on our early work (Ye et al., 2007), illustrates an approach to measuring privacy inference risks on the k-anonymized dataset in a quantitatively manner. Through the data anonymity analysis, we discover four main factors that may incur various privacy violations. Further, we propose two effective *anti-inference* privacy policies for access-unrestricted data publication. In this chapter, k-anonymity is used as a scenario model to perform data anonymity. In the experiments studied, we have proven the existence of potential privacy inference violations and the efficiency of our anti-inference policies.

The rest of this chapter is organized as follows. Section 2 discusses the related work. Section 3 defines some useful concepts and notations for the data anonymity analysis; Section 4 discusses the anonymity principles and anonymization rules that are explicitly or implicitly used during data anonymity; Section 5 illustrates the formal model of access-unrestricted data anonymity; Section 6 analyzes the data anonymization process and the corresponding privacy inference attacks with the k-anonymity model, in which experimental findings are shown for discovering and removing potential various privacy violations and convincing the advantages of our proposed anti-inference policies over the existing ones; Finally, Section 7 gives a short conclusion for the chapter.

RELATED WORK

In access-unrestricted microdata publication applications, all privacy attacks have the common feature: to find the exact or enough-precise mapping between individual identities (*UI*) and sensitive values (*SA*) (Sweeney, 1997; Meyerson & Williams, 2004; Staddon, 2003; Rastogi et al., 2007; Mcguckin & Nguyen, 1990). From the perspective of privacy protection, data publisher should prevent users from them through some special data processing techniques.

Practical strategies towards this goal can be roughly classified into two types: *separation* and *distortion*. The separation strategy is to separate *UI* from *SA* and guarantee users cannot fetch and map both of them. De-identification (Sweeney, 1997; Samarati & Sweeney, 1998) is a typical separation strategy, which is not always effective, since users can always fetch both *UI* and *SA*, and map them high-precisely by the re-identification capability of *QI*. The separation strategy can be used as a fundamental policy for privacy protection.

Based on the separation method, the distortion strategy is to change the original *UI*, *QI*, or *SA*, for decreasing the likelihood of finding the real *UI-SA* mapping in a population. Many data processing techniques can be used for distortion, including *recording* (Adam & Wortmann, 1989), *swapping* (Dalenius & Reiss, 1982), *suppression* (Samarati & Sweeney, 1998), *micro-aggregation* (Domingo-Ferrer et al., 2002), *generalization* (Sweeney, 2002a), etc. Recording is to reduce the granularity of the microdata. Swapping involves moving data entries from one cell to another in the contingency table in a manner that is consistent with the set of published margins. Suppression just deletes the sensitive cells when answering the queries for them. Micro-aggregation is to make data vectors for publication correspond to groups of *k* or more individuals, where no individual dominates (Domingo-Ferrer et al., 2002). Generalization involves replacing (or recording) a value with a less specific but semantically consistent value. General anonymity policy is to use one or more of the above methods on *QI* or *SA* values, then publish the distortion result. K-anonymity is a typical model (Sweeney, 2002b).

Even the original data can be processed with either separation or distortion or both, it is not always "secure" from privacy inference violations. After long-history research, inference problem is still an open challenge in many research areas. As in commercial database (a typical access-restricted data management application), privacy inference control is implemented during data designing or query processing phase (Farkas & Jajodia, 2002; Staddon, 2003). For static data designing, it is easy to decrease data usability when security levels on data are complicated and application tasks are changeable. For dynamic query processing, query performance can be degraded as to check all dynamic inference channels in the current database status is a *NP* problem (Yip & Levitt, 1998).

Privacy inference control (Farkas & Jajodia, 2002) can prevent mutual linking among data, users by eliminating inference channel violations. The *dynamic inference control* schemes (Staddon, 2003) provide collusion resistance and have a query processing time that depends only on the length of the inference channels (not on the length of user query histories). All of them can only be used for access-restricted data management, such as on multi-level secure data models (Morgenstern, 1987; Sandhu & Chen, 1998) and database applications.

Inference control policies and strategies for access-unrestricted data publication are different than the above one, since there are much richer information can be freely fetched by users that can be available for privacy inference violations on published data. Anonymity policies with data processing methods just focus on data, but not the complicated application environment, where knowledge used for re-mapping *UI* and *SA* are not considered fully in general data anonymity and publication applications.

As in many k-anonymity applications, privacy disclosure may occur when a value can be determined to be among less than *k* possibilities by using the views together with just the schema information of the table. To check whether a set of views violates the k-anonymity requirements is a computationally *hard* problem (Yao, 2005). A conservative checking algorithm is proposed in (Yao, 2005), but still, there are other table-inside information (e.g., hierarchical anonymity structure on attributes or values) and table-outside information (e.g., value inference relations in commonsense) available for privacy disclosure detection not discussed in the chapter. Some privacy violations on k-anonymized microdata are also discussed in (Machanavajjhala et al., 2005). The *ℓ-diversity* ℓmechanism can prevent some background knowledge-based privacy attacks. It guarantees the diversity on each cluster is enough to prevent *QI-SA* matching by users.

Further, the anonymity cost metrics used in the current k-anonymity models cannot exactly indicate the real information loss on data after publication (Li et al., 2006b). While to maintain information on data is the most critical to data usability. Information loss is the quantity of information that existed in the original microdata but does not occur in anonymized microdata because of disclosure control methods (Winsborough & Li, 2004; Truta et al., 2003).

There are many information loss metrics proposed in literature, such as the five kinds of *information loss metrics* summarized in (Truta et al., 2003). (Domingo-Ferrer et al., 2002) proposes a disclosures risk measure that is compatible with some disclosure control methods in literature, based on data processing techniques. In (Winsborough & Li, 2004; Truta et al., 2003), the record-level disclosure risk is analyzed through computing the probability of matching an exact individual to any record in the original microdata under many assumptions.

Overall, one disadvantage of these technologies is that they do not consider some other influence out of the microdata, especially in access-unrestricted data publication applications. In k-anonymity models, anonymity cost metrics should measure the real information loss under the assumption that users can utilize some knowledge, which should be directly relative with the strength of preventing inference attacks (Li et al., 2006b). It is the main idea of one of the above proposed anti-inference policies.

Beside, current k-anonymity models are really hard-shelled: the k is defined on the table level, i.e., it cannot support *individual-specific k* on each corresponding data (Li et al., 2006a). Some other aspects should be considered such as *personalized* privacy protection policies (Xiao & Tiao, 2006), etc.

Martin et al. (Martin et al., 2007) considered the worst case of inference based on background knowledge in the (c, k)-safety model. In the model, attackers are supposed to have at most k pieces of information about the dataset; this information is expressed with a formal language. Later, Chen et al. (Chen et al., 2007) proposed the skyline privacy model. This model considered external background knowledge. Users are allowed to set parameters for this model in order to anonymize the dataset.

The privacy inference theory is introduced in our previous work (Ye et al., 2007), which defines the framework of measuring privacy inference risks based on k-anonymized dataset. In this chapter, we extend the theory into extended anonymity modeling, and give k-anonymity scenario as a typical application with this theory.

The above discussed k-anonymity models can be used on only static mircodata publication. Byun et al. (Byun et al., 2006) firstly proposed the incremental k-anonymity model for continuously growing datasets. Later, Xiao and Tao (Xiao & Tao, 2007) proposed a dynamic k-anonymity model (termed m-Invariance) that supports both insertions and deletions of the subject datasets.

CONCEPTS AND NOTATIONS

Suppose individual data are collected into a relational table in database, where each record maps to an individual in reality. For such a special table, the possible data attributes can be divided into three classes (similar to (Domingo-Ferrer, 2002)): (1) *Unique identifier* attributes (*UI*), where each value refers to an individual in a scope; (2) *Quasi-identifier* attributes (*QI*), where each value can be joined with other values to re-identify an individual in the scope with high precision; and (3) *Sensitive* attributes (*SA*), where each value should not be accessed by any users except the corresponding individual.

We define the *schema* of table T as <UI, QI, SA>, where UI = <UI^1, UI^2, ..., UI^u>, QI = <QI^1, QI^2, ..., QI^q>, SA = <SA^1, SA^2, ..., SA^s>, and $u + q + s = m$. An instance of T is a set of records. The ith record r_i in the T can be specified as: r_i = < ui_i, qi_i, sa_i>, where ui_i, qi_i, and sa_i denote a value tuple of UI, QI, and SA, respectively. $ui_i = <ui_i^1, ui_i^2, \ldots, ui_i^u>, qi_i = <qi_i^1, qi_i^2, \ldots, qi_i^q>, sa_i = <sa_i^1, sa_i^2, \ldots, sa_i^s>$. In this chapter, we simply take T as a table instance without misunderstanding. In T, the value *domain* of attribute A is written as D_A. We define a *population* Ω over T as a cluster of elements. Generally a population is a set of individual information where all individuals belong to a group (such as a social community). It holds: T □ Ω. For convenience, we define the following six functions and two concepts on T (N is set of natural number):

Definition 1 (Information Capacity) The *information capacity* (*IC*) on value v in Ω, denoted as $IC_\Omega(v)$, is a quantitative value (within [0, 1]) indicating the specification degree of the information beyond v in Ω.

The *IC* of v on attribute A in Ω is directly relative with the properties of A. For instance, suppose the value format on attribute *BirthDate* is *mm-yy* (i.e., *month-year*), then IC_Ω(09-06) =1, which indicates 09-06 is the most specific value on this attribute in Ω. But if the value format is *dd-mm-yy* (i.e., *day-month-year*), then IC_Ω(*-09-06) $=\frac{1}{30}$, since the "*day*" sub-item is suppressed, and there exist 30 more specific values (1st-30th in Sep., 2006) with larger *IC* values. From a computable perspective, $IC_\Omega(v)$ is related with Ω. If there exist no values in [01, 15]-09-06 in Ω, then we can derive IC_Ω(*-09-06) $=\frac{1}{15}$, which is closer to the *IC* value of any specific value in [16, 30]-09-06. Generally, the bigger of $IC_\Omega(v)$, the more specific of v in Ω, i.e., the better usability of v (but exist some exceptions on this statement, discussed later on).

Data anonymity is the main privacy technique used to decrease the potential disclosure risks beyond data. It derives *anonymity forms* of the original data, where specific privacy-aware information is generalized.

Definition 2 (Anonymity Form) An *anonymity form* of data v on attribute A in T, denoted as $AF(v)$, is a *generalized* value on A than v, satisfying $AF(v) \in \Omega$and $IC_T(AF(v)) \leq IC_T(v)$.

According to the above definition, we can infer the usability of $AF(v)$ is less than that of v. In Ω, v may have multiple anonymity forms, indicating different usability levels. It is an *anonymity relation* between v and one of its anonymity forms $AF(v)$, denoted as $v \prec AF(v)$. Obviously, this is a *partial order* relation, and the following transitive anonymity relations hold: $v \prec AF(v) \prec AF(AF(v))$. We define: (*i*) $v_i \prec^+ v_j \Leftrightarrow \exists\, v_k$ ($k \neq i$, $k \neq j$): $v_i \prec v_k$, $v_k \prec v_j$; (*ii*) $v_i \prec^* v_j = v_i \prec^+ v_j \cup v_i \prec v_j$. Beside, an anonymity form of value tuple $<v_i, v_{i+1}, \ldots, v_{i+k}>$, written as $AF(<v_i, v_{i+1}, \ldots, v_{i+k}>)$, satisfies: (1) $v_{i+j} \prec AF(v_{i+j}) \in \Omega$ ($0 \leq j \leq k$); (2) $IC_\Omega(AF(<v_i, v_{i+1}, \ldots, v_{i+k}>)) \leq IC_\Omega(<v_i, v_{i+1}, \ldots, v_{i+k}>)$.

Definition 3 (Inference Relation) An *inference relation* from value v_i to v_j holds, as $v_i \mapsto v_j$ [p], if v_j can be ascertained according to v_i with the *inference probability* p (within [0, 1]).

Specifically, the inference probability from $AF(v)$ to v in Ω_v (a cluster in Ω containing all elements on the attribute v occurs), as $P_{\Omega_v}(AF(v) \mapsto v)$, indicates the likelihood of correctly ascertaining v in Ω_v with $AF(v)$, as:

$$P_{\Omega_v}(AF(v) \mapsto v) = \frac{|V_{\Omega_v}(v)|}{|V_{\Omega_v}(AF(v))|} \tag{1}$$

Where $V_{\Omega_v}(v) = \{v' \mid v' \in \Omega_v, v' \prec^* v\}$, $V_{\Omega_v}(AF(v)) = \{v' \mid v' \in \Omega_v, v' \prec^* AF(v)\}$. Similarly, we define the *inference probability* within the domain of *v*, D_v ($v \in D_v \subseteq \Omega_v$), as: $P_{D_v}(AF(v) \mapsto v) = \frac{|V_{D_v}(v)|}{|V_{D_v}(AF(v))|}$. It holds:

$$P_{\Omega_v}(AF(v) \mapsto v) \le P_{D_v}(AF(v) \mapsto v) \quad (2)$$

Evaluating the *inference probability* is similar to *information capacity*. But *inference probability* can be used between any two values on different attributes, which is more generic than *information capacity*. It also depicts the privacy inference risk from a known value to the target value. *Information capacity* refers to the information maintenance on the anonymity form of an attribute value in comparison with the original one.

In scope *T*, we define an *attribute inference* relation from attribute A_i to A_j, if for any value v_j on A_j, there exists value v_i on A_i, satisfying: $P_T(v_i \mapsto v_j) > \beta$ (β is a predefined *threshold* value). For example, the following attribute inference relation holds obviously: *BirthDate* $\mapsto$ [100%]. Simply, we also say there exists a *value inference* relation from v_i to v_j, if $P_T(v_i \mapsto v_j) > \beta$.

Definition 4 (Anonymity Degree) The *anonymity degree* on $v \prec AF(v)$ in Ω_v (or D_v), denoted as $AD_{\Omega_v|D_v}(v \prec AF(v))$, refers to the coverage ratio of the anonymity relation in the whole domain:

$$AD_{\Omega_v|D_v}(v \prec AF(v)) = \frac{|V_{\Omega_v|D_v}(AF(v))|}{|\Omega_v \mid D_v|} \quad (3)$$

With Formula (1) and (3), we can derive:

$$P_{\Omega_v|D_v}(AF(v) \mapsto v) = \frac{\frac{|V_{\Omega_v|D_v}(v)|}{|\Omega_v \mid D_v|}}{|AD_{\Omega_v|D_v}(v \prec AF(v))|} \quad (4)$$

It is obvious there is an *inverse proportion* relationship between the inference probability and the anonymity degree on an anonymity relation (since $\frac{|V_{\Omega_v|D_v}(v)|}{|\Omega_v \mid D_v|}$ is always a constant). We can derive: To increase $P_{\Omega_v|D_v}(AF(v) \mapsto v)$, it is necessary to decrease $AD_{\Omega_v|D_v}(v \prec AF(v))$, i.e., to increase $IC_{\Omega_v|D_v}(AF(v))$.

Definition 5 (Diversity Degree) The *diversity degree* of cluster *C*, denoted as *DD*(*C*), refers to the diversity level (i.e., the *information entropy*) of elements in *C*:

$$DD(C) = -\sum_{v \in C} \frac{V_C(v)}{|C|} \times \log \frac{V_C(v)}{|C|} \quad (5)$$

ANONYMITY PRINCIPLES AND RULES

Because each individual data is organized as a record in a relational table, such a tabular structure can be easily understood by users to match *UI* and *SA* values after data publication, which in fact is a typical privacy disclosure situation. Anonymity on such a dataset aims to break-downing the direct connection between *UI* and *SA*. Some data processing techniques adopted by anonymity including hiding *UI*, generalizing *QI*, clustering *SA*, etc. (Winsborough & Li, 2000; Truta et al., 2003; Domingo-Ferrer, 2002). The ultimate purpose is to guarantee the inference probability from *UI* to *SA*, or from *SA* to *UI* under the "safe" *threshold*. We cannot hope to publish an enough anti-inference dataset by just hiding *UI*, since *QI* values may re-identify *UI* values with enough precision. For instance, the inference probability from *QI* to *UI* is close to 100% in many real applications (See the example in (Sweeney, 1997)).

The cause for this problem is that users can find and utilize some <*UI, QI*> data pairs in other public datasets to help infer the exact *UI-SA* mappings. So even on <*QI, SA*>, we need to reduce the inference probability from *QI* to *SA*, and from *SA* to *QI*. As well, it must consider data usability for practical applications. Overall, it is a problem how to achieve or balance two main goals: (1) Guaranteeing the inference probability on *UI-SA* mapping under the predefined "safe" threshold; and (2) Keeping information capacity on published data as close to that on the original data as possible.

Together with other requirements from practice, such as personalized anonymity (Xiao & Tiao, 2006) and most data usability, we propose the following *five anonymity principles* as the criteria for access-unrestricted data anonymization.

- *Enough anonymity*: Published data should satisfy the anonymity requirement;
- *Personalization*: Individuals should be able to specify personalized privacy preferences levels on their data records;
- *Information maintenance*: Data publisher should guarantee the published data as specific as possible;
- *Information preciseness*: Information on published data should be accurate, avoiding misunderstanding; and
- *Enough usability*: Published data should be the most "meaningful" in real practice.

The first principle requires that the used data anonymization techniques should make the inference probability on *UI-SA* mapping under a predefined threshold. The second principle requires data anonymization to support individual-specific privacy policies since individuals are the ultimate victims on privacy disclosure and they have the right (in society) to require this feature. The following three principles focus on data usability and quality, where the third one requires data anonymization to guarantee information capacity on published data from a general perspective. The fourth one requires the anonymity methods (such as suppression, aggregation, etc.) should not create wrong or misunderstand-prone data for publication. The last one requires data anonymization should make data real-most-available in public.

Information capacity on data may not always represent the data usability from a social perspective. For instance, on *BirthDate* attribute with *dd-mm-yy* format in Ω, it is obvious that IC_Ω(*-09-06) $\leq IC_\Omega$(01-*-06) $\leq IC_\Omega$(01-09-06). But we should not choose 01-*-06 instead of *-09-06 as an anonymity form of 01-09-06 for publishing even both of them satisfy the anonymity requirement, since *-09-06 is more widely used and meaningful than 01-*-06 in a social viewpoint.

The above five anonymity principles indicate the goals of data anonymization. To be more concretely, we propose *six anonymization rules* to guide the data anonymization practice:

(1) Data anonymization should conform to the individual-specific privacy preferences including the privacy protection degree and anonymization granularity;
(2) Data anonymization should base on data characteristics, including data format, semantic structure, value domain, and distribution in both technique and social viewpoints;
(3) Data anonymization should prefer the more-available-data choice when facing multiple anonymization plans;
(4) Data anonymization should balance the anonymity degrees on different values in a record, avoiding non-proportioned data anonymization (i.e., make some data loss much information while others loss little);
(5) Data anonymization priority on different values in a record may base on the ascend-

ing order of information capacity on them, or conform to individuals' preferences;

(6) Data anonymization within the same attribute should conform to the same rules if no individual-specific requirements existed.

The above six anonymization rules illustrate a framework for data anonymity implementation in access-unrestricted data publication practice.

DATA ANONYMITY ANALYSIS

In this section, we formally analyze data anonymity, mainly focusing on *knowledge* definition, classification for privacy inference control, and the three anonymity levels on data anonymization.

Level 1 Anonymity

Firstly, we define the notion *anonymity schema* of an anonymity table.

Definition 6 (Anonymity schemaS^A) The *anonymity schema* S^A is a *schema* used for the construction of an anonymity table on T, like <*UI, QI*, β_{AD}, *SA*, β_{DD^*} >, where $\beta_{AD} = \{\beta_{AD}, \beta_{AD}, \ldots, \beta_{AD}\}$; $\beta_{DD^*} = \{\beta_{DD^1}, \beta_{DD^2}, \ldots, \beta_{DD^s}\}$; $s=|SA|$. $\beta_{AD}(1 \le i \le s)$ is the threshold value of *AD* on *QI* mapping to SA^i, β_{DD^i} is the threshold value of *DD* on SA^i.

In S^A, $\beta_{AD}$and $\beta_{DD_i^j}$ are individual-specific *attribute-level* thresholds on qi_i and sa_i^j, indicating the specific privacy protection level on them, which can be seen as a kind of personalized privacy preference. Suppose $\overline{UI}$, $\overline{QI}$, and $\overline{SA}$ are the anonymity forms of *UI*, *QI*, and *SA*, respectively; $\sum(\overline{SA})$ is the *power set* of $\overline{SA}$ except the empty set. Based on S^A, the anonymity table on T for publication is defined as follow.

Definition 7 (Anonymity TableT^A) The *anonymity table* T^A is an anonymized table constructed based on both T and S^A, like < <$\overline{QI}$, $\overline{SA}^{(*)}$ > >, where (*)$\in \sum(\overline{SA})$.

In T^A, *AD* on $\overline{qi}_i$ mapping to $\overline{sa}_i^j$, as AD_i^j, and *DD* on $\overline{sa}_i^j$ mapping to $\overline{qi}_i$, as DD_i^j, can be calculated as follows:

(1) $AD_i^j = AD_T(qi \prec SQ_{T^A}(\overline{sa}^j))$; (2) $DD_i^j = DD_{T^A}(QS_{T^A}(QCluster_i, j))$.

Definition 8 (Level 1 Anonymity)T^A is of *Level 1 anonymity*, if and only if they hold ($1 \le i \le n$, $1 \le j \le s$): (1) $AD_i^j > \beta_{AD_i^j}$; (2) $DD_i^j > \beta_{DD_i^j}$.

From a data publisher perspective, individual record anonymity includes three steps: concealing *UI*, anonymizing *QI* and clustering *SA*. So the privacy protection level is indicated by *AD* and *DD*. Through setting the thresholds on them, the goal can be achieved by the data anonymization process. While in users' viewpoint, the above *Level 1 anonymity* definition can be expressed in the following way.

Definition 9T^A is of *Level 1 anonymity*, if and only if they hold hold ($1 \le i \le n$, $1 \le j \le s$): (1) $P_{T^A}(\overline{ui}_i \mapsto \overline{sa}_i^j) \le \beta_{\overline{us}_i^j}$; (2) $P_{T^A}(\overline{sa}_i^j \mapsto \overline{ui}_i) \le \beta_{\overline{su}_i^j}$. The above conditions can be expressed in a computable manner:

(1) $P_{T^A}(\overline{ui}_i \mapsto \overline{qi}_i^j) \le \beta_{\overline{uq}_i}$; (2) $P_{T^A}(\overline{qi}_i \mapsto \overline{sa}_i^j) \le \beta_{\overline{qs}_i^j}$;

(3) $P_{T^A}(\overline{sa_i}^j \mapsto \overline{qi_i}) \leq \beta_{\overline{sq_i}^j}$; (4) $P_{T^A}(\overline{qi_i} \mapsto \overline{ui_i}) \leq \beta_{\overline{qu_i}}$.

Generally, individuals do not need know the detail anonymity policy on *T*, but they care about inference relations and probabilities on the disclosure of their sensitive information (i.e., *UI-SA mappings*). So they need to define the threshold on these inference probabilities as extra constraints for data anonymization.

Theorem 1 To create a usability-optimal T^A of Level 1 anonymity is *NP-hard*.

Proof Sketch: T^A is a set of sub-tables: $\{<\overline{UI}, \overline{SA}^j> \mid 1 \leq j \leq s, |SA| = s\}$. Records in $T^j = <\overline{UI}, \overline{SA}^j>$ are divided into a set of $QCluster_i$-$SCluster_i^j$ mappings. Suppose $\overline{qi_i}$-$\overline{sa_i}^j$ is in $QCluster_i$-$SCluste_i^j$, according to Definition (9), $P_{T^A}(\overline{qi_i} \mapsto \overline{sa_i}^j) \leq \beta_{\overline{qs_i}^j}$ and $P_{T^A}(\overline{sa_i}^j \mapsto \overline{qi_i}) \leq \beta_{\overline{sq_i}^j}$ hold. If we define: $\forall i: \beta_{\overline{qs_i}^j} = \frac{1}{k}$, $\beta_{\overline{sq_i}^j} = 1$, this problem is equivalently transformed to a general optimal k-anonymity problem, which has been proven as a *NP-hard* in (Meyerson & Ryan, 2004) and (Aggarwal et al., 2005). While, if we define: $\forall i: \beta_{\overline{qs_i}^j} = \frac{1}{k}$, $\beta_{\overline{sq_i}^j} = \frac{1}{\ell}$, then it is changed to be an optimal *ℓ-diversity* or an *optimal (k, ℓ)-anonymity* problem, which are discussed in our previous chapter (Li et al., 2006a). So the statement holds.

KNOWLEDGE-BASED PRIVACY INFERENCE

If one of these threshold constraints defined in S^A is not satisfied, a *privacy inference violation* may occur.

Definition 10 *Privacy inference violation* is a kind of privacy disclosure, which occurs on T^A, if at least one of the following two conditions holds: (1) $P_{T^A}(\overline{ui_i} \mapsto \overline{sa_i}^j) \leq \beta_{\overline{us_i}^j}$; (2) $P_{T^A}(\overline{us_i}^j \mapsto \overline{ui_i}) \leq \beta_{\overline{su_i}^j}$. Or, being more practicable, it occurs if one of the following two conditions holds:

(1) $P_{T^A}(\overline{qi_i} \mapsto \overline{sa_i}^j) \leq \beta_{\overline{qs_i}^j}$; (2) $P_{T^A}(\overline{sa_i}^j \mapsto \overline{qi_i}) \leq \beta_{\overline{sq_i}^j}$.

In practice, if all of the above *AD* and *DD* thresholds are given in S^A, it is easy to implement the corresponding threshold constraint in data anonymization process. What we need to concern more is some special information in practical application context (usually the social environment) that is related to the original data and the anonymization process, which can incur potential privacy inference violations after data publication. This information used for privacy inference attacks is called (inference) *knowledge*.

Definition 11 (Knowledge) *Knowledge* is special information used to more precisely ascertain an inference relation in T^A.

Any information may have the possibility to become knowledge. For instances, knowledge may be anonymity principles, anonymity relations, attribute or value inference relations, even the structure of individual data and a $\overline{qi}$ - $\overline{sa}$ mapping on a record.

Proposition 1 T^A of *Level 1 anonymity* is *secure* from privacy inference violations, if *no knowledge* is available.

Proof: It is easy to prove the statement according to Definition (9).

Definition 12 A *knowledge-based (privacy) inference* relation refers to an inference relation with inference probability increased based on knowledge.

To various users, the knowledge scope may be different. It is obvious that knowledge may result in a privacy inference violation on T^A, named *knowledge-based (privacy) inference violation*. To avoid these violations, it needs to discover all possible knowledge available in access-unrestricted data publications. But it is hard to ascertain or suppose what knowledge the attacker may have. There are three main reasons to this statement: (1) Attackers having diverse knowledge are unpredictable; (2) Knowledge available for attackers cannot be covered fully since the application environment is changeable and complicated; (3) To prevent inference attacks is not always consistent with the anonymity principles. We conclude these features of knowledge as its *global unpredictability*.

While on a concrete table that is published with anonymity techniques conforming to the above anonymity principles and anonymization rules, we can capture most of knowledge used for knowledge-based inference attack, called *core knowledge*. Core knowledge on access-unrestricted data with tabular structure can be captured before the data anonymization process, because of the structure of data, anonymity principles and anonymization rules. Such a feature of knowledge is called *local coverability*.

Generally, there are four kinds of core knowledge on an anonymity table. We discuss privacy inference violation problems on T^A with the assumption of these four kinds of core knowledge one by one (labeled from (**I**) to (IV) in the following content).

- <*UI, QI*> in *T* or Ω

It is obvious this type of knowledge is indispensable for privacy disclosure on access-unrestricted data publication with anonymity. Mostly, attackers may not know the whole <*UI, QI*> in *T* or Ω, but they always capture a part of it, which is enough as the base of some privacy inference attacks and violations.

- Value domains and anonymization rules

If users can capture the *QI* value domains in T^A, to infer the original form from the anonymity form may be easier (see Equation (2)). While if users also capture anonymization rules (discussed above), it is more convenient for them to make inference violations, as the anonymization rules reveal the implementation details to some extent, such as rules of choosing the most available anonymity forms.

- Attribute/value inference relations

Some classical association rule algorithms can be used to discover most attribute/value inference relations, such as the *Apriori* algorithm (Han, 2001). Now we take two instances to explain how to use such knowledge for inference attacks. Suppose in a *QCluster-SCluster* mapping, if exist one or more high-precise value inference relations, the probabilities on other inference relations can be easily increased consequently. For instance, suppose exist $\{A\}$-$\{e\}$ on *QI-SA*^{i} and $\{A, B\}$-$\{a, b\}$ on *QI-SA*^{j} (I ≠ j), if $P(e \mapsto a |_{\Omega}) = 100\%$, then we can infer $\{A\}$-$\{a\}$ and $\{B\}$-$\{b\}$ hold on *QI-SA*^{j}. Further, we suppose a *QCluster-SCluster* mapping, $\{A, B, C\}$-$\{a, b, c\}$, indicates there exist three inference relations from $\{A, B, C\}$ to $\{a, b, c\}$. If there is no knowledge available, we can subjectively ascertain: $P(A \mid B \mid C \mapsto a \mid b \mid c) = \frac{1}{3}$. But if exists an inference relation *ir*: $A \mapsto$ a $[\frac{2}{3}]$ (i.e., $P(A \mapsto a) = \frac{2}{3}$), then we can derive based on knowledge *ir*: $P(A \mapsto b \mid c \,||\, ir) = \frac{1}{6}$, $P(B \mid C \mapsto a \,||\, ir) = \frac{1}{6}$ and $P(B \mid C \mapsto b \mid c \,||\, ir) = \frac{5}{12}$, where "|| *ir*" indicates the inference relation holds based on *ir*. We define this kind of inferences as *conditional inference*.

Definition 13 *Conditional inference* relation refers to an inference relation with more precise probability based on other inference relations. The probability on a conditional inference relation is named *conditional probability*.

Based on some known inference relations, denoted as $\{ir\}$, to infer a new inference relation *ir'* can be written as:

$$ir' \,||\, \{ir\} \tag{6}$$

Conditional inference is a special kind of knowledge-based inference, because of the special knowledge. Conditional inference probability is different from general conditional probability discussed in the probability theory. Its calculation is more complicated. As in the above example, the inference relation from the left side to the right side is a *more-to-more* relation with different inference probabilities. The probability on an inference relation used as knowledge can be derived from other kind of knowledge, such as value domain, attribute/value inference relations.

The real probability of an inference relation is never exactly known in T^A, since cluster mapping is *more-to-more* (i.e., it conceals the *one-to-one* inference relation information between two single elements). To infer on a conditional inference relation is a process towards the exact value inference relation with full probability. It is known that each value inference relation in T is 100%, while in T^A, the probability should be decreased. This is the essence of anonymity being a way to prevent privacy disclosure, i.e., it can make the inference risk on T^A much lower. But if the conditional inference probability on an inference relation is too highly precise to be tolerated by individuals whose private information is recorded in T, it should be a kind of knowledge-based privacy inference violations too, although the probability is less than 100% (close to 100%, or over the predefined threshold).

It should be noted the criteria (i.e., inference probability threshold) to determine whether an inference relation is privacy-violated or not should be defined by individuals who are responsible for the records in T. An individual may define the threshold of inference risks requiring the data publisher to control the inference probability on his/her record under the threshold. On a *QCluster-SCluster* mapping in T^A, written as *Q-S*, they hold obviously:

$$\forall q \in Q : \sum_{s \in S} P_{T^A}(q \mapsto s) = 1;$$

$$\forall s \in S : \sum_{q \in Q} P_{T^A}(q \mapsto s) = 1.$$

$$\text{If} \exists q, S' : q \in Q, S' \subseteq S, \sum_{s \in S'} P_{T^A}(q \mapsto s) > \beta$$

(β is a threshold on the record containing q), we call it *aggregation effect* of conditional inference on *Q-S*. For instance, on the above example, if the knowledge is *ir*: $A \mapsto a\ [\frac{1}{100}]$ (i.e., $P_{T^A}(A \mapsto a) = \frac{1}{100}$), then we can derive: $P_{T^A}(A \mapsto \{b, c\} \,||\, ir) = \frac{99}{100}$. It can be seen as an aggregation effect. Following the example, if exists $ir_1 : A \mapsto a\ [\frac{2}{3}]$, we can derive $ir_2 : A \mapsto b\ [\frac{1}{6}]$. But if exists $ir_3 : B \mapsto b\ [\frac{3}{4}]$, then we derive $A \mapsto b\ [\frac{1}{15}]$, and then $A \mapsto c\ [\frac{4}{15}]$, finally, $B \mapsto a\ [\frac{5}{64}]$. We call it a *domino effect* of conditional inference, since a known inference relation can impact all other inference relations in the same *Q-S* mapping.

Theorem 2 T^A of Level 1 anonymity is *not secure* from knowledge-based privacy inference violations if some of knowledge **(I-III)** exists.

Proof: Suppose on T^A: $\exists i, j$, $QCluster_i$-$SCluster_i^j$ holds. To prove this statement, we only need to prove the real *AD* on $QCluster_i$ or *DD* on $SCluster_i^j$ is decreased on knowledge-based inference relations. From the above discussion, knowledge (I) is the base for privacy inference attacks, knowledge (II) can decrease *AD* on $QCluster_i$, and knowledge (III) can make both *AD* on $QCluster_i$ and *DD* on $SCluster_i^j$ decreased in published table to incur knowledge-based privacy inference violations. So the statement holds.

Level 2 Anonymity

T^A of *Level 2 anonymity* can prevent privacy inference violations based on attribute/value inference relations, which is defined as follows:

Definition 14 (Level 2 anonymity) T^A is of *Level 2 anonymity*, if and only if the following two conditions hold: (1) T^A is of Level 1 anonymity; (2) $\forall i, \exists\{ir\}$ on $QCluster_i$-$SCluster_i^j$ as knowledge, then the following two conditions hold:

(2.1) $P_{T^A}(\overline{qi}_i \mapsto \overline{sa}_i^j \,||\, \{ir\}) \le \beta_{\overline{qs}_i^j}$; (2.2) $P_{T^A}(\overline{sa}_i^j \mapsto \overline{qi}_i \,||\, \{ir\}) \le \beta_{\overline{sq}_i^j}$.

Theorem 3 T^A of *Level 2 anonymity* is *secure* from knowledge-based privacy inference violations if kinds of core knowledge (I-III) exist.

Proof: According to Definition (14), to prove this statement, we need to transform knowledge (I-II) to inference relations forms for conditional inference attacks. First, knowledge of <*UI*, *QI*> can be represented as $\forall i, \exists ir : qi_i \mapsto ui_i[\frac{1}{V_T(qi_i)}]$. Second, knowledge of QI^j value domain can be represented as $\forall i, \exists ir : \overline{qi}_i^j \mapsto qi_i^j[\frac{V_T(qi_i)}{V_T(\overline{qi}_i)}]$. The above defined anonymization rules are essentially to judge and choose anonymity forms and anonymity relations based on information capacity, individuals' preferences, etc., which can be represented as inference relations too.

- Relationship among published tables

Suppose T^A with $|SA| > 1$. According to Definition (7), T^A should contain several sub-tables < $\overline{QI}$, $\overline{SA}^{(*)}$ >, $(*) \in \sum(\overline{SA})$, based on *T*. Through analysis of these sub-tables, some privacy inference violations may occur in them. Suppose $T^{(I)} = <\overline{QI}^{(I)}, \overline{SA}^{(I)}>$ and $T^{(J)} = <\overline{QI}^{(J)}, \overline{SA}^{(J)}>$, where $\overline{QI}^{(I)|(J)}$ is the corresponding $\overline{QI}$ mapping to $\overline{SA}^{(I)|(J)}$. $QU_{T^{(I)}}(\overline{QI}^{(I)}) = QU_{T^{(J)}}(\overline{QI}^{(J)})$. If $qi_i \prec \overline{qi}^{(I)} \in \overline{QI}^{(I)}, qi_i \prec \overline{qi}^{(J)} \in \overline{QI}^{(J)}$, s.t., $QU_{T^{(I)}}(\overline{qi}_i^{(I)}) = QU_{T^{(J)}}(\overline{qi}_i^{(J)})$, $\overline{qi}_i^{(I)} \neq \overline{qi}_i^{(J)}$, we derive $qi_i \in \overline{qi}_i^{(I)} \cap \overline{qi}_i^{(J)}$, $IC_T(\overline{qi}_i^{(I)} \cap \overline{qi}_i^{(J)}) \ge IC_T(\overline{qi}_i^{(I)}) \,|\, IC_T(\overline{qi}_i^{(J)})$, $(\overline{qi}_i^{(I)} \cap \overline{qi}_i^{(J)}) \le AD_{T^{(I)}}(\overline{qi}_i^{(I)}) \,|\, AD_{T^{(J)}}(\overline{qi}_i^{(J)})$.

Further there are two situations of privacy inference violations between (*I*) and (*J*): First, if $(I) \cap (J) \neq \phi$, or more simply, suppose $(I) \subseteq (J)$, then both $QCluster_i^{(I)}$-$SCluster_i^{(I)}$ in $T^{(I)}$ including $\overline{qi}_i^{(I)}$ - $\overline{sa}_i^{(I)}$ and $QCluster_i^{(J)}$-$SCluster_i^{(J)}$ in $T^{(J)}$ including $\overline{qi}_i^{(J)}$ - $\overline{sa}_i^{(J)}$ can be inferred as $QCluster_i^{(I)} \cap \mathrm{QCluster}_i^{(J)}$ - $SCluster_i^{(I)} \cap \mathrm{SCluster}_i^{(J)}$. So, *AD* on $\overline{qi}_i^{(I)} \cap \overline{qi}_i^{(J)}$ is decreased obviously, and *DD* should also be possible to be decreased since $|\,SCluster_i^{(I)} \cap \mathrm{SCluster}_i^{(J)} \le |\,|SCluster_i^{(I)|(J)}.|$

Second, if (I) $\cap$ (J) = ϕ, then it can be inferred: $QCluster_i^{(I)} \cap QCluster_i^{(J)}$ - {$SCluster_i^{(I)}$, $SCluster_i^{(J)}$}, i.e., $\overline{qi}_i^{(I)} \cap \overline{qi}_i^{(J)}$ - {$\overline{sa_i}^{(I)}, \overline{sa_j}^{(J)}$}. So its *AD* is decreased. Beside, there is a special situation on $\overline{SA}$: $\exists k,\ ir: \overline{sa_k}^{(I)} \mapsto \overline{sa_k}^{(I)}$ with enough precision, suppose $\overline{qi}_k^{(I)} \mapsto \overline{sa_k}^{(I)}$ in $QCluster_k^{(I)}$-$SCluster_k^{(I)}$ and $\overline{qi}_k^{(J)} \mapsto \overline{sa_k}^{(J)}$ in $QCluster_k^{(J)}$-$SCluster_k^{(J)}$, if $DD_{T^{(I)}}(SCluster_k^{(I)}) \le DD_{T^{(J)}}(SCluster_k^{(J)})$, then $P_{T^{(J)}}(\overline{qi}_k^{(J)} \mapsto \overline{sa_k}^{(J)}) \le P_{T^{(J)}}(\overline{qi}_k^{(J)} \mapsto \overline{sa_k}^{(J)} \| ir)$ holds, which indicates a privacy inference violation to some extent.

Theorem 4 T^A of Level 2 anonymity is *not secure* from knowledge-based privacy inference violations when knowledge (IV) exists.

Proof: Suppose on T^A: $\exists i, I, J, QCluster_i^{(I)}$-$SCluster_i^{(I)}$ and $QCluster_i^{(J)}$-$SCluster_i^{(J)}$ hold. To prove this statement, we need to prove the real *AD* on $QCluster_i^{(I)}$ or *DD* on $SCluster_i^{(I)}$ is decreased when $QCluster_i^{(J)}$-$SCluster_i^{(J)}$ exists. While the above discussed three situations have explicitly proven the statement.

Level 3 Anonymity

Definition 15 (Level 3 Anonymity) T^A is of *Level 3 anonymity*, if and only if the following two conditions hold: (1) T^A satisfies the Level 2 anonymity requirement; (2) $\forall (I) \in \sum(\overline{SA})$, $T^{(I)} \subseteq T^A$, *i*: $P_{T^{(I)}}(\overline{qi}_i^{(I)} \mapsto \overline{sa_i}^{(I)} \| \{\overline{qi}_i^{(J)} \mapsto \overline{sa_i}^{(J)}, \overline{sa_i}^{(J)} \mapsto \overline{sa_i}^{(I)} | (J) \in \sum(\overline{SA}),\ T^{(J)} \subseteq T^A\}) \le \beta_{\overline{qs}^{(I)}}$.

Proposition 2 T^A of Level 3 anonymity is *secure* from knowledge-based privacy inference violations when multiple relative tables are published.

Proof: It is obvious by Definition (15).

The above four kinds of core knowledge can be composed for more complicated knowledge-based privacy inference attacks or violations. The *Level 1 anonymity* can be satisfied easily through general data processing, but its strength on privacy enhancing is not enough. The *Level 3 anonymity* is hard to achieve, because the relationships between relative tables is hard to predict during the data anonymization process. But its strength on privacy enhancing is enough to prevent most of existed observed privacy inference violations.

DATA ANONYMITY VS. PRIVACY INFERENCE

In this section, we take k-anonymity as a scenario model for illustrating knowledge-based privacy inference attacks, violations, and anti-inference policies on k-anonymized tabular datasets.

Modeling K-Anonymity

An individual that is represented as a record in a database might be re-identified by joining the released data with another public database. To reduce the risks of this privacy inference attacks type, *k-anonymity* is proposed (Samarati & Sweeney, 1998; Sweeney, 2002b).

In practice, k-anonymity is to anonymize table *T* with <*UI, QI, SA*> to table T^K with <$\overline{QI}$, *SA*>. T^K is a k-anonymized version of *T* for publication, if T^K satisfies the following k-anonymity definition.

Definition 16 (K-Anonymity) T^K satisfies k-anonymity, if for every record t_1, exist *k-1* other records $t_2, \ldots, t_k \in T^K$, satisfying $t_1[\overline{QI}] = t_2[$

$\overline{QI}$]=…=t_k[$\overline{QI}$], where t_i[$\overline{QI}$] ($1 \le i \le$ k) denotes the projection of t_i on $\overline{QI}$.

K-anonymity is just to generalize or suppress *QI* values for anonymity with the static predefined table-level *k* constraints that are specified by data publishers.

As a set of original individual microdata, Table 1 shows a set of health-care records from a fictitious hospital, in which *UI* = {*Name*}, *QI* = {*BirthDate, Sex, Zipcode*}, *SA* = {*Disease*}. We firstly remove the *Name* (identity) attribute column and then use k-anonymity on Table 1 with *data generalization* approach (Sweeney, 2002a) and derive a 2-anonymized table (See Table 2, * denotes a suppression of the corresponding data value, B.Cancer stands for Breast Cancer).

$$UI \underbrace{\xleftarrow{P_T(QI \mapsto UI)}}_{In\ T} QI \underbrace{\xleftarrow{P_T(\overline{QI} \mapsto QI)}}_{Anonymization} \overline{QI} \underbrace{\xrightarrow{P_{T^K}(\overline{QI} \mapsto SA)}}_{In\ T^K} SA \qquad (7)$$

Table 1. A table of health-care data

	Name	BirthDate	Sex	Zipcode	Disease
1	Lucy	11-12-39	F	13068	Flu
2	Lily	11-02-39	F	13068	Flu
3	Alice	08-24-57	F	14092	B.Cancer
4	Bob	08-02-57	M	13053	Stoke
5	Frank	08-02-42	M	13053	Stoke
6	Jack	11-22-42	M	13053	No
7	Michael	07-25-42	M	13053	AIDS

Table 2. A 2-anonymized table

	BirthDate	Sex	Zipcode	Disease
1	11-*-39	F	13068	Flu
2	11-*-39	F	13068	Flu
3	08-*-57	*	1****	B.Cancer
4	08-*-57	*	1****	Stoke
5	*-*-42	M	13053	Stoke
6	*-*-42	M	13053	No
7	*-*-42	M	13053	AIDS

$$P_{T\&T^K}(UI \leftrightarrow SA) = P_T(QI \mapsto UI) \times P_T(\overline{QI} \mapsto QI) \times P_{T^K}(\overline{QI} \mapsto SA) \qquad (8)$$

The principle for k-anonymity is to prevent individual re-identification, and ultimately, to prevent the successful (privacy) inference attack on *UI-SA* matching in T^K after publication. A k-anonymized table T^K is of *Level 1 anonymity* if and only if the two conditions holds (on the *i*th record): (1) $P_T(\overline{qi}_i \mapsto qi_i) \le \beta_{qq}$, and (2) $P_{T^K}(\overline{qi}_i \mapsto sa_i) \le \beta_{qs}$.

The k-anonymization process can be modeled as Formula (7), mainly including three inference relations, *UI-QI*, *QI*-$\overline{QI}$, and $\overline{QI}$-*SA*. With the above theoretical foundation, we can conceptually describe the generic k-anonymity model in a measurable view (See Formula (8)):

First, $P_\Omega(QI \mapsto UI)$ is calculated within Ω. For instances, in Cambridge, Massachusetts in 1997, the following hold (Ω may be the whole Cambridge city, *UI* = {*Name, Address*}, *QI* = {*BirthDate, Sex, Zipcode*}): P_Ω({*BirthDate, Sex, Zipcode*}↦{*Name, Address*}) ≥ 0.97 (Sweeney, 1997). To make it more significant and operable in practice, we define Ω = *T*. It is reasonable according to Equation (2). Within the *i*th record of T, $P_T(qi_i \mapsto ui_i)$ can be calculated as follows:

$$P_T(qi_i \mapsto ui_i) = \frac{1}{|V_T(qi_i)|} \qquad (9)$$

In *T*, $V_T(qi_i)$ equals to the set of records with the same value on *QI* to qi_i. $P_T(qi_i \mapsto ui_i)$ is the re-identification probability in *T*. For instances, as in the above Table 1, *P*({*Sex=F*}↦{*Name=Lucy*}) $= \frac{1}{3}$, *P*({*Zipcode*=13053} ↦ {*Name=Frank*}) =

$\frac{1}{4}$, and $P(Sex \mapsto \text{Name}) = \frac{2}{7}$. We can define: $P_T(QI \mapsto UI) = \frac{1}{n} \times \sum_{i=1}^{n} P_T(qi_i \mapsto ui_i)$. As in Table 2, $P(\{Sex, Zipcode\} \mapsto Name) = \frac{7}{12}$, and $P(\{BirthDate, Sex, Zipcode\} \mapsto Name) = 1$. As $\overline{qi_i}$ is an anonymity form of qi_i in *T*. On $qi_i \prec^* \overline{qi_i}$, the inference probability $P_T(\overline{qi_i} \mapsto qi_i)$ is calculated:

$$P_T(\overline{qi_i} \mapsto qi_i) = \frac{|V_T(qi_i)|}{|V_T(\overline{qi_i})|} \tag{10}$$

$P_T(\overline{QI_i} \mapsto QI_i)$ is the mean of $P_T(\overline{qi_i} \mapsto qi_i)$ on the whole *T* scope. It is noted the above formula is to calculate the real inference probability when users capture knowledge *QI* value domains in *T*. For instances, $P(BirthDate{=}11\text{-}*\text{-}39 \mapsto BirthDate{=}11\text{-}12\text{-}39) = \frac{1}{2} \neq \frac{1}{30}$, $P(\{Sex{=}*, Zipcode{=}1****\} \mapsto \{Sex{=}F, Zipcode{=}14092\}) = \frac{1}{7}$. As $\overline{qi_i}$ in T^K has at least *k-1* duplicated tuples, $P_{T^K}(\overline{qi_i} \mapsto sa_i)$, can be calculated on the $SCluster_i$ size. $P_{T^K}(\overline{QI} \mapsto SA)$ is its mean value in scope T^K.

$$P_{T^K}(\overline{qi_i} \mapsto sa_i) = \frac{|V_{SCluster_i(sa_i)}|}{|SCluster_i|} \tag{11}$$

As in Table 2, $P(\{BirthDate{=}08\text{-}*\text{-}57, Sex{=}*, Zipcode{=}1****\} \mapsto Disease{=}Stoke) = \frac{1}{2}$, $P(\{BirthDate{=}*\text{-}*\text{-}42, Sex{=}M, Zipcode{=}13053\} \mapsto Disease{=}Stoke) = \frac{1}{3}$. It is obvious that $P_{T^K}(\overline{qi_i} \mapsto sa_i)$ is relative with $DD_{T^K}(SCluster_i)$, see Definition (5) for detail. The purpose for k-anonymity is to make $P_{T\&T^K}(ui_i \leftrightarrow sa_i)$ less than a predefined threshold β_i. Based on the analysis in the above section, the way to reach it is to keep each "middle" inference probability between ui_i and sa_i under a threshold, including $P_T(\overline{qi_i} \mapsto qi_i)$ and $P_{T^K}(\overline{qi_i} \mapsto sa_i)$.

Privacy Inference Attacks

As the above analyzed, when no knowledge considered, general k-anonymity models can make them under a controllable scope except some special situations (defined as *simple privacy inference attacks*). But when considering some knowledge, some potential privacy inference violations may occur, defined as *knowledge-based inference attacks*.

Due to the defects of practical k-anonymity algorithm, simple privacy inference attacks are existed to re-identify individual identity on k-anonymized *QCluster* tuples, or to infer some *QCluster-SCluster* matches with enough precision. In short, it is to discover some information that should be concealed by k-anonymization. For instance, in Table 2, it is easily discovered that the exact *Disease* value on the 1st record mapping to <11-*-39, F, 13068> is *Flu*, because of the mapped $SCluster_1$ just contains this single element, which incurs $P_{T^K}(\overline{qi_i} \mapsto sa_i) = 1$. The main reason is the diversity degree of $SCluster_1$ is not enough to prevent inferring an exact value in it. To avoid this attack, we can define a threshold β_{DD}, and take $DD(SCluster) \geq \beta_{DD}$ as an anonymization rule effectively in the k-anonymization process (similar to ℓ-*diversity* (Machanavajjhala et al., 2005) and (*k*, ℓ)-anonymity in (Li et al., 2006a). In short, although simple privacy inference attacks are caused by some anonymization defects, they can be easily discovered and prevented when adjusting the k-anonymity algorithm to consider these special requirements.

According to the core knowledge discussion in the above section, we propose three kinds of core knowledge on a k-anonymized tabular microdata set: (1) *QI* value domains in *T*. In a vertical view, *QI* value domains should be helpful for inferring *qi* from its anonymity form $AF_{D_{qi}}(qi)$; (2) Value inference relations in *T*. In a horizontal view, value inference relations with precise probability maintained in T^K can be available for increase the precision on other inference relations dominated in same *QCluster-SCluster* mapping; (3) Anonymity rules and relations.

We divide knowledge-based privacy inference attacks on k-anonymized table into three classes according to the above three kinds of core knowledge.

1. *Value inference attack under value domains known*, which is to infer the exact value on an attribute when knowing *QI* value domains;
2. *Conditional inference attack under value associations known*, which is to infer the more precise value associations when knowing some value associations;
3. *QI tuple inference attack under anonymization rules and relations known*, which is to infer the original or more specific *QI* tuples when knowing some anonymization rules and relations.

For the first class, it is obvious that the probability of inferring *v* in D_v is much larger than that in Ω, as Equation (2). The essential reason for this type attacks is $D_v \subseteq \Omega$ and D_v is tractable in most situations by users. For example, suppose the attackers have known: $D_{BirthDate}$ = {11-12-39, 11-02-39, 08-24-57, 08-02-57, 08-02-42, 11-22-42, 07-25-42} in *T*. The following inference probabilities (on *BirthDate* in T^K) hold to the attacker: $P(11\text{-}*\text{-}39 \mapsto 11\text{-}12\text{-}39 | 11\text{-}02\text{-}39) = \frac{1}{2} \neq \frac{1}{30}$; $P(*\text{-}*\text{-}42 \mapsto 08\text{-}02\text{-}42 | 11\text{-}22\text{-}42 | 07\text{-}25\text{-}42) = \frac{1}{3} \neq \frac{1}{12 \times 30}$. In short, to know the value domain on an attribute is helpful to narrow the scope used to infer the exact original value.

For the second class of knowledge-based inference attacks, to capture all value inference relations with enough precision among two or several attributes is hard. Same to destroy or conceal all of them in all *QCluster-SCluster* mappings in T^K, as it needs the anonymization process to find them while they are not appeared automatically in data anonymization. For example, it is hard to discover the value inference relation *Disease=B. Cancer* ↦ Sex=F [100%] in {08-*-57, *, 1****}-{*B.Cancer, Stoke*}. Even if it is discovered before anonymization, it is hard to guarantee its inference probabilities in the matching smaller than the threshold (as in this instance, the inference probability is inherited high unless the anonymized *Sex* value in *QCluster* is *F*).

For the third class of knowledge-based inference attacks, since anonymization rules and relations for k-anonymity models are always static and simple, they can be easily captured by attackers. For example, it is easy to know the k value, the generalization hierarchy or some suppression techniques on some attributes used for the anonymization process through observing and analyzing the published microdata.

If attackers capture several kinds of knowledge for inference, such as capturing the value domain on an attribute, some value inference relations, and some anonymization rules, etc., the inference attacks may be more complicated to be prevented. For instance, if $D_{Zipcode}$ and the anonymization rules on it are known, it can be inferred the original values from 1**** are {14092, 13053}, but not 13068, since there are just two 13068 in the value domain, and they have been appeared in T^K, i.e., $P_T(1^{****} \mapsto 14092|13053) = \frac{1}{2}$. So if the attacker knows *Disease=B.Cancer* $\mapsto$ *Sex=F* [100%] and P_T *(Zipcode=*14092 $\mapsto$ *Name=Alice)* = 1, then P_{T^K} *(Disease=B.Cancer* $\mapsto$ *Name=Alice)* $= \frac{1}{2}$ holds. With similar knowledge, the attacker can infer: $P_{T\&T^K}$ *(Name=Lucy|Lily* $\mapsto$ Disease=Flu) $= \frac{1}{2}$.

Anti-Inference Privacy Policies

K-anonymity takes an anonymity cost metric to measure the information loss on anonymity forms of values, records, and the whole table. So the anti-inference capability on k-anonymized table is tightly related with such an anonymity cost metric. The inference probability from $\overline{qi}$ to qi reflects the specific information loss on $\overline{qi}$ than qi, and ultimately, influences the data usability. The lower of the information loss on k-anonymized data, the higher of its usability, to attackers, the smaller of the inference probability on an anonymity relation, the larger of the anonymity cost on the anonymity form.

Together the above analysis, we propose two anti-inference policies for k-anonymization on microdata:

- *Policy 1*: Take $P_T(\overline{qi}_i \mapsto qi_i)$ and $P_{T^K}(\overline{qi}_i \mapsto sa_i)$ in the anonymity cost metric: Greatly increase the metric value when the inference risk is enough high, while keep it uninfluenced when the risk is tolerable.
- *Policy 2*: Require anonymity relations satisfy $P_T(\overline{qi}_i \mapsto qi_i) \leq \beta_{qq}$ and $P_{T^K}(\overline{qi}_i \mapsto sa_i) \leq \beta_{qs}$ constraints before measuring the anonymity cost.

For convenience, we define *Policy 0* as the anonymity cost metric used in general k-anonymity models, which is to compute data anonymity cost based on its generalization height or the height ratio in the data generalization hierarchy. In the following subsection, we will implement these three policies on an individual mircodata set for explicitly describing the defects of *Policy 0* on knowledge-based privacy inference violations and illustrating the anti-inference capability of both *Policy 1* and *Policy 2*.

Experimental Study

Same to (Lefevre et al., 2005), the initial microdata set in the experiment is the Adult database supported by UCI (UCI, 2004). We adopt the training dataset containing 4 *QI* attributes (*Age, Sex, Education, and Native country*) and 2 *SA* attributes (*Salary class and Occupation*).

As the above discussed, the core knowledge available for privacy inference attacks on k-anonymized microdata include value domains, value associations, and anonymization rules and relations. We suppose the population is the Adult Database (*T*). Beside, we take an *Apriori* algorithm to create a set of value inference relations in *T* with high inference probability (≥ 90%). There are two purposes of the experiments:

(1) To discover potential existed privacy inference violations on general k-anonymized microdata without special anti-inference consideration (i.e., Policy 0), and
(2) To prevent these behaviors by taking our proposed anti-inference policies (i.e., Policy 1 and Policy 2).

Table 3 describes some useful information derived from experiments. $p^0_{s_0}$ - $p^0_{s_3}$, $p^1_{s_0}$ - $p^1_{s_3}$, and

$p^2_{s_0}$ - $p^2_{s_3}$ stand for $P_{T\&T^K}(UI \leftrightarrow SA)$ in *Policy 0, Policy 1*, and *Policy 2* under four kinds of different knowledge assumptions: under no knowledge (s_0), knowledge of value domains (s_1), value domains and anonymization rules and relations (s_2), and all core knowledge (s_3), respectively, in a 2-anonymized table (T^K) when β_{us}= 0.1, 0.2, 0.3 (for all records in T^K).

Each data item in Table 3 is pair of average inference probability and the amount of successful inference attacks (with inference probability larger than β_{us}). For instances, 0.185(238) indicates there are privacy inference violations existed on 238 records in $T^K$with average inference probability 0.185 when the predefined table-level threshold β_{us} is 0.1. From Table 3, we can derive:

(1) There are large amount of privacy inference violations potentially existed on T^K, as there exist 1461 records may occur privacy disclosure when users utilize all core knowledge;
(2) Both *Policy 1* and *Policy 2* are effective for preventing these behaviors.

Table 3: Anonymization result when k = 2

(k, β_{us})	(2, 0.1)	(2, 0.2)	(2, 0.3)
$p^0_{s_0}$	0.185(238)	0.284(74)	0.406(16)
$p^0_{s_1}$	0.186(339)	0.282(114)	0.433(20)
$p^0_{s_2}$	0.234(1292)	0.285(863)	0.476(135)
$p^0_{s_3}$	0.241(1461)	0.298(940)	0.489(194)
$p^1_{s_0}$	0(0)	0(0)	0(0)
$p^1_{s_1}$	0(0)	0(0)	0(0)
$p^1_{s_2}$	0.207(82)	0.261(47)	0.5(2)
$p^1_{s_3}$	0.206(83)	0.261(47)	0.5(2)
$p^2_{s_0}$	0(0)	0(0)	0(0)
$p^2_{s_1}$	0(0)	0(0)	0(0)
$p^2_{s_2}$	0.325(59)	0.418(43)	0.491(26)
$p^2_{s_3}$	0.326(61)	0.418(43)	0.491(26)

The amount of privacy-violated records in T^K with either of two anti- inference policies is greatly decreased under four kinds of knowledge assumptions. While, we also need to interpret that the reason of there still exist some privacy inference violations is the special properties of the original microdata set: several records are greatly distinguished from others, which make privacy inference much easier in 2-sized *QCluster-SCluster* when considering information maintenance on data anonymity.

When parameter k increases, the average relative inference probability ratios of inference probabilities under different knowledge (s_1, s_2, and s_3 as above) to that under *empty* knowledge (s_0), as $\frac{P_{s_{1|2|3}}(UI \leftrightarrow SA) - P_{s_0}(UI \leftrightarrow SA)}{P_{s_0}(UI \leftrightarrow SA)}$, are illustrated in Figure 1.

For instance, when k = 3, the average inference probability ratio with s_2 is 15 times larger than that with s_0. Along with k increases, such ratios with s_1, s_2, and s_3 to that with s_0 is decreased greatly. But we should note general k-anonymity applications prefer k = 2 or 3 (Aggarwal et al., 2004&2005), in Figure 1, is the most sensitive to knowledge-based privacy inference.

Figure 2 and Figure 3 illustrate such average inference probability ratio with s_1, s_2, and s_3 to that with s_0 on k-anonymized table with *Policy 1* and

Figure 1. The relative ratio of $P_{T\&T^K}(UI \leftrightarrow SA)$ *with Policy 0*

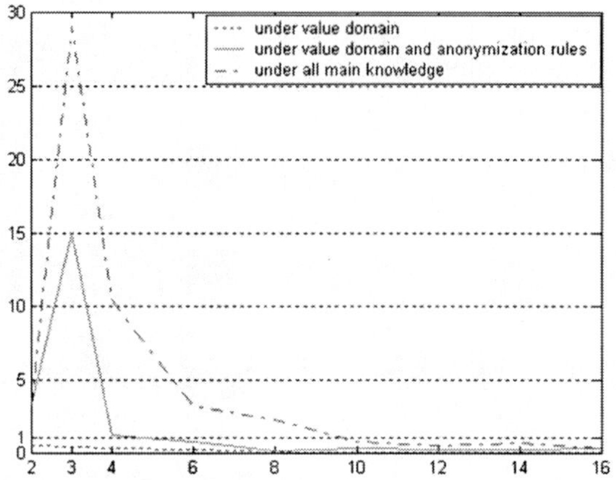

Policy 2. It is obvious that both policies make the ratio decreased much. Especially, *Policy 2* make it never larger than 1 (see Figure 3).

Figure 4 illustrates the $P_{T\&T^K}(UI \leftrightarrow SA)$ distribution of same records in 2-anonymized and 3-anonymized tables with the inference probability larger than β_{us} ($\beta_{us} = 0.1, 0.2, 0.3$).

Along with β_{us} increases, it is easy to observe that the amount of same records with $P_{T\&T^K}(UI \leftrightarrow SA)$ larger than β_{us} in both two anonymized microdata is decreased to some extent. But it can prove an important problem in this microdata set (may exist in other microdata sets): there exist some records with distinguished values comparing to others, i.e., $P_T(qi_i \mapsto ui)$, $P_T(\overline{qi_i} \mapsto qi_i)$, or $P_{T^K}(\overline{qi_i} \mapsto sa_i)$ on them are always much higher than others. It is an inherited knowledge available for incurring privacy disclosure in the *QCluster-SCluster* they are in after anonymization.

From Figure 4, even with different *k* for anonymity, privacy violations on these records are surely to occur. Such a problem also exists in the data anonymization process with our defined

Figure 2. The relative ratio of $P_{T\&T^K}(UI \leftrightarrow SA)$ *with Policy 1*

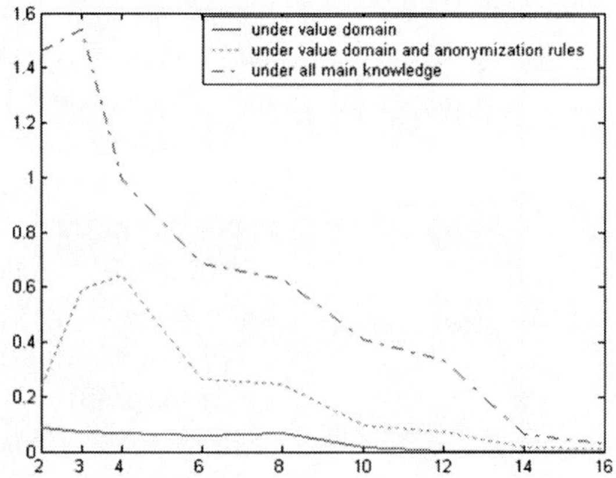

Figure 3. The relative ratio of $P_{T\&T^K}(UI \leftrightarrow SA)$ *with Policy 2*

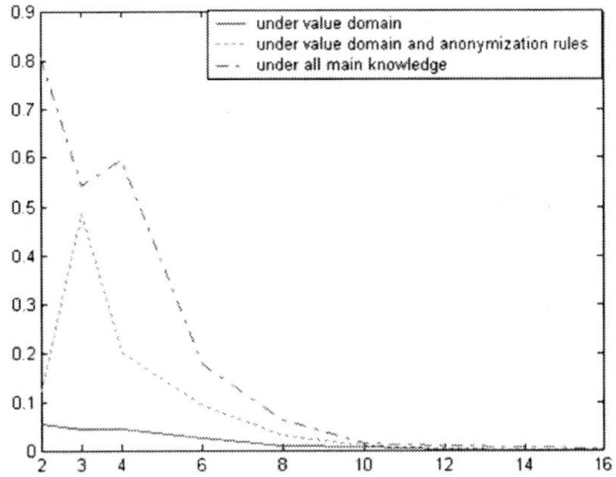

Figure 4. (β_{us} *= 0.1, 0.2, 0.3) on same records (*$\geq \beta_{us}$*) when k = 2 and k= 3*

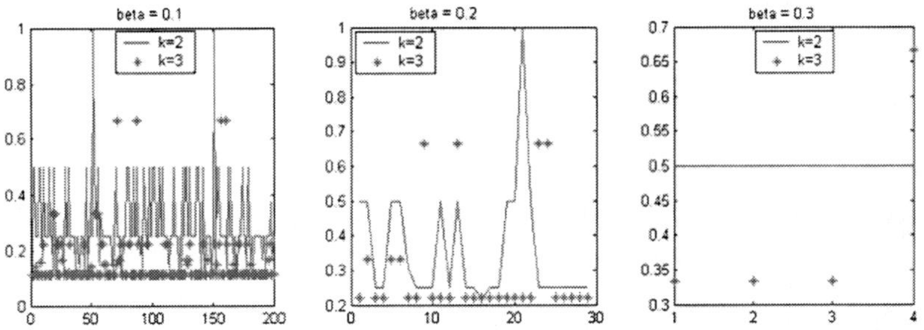

anti-inference policies. The only way to avoid it is to increase *AD* and *DD* (i.e., to increase the corresponding threshold values) on the corresponding *QCluster-SCluster* mapping, which in another side, may greatly decrease the information capacity on the relative data, which is not consistent with the information maintenance principle for anonymity.

CONCLUSION

In conclusion, we give a formal research on access-unrestricted data anonymity and the relative privacy inference disclosure control, mainly knowledge-based inference violation prevention. The core policy of anonymity is to generalize specific information on data for avoiding privacy inference disclosure. The essential idea to prevent diverse knowledge-based inference violations is to consider the related knowledge in anonymization before data publication. Through the experiments with k-anonymity model, privacy inference violations are explicitly illustrated and our two anti-inference policies are proven to be efficient for privacy inference control.

Future work will focus on the exploration of various relationships in relative tables, because these relationships may incur newly potential privacy violations.

ACKNOWLEDGMENT

I would like to thank Prof. Xiaojun Ye from Tsinghua University, Beijing, China.

REFERENCES

Adam, N., & Wortmann, J. (1989). Security-control methods for statistical databases: A comparative study. *ACM Computing Surveys*, *21*, 515–556. doi:10.1145/76894.76895

Aggarwal, C. C. (2005). On k-anonymity and the curse of dimensionality. In *Proc. of the 31st very large data bases conf. (VLDB'05)*. Trondheim, Norway.

Aggarwal, G., Feder, T., & etc. (2004). Anonymizing tables for privacy protection. Retrieved from http://theory.standford.edu/ rajeev/privacy.html.

Agrawal, R., Kiernan, J., Srikant, R., & Xu, Y. (2002). Hippocratic databases. In *Proc. of the 28th very large data bases conf. (VLDB'02)*. Hong Kong, China.

Anton, A. I., Bochini, D., & He, Q. (2004, May). The use of goals to extract privacy and security requirements from policy statements. In *Proc. of the 26th IEEE intl. conf. on software engineering*. Scotland, UK.

Byun, J. won, Bertino, E., & Li, N. (2005, Jun.). Purpose-based access control of complex data for privacy protection. In *Proc. of the 10th ACM symposium on access control models and technologies (SACMAT'05)*. Stockholm, Sweden.

Byun, J., Sohn, Y., Bertino, E., & Li, N. (2006) Secure anonymization for incremental datasets. In *Proc. of the 3rd VLDB Workshop on Secure Data Management*. Korea

Chen, B., LeFevra, K., & Ramakrishnan, R. (2007) Privacy skyline: Privacy with multidimensional adversarial knowledge. In *Proc. of the 3rd Int'l Conf. on Vary Large Data Bases (VLDB'07)*. Vienna, Austria

Crook, R., Ince, D., & Nuseibeh, B. (2005). On modelling access policies: Relating roles to their organisation context. In *Proc. of the 13th IEEE Int'l conf. on requirements engineering* (p.157-166). IEEE.

Dalenius, T., & Reiss, S. (1982). Data swapping: A technique for disclosure control. *Journal of Statistical Planning and Inference*, 6.

Domingo-Ferrer, J., Oganian, A., & Torra, V. (2002). Information-theoretic disclosure risk measures in statistical disclosure control of tabular data. In *Proc. of the 14th international conf. on scientist and statistical database management* (pp. 227-231). IEEE.

Farkas, C., & Jajodia, S. (2002). The inference problem: A survey. *SIGKDD Explorations*, *4*(2), 6–11. doi:10.1145/772862.772864

Han, J. (2001). *Data mining concepts and techniques*. Magan Kanfmann Publishers.

Lefevre, K., & DeWitt, J. D., & Ramakrishnan, R. (2005, June). Incognito: Efficient full-domain k-anonymity. In *Proc. Of the 24th ACM Intl. Conf. on Management of Data (SIGMOD'05). Maryland, USA.*

Li, Z., Zhan, G., & Ye, X. (2006a). Towards an anti-inference (k, l)-anonymity model with value association rules. In *Proc. Of the 17th Intl Conf. on Database and Expert System Applications (DEXA'06)*. Krakow, Poland.

Li, Z., Zhan, G., & Ye, X. (2006b). Towards a more reasonable generalization cost metric for k-anonymity. In *Proc. Of the 23rd British National Conf. on Database (BNCOD'06)*. Northern, Ireland.

Machanavajjhala, A., Gehrke, J., & Kifer, D. (2006). l-diversity: Privacy beyond k-anonymity. In *Proc. of the 22nd international conference on data engineering (ICDE'06).*

Martin, D., Kifer, D., Machanavajjhala, A., Gehrke, J., & Halpern, J. (2007). Worst-case background knowledge for privacy-preserving data publishing. In *Proc. of the 23nd international conference on data engineering (ICDE'07).*

Mcguckin, R. H., & Nguyen, S. V. (1990). Public use microdata: Disclosure and usefulness. *Journal of Economic and Social Measurement, 16*, 19–39.

Meyerson, A., & Williams, R. (2004). On the complexity of optimal k-anonymity. In *Proc. of the 23th ACM SIGMOD principles of database systems (PODS'04).* France.

Morgenstern, M. (1987, December). Security and inference in multilevel database and knowledge-based systems. [ACM.]. *SIGMOD Record, 16*(3), 357–373. doi:10.1145/38714.38752

Rastogi, V., Hong, S., & Suciu, D. (2007). The boundary between privacy and utility in data publishing. In *Proc of the 33rd Intl. Conf. on Very Large Data Bases (VLDB'07).* Vienna, Austria.

Samarati, P., & Sweeney, L. (1998). *Protecting privacy when disclosing information: K-anonymity and its enforcement through generalization and suppression. Technical Report.* SRI Computer Science Lab.

Sandhu, R., & Chen, F. (1998). The multilevel relational (MLR) data model. *ACM Transactions on Information and System Security, 1*(1), 93–132. doi:10.1145/290163.290171

Staddon, J. (2003, June). Dynamic inference control. In *Proc. of the 8th ACM SIGMOD workshop on research issues in data mining and knowledge discovery (DMKD'03)* (p. 94-100). San Diego, CA, USA: ACM.

Sweeney, L. (1997). Guaranteeing anonymity when sharing medical data, the datafly system. *Journal of the American Medical Informatics Association.*

Sweeney, L. (2002a). Achieving k-anonymity privacy protection using generalization and suppression. *Intl. Journal on Uncertainty, Fuzziness and Knowledge -based Systems, 10*(5), 571-588.

Sweeney, L. (2002b). K-anonymity: A model for protecting privacy. *Intl. Journal on Uncertainty, Fuzziness and Knowledge-based Systems, 10*(5), 557–570. doi:10.1142/S0218488502001648

Truta, T. M., Fotouhi, F., & Barth-Jones, D. (2003). Disclosure risk measures for microdata. *Proc. of the 15th Intl. Conf. on Scientific and Statistical Database Management* (pp. 15-22).

UCI. (2004). U.C. Irvine machine learning repository. Retrieved from http://www.ics.uci.edu/mlearn

Westin, A. F. (1976). *Privacy and freedom.* New York: Atheneum.

Winsborough, W. H., & Li, N. (2000). Safety in automated trust negotiation. In *Proc. of 2004 IEEE symposium on security and privacy.*

Xiao, X., & Tao, Y. (2006). Personalized privacy preservation. In *Proc. Of the 25th ACM Intl. Conf. on Management of Data. (SIGMOD'06).* Illinois, USA.

Xiao, X., & Tao, Y. (2007). m-Invariance: Towards privacy preserving re-publication of dynamic. In *Proc. Of the 26th ACM Intl. Conf. on Management of Data. (SIGMOD'07).* Beijing, China.

Yao, C., Wang, X. S., & Jajodia, S. (2005). Checking for k-anonymity violation by views. In *Proc. Of the 31st Very Large Data Bases Conf. (VLDB'05).* Trondheim, Norway.

Zhan, &Li, Y (2007). Capture inference attacks for k-anonymity with privacy inference logic. In *Proc. Of the 12th Intl Conf. on Database System for Advanced Applications (DASFAA'07).* Bangkok, Thailand.

ENDNOTE

[1] This chapter is an enhanced version of an earlier paper: Zude Li and Xiaojun Ye: On Access-Unrestricted Data Anonymity and Privacy Inference Disclosure Control. *International Journal of Information Security and Privacy*, *2*(4), 1–21.

Chapter 11
Designing Efficient Security Services Infrastructure for Virtualization Oriented Architectures*

Syed Naqvi
Senior Member IEEE

ABSTRACT

Virtualization technologies are emerging as a promising solution for managing the rapidly growing complexities of modern distributed ICT infrastructures. However, a mainstream operational concern for these virtualization oriented architectures is to provide efficient security services. Establishment of in-depth security services and trust relationships are the most desirable features for the effective functioning of these systems. This chapter presents a security architecture to address the comprehensive security needs of today's virtualization oriented architectures. The concept of virtualization of security services is introduced so as to have absolute freedom to choose the underlying security mechanisms. This concept of virtualization of security services is realized through distributed virtual engines that enable unification of security service calls according to requirements and not according to the underlying technologies. A configurable mechanism for the invocation of security services is proposed to address the security needs of different kinds of users. This approach permits the evolution of efficient security infrastructure with minimal impact on the resource management functionalities. In this way, users and resource providers can configure the security services according to their requirements and satisfaction level. The resulting extensible set of security services include both core security services and contemporary security services required for the modern virtualization oriented infrastructures.

INTRODUCTION

The concept of virtualization in information technology dates back to the development of programming language compilers that were designed to virtualize the object code. Later virtualization emerged as cutting-edge technology not only for cost reduction of IT operations and support but also for ultimate simplicity, flexibility and agility of the underlying infrastructures. Server virtualization and consolidation are now regarded

DOI: 10.4018/978-1-61692-000-5.ch011

as top cost containment strategies by the majority of data center managers (Symantec, 2007). However, these emerging virtualization-oriented architectures give birth to several challenges for its deployment including a lot of uncertainty as to how and where to implement security (Adhikari, 2008). Although, initially security issues were not seen as bottleneck for the widespread adoption of virtualization-oriented architectures as overall security concerns are outweighed by the cost savings and operational benefits (Dignan, 2008). However, now it is expected that more attacks on the IT resources will take place with the increase in the number of virtualization-oriented architectures; and these architectures are going to face non-classical threats model that will require novel protection mechanisms for assuring the smooth running of their security operation procedures (Lewis, 2008).

In the recent years, various research funding agencies emphasized the need for a comprehensive research efforts of building scientific and technical excellences in security, dependability and resilience of systems, services and infrastructures, whilst meeting demands for privacy and trust (EU-IST; NSF). The work presented in this chapter has got direct and indirect support from these research funding agencies.

PROPOSED ARCHITECTURE

Overview

In the large scale distributed systems, such as computational Grids, Clouds, etc., the need for efficient and secure data transportation over potentially insecure channels creates new security and privacy issues, which are exacerbated by the heterogeneous nature of the collaborating resources. Traditional security approaches require adequate overhauling to address these paradigms. The two-pronged approach proposed in (Naqvi, 2004) to address these security issues is elaborated in this section. The proposed model is called VIPSEC: Virtualized and Pluggable Security Services Architecture. In this model, first, the virtualization of security services provides an abstraction layer on top of the security infrastructure that harmonizes the heterogeneity of underlying security mechanisms. Second, the configurable/pluggable nature of various security services permits the users and resource providers to configure the security architecture according to their requirements and satisfaction level. This approach allows the security infrastructure to be developed with minimal impact on the resource management functionalities.

Since security implementations are more and more numerous and complex, it has become almost impossible for an inexperienced user to understand their meaning and especially how they should be used. Additionally, the heterogeneity of networks does not simplify the understanding and definition of a security system. Therefore, it is currently impossible to establish a security policy for a communication by using the low level properties of the different networks that are being crossed. The classical solution to this problem consists in setting up a secured high-level ciphered tunnel from end to end. This is acceptable in some situations, but it may not satisfy future evolutions of networks. The goal of virtualization is to reinstate security principles (transparency, responsibility, traceability, etc.), security objectives (integrity, availability, confidentiality, etc.), security policies (protection, deterrence, vigilance, etc.) and security functions (identification, authentication, access control, management of secret elements, privacy, etc.) in their rightful place. Virtualization aims at describing a policy and at refining it. Actually, a unique security policy cannot be implemented on several heterogeneous networks, architectures or environments. The current complexity of networks comes from the fact that on the one hand each element defines its own security policy in accordance with the security domain to which it pertains (a priori…), and on the other hand each security domain has its own security policy.

In the virtual paradigm, the policy of the element (wherever it may be) shall be merged with the policy of the domain to which it belongs. Then, this policy will be automatically implemented depending on the available security functions.

VIRTUALIZATION

The concept of virtualization in information technology finds its roots in the very earliest software. The first generation of digital computers employed the digital logics that dealt with the 0s and 1s (0: switch OFF ; 1: switch ON). Therefore, software programs consisted of 0s and 1s, and their output also consisted of 0s and 1s. This situation made computer programming very difficult. Then the compiler programs came into existence that let programmers work with English-like (high-level) languages such as COBOL. The compiler programs were designed to virtualize the object code by transforming the high-level instructions into the 0s and 1s object code understood by the digital computers. Graphical user interface (GUI) is another classical example of virtualization that provides virtual access to a system's underlying resources. In other words, the objective of virtualization technologies is to simplify the complexities of a system.

Virtualization in the Context of Security Architecture

The concept of virtualization of security services is needed to have the absolute freedom to choose the underlying security mechanisms. It could be extended to adapt country-specific legal requirements, population-based ethical issues, and the business-oriented interests. Moreover, virtualization could be used to achieve the best trade-off between security guarantees and processing capabilities.

A security services handler is shown in Figure 1. It depicts the capacity of absorbing the underlying heterogeneity of the security services by providing a uniform interface to the applications running on top of it.

Advantages of Virtualization

Usability is a decisive factor from a user's point of view (Kishimoto, 2003). Hence, security architecture must provide a way that the user merely needs to have technical knowledge of the underlying security infrastructure. Virtualization facilitates the development of flexible, custom-designed hierarchical security model that responds to the security needs. It dynamically delivers high performance and statistically consistent level of the accessibility to security services. Virtualization introduces an abstraction layer to manage the complexity by defining security rules for

Figure 1. Security Services Handler

the system that are applied whenever needed. In a virtualized security services environment, a user does not need to know which authentication mechanism (such as X.509 certificate, Kerberos ticket, etc.) he requires for accessing a remote node. If he is authorized to perform some operation(s) at the remote site then the security architecture will take care of the interpretation of credentials for the various sites.

Virtualization is a gimmick, an artifice enabling to keep the standards in place and improve them by providing additional functions, without, however changing anything in the existing setup. Virtualization makes it possible to coexist with the pre-existing structures that will be utilized by default, utilizing the diverted systems only by request. Virtualization also enables to act in the opposite way, by subordination to the additional functions by default, while relegating the normal and standard functions to exception status, somewhat like traffic relief using a bypass road on a holiday route on which those in seasonal migration naturally follow the added traffic signs to avoid traffic jams, while the natives follow the normal signs to travel locally. The existing architecture, still operating by default, is thus preserved while being outfitted with an add-on smart design and not with an add-on module. Virtualization enables slipping into any system irrespective of its architecture without defacing it. It is a guarantee of fluidity and upgradeability. If more ambitious, this approach is not incompatible and does not conflict with the previous approaches based on addition of specific devices (firewall, intrusion detection, specific cryptographic module) intended to take charge of certain security functions within a known domain or to perform a very specific and proprietary security function of a component (secure operating system, GSM security, etc.).

It is clear that large-scale virtualization is an important lock. If classical abstraction such as XML is rapidly deployable, the case of virtualization is different, as it requires installation of specific virtual machines for all of those heterogeneous environments. Universal types of virtualization (Java Virtual Machine, etc.) are usually heavy. The internal mechanism required to perform such virtualization may be quite voluminous (VPNs, distributed operating systems, etc.). The cost of this approach undoubtedly involves additional computing resources needed to implement the virtualization engines in all the existing hosts (terminals, servers, routers, gateways). Moreover, additional entities will ultimately have to be created to perform this virtualization by means of virtualization gateways between two widely different adjoining networks. Virtualization is abstract, but each time it is the implemented mechanism and not the concept that will have to prove itself. Virtualization must remain simple and effective.

PLUGABILITY

A *configurable/pluggable* mechanism is proposed for the invocation of security services to address security needs of the different kinds of users. This approach permits the evolution of security infrastructure with less impact on the resource management functionalities, which are still on the verge of evolution. Moreover, it permits the users and resource providers to configure the security architecture according to their requirements and satisfaction level. The set of these security services include core security services (such as authentication, authorization, identity mapping, audit, etc.) as well as contemporary security services (such as mobile access control, dynamic digital signature, etc.).

Pluggable Security Services (PSS)

Authentication Service is concerned with verifying proof of an asserted identity.

Authorization Service is concerned with resolving a policy based access control decision.

Identity Mapping Service provides the capability of transforming an identity that exists in one identity domain into an identity within another identity domain. This service is not concerned with the authentication of the service requestor; rather it is strictly a policy driven name mapping service.

Credential Conversion Service provides credential conversion between one type of credential to another type or form of credential. This service facilitates the interoperability of differing credential types, which may be used by services.

Policy Service is concerned with the management of policies. The policy service may be thought of as another primitive service, which is used by the authorization, audit, identity mapping and other services as needed.

Audit Service is responsible for producing records, which track security relevant events. The resulting audit records may be reduced and examined to determine if the desired security policy is being enforced.

Profile Service is concerned with managing service requestor's preferences and data which may not be directly consumed by the authorization service. This data will primarily be used by applications that interface with a person.

Privacy Service is primarily concerned with the policy driven classification of personally identifiable information (PII). Such a service can be used to articulate and enforce a privacy policy.

Encoded Communication Service is concerned with ciphering the data before it leaves the resource terminal. Participating nodes mutually agree on the encoding technique being employed to encrypt the data.

Nonrepudiation Service ensures that a party to a contract or a communication cannot deny the authenticity of their signature on a document or the sending of a message that they originated.

Encrypted Data Storage Service is concerned with ciphering the data before it is stored in a storage repository. Authorized retrieving entities are aware of the encoding technique being employed to encrypt the data.

Communication Channels Service facilitates lossless transfer of communication parameters to a stand-by or a secondary communication channel in the case of main communication channel failure.

Requirements for Configurable/ Pluggable Services

Definition of standard and flexible interfaces: To assure its proper functioning in various heterogeneous environments, PSS should define a set of standard, yet flexible, interfacing protocols. Most applications need not use these interfaces directly. Therefore, the PSS protocols interface should be exposed only to application developers interested in defining new protocols or in configuring them in novel ways.

Integration at application layer: To enable a user to invoke his desired set of security services in the beginning of a task session, PSS should offer its integration at application layer without requiring any low-level programming knowledge.

Coordinated invocation of Services: To assure secure links among the various nodes, there is a need for an identical set of invoked security services at all the ends. This coordination is important for flushing out any vulnerability that may be introduced due to mismatched services.

Usability by users and services: To assure the adaptability of the security architecture in the various operational situations, the security services invocation should be possible by not only the users but also by the appropriate computing services.

Simultaneous use of multiple services: To assure security in depth, various security services are needed simultaneously. Moreover, certain security services are dependent on other services; e.g., Invocation of Authorization service requires invocation of Authentication service. In such a situation, prerequisite services of a certain invoked service should be activated automatically.

Support for future enhancements: To maintain its usability, PSS should accommodate forthcoming enhancements to the Grid security infrastructure, such as allowing integration of semantic firewall, etc.

Optimization for various communication links: To ensure sustained communications, PSS should automatically adapt the current communication channel like wired network or wireless network (Bluetooth, 802.11, …).

Providing real-time invocation features: To make the security architecture flexible and adaptable to the needs of Grid users and services, real-time invocation and de-invocation of security services are indispensable.

Using standard programming interface: To allow the integration of security services with the various Grid systems, it should use standard programming interface. It will further enable the Grid programmers to extend it to meet their specific needs.

DESCRIPTION OF THE ARCHITECTURE

To meet the requirements outlined in the previous section, we identified logical components, factored out common features, and defined general framework interfaces. Figure 2 depicts the PSS framework archi-tecture. The security policy provides the fundamental guidelines for the various security operations. The Security Broker interacts between applications (more precisely the distributed applications) and the security services. The security broker has a security services handler (cf. Figure 1), which is employed to absorb the heterogeneity. The layered architecture of the security broker is presented in the corresponding section. Invocation of various security services at one site requires coordination with the set of the services invoked at the other sites. This coordination is carried out by a special service called Coordination Service.

The functioning of the coordination service is elaborated in the next section. However, it is worth mentioning here that the coordination service is guided by the security policy especially for the resolution of conflicts that arises if various grid sites try to invoke a dissimilar set of security services.

Figure 2. PSS Framework Architecture

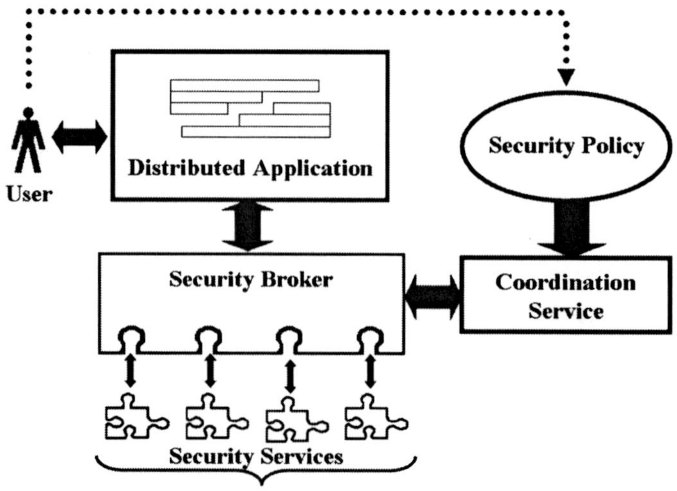

Addition and Deletion of Security Services

Adding New Services Transparently

To add new services without making any changes to the rest of the system, a *registry* is used to maintain a collection of abstract factories. In the abstract factory pattern, a single class defines an interface for creating families of related objects, without specifying their concrete types. Subclasses of an abstract factory are responsible for creating concrete classes that collaborate among themselves. In the context of pluggable services, each abstract factory can create the *Connector*, *Acceptor*, *Profile*, and *Transport* classes for a particular service.

Adding New Services Dynamically

To configure new services dynamically, even while the system is running, a *configurator* is used that can decouple the implementation of a service from its configuration into the application. This configurator can be applied in either of the following ways:

a. The configurator can be used to dynamically load the registry class. This facade knows how to configure a particular set of services. To add new services, one must either implement a new registry class or derive from an existing one.
b. The configurator can be used to load the set of entries in a registry dynamically. For example, a registry can simply parse a configuration script and link the services listed in it dynamically. This is the most flexible design, but it requires more code to parse the configuration script and load the objects dynamically.

Services Synchronization

If the various communicating nodes invoke different sets of security services, a coordination service is used to resolve the invocation of unpaired services. A *log* of the services invoked at the various nodes is maintained and if a conflict is found in the set of invoked services in the collaborating nodes then it is resolved according to the rules set forth in the security policy. For example, if the security policy permits the automatic invocation of corresponding services then if a VO member wishes to invoke *Audit Service* whereas the other prefers *Profile Service* then the coordination service will invoke the two services at both places.

SECURITY BROKERING

The Security Broker mediates between applications (more precisely the distributed applications) and the security services. The security broker has a security services handler, which is employed to absorb the heterogeneity of the underlying security services and to provide a homogeneous interface to the upper layer.

An Analogy of Resource Brokering

The idea of introducing a security services broker is actually inspired by the utilization of a brokering agent for the exploitation of suitable computing/ storage resource (also known as the resource broker) in distributed applications.

The virtual security services handler shown in Figure 1 could be seen as a part of the security broker (as shown in Figure 3) that interacts between the core security architect-ture and the applications. Such an arrangement with no direct interaction between applications and core security architecture will raise the protection level of the security infrastructure from the malevolent applications.

Figure 3. Induction of a Security Broker

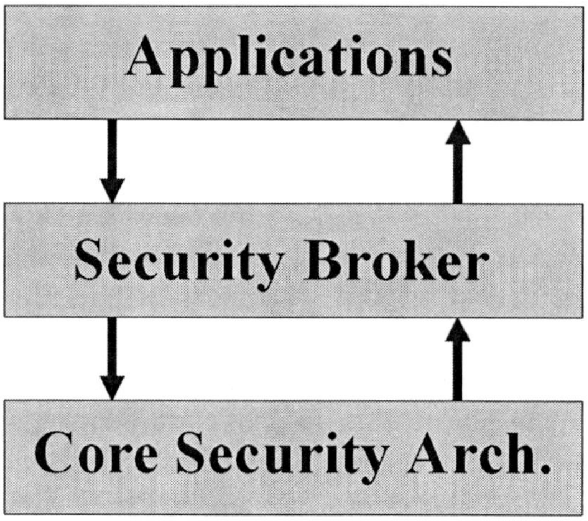

This concept of virtualization of security services can be realized through distributed virtual engines that will enable security service calls to be unified according to requirements and not according to the technologies to be supported. Distributed virtual engines are implemented by using brokering agents for the security services.

Coordination between Applications and Core Security Architecture

This component of PSS is responsible for the surety that a coordinated set of security services are invoked at the various sites of the VO. It contains traces of all the services invoked at the various nodes (cf. Figure 4). When a user invokes a set of services (default or user-defined) and it does not match with the set of services invoked at the other nodes then the mismatch is identified as *conflict* in the invoked services which is managed in the light of the security policy. Once the conflict is resolved, security services invocation is made to the security broker.

The security broker is not involved in the conflict management itself, rather it forwards the service invocation, made by a user/service, to the coordination service for its mapping and to look for any conflict(s) with the security services invoked at the other nodes. The security services are invoked by the security broker only when it receives a command from the coordination service.

In the arrangement shown in Figure 2, the security broker is deliberately placed between the application and the coordination service so as to isolate the latter from the former. One of the objectives of the security broker is to isolate the core security architecture from the applications running over the grid to increase the protection level. The need to protect the coordinated service is evident from the fact that if some malicious user/application succeeds in influencing it then the mismatch of the various security services invoked at the various nodes will cause the self-destruction of the entire security architecture.

Layered Architecture

The layered architecture of the proposed security broker is shown in Figure 5. The functionalities associated with these layers are:

1. **Application/Client Interface** authenticates the user/application and provides the glue between the user/application and the un-

Figure 4. Coordination Service Architecture

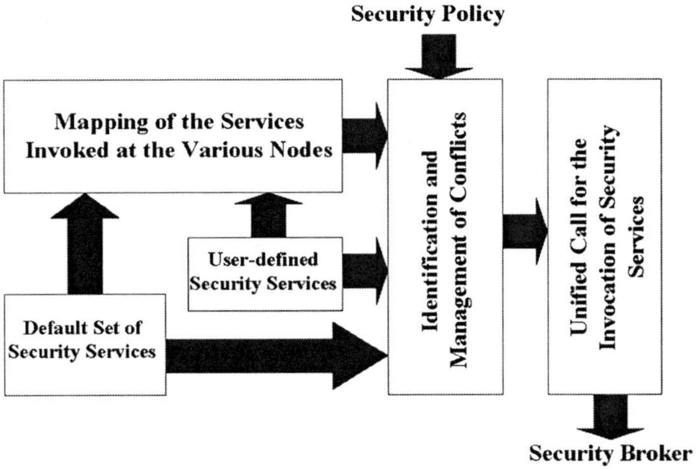

Figure 5. Security Broker Architecture

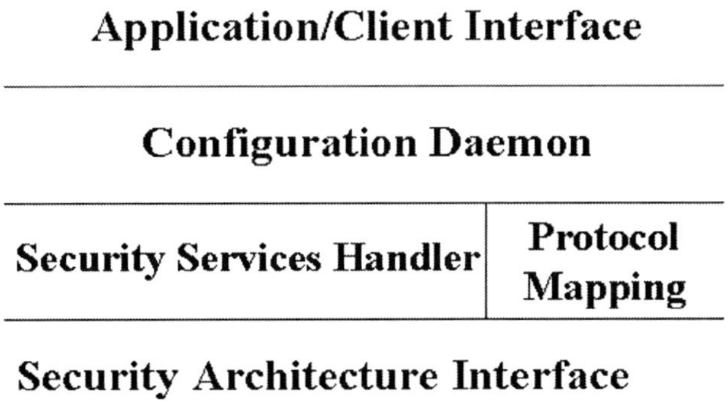

derlying security broker infrastructure to facilitate communications between them.

2. **Configuration Daemon** is a configuration server. It accepts a machine independent, abstract configuration request and then interacts with the coordination service through a secure channel. It notifies when the coordination service approves the security service configuration. It can keep a log of configurations done or even a complete backup configuration.
3. **Security Services Handler** absorbs the diversity of the security mechanisms to enable security service calls to be unified according to requirements and not according to the technologies to be supported (cf. Figure 1).
4. **Protocol Mapping** contains a comprehensive list of the protocols supported by the security architecture through the Security Services Handler.
5. **Security Architecture Interface** consists of *socket* modules to plug various security services. Call for a particular security service is

sent to the security services handler through the Configuration Daemon. The security services handler checks the existence of such a security service from the security protocol mapping and if it exists then an instance is invoked to hook the corresponding security service to the security architecture interface.

6. **Real-Time Algorithms**, similar to real-time operating system, are used to address the performance concerns. When building components in a layered architecture, efficiency of interactions among the various layers is of prime importance. These algorithms assure that the entire processing of the security broker takes place in real time and the users/ services can invoke these security services at the application layer. These real-time features are implemented at each layer.

HOW VIPSEC RESPONDS THE SECURITY REQUIREMENTS

Figure 6 shows various entities of a user and his target domains including auxiliary pluggable security services. ***User domain*** consists of *user*, *local resources* (both computational and storage), an *authentication server* (that authenticates the user and delivers credentials), and an *attribute server* (that delivers user's privilege attributes and sends the assertions with service requests). ***Target domain*** consists of *target resources* (both computational and storage), an *authorization server* (that validates the certificates), a local CA, and *Access policy* (that makes authorization decisions).

Any interaction between the user domain and his target domain requires some intermediary architecture that can convert the assertions into a form understood by the target domain – for example conversion of authentication credentials (e.g. Kerberos ticket) into a credential form that target domain can work with (e.g. X.509 certificates). This intermediary architecture can also offer a number of pluggable security services to the user. These services are discussed in detail in the second part of this document. Moreover, this intermediary architecture can honor a set of policies when forwarding the request (including the mapping rules and delegation policies. This mapping server, pluggable security services and the various security units of user and his target

Figure 6. Virtual security interaction between a user and his target domains

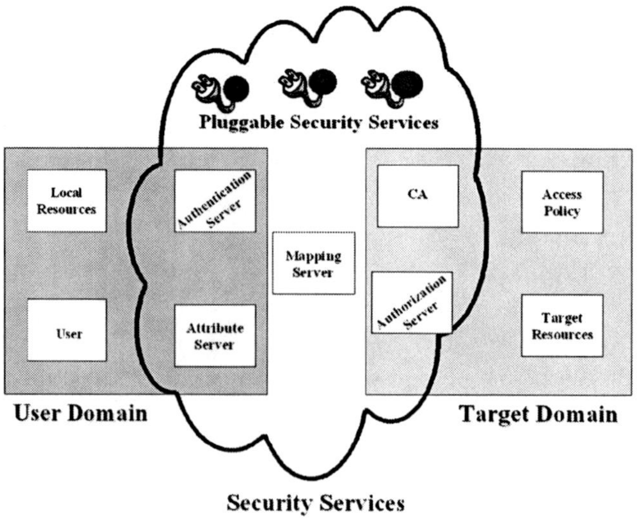

domains are grouped together in Figure 6 as ***Security Services*** that virtualizes the security dialogues between these domains.

In such architecture

- New users or groups may be introduced quickly (scalability) as the security services layer harmonizes (virtualizes) the diverse security mechanisms of participating nodes and there is no restriction of specific communication or security requirement.
- The handling of privileges provided to a group or individual can be easily managed as it employs role based access control (RBAC).
- Isolation of applications layer from the core security architecture layer (Figure 3) enhances the protection of the private data including authentication data.
- Agreed security features could be implemented by making corresponding adjustments in the security broker layer (Figure 3).
- The intermediary architecture (Figure 6) could be employed to delegate actions; however, there is a need to shun the cloning of credentials as they could be exploited.
- The attribute server (Figure 6) could be employed to place limits on the overall amount of resources consumed by particular user or group. These limits are generally defined in the access policy of the target domain (Figure 6).
- The confidence of the resource providers can be gained by offering them a number of pluggable security services. They can easily incorporate additional security features that assure them that their resources could neither be exploited nor be misused; and in the case of any misuse a chain of accountability could be established.

Performance Evaluations

Validation of the VIPSEC is the focal point of its applicability. A pervasive grid prototype is developed to validate these propositions. An example scenario of the system is described in this section:

All the teachers and students of our department are supposed to use their PDAs to gain access to the pedagogic resources. Wireless access points are provided in every room of the department. These access points are also used to determine the context of the users. In the library, students can read e-books but can not read their examination paper; whereas in the examination hall, from 9 am to noon, the students can read the examination paper, write the answers file, but can not read books. The teachers can read and write the examination paper from both library and from the exam hall. A PDA is placed in the quarantine zone if its user:

1. tries more than three unsuccessful log-in attempts as student or more than two unsuccessful log-in attempts as teacher, as he/she may be a potential intruder;
2. is using too much bandwidth, as he/she may be trying to cause the Denial of Service (DoS) attack;
3. is seeking unauthorized privileges.

Placement in a quarantine zone implies that:

1. other users are informed of his/her presence, as a troublemaker;
2. he/she is asked to behave normally otherwise he/she will be expelled;
3. after some time Δt it is evaluated whether to clear him/her out the quarantine zone or disconnect him/her from the system. This decision will be based on the close observation of his/her activities during the quarantine period Δt.

As shown in Figure 7, two different Wi-Fi access points at our department building are used to model library and exam hall. PDAs with embedded Wi-Fi card are used to model students (S), teacher (T) and potential attacker (encircled). One PC is used (to be connected from the third Wi-Fi access point) to act as the CA. The overall happening of the system is displayed on its screen including the log of the various actions taken by these PDAs and the time taken by each operation.

We consider a bunch of heterogeneous nodes containing some malicious nodes. These nodes are considered mutually trusted until an attack is detected. A malicious node regularly tries to attack the other nodes. Each attack has a probability p of success. This probability depends on the target node type. A successful attack turns the victim node into a new attacking node for the others. However, in the contrary case the attacker is blocked in its firewall and an alert concerning this node is transmitted in the system.

The results obtained from this grid setup show that there is no considerable overhead on the overall performance of the system due to the consideration of context and state of the mobile nodes. Figure 8 shows few screen shots of this experimental set-up:

The performance of the entire system is of prime consideration. We carried out a study to observe the impact of the dynamic consideration of the access privileges. Figure 8(a) provides a screen shot of the CPU performance graph.

Figure 8(b) shows a screen shot of the CPU load monitoring. It shows the normal behavior of CPU even during the execution of dynamic privileges management.

Figure 8(c) shows a screen shot of the network confi-guration settings of a mobile node.

As shown in Figure 8(d), a mobile user can activate or deactivate the encrypted communications mode. One of the objectives of providing this feature was to study the impact of encryption on the overall performance of the system.

Figure 8(e) provides the screen shot of the mobile node which is in the process of establishing connection with the Grid.

Figure 8(f) shows a glimpse of the log file that contains all the actions took place with the entire Grid system.

Figure 8(g) shows the screen shot of the options of modifying the security parameters in the real time.

Figure 7. Prototype setup

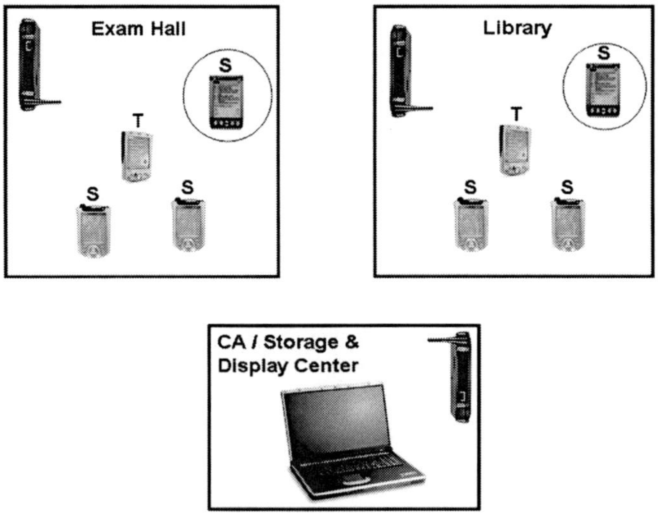

Figure 8. (a) CPU performance, (b) CPU load monitor, (c) A mobile node, (d) Selection of encrypted communication, (e) Connection to the grid, (f) Log file, (g) Modifications of the security parameters in the real time

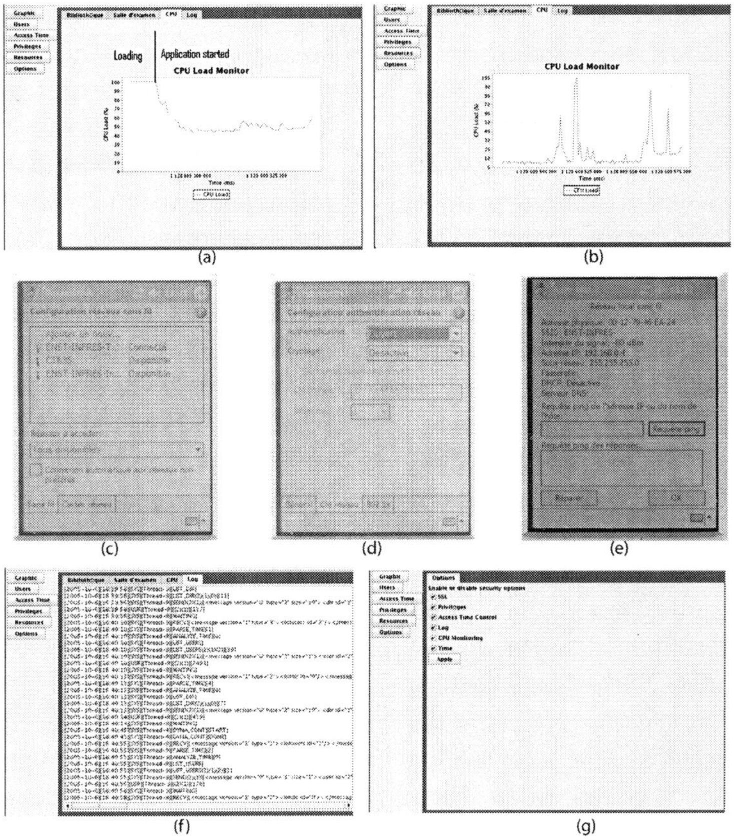

TRUST ESTABLISHMENT

Establishment of security services and trust relationships are the most desirable features for large scale open heterogeneous systems. They are in need of a consistent security architecture that is sufficiently efficient and scalable. Security of large numbers of users and resources, without relying on a centralized and integrated security infrastructure, should have to move towards the concepts of spontaneous recognition and dynamic trust establishment.

The problem addressed in this section is *how different nodes can trust unknown infrastructure with their private data and vice versa i.e. how a computing infrastructure can trust a node which is seeking access to its resources.* In particular, our trust establishment approach gives greater protection to data and information that may need to be revealed to a third party's computing platform during the process of a transaction. In such computing system, certain number of mutually distrustful participants with varying degree of priori relationships (or perhaps not at all) may want to share resources in order to perform some tasks. In this situation we envisage the trust reflection will provide a means for these participants to evolve the trust relationships quickly and accurately via limited iterations. This will provide an opportunity for the collaborating participants to either achieve full cooperation or to remove untrustworthy participants depending on the result of trust evaluation.

The collaborating members are from different security domains, they may not adhere to the same security policy. The decentralized nature of administration makes it difficult to establish and propagate trust. A distributed trust evaluation scheme is therefore required for these environments. Trust can be based on a history of interactions where credentials can be shown to demonstrate some previous relationships. Such mechanism is called *history-based trust establishment* (Bussard, 2004). However, a challenging situation arises if there is no trust among parties and there is no mechanism to build some trust based on a history of previous interactions. New solutions addressing these issues are required both for the protection of users, including privacy measures, and for controlling the access to valuable resources like commercial services.

We propose dynamic distribution of trust. Our proposed scheme provides a mechanism for delegation of trust and continuous monitoring of the changes to the level of trust of each node. It has the advantage of decentralized hierarchical administration, scalability of certificate issuing capacity and the flexibility of delegation. Since the open services are not limited to a specific range of domains and organizations, we propose a distributed, flexible (adaptable to different security domains) and general-purpose trust management for establishing a trust relationship among entities, which have no previous interaction, to provide a scalable and decentralized access-control mechanism over the Internet.

To establish trust among the different nodes, we show that instead of having a single value representing the *trustworthiness of a node*, the value should be broken into separate attributes. These attributes are presented to exemplify how to break trust into separate *confidences*. Each attribute represents a *confidence,* and each *confidence* represents a characteristic of a node from which trust can be synthesized. There are varying forms of trust. We can *trust* a node to be accurate (this is important for data integrity). We can *trust* a node to complete tasks reliably. We can *trust* nodes to return data quickly, or always in a guaranteed time, so on and so forth. Like people *trust* physicians for medical advice and stock brokers for financial advice, attributes should be viewed as foundational characteristics used to build particular types of *trust*.

These attributes form a virtual plane to link the resources, users (individuals and services), and the applications. This relationship signifies that there is not a fix form of trust among the various entities. Using a virtual and extensible basis for synthesizing varying types of trust allows for the greatest flexibility from one entity to the other. Flexibility is essential as anything too rigid cannot be easily adopted in a grid environment.

From the functional point of view, these attribute certificates will be used in compliment with identity certificates provided by the existing infrastructure (Foster, 1998; Pearlman, 2002). Where the identity certificates are used to verify the identity of an entity in a highly anonymous environment (e.g. the Internet), the attribute certificates will be used to determine the trustworthiness of an entity in an uncertain environment (such as the pervasive grid).

Our proposed model comprises of:

1. Definition of trust relationships between two nodes when there exist:
    a. direct trust relationships within a single domain – although there exist a unique CA or authorization policy, still an invalid proxy certificate generated by a malicious host can run a faked sub-process to accomplish a malicious task. So a node should estimate the *trustworthiness* of the node it is going to interact. Our trust model handles this scenario by using the centralized credentials (X509 or Kerberos) architecture to determine the trust values of the individual nodes by maintaining a trust table of the domain.

b. indirect trust relationships across multiple domains – crossing the domains further complicate the problem described in part 1a. In this scenario, for a successful interaction, a node has to trust all the intermediate hosts along the path that it has traversed before arriving to the second node (with which it will interact). Our trust model evaluates the trust degree along the whole path keeping in mind that the security policies in different domains as well as in VO may be different. Thus, the trust relationship between a set of nodes is establish.

2. Dynamic establishment of trust relationships (using intermediaries in a distributed means) where any node can join and leave anytime and anywhere. As the nodes may belong to different security domains, they may not share the same security policy. The decentralized structure makes it difficult to establish trust in the grid. Our trust model employs a distributed trust evaluation scheme to fit the large scale heterogeneous distributed environments and also supports the basis for satisfying the security requirements to achieve single sign-on and delegation.

ASSESSMENT OF VIPSEC FUNCTIONALITIES

The range of available grid simulation tools, such as Bricks (Takefusa, 1999), SimGrid (Legrand, 2003), GridSim (Buyya, 2002), GangSim (Dumitrescu, 2005), OptorSim (Cameron, 2004) etc., does not provide any support for the simulations of grid security functions. The deployment of a grid infrastructure without proper simulations of its various defense capabilities will certainly be an invitation to disaster. One can not remove all the vulnerabilities from a design, no matter how analytically good it is, unless the design has undergone a series of 'real-application-specific' tests. In the absence of a proper validation mechanism, security designers risk wasting time and effort implementing safeguards that do not address any realistic threat to the grid. Or, just as dangerously, they run the risk of concentrating their security measures on one threat while leaving the grid architecture dangerously exposed to others. We have faced the same problem while working on the virtualization of security services for the grid. This situation obliged us to develop a tool to perform grid security simulations – Grid Security Services Simulator (G3S).

G3S: Grid Security Services Simulator

Motivations The prime motivations behind the design and development of G3S was to lay the foundation of a simulations tool for the grid security services as none of the existing grid simulators provides any support for the security functionalities. It was felt imperative to provide a *Graphics User Interface (GUI)* so that even non-computer professionals (such as health grid users) can benefit from this tool by interactively simulating various grid security features (such as secure exchange of documents, attack patterns, etc).

Principles G3S models security functionalities of a grid. The grid nodes may be static or mobile. For the mobile nodes, it also considers the mobility-related security issues such as security gaps. It is designed to support multiple authentication mechanisms such as X.509 certificates and Kerberos tickets. Role-Based Access Control (RBAC) is used for the authorization purposes – work is underway to support the Community Access System (CAS) (Pearlman, 2002). G3S supports Bell-LaPadula Model for the assurance of grid data confidentiality and the Watermarking technique is used to assure the integrity of the data flowing across the grid resources. G3S is designed in user-friendly way, so that even a user with a shallow knowledge of security services may equally use it. For example, a user may intend to simulate confidentiality features without knowing that confidentiality requires access control mechanism. G3S automatically invokes the *prerequisite* security services so that a true scenario can be simulated even if its user does not know all of its parameters.

Simulations of different *attack patterns* is provided so that the designers can see if their design can deter the security threats and can survive after the attack. G3S has a mechanism for threats dissemination. If a node attempts to cross its defined privileges then an alert signal about the presence of a malicious node is sent to all the relevant nodes.

Implementation G3S is written in Java. It is lightweight and can be installed and executed from a single PC. An easy-to-use graphics user interface (GUI) is provided. Detailed log of the various operations is maintained to facilitate the auditability. This log file can be accessed by any querying program for swift access to some particular event as it is very difficult to find the trace of certain activity by general observation of a huge audit trail. Nodes have different geometrical shapes (such as circular, square, triangular, etc.) to graphically exhibit their heterogeneous nature. These shapes correspond to the nature of the participating nodes (e.g. their communication mechanisms, their static or mobile nature, etc.) These nodes can be grouped together to form virtual organizations (VOs) at any instant. A number of VOs may be created simultaneously and their transactions are consequently simulated. A different color is allocated for each VO.

Applications G3S can be used to simulate the working and efficiency of a grid security model. The alpha version of G3S can simulate the security services of a grid of maximum 100 nodes; however, the next release will be able to handle 1000 nodes. These nodes are not necessarily the fixed resources – mobile grid nodes can also be simulated with their corresponding mobility features and constraints.

G3S Structure

G3S is composed of five main components (as shown in Figure 9): *Core*, *DocumentExchange*, *SecurityPolicy*, *TrustManger* and *Attack.*

Interdependencies of various G3S components are summarized in the Table 1. The *Core* uses *Security Policy*; the *Document Exchange* uses Core, *Security Policy*, and *Trust Management*; the *Security Policy* is totally independent of other

Figure 9. G3S main components

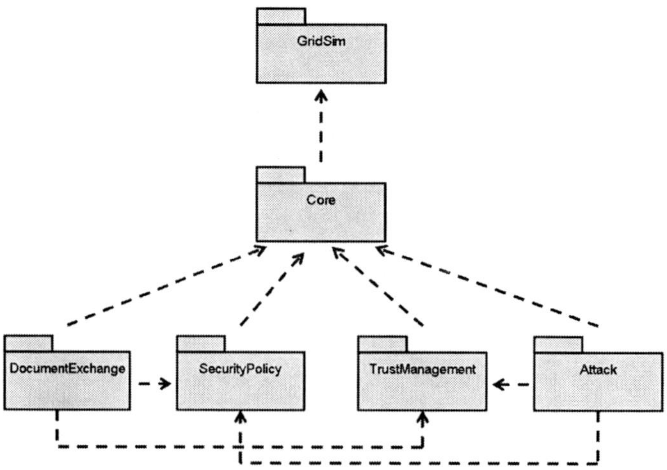

Table 1. Interdependencies of the G3S components

	Core	Document Exchange	Security Policy	Trust Management	Attack
Core		X	• R1.3	X	X
Document Exchange	• R2.1		• R2.3	• R2.4	X
Security Policy	X	X		X	X
Trust Management	• R4.1	X	X		X
Attack	• R5.1	X	• R5.3	• R5.4	

• (left) uses (up)
X (left) doesn't use (up)

components; the *Trust Management* uses *Core*; and the *Attack* uses *Core*, *Security Policy*, and *Trust Management*.

Various relationships of the Table 1 are described below. These relationships of the various components are not rigid. The existing functions can be easily extended. Likewise, more security functions can also be easily added.

R1.3: A VO has one Security Policy (characteristic feature of VOs)
R2.1: A document exchange requires 2 G3SNodes (which exchange the document)
R2.3: A document exchange takes place according to the rules set forth in the security policy of the VO
R2.4: A document exchange needs to check the current trust value of the sending and receiving nodes
R4.1: Trust Management deals with the trust level of each node
R5.1: An attack may result in several victim nodes
R5.3: Success or failure of an attack depends on the strength of the VO security policy
R5.4: If an attack is detected, the trust levels of the attacker and the attacked nodes are changed

Graphics User Interface (GUI) of G3S

G3S has a graphics window for user interaction. As shown in Figure 10, buttons for various simulation features (such as adding new users, resources, creation of VOs, security policy configuration, documents exchange, attack pattern, etc.) are provided on the left side of the window beside in the pull-down menus. The central zone is the area where the results of simulations are graphically displayed. A list of different symbols used by the G3S is given on the right side.

In the G3S Graphics User Interface (GUI), new users (individuals or groups) can be dynamically introduced (cf. Figure 10) at any time instant. Apart from the fundamental parameters, such as *name*, *confidentiality level*, etc., specific

Figure 10. G3S Graphics User Interface (GUI)

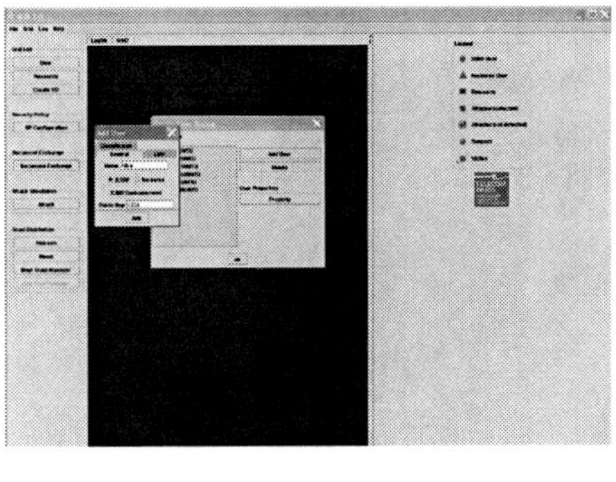

authentication parameters can be provided after choosing the desired authentication mechanism (Kerberos ticket or X.509 certificate). As soon as a certain authentication mechanism is chosen, G3S GUI automatically asks for the corresponding parameters. Similarly, new computing resources can be dynamically added (cf. figure 11).

New VOs can be created anytime by choosing the participating nodes (users and resources). A unique name is required for each VO (cf. Figure 12) and the security policy for each VO is configured. A number of VOs may be created simultaneously and their transactions are consequently simulated. As shown in Figure 13, various nodes have different colors and geometrical shapes (such as circular, square, triangular, etc.) to graphically exhibit their heterogeneous nature. These shapes correspond to the nature of the participating nodes (e.g. their communication mechanisms, their mobility mechanism, etc.). A different color is allocated for each VO. The various nodes of these VOs can collaborate and share resources according to their roles and privileges. All the exchange of data is recorded and the current status of each transaction is graphically displayed. Apart from the collaborations among a VO's nodes, the VOs themselves can collaborate for certain jobs.

Simulations of Attacks Patterns

G3S can also simulate various attack situations such as denial-of-service, man-in-the-middle, relay, and wormhole (cf. Figure 14). It can also simulate the survivability feature of a security design.

As an example of attack simulations, we consider a bunch of heterogeneous nodes containing some malicious nodes. These nodes are mutually trusted nodes until an attack is detected. A malicious node regularly tries to attack the other

Figure 11. Adding new resources

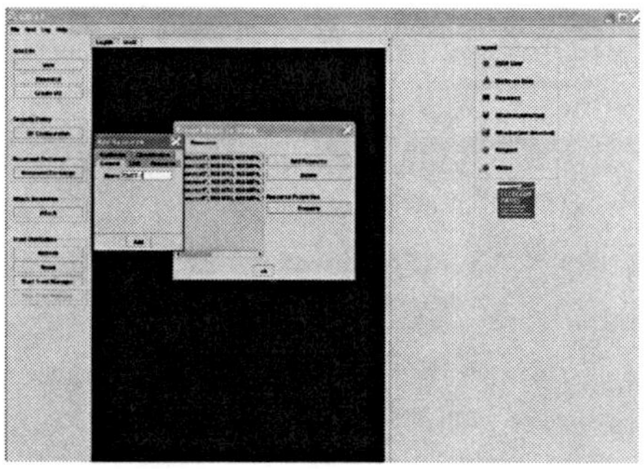

Figure 12. Creation of a VO

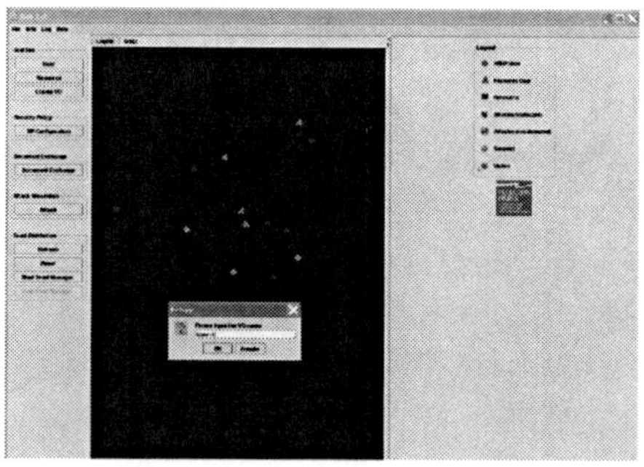

Figure 13. G3S entities

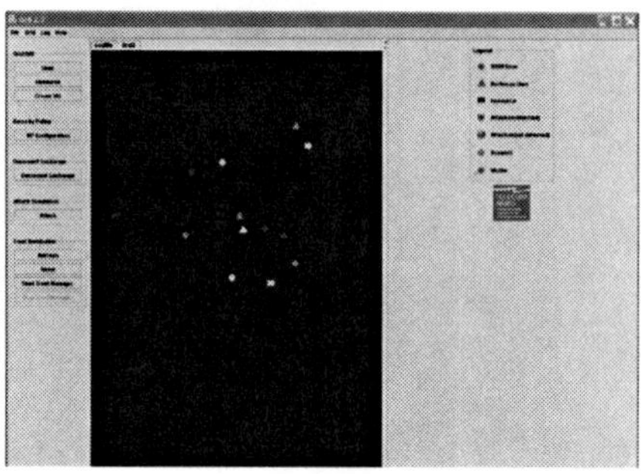

Figure 14. Attacks simulations

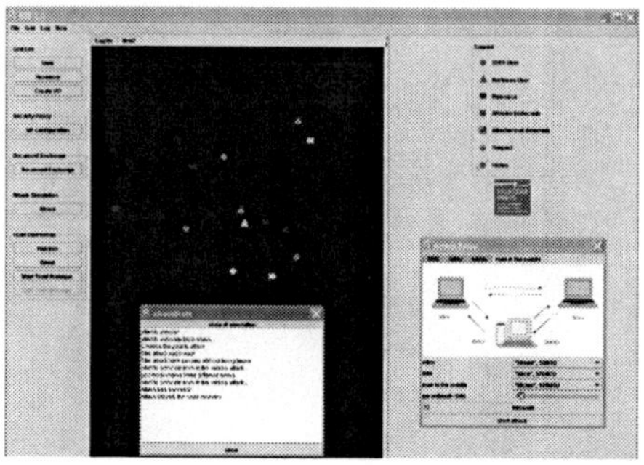

nodes. Each attack has a probability p of success. This probability depends on the target node type. A successful attack turns the victim node into a new attacking node for the others. However, in the contrary case the attacker is blocked in its firewall and an alert concerning this node is transmitted in the system. Figure 15 shows a graphics display of the Grid nodes. Each node has *xy*-coordinates and its class is determined by its shape (e.g. a triangular shape corresponds to a PDA; a round shape corresponds to a PC, etc.). The color coding used in this scheme is as follows: A node is gray if it does not know about the presence of the malicious node, blue if it is informed of malicious node, and white if it knows all the malicious nodes in the system.

A red halo around a node indicates that it is a victim node (which has become a malicious node itself), blue if the attack was foiled by the security architecture and yellow if the attack failed due to some other reason. The triangles in the display show the attack propagation whereas the arrows correspond to the distribution of trust among the nodes. The calculation of the distribution of trust is based on a *trust table*. A trust table is shown in Figure 16. The left entry A is the node that evaluates the entry of the node B from the top side. A color code is employed to quickly determine if there remains a danger of attack in the system: green, if A relies on B, and that A and B are indeed trustworthy; red, if A relies on B, and that B belongs to the attacker or is an attack victim; blue, if A does not rely on B and that B is indeed untrustworthy due to the reasons described in the previous case; white, if A's confidence in B has no importance.

Figure 17 presents the collective defense behavior of the nodes with the described infrastructure of confidence. If the attacker fails in its first attempt, it will be difficult for it to take control of the other nodes. Here node 0 escapes an attack from node 1 and blocks its transmissions. The other nodes are promptly informed of the threat so that they do not remain confident in node 0; and hence the overall system is protected (cf. corresponding values in the trust table). But if the node 0 fell prey to the attack of node 1 (Figure 18) and then manages to take control of node 3 all the other nodes will soon be affected resulting in the successful endeavor of the attacker.

Figure 15. Grid nodes

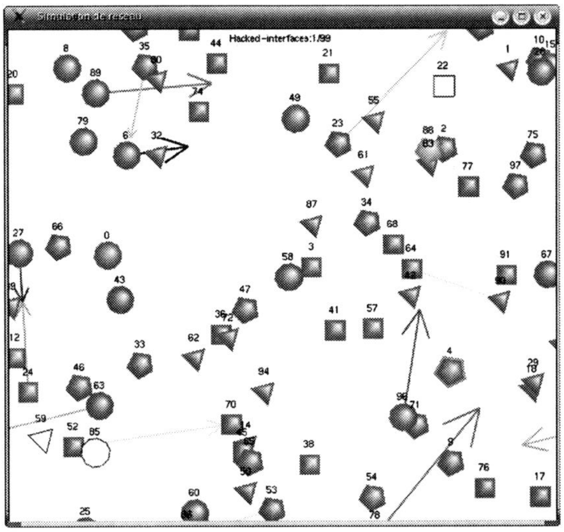

Figure 16. Trust table

	0	1	2	3	4	5	6	7	8	9
0	1.0	2.0	-6.8	2.0	2.0	2.0	2.0	2.0	1.8	1.9
1	1.0	1.0	1.0	1.0	1.0	1.0	1.0	1.0	1.0	1.0
2	1.1	1.1	1.0	1.1	1.1	1.1	1.1	0.6	1.1	1.1
3	-17.7	3.3	3.3	1.0	3.3	3.3	3.3	3.3	3.3	3.3
4	2.5	2.5	-10.4	2.5	1.0	2.5	2.5	2.0	2.5	2.5
5	2.5	2.5	-10.4	2.5	2.5	1.0	2.5	2.0	2.5	2.5
6	-10.4	2.5	2.5	2.0	2.5	2.5	1.0	2.5	2.5	2.5
7	-5.0	3.7	-12.8	3.3	3.7	3.7	3.7	1.9	3.7	4.2
8	2.5	2.5	-10.4	2.5	2.5	2.5	2.5	2.0	1.0	2.5
9	-10.4	2.5	2.5	2.0	2.5	2.5	2.5	2.5	2.5	1.0

Figure 17. Failed attack paradigm

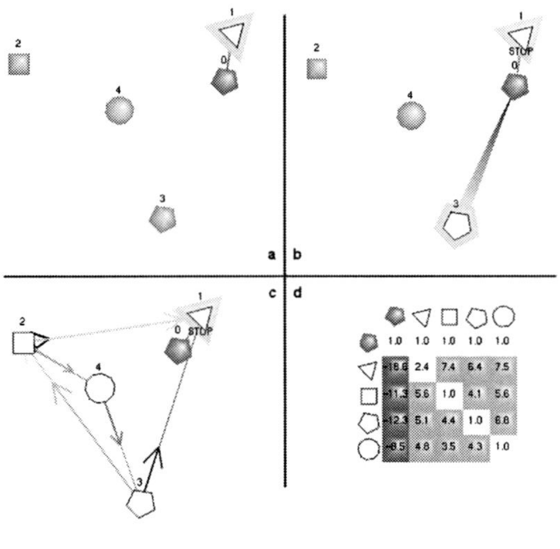

Figure 18. Successful attack paradigm

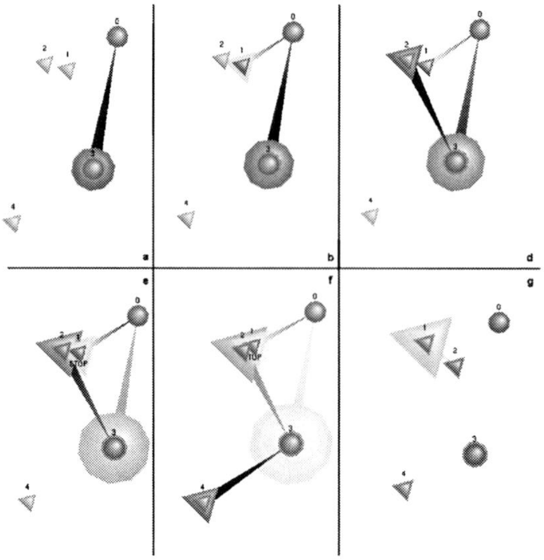

Integration of G3S with GridSim

We interacted with the developers of GridSim during the development phase of the alpha version of G3S so as to give a broader scope to G3S. Moreover, the integration of security services simulations into GridSim will provide a comprehensive simulations tool for the grid community; and hence the users of GridSim can also simulate security functionalities beside scheduling and resource management parameters.

The users and resources defined for the G3S are the *GridUser* and *GridResource* of GridSim, and the actions (such as the exchange of document) are *Gridlets* of GridSim. A *Gridlet* is a package that contains all the information related to the job and its execution management details such as job length expressed in MI (Millions Instruction). For example, the exchange of document is defined as a DocumentGridlet which extends to gridsim. GridSim class.

In the G3S *Core* module, we have defined G3SUser, G3SResource, and G3Slink classes. These classes inherit (*extend*) following GridSim functions:

G3SUser extends gridsim.GridUser
G3SResource extends gridsim.GridResource
G3Slink extends gridsim.net.Link

These classes are harnessed together by a superclass called G3SNode.

There exist some redundancies of code between G3S and GridSim, such as simJava classes. It is in fact required so that G3S can be executed independently without GridSim.

Upgrading of G3S

The G3S has been extended to incorporate the simulations of security features of service oriented architectures (SOA). The upgraded version better represents the intrinsic characteristic of virtualization oriented architectures by offering user defined dynamic composition of services and resources. The simulator is already used for the validation of security services architecture for a regional collaborative project dealing with the highly decentralized service oriented architecture. A screenshot of the upgraded version of the G3S is shown in the Figure 19.

Figure 19. Upgraded version of G3S

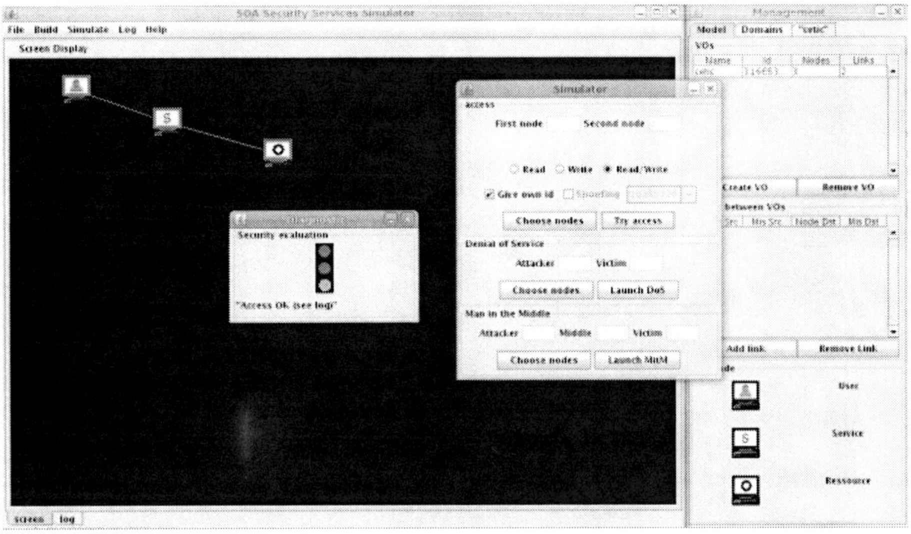

CONCLUSION

Managing security in large scale heterogeneous distributed computing systems such as virtualization-oriented architectures is a non-trivial problem. In such systems, the relationships are dynamic in nature which requires dynamic and adaptable security modules. Due to these reasons, currently available solutions usually lead to heavy administrative burden or weak security.

In this chapter, a new approach is proposed to deal with a number of security challenges presented by large scale, open, distributed heterogeneous systems. The most salient feature of the proposed approach is the flexible and adaptive nature of security services. Virtualization is used to provide standardized ways of enabling the federation of multiple heterogeneous security mechanisms. To have minimal reliance on the emerging resource management functionalities, and to make the proposed model more adaptive, the concept of security as a service is extended to security as a set of pluggable services. The other features are the self-security of the security architecture; use of security broker that negotiates for security services; description of security ontology to enable standard protocol interactions of core security bootstrapping services; and user-centered security services where usability is the prime motivation.

This research work has been a first step to come towards a systematic approach in the design process of security architecture for large scale, open, distributed heterogeneous systems. Although a wide variety of complex systems are considered but more consideration is given to the virtualization-oriented architectures. This work can be continued to explore more specific security solutions for other highly distributed and decentralized systems. Moreover, the concept of virtualization could be extended to adapt country-specific legal requirements, population-based ethical issues, and the business-oriented interests. Furthermore, virtualization could be used to achieve the best trade-off between security guarantees and processing capabilities.

It is important to remember that security is a process, the threat picture is always changing, and threat analysis needs to be continuously updated. In other words, virtualization-oriented architectures should be subject to constant review and upgrade, so that any security loophole can be plugged as soon as it is discovered. The growth in the users community should lead to improvements as larger number of users will find the loopholes faster, and more developers will be available to fix them and release patches.

REFERENCES

Adhikari, R. (2008). The Virtualization Challenge, Part 5: Virtualization and Security. *TechNewsWorld*

Bussard, L. (2004). *Trust Establishment Protocols for Communicating Devices*. PhD Thesis, October 2004

Buyya, R. & Murshed, M. (2002). GridSim: A Toolkit for the Modeling and Simulation of Distributed Resource Management and Scheduling for Grid Computing. *The Journal of Concurrency and Computation: Practice and Experience,* 1-32.

Cameron, D., Carvajal-Schiaffino, R., Millar, P., Nicholson, C., Stockinger, K., & Zini, F. (2004). OptorSim: A Grid Simulator for Replica Optimisation. In *Proceedings of the UK e-Science All Hands Conference*. ISBN 1904425216.

Dignan, L. (2008). Virtualization: What are the security risks? *ZDNet Blogs*. Retrieved from http://blogs.zdnet.com/security/?p=821

Dumitrescu, C., & Foster, I. (2005). Gangsim: A Simulator for Grid Scheduling Studies. *Proceedings of the IEEE International Symposium on Cluster Computing and the Grid (CCGrid'05)*. (pp 1151-1158).

EU-IST. (n.d.). European Union Information Society Technologies. *A thematic priority for Research and Development under the Specific Program "Integrating and Strengthening the European Research Area" in the Community sixth Framework Program*. Retrieved from http://www.cordis.lu/ist

Foster, I., & Kesselman, C. (1998). *The Grid: Blueprint for a New Computing Infrastructure*. Morgan Kaufman Publishers.

Foster, I., Kesselman, C., Tsudik, G., & Tuecke, S. (1998). A Security Architecture for Computational Grids, *Proceedings of the 5th ACM Conference on Computer and Communications Security* (pp 83-92).

GridBus. (n.d.). The GRIDBUS Project. Retrieved from www.gridbus.org

Kishimoto, H., Savva, A., & Snelling, D. (2003). *OGSA Fundamental Services: Requirements for Commercial GRID Systems. Technical Report, Open Grid Services Architecture Working Group*. OGSA-WG.

Legrand, A., Marchal, L., & Casanova, H. (2003). Scheduling Distributed Applications: The SimGrid Simulation Framework. In *Proceedings of the 3rd IEEE/ACM International Symposium on Cluster Computing and the Grid 2003 (CCGrid2003)* (pp 138-145).

Lewis, N. (2008). Virtualization Security. *IT Security Magazine*. http://www.itsecurity.com/features/virtualization-security-061708/

Naqvi, S., & Riguidel, M. (2004). Security Architecture for Heterogeneous Distributed Computing Systems. *Proceedings of the IEEE International Carnahan Conference on Security Technology 2004 (IEEE ICCST2004)*. (pp. 34-41).

NSF. (n.d.). National Science Foundation. Retrieved from http://www.nsf.gov

Pearlman, L., Welch, V., Foster, I., Kesselman, C., & Tuecke, S. (2002). A Community Authorization Service for Group Collaboration. In *Proceedings of the 3rd International Workshop on Policies for Distributed Systems and Networks (POLICY '02)* (pp 50-59).

Schneck, P. A., & Schwan, K. (1998). *Dynamic Authentication for High-Performance Networked Applications*. (Technical Report GIT-CC-98-08). Georgia Institute of Technology, College of Computing.

Symantec Corp. (2007). State of the Data Center Report 2007. *Overview of the findings of world-wide State of the 2007 Data Center Research report*. http://www.symantec.com/content/en/us/about/media/SOTDC_report_2007.pdf

Takefusa, A., Matsuoka, S., Aida, K., Nakada, H., & Nagashima, U. (1999). Overview of a Performance Evaluation System for Global Computing Scheduling Algorithms. In *Proceedings of the Eighth IEEE International Symposium on High Performance Distributed Computing (HPDC'99)* (pp 97-104).

ENDNOTE

* This chapter is an enhanced version of the paper published by the author in IJISP in 2008

Chapter 12
Architectural Support for Enhancing Critical Secrets Protection in Chip-Multiprocessors

Lu Peng
Louisiana State University, USA

Li Yang
University of Tennessee at Chattanooga, USA

Balachandran Ramadass
Louisiana State University, USA

ABSTRACT

Security has been considered as an important issue in processor design. Most of the existing designs of security handling assume the chip as a single secure unit. However, such assumption is vulnerable to exposure resulted from a central failure point. This chapter proposes a secure Chip-Multiprocessor architecture (SecCMP) to handle security related problems such as key protection and core authentication in multi-core systems. Matching the nature of multi-core systems, a distributed threshold secret sharing scheme is employed to protect critical secrets. A critical secret (e.g., encryption key) is divided into multiple shares and distributed among multiple cores instead of being kept a single copy in one core that is sensitive to exposure. The proposed SecCMP can not only enhance the security and fault-tolerance in secret protection but also support core authentication. SecCMP is designed to be an efficient and secure architecture for CMPs.

INTRODUCTION

Computer networking makes every computer component vulnerable to security attacks. Examples of such attacks include injection of malicious codes (e.g., buffer overflow), denial of service (DoS) attacks, and passive eavesdropping between CPU cores and off-chip devices. Also off-chip or on-chip devices taken over by an adversary can launch attacks to other components of a computer. Pure software solutions itself can not counter all attacks,

DOI: 10.4018/978-1-61692-000-5.ch012

therefore, enforcing security in processor design has drawn more and more attention. Currently many proposed works focus on encryption and authentication of hardware memory in single-core systems (Gassend, Suh, Clarke, Dijk, & Devadas, 2003; Lee, Kwan, McGregor, Dwoskin, & Wang, 2005; Shi, Lee, Ghosh, Lu, & Boldyreva, 2005; Yan, Rogers, Englender, Solihin, & Prvulovic, 2006; Yang, Zhang, & Gao, 2003). They usually assume the processor core as a safe and secure unit. When Chip-Multiprocessors (CMPs) have become mainstream products, applying encryption scheme of existing works to each core independently is one possible solution to enforce security in CMPs. The weakness of this solution is that the critical secrets (e.g. encryption key) stored or processed by one processor core can be easily exposed to adversaries through remote exploit attacks such as buffer overflow or Trojan horse, which leads to a central failure point. Once a core is compromised or taken over, the adversary could either access the critical secrets or wait until the compromised thread migrating onto another clean core then access unauthorized critical secrets. Therefore, this is not an effective approach to protecting shared critical secrets for CMPs.

Utilizing the distributed nature of CMPs is an alternative solution to reinforce the security of CMPs. Not only the computation load but also the security risks are distributed among multiple processor cores that are designed to collaboratively protect and access the critical secret. No individual core is possible to access the critical secret alone. We proposed a novel *Secure Chip-Multiprocessor (SecCMP)* architecture (Yang, & Peng 2006) to protect critical secrets based on a distributed *Secret Sharing* (Pedersen 1991). Instead of protecting a secret in one processor core, *Secret Sharing* is employed to distribute the secret among multiple cores that protect the secret collaboratively. The distributed security management matches the nature of multi-core architecture in CMPs. By employing a threshold *Secret Sharing* scheme, critical secrets are protected safely in a CMP processor even when one or more processor cores are compromised. In this chapter we integrate the SecCMP architecture with identity based cryptography to support remote information access and sharing. The performance degradation of our approach is studied through simulation. Low overheads and improved fault-tolerance are two major features of our approach. Low overhead is achieved via distributing the encryption and decryption load among multiple cores. Fault-tolerant is achieved via *(k, n)* secret sharing where at least k out of n cores are required to recover the secret. From a secret protection point of view, fewer than *k-1* cores are not able to recover the secret (i.e., the encryption key) such that our solution is resistant to the compromise of fewer than *k-1* cores. From a service protection point of view, k cores are able to provide the secret recovery service (i.e., retrieve the encryption key) such that our solution is tolerant to failure (i.e., hardware failure, DoS attacks) of up to *(n-k)* cores. Moreover, confidentiality and authentication among cores are supported through core authentication in *SecCMP*. Core authentication, which identifies whether a core is compromised, could be performed during critical information reconstruction or periodically. If not enough authenticated cores available, a system error will be called. The user may restart the system and reconstruct the critical secrets.

We use an application to demonstrate secure and remote critical information access and sharing supported by our *SecCMP*. Integrated with identity based cryptography (Bonh, & Franklin, 2003), the *SecCMP* provides a secure and reliable way to generate and distribute encryption keys between local host and remote site when prior distribution of keys is not available. Each local host has a pair of *master public key* (*MUK*) and *master private key* (*MRK*). In addition, each account has a pair of *account public key* (*AUK*) and *account private key* (*ARK*). In the local host which contains a multi-core processor, the MRK is divided and distributed among multiple cores and the ARK is generated from the MRK. On the remote site, the MUK and an Account ID will generate an AUK, which is used to encrypt the requested critical information. After receiving the encrypted critical information, k authenticated cores in the local host involve in generating the ARK, which finally decrypts received information.

To support critical information protection on CMPs, each processor core maintains two registers for the secret share and a public/private key pair for core authentication. These registers can only be accessed by a trusted application which constructs an account private key and decrypts the information. An encrypted I/O channel is employed to support user input and critical information receiving. To avoid bus or interconnection eavesdropping, all critical information related cache blocks are encrypted.

The rest of this chapter is organized as follows. Section 2 introduces related work. Section 3 states the attack and thread model of this chapter. We discuss the *master private key* protection and core authentication of the *SecCMP* architecture in Section 4. An application of the *SecCMP* is to support critical information remote access and sharing is given in Section 5, followed by security and computational complexity analysis in Section 6. Section 7 introduces performance evaluation. Finally, we summarize this chapter in Section 8.

RELATED WORK

Lee et al. (Lee et al., 2005) proposes a *"secret-protected (SP)"* architecture focusing on key protection and management, featured by secure on-line access of users' keys from different network computing devices. The keys are organized as a tree structured key chain rooted at a secret *"User Master Key"*. With helps from additional hardware features supporting *Concealed Execution Mode (CEM)* and *Trusted Software Module (TSM)*, the *SP* architecture protects confidentiality and integrity of sensitive data transmitted between processor chip and off-chip devices. Our proposed mechanism can enhance the security for the SP processor architecture working on a CMP. With a threshold distributed secret sharing, even if one or more pieces of critical secrets are released, the adversaries still cannot obtain the secrets as long as the number of released pieces is less than the threshold.

In Shi et al. (Shi, Lee, Falk, & Ghosh, 2006) the authors present an integrated framework utilizing multi-core processors to detect intrusions and recover from infected states. The processor cores are divided as resurrectors and resurrectees and memory space is also insulated. Resurrectees cannot access resurrectors' memory but resurrectors can access all the memory space. Fine grain internal state logging for low privileged cores, resurrectees, is employed. Resurrectors dynamically check the states of resurrectees. If any suspicious intrusions are detected, a logged state will be recovered. This design presumes that there are one or more master cores which are immune to attacks. In our scheme, we assume that all cores inside a chip are organized to a peer-to-peer relationship. Any cores could be compromised. However, if there are not enough authenticated cores, the system can be recovered by restarting.

There are two schemes to protect memory integrity and confidentiality for symmetric shared memory multiprocessor systems (SMP) proposed in (Shi, Lee, Ghosh, & Lu, 2004; Zhang, Gao, Yang, Zhang, & Gupta, 2005). In Shi et al. (Shi et al., 2004), the authors propose a one-time-pad based memory encryption scheme and an SHA256 hash function based authentication approach to protect bus communication. The scheme proposed in (Zhang et al., 2005) further improves security by generating a Cipher Block Chaining (CBC) encryption pad from snooped data. In (Rogers, & Solihin, 2006) the authors propose a memory encryption and authentication mechanism for Distributed Shared Memory (DSM) systems. All above proposals assume that a processor is a single secure unit. In this chapter, we assume that adversaries could intrude one or more cores in a multi-core processor and they need more efforts and time to intrude more than one core than they compromise a single core.

Sophisticated cache design to prevent software cache-based side channel attacks has been also proposed as important topic in secure processors. Wang. et. al (Wang, & Lee, 2007) analyzes these attacks and proposed two new security aware cache designs: a partition-based solution (Partition-Locked cache) and the randomization-based solution (Random Permutation cache). Both caches can thwart side channel attacks while keep very low performance degradation. Recently, Kong et. al. (Kong, Acricmez, Seifert, & Zhou, 2009) further improve the security of the Partition-Locked cache and the Random Permutation cache by employing preloading, informing load (a special load instruction) and software permutation techniques. The enhancements prevent the processor from advanced cache attacks.

Many other works (Gassend et al., 2003; Lee et al., 2005; Shi et al., 2005; Yan et al., 2006; Yang et al., 2003) emphasize on memory encryption and authentication by efficient hardware approaches in single-core systems. Our proposed scheme focuses on on-chip secret protection in multi-core processors. Additionally, there are a few proposals provide efficient bus or interconnection protection (Gao, Yang, Chrobak, Zhang, Nguyen, & Lee, 2006; Zhuang, Zhang, & Pande, 2004). Incorporating with the above memory/bus protection and recovery schemes, our proposal will further enhance security and fault-tolerance of CMP systems.

ATTACK MODEL

In this chapter, we focus on vulnerabilities resulted from the remote exploit network attacks The system suffers attacks at different points including memories, crossbar switch or one or more cores on multi-core processors, as shown in Figure 1. A firewall or intrusion detection system (IDSs) helps to protect users from attacks such as known virus or denial-of-service. However, they cannot protect the system once attacks have bypassed firewalls or IDSs. Therefore, the system suffers from a variety of vulnerabilities and attacks that expose critical secrets of a core to adversaries. The

Figure 1. Attack model

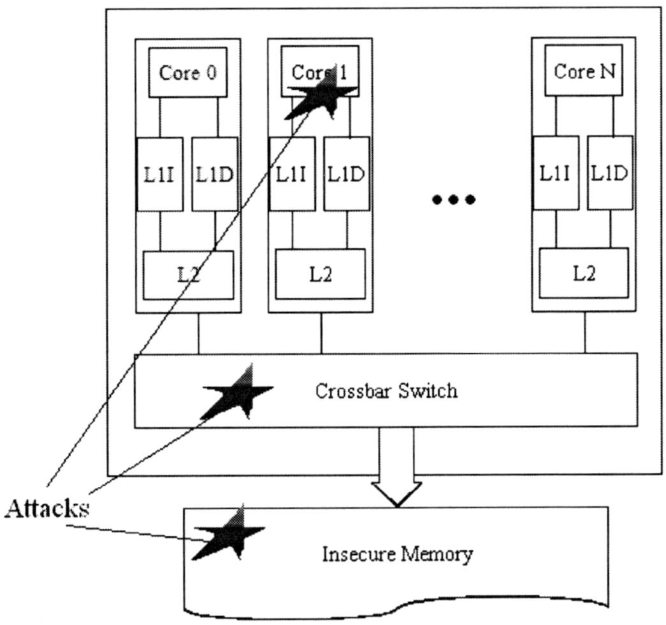

different kinds of attacks include eavesdropping, buffer overflow, cascading breakdowns, message spoofing, and message blocking.

The adversaries can access critical secrets by *eavesdropping* the crossbar switch connecting cores. During an eavesdropping attack, an attacker tries to learn critical secrets such as the root password that he/she was unable to access. *Buffer overflow* attack happens when a process attempts to store data beyond the boundaries of a fixed length buffer, and the extra data overwrites adjacent memory locations. By doing so, possible malicious code is injected into an execution path. If executed, the injected malicious code grants attackers unauthorized privileges to access critical secrets. Therefore, both eavesdrops and buffer overflow try to increase privileges of an attacker and expose the critical information to unauthorized attackers. Adversaries can break down other processors in a cascading manner if one core's critical secret is compromised. *Message spoofing* can occur if a fake message is generated and attributed to other senders. Examples are message insertion or replaying. Also, a message destined to a processor can be *blocked illegally* when it is transmitted through the crossbar switch.

A single authentication or encryption algorithm is not able to counter all aforementioned attacks. Section 4 describes how SecCMP enhances confidentiality by distributing secrets among multi-cores and augments authentication by employing digital signature from each core. Section 5 demonstrates countermeasures proposed in SecCMP by using an example. The example shows how to support access of the critical information shared remotely

SECURE CHIP MULTIPROCESSOR ARCHITECTURE

Each processor has a pair of *master public key* and *master private key*. The *master public key* is available to all other network devices and hosts. The *master private key* is protected by a threshold secret sharing scheme in the *SecCMP*. Protection of the *master private key* allows critical information related to an application to be stored and accessed over public network. Each core in the processor also employs new hardware features to support trusted applications. Trusted applications generate and distribute application related keys and encrypt, decrypt the remotely shared critical information. Because all activities including key computations, distributions and critical information encryptions and communications are protected by the trusted applications, an adversary cannot observe other cores' secret share even when one or more cores are compromised. Most importantly, the *SecCMP* comprises two unique components: protection of a master (chip) private key and core authentication.

Protection of the Master Private Key

The *master private key* is used to generate an application private key (we call it as an *account private key* in Section 5) and decrypt the application related critical information stored on-line and accessed over public network, i.e., banking PIN, PGP keys. To avoid a central failure point in a single processor system, the *master private key* is divided and distributed among multiple CPU cores in a processor chip. In a (k, n)-threshold secret sharing, each core holds a secret share, and any k of these n cores can reconstruct the *master private key*. Any collection of less than k partial shares can not get any information about the *master private key* (Pederson, 1992). Here, k is the threshold parameter such that $1 \leq k \leq n$. Each processor will authenticate itself in fine grained intervals. Therefore, it is difficult for adversaries to obtain k or more pieces of secret shares during a short time.

Figure 2 is an example of a (2, 4)-threshold scheme among 4 cores where a *master private key MRK* is divided into 4 unique pieces (Secret Share 0, 1, 2, 3), such that any 2 of them can be used to reconstruct the *master private key MRK*.

Figure 2. A Secure Chip Multiprocessor (SecCMP) with Master Private Key MRK Sharings

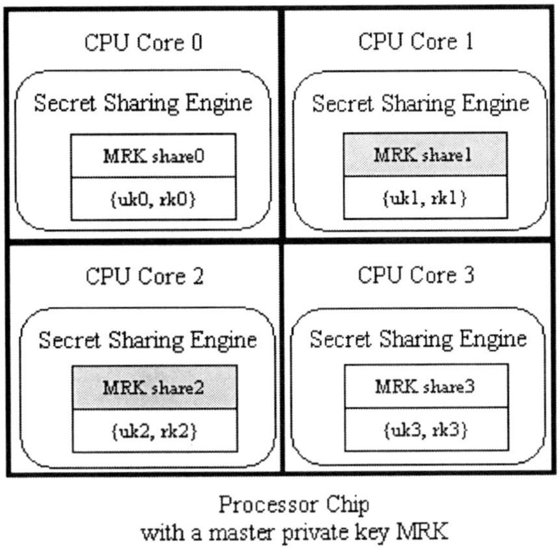

Traditional Shamir's secret sharing scheme suffers from the requirement of a trust authority and the absence of share verification. We employ a scheme based on (Pederson, 1991), which is an extension to Shamir's secret sharing without the support of a trust authority. We also deploy the verifiable secret sharing (Pederson, 1992) to detect the invalid share that some shareholders generate to prevent reconstruction of the *master private key*.

For each core in a processor supporting SecCMP, we have a 128 bit register for the secret share and a 128 bit register for the public/private key pair which is used for core authentication. These dedicated registers can only be accessed by the trusted application which generates the account private key. To ensure security, a thread running on one core cannot access another core's secret share register and public/private key register. During the procedure of critical information generation, threads involved are not allowed to migrate to another core. Besides, an encrypted I/O channel is required for a user to input the AcctId and receive the decrypted critical information. To keep confidentiality of the decrypted critical information, we employ an AES based encryption (Menezes, Oorschot, & Vanstone, 1997) scheme to encrypt a related cache block before it goes outside of the core. This can prevent eavesdropping attacks over the crossbar interconnection. For performance consideration, however, other cache blocks are not necessary to be encrypted. To identify critical information related cache blocks, we add one bit into all on-chip cache tags. This bit is set if a cache block contains part critical information.

Core Authentication

One or more cores may be compromised and their secret share is exposed to an attacker. We assume that an honest core will present the correct share to authenticate itself, and a compromised core will present a random number instead of the correct share. The attacker learns the secret shares from compromised cores and interrupts the *master private key* reconstruction. Failure of *master private key* reconstruction will result in a denial of service (DoS) attack. In order to exchange secret share securely, each core holds a public/private key pair {<, >} to encrypt the secret share and authenticate each other. Each core signs its secret share and hash code with its private key (digital signature). Then the signed message is encrypted with requesting core's public key. The requesting core decrypts the message with its private key. Then requesting node checks the signature to authenticate the sender, checks the hash code to make sure the integrity of the secret share. The key pair is created during core installation based on each core's identity. An adversary is not able to observe the encrypted share without a correct key pair. The private key is also used to encrypt the critical key sent to and from off-chip devices. In our design, core authentication also allows that a sender core to authenticate a receiver core. This is implemented by sending an authentication request message to the receiver core and checking the returned signature.

We not only passively protect the *master private key*, but also actively detect the compromised core. In order to detect the compromised cores, we design a series of *m master private keys* such that none of the participants knows beforehand which is correct. The *master private keys* are ordered incrementally based on their values, except for the real key. The participants combine their shares to generate one key after the other, until they create a correct key that is less than the previous key. This helps us to detect the compromised core before the master private key is exposed to the cheating compromised core. The detection and prevention of cheaters in threshold schemes (Lin, & Harn 1991) is adopted in our approach. Once the compromised core is detected, we isolate the compromised one. The work in (Martin, 1993) allows a new sharing scheme to be activated instantly once one of the cores becomes untrustworthy.

SECCMP SUPPORTED CRITICAL INFORMATION ACCESS AND SHARING

An application of *SecCMP* is to support critical information remote access and sharing. *SecCMP* provides secure channels to generate, store and exchange encryption keys for a local host and remote sites to share critical information associated with a specific account (i.e., a bank account, an email account). Each local host has a pair of *master public key* (*MUK*) and *master private key* (*MRK*). In addition, each account has a pair of *account public key* (*AUK*) and *account private key* (*ARK*). Based on the identity-based cryptography (Bonh & Franklin 2003), a user *account public key* can be any arbitrary string. In other words, users may use some well-known information, such as email address, IP address, URL as their *account public key*.

When a local host tries to retrieve critical information from a remote site, it creates a pair of MUK and MRK. The MUK is available to the remote site, and the MRK is distributed and stored in multiple cores of the local host. Such distributed design of the *master private key* is resistant to eavesdropping since at least k cores need to be compromised in the active session to reconstruct *master private key*. Moreover, *k* out of *n* cores need to be contacted in order to create an *account private key* based on the account ID. A buffer overflow attack may expose secret share of a core or interrupt private key generation from a core. Our core authentication service could detect such attacks. Because an attack from network exploits usually cannot be performed in a very short duration, the undergoing attack can be reported and blocked before *k* cores are compromised.

Identity-Based Cryptography

Identity-based systems 0 allow any party to generate a public key from a known identity value such as an ASCII string. The Private Key Generator (PKG) generates the corresponding private keys. To operate, the PKG first publishes a *master public key*, and keeps the corresponding *master private key*. Given the *master public key*, a public key can be generated corresponding to the identity by any party. To obtain a corresponding private key, PKG is contacted to generate the private key using *master private key* based on the identity. As a result, messages may be encrypted without prior key distribution between individual participants. Such solution is helpful when the pre-distribution of the authentication keys is not available. A major challenge of this approach is that the PKG must by highly trusted since it generates any user's private key and thus decrypt messages. In the *SecCMP*, multiple cores work together to provide secure private key generation service when there is no prior distribution of keys.

Remote Information Access and Sharing

When a local host, a multi-core processor system, tries to access or retrieve critical information from a remote site. The local host and remote site need to authenticate and exchange *master public key* with each other. How two remote hosts authenticate each other and how local host authenticate the current user is valid or not are out of the scope of this chapter. The former can be accomplished either by a Certificate Authority (CA) or a trusted third party. The later can be achieved through access control or biometrics. We focus on how remote site and local host generate, store and distribute *master public key*, *master private key*, *account public key* and *account private key*.

Figure 3 shows the general procedure of remote information access and sharing. To initiate the remote critical information access, the local host sends its *master public key* to remote site in the step 1. In step 2, the local host also sends account ID (*AcctID*) whose critical information the user would like to access. The remote site computes an *account public key* based on a *master public key* and an account ID (*AcctID*). After that, the remote host encrypts requested critical information with the *account public key* and transmits requested critical information to public networks in step 3. In step 4, the local host computes *account private key* by interacting at least *k* out of *n* cores in the local host processor, and uses the *account private key* to decrypt the received encrypted critical information.

The remote site needs to generate an *account public key* to encrypt requested critical information

Figure 3. Remote Critical Information Access

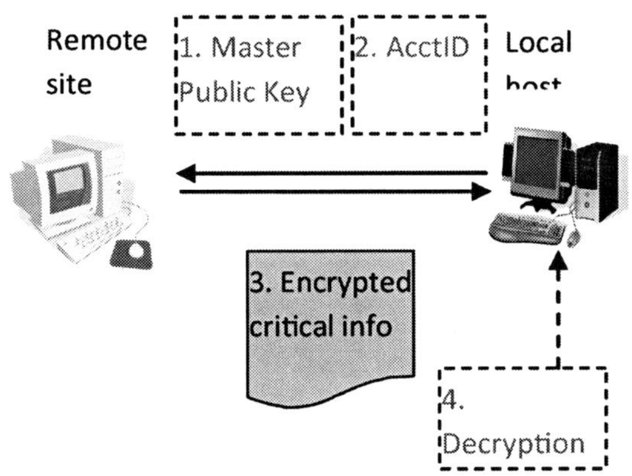

and send it to public networks. The local host will obtain *account private key* to decrypt the critical information encrypted using its corresponding public key. Figure 4 shows the *account public key* generation and critical information encryption in the remote site. There is a generator producing an account public key from the master public key and the input account id. The critical information is encrypted by the account public key and transmitted back to the requestor over a public network.

The local host generates an *account private key* by which the encrypted critical information is decrypted, as shown in Figure 5. The method to obtain the *account private key* is to contact at least *k* cores, present the account identity and request private key generation service.

The trusted application is the only application that can access the secret share in registers of a core. It can be implemented as a system call which can only be executed by a thread with a special privilege. This application is called when an authenticated local user sends a request for critical information to a remote site. When a trusted application sends the *MUK* to a remote site, the *MRK* was divided into *n* secret shares and kept in *n* cores respectively. The distributed secret shares

Figure 4. Account Public Key Generation and Encryption in the Remote Site

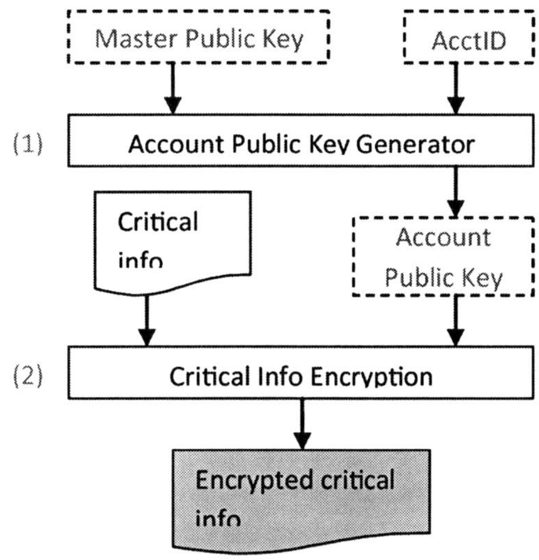

Figure 5. Account Private Key Generation and Decryption in the Local Host

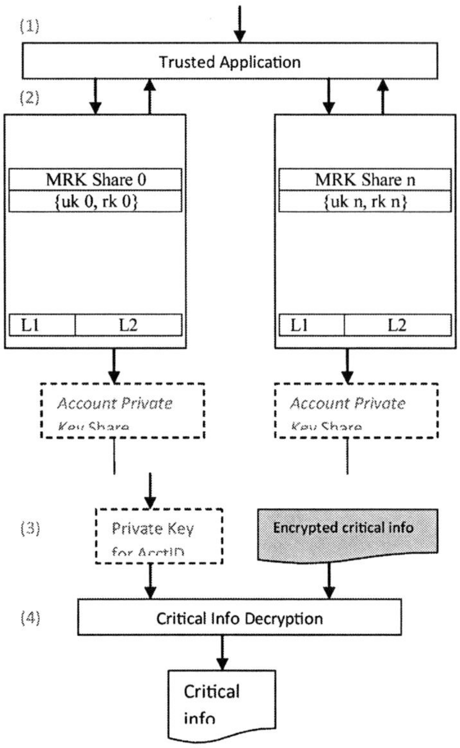

and account public key pairs are stored in a set of special registers which can only be accessed by the trusted application. The general procedure of the account private key generation and decryption in the local host is listed as follows:

(1) Authenticated users input to the trusted application and *AcctID* whose related critical information will be retrieved through public network by an authenticated user.
(2) Each core generates an *account private key* share.
(3) *K* shares of the *account private key* construct the corresponding *ARK* for *AUK* of the *AcctID*. Note each core only constructs its own share and then sends the output results to the next core. Therefore the confidentiality of the *MRK* is preserved here. Meanwhile, during the transmission, core authentication will also be preformed. To keep the constructed Account Private Key secret, the *kth* core has to be authenticated. Before the *(k-1)th* core sends out its decrypted secret share, it has to authenticate the last receiver. If no available authenticated cores exist, there are no enough cores to decrypt requested information. It will report this failure to the user. The user may require restart the system. Details of core authentication refer to section 4.2.
(4) The constructed *ARK* is used to decrypt the cipher text encrypted by the corresponding *AUK*. We will perform this step on an authenticated core identified by step 3. By doing so, a user can securely access and share critical information remotely. The received critical information can be protected until the user session finished.

SECURITY AND COMPLEXITY ANALYSIS

Confidentiality and *Integrity* are taken care of by the encryption and hash function performed on secret share. Hash function guarantees that a share being transferred is never corrupted due to non-benign failure. *Availability* ensures the survivability of a processor chip despite denial of service attack. In our schema we take care of this problem by making use of *(k, n)* threshold secret sharing algorithm, as any *k* out of *n* cores work together for critical master key reconstruction. Thus our security solution is tolerant to *k-1* compromised cores. *Authentication* is taken care of by digital signature that enables a core to ensure the identity of the peer core it is communicating with. During a network exploit attack, the adversary compromises the processor cores one by one through attacks such as buffer overflow. We authenticate cores during the critical information generation or actively perform authentication periodically. If there are no enough cores for decryption, a system error will be triggered. Therefore, our secret sharing mechanism prevents an adversary

Table 1. Processor and Memory Hierarchy Parameters

Processor	8 cores, 5GHz, Out-of-order execution pipeline. Issue width: 1.
L1 instruction cache	Two way set-associative. 32 bytes block size. 32KB inst. cache for each core. 1 cycle access delay.
L1 data. cache	Four way set-associative. 32 bytes block size. 32KB inst. cache for each core. 2 cycles access delay.
L2 shared unified cache	16-way set-associative. 128 bytes block size. 1MB. 9 cycles access delay. MESI protocol for L1 cache coherence.
Memory latency	500 cycles.
Secret Sharing Engine latency	80 cycles.

from spoofing a secure core and gaining unauthorized secret share. Main computations in our approach come from secret share reconstruction and encryption. The reconstruction *computational complexity* depends on the number of thresholds. The encryption *computational complexity* is same as the traditional schemes and depends on the size of shares. The shorter length of a share results in less resource consumption. The computations will be accelerated by involvement of multi-cores.

PERFORMANCE ANALYSIS

User activities can either not involve the request of remote critical information (e.g., browse unclassified daily news) or involve the request of remote critical information (e.g., access bank account). The former one does not call secret protection procedure so that the system performance is not affected. However, the latter one calls the secret protection procedure demonstrated in Figure 5, resulting in performance degradation . The performance degradation in the local host of Figure 3 is considered and simulated in this chapter. We evaluate the proposed architecture by a multi-core processor based on SESC (http://sesc.sourceforge.net/) with SPLASH2 applications and kernels. The setting of the processor simulator is listed in Table 1. For each program, we skip the first 100 million instructions and collect statistics for the next 200 million instructions. Three schemes are evaluated: the baseline machine has no secret sharing mechanism; a CMP triggering a secret sharing engineering every 1 million cycles and a CMP triggering a secret sharing engineering every 100 thousands cycles. We measure the average number of cycles executed for all eight cores. We calculate performance degradation by comparing average execution cycles for each scheme. The simulated secret sharing engine is a

Figure 6. Performance Degradation for a SecCMP Triggering a Secret Sharing Engine Every 1 Million and Every 100 Thousand Cycles

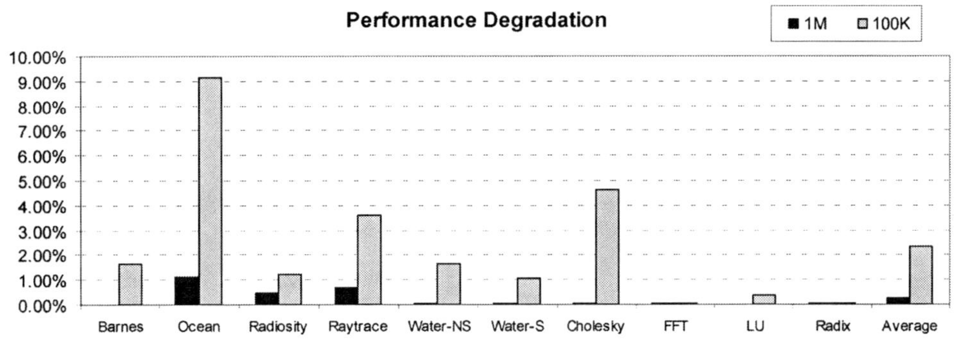

(4, 8) scheme which means that a secret can only be reconstructed by at least 4 correct sharing. During secret generation, the current executed program will be paused and the secret sharing engine will use caches for it calculation. Performance degradation comes from the latency of the secret sharing engine and cache pollution. Figure 6 shows performance degradation for simulated programs. We can see that average performance degradation are 0.26% and 2.35% respectively when the secret sharing scheme is triggered every 1 million cycles and every 100 thousands cycles. Ocean, which is sensitive to cache misses, slows down 9.17% when the secret sharing engine is triggered every 100 thousand cycles.

CONCLUSION

In this chapter, we introduce a low cost secure architecture design for CMPs. The proposed architecture employs a threshold *Secret Sharing* scheme to protect critical secrets and support core authentication for a CMP system. It supports online critical information retrieval and protection in a local host. Instead of keeping a whole copy of a critical secret, the secret is divided and distributed among multiple cores. A user can only reconstruct the secret if the number of authenticated cores is equal to or larger than the threshold. Compared with existing mechanisms, the proposed scheme is more secure and fault-tolerant.

REFERENCES

Bonh, D., & Franklin, M. (2003). Identity-based encryption from weil pairing. *SIAM Journal on Computing*, *32*(3), 586–615. doi:10.1137/S0097539701398521

Gao, L. Yang, J., Chrobak, M., Zhang, Y., Nguyen, S. & Lee, H.-H. (2006). A low-cost memory remapping scheme for address bus protection. In *Proceedings of the 15th International Conference on Parallel Architectures and Compilation Techniques (PACT)* (pp. 74-83).

Gassend, B., Suh, G., Clarke, D., Dijk, M., & Devadas, S. (2003). Caches and hash trees for efficient memory integrity verification. In *Proceedings of the 9th International Symposium on High-Performance Computer Architecture (HPCA)* (pp. 295-306).

Kong, J., Acricmez, O., Seifert, J., & Zhou, H. (2009). Hardware-Software Integrated Approaches to Defend Against Software Cache-based Side Channel Attacks. In *Proceeding of the 15th IEEE International Symposium on High-Performance Computer Architecture (HPCA)* (pp. 393-404).

Lee, R. B., Kwan, P. C. S., McGregor, J. P., Dwoskin, J., & Wang, Z. (2005). Architecture for protecting critical secrets in microprocessors. In *Proceedings of the 32nd International Symposium on Computer Architecture (ISCA)* (pp. 2-13).

Lie, D., Thekkath, C., Mitchell, M., Lincoln, P., Boneh, D., Mitchell, J., & Horowitz, M. (2000). Architectural support for copy and tamper resistant software. In *Proceedings of the 9th International Conference on Architectural Support for Programming Languages and Operating Systems (ASPLOS)* (pp. 168-177).

Lin, H.-Y., & Harn, L. (1991). A generalized secret sharing scheme with cheater detection. In *Proceedings of the International Conference on the Theory and Applications of Cryptology: Advances in Cryptology* (LNCS 739, pp. 149-158).

Martin, K. M. (1993). Untrustworthy participants in perfect secret sharing schemes . In Ganley, M. J. (Ed.), *Cryptography and Coding III* (pp. 255–264). Oxford University Press.

Menezes, A. J., van Oorschot, P. C., & Vanstone, S. A. (1997). *Handbook of applied cryptography. CRC Press*. LLC.

Pedersen, T. P. (1991). A threshold cryptosystem without a trusted party. In *Proceedings of EURO-CRYPT* (pp. 522-526).

Pederson, T. P. (1992). Non-interactive and information-theoretic secure verifiable secret sharing. In *Proceedings of the 11th Annual International Cryptology Conference on Advances in Cryptology* (LNCS 576, pp. 129-140).

Rogers, M. P., & Solihin, Y. (2006). Efficient data protection for distributed shared memory multiprocessors. In *Proceedings of the 15th International Conference on Parallel Architectures and Compilation Techniques (PACT)* (pp. 84-94).

SESC. (n.d.). Retrieved from http://sesc.sourceforge.net/

Shi, W., Lee, H.-H., Falk, L., & Ghosh, M. (2006). An integrated framework for dependable and revivable architectures using multicore processors. In *Proceedings of the 33rd International Symposium on Computer Architecture (ISCA)* (pp. 102-113).

Shi, W., Lee, H.-H., Ghosh, M., & Lu, C. (2004). Architectural support for high speed protection of memory integrity and confidentiality in multiprocessor systems. In *Proceedings of the 13th International Conference on Parallel Architectures and Compilation Techniques (PACT)* (pp.123-134).

Shi, W., Lee, H.-H., Ghosh, M., Lu, C., & Boldyreva, A. (2005). High efficiency counter mode security architecture via prediction and precomputation. In *Proceedings of the 32nd International Symposium on Computer Architecture (ISCA)* (pp. 14-24).

Suh, G. E., Clarke, D., Gassend, B., van Dijk, M., & Devadas, S. (2003). AEGIS: architecture for tamper-evident and tamper-resistant processing. In *Proceedings of the 17th International Conference on Supercomputing (ICS)* (pp. 160-171).

Wang, Z., & Lee, R. B. (2007). New Cache Designs for Thwarting Software Cache-based Side Channel Attacks. In *Proceedings of the 34th International Symposium on Computer Architecture (ISCA)* (pp. 494-505).

Yan, C., Rogers, B., Englender, D., Solihin, Y., & Prvulovic, M. (2006). Improving cost, performance, and security of memory encryption and authentication. In *Proceedings of the 33rd International Symposium on Computer Architecture (ISCA)* (pp. 179-190).

Yang, J., Zhang, Y., & Gao, L. (2003). Fast secure processor for inhibiting software piracy and tampering. *Proceedings of the 36th International Symposium on Microarchitecture (MICRO)*, 351-360.

Yang, L., & Peng, L. (2006). SecCMP: a secure chip-multiprocessor architecture. In *Proceedings of the first Workshop on Architectural and System Support for Improving Software Dependability (ASID)* (pp. 72-76).

Zhang, Y., Gao, L., Yang, J., Zhang, Z., & Gupta, R. (2005). SENSS: Security Enhancement to Symmetric Shared Memory Multiprocessors. *Proceedings of the 11th Intl. Symposium on High-Performance Computer Architecture (HPCA)*, 352-362.

Zhuang, X., Zhang, T., & Pande, S. (2004). HIDE: An Infrastructure for Efficiently Protecting Information Leakage on the Address Bus. *Proceedings of the 11th International Conference on Architectural Support for Programming Languages and Operating Systems (ASPLOS)*, 72-84.

Chapter 13
Memory Based Anti-Forensic Tools and Techniques

Hamid Jahankhani
University of East London, UK

Elidon Beqiri
University of East London, UK

ABSTRACT

Computer forensics is the discipline that deals with the acquisition, investigation, preservation and presentation of digital evidence in the court of law. Whereas anti-forensics is the terminology used to describe malicious activities deployed to delete, alter or hide digital evidence with the main objective of manipulating, destroying and preventing the creation of evidence .Various anti-forensic methodologies and tools can be used to interfere with digital evidence and computer forensic tools. However, memory-based anti-forensic techniques are of particular interest because of their effectiveness, advanced manipulation of digital evidence and attack on computer forensic tools. These techniques are mainly performed in volatile memory using advanced data alteration and hiding techniques. For these reasons memory-based anti-forensic techniques are considered to be unbeatable. This chapter aims to present some of the current anti-forensic approaches and in particular reports on memory-based anti-forensic tools and techniques.

INTRODUCTION

The advent of Information Technology and personal computers has transformed significantly our way of living. Most of our day to day activities rely heavily upon the use of electronic devices and digital communications. More people are relying on these technologies to learn, work and entertain. In 2003, USA Census Bureau estimated that sixty-two percent of the households had access to a personal computer while fifty-five percent had access to the internet (Census, 2003). Without doubts digital communications can be considered as one of the greatest inventions of the last century because of its impact and benefits on the society.

On the other hand, digital communications have provided new opportunities for criminals

DOI: 10.4018/978-1-61692-000-5.ch013

and shaped the ways they commit crime (Shinder, 2002). Criminals are exploiting now digital communications to commit a wide range of crimes such as identity theft, online piracy, financial fraud, terrorism and pornography distribution. Furthermore the incidences of some of types of crimes increased significantly with the introduction of digital communications and personal computers. For example, internet communications have escalated the problem of child pornography by increasing the amount of material available, the efficiency of its distribution and the ease of its accessibility (Wortley, 2004).

According to Bruce Schneier, electronic crime is flourishing because of three main reasons: a). automation, b) action at distance, c) technique propagation (Schneier, 2000).

a) Automation: Software packages are used to perform repetitive tasks and cross reference more and more data.
b) Action at distance: We live in a global digital communication era. Criminals perform electronic crimes in distance and with a high rate of anonymity.
c) Technique Propagation: successful electronic crime techniques and malicious software is propagated easily through the internet.

Law enforcement agencies have started dealing with crimes involving electronic devices and communications since the 1970's when these technologies were introduced. These were coined as electronic crimes since electronic devices and digital communications were used to commit them; while electronic evidence was defined as information or data of investigative value that is stored or transmitted by electronic devices (Ashcroft, 2001).

Law enforcement investigators initially considered electronic evidence as any other type of evidence; however they realised soon that this was not the case and that the conventional approach was not suitable to collect, preserve and analyze electronic evidence. This is because "conventional evidence lives in an analog world, whereas computer-derived evidence comes from a digital world and the transition between these worlds is not always as smooth as one would hope" (Johansson, 2002).

Computer forensics was then established as a discipline to support law enforcement agencies in their fight against electronic crime. Computer forensics deals with the acquisition, investigation, preservation and presentation of digital evidence in the court of law with the final objective of finding evidence that would lead to prosecution. Computer forensics is also known as cyber forensics since it deals with crimes committed in the cyber world (electronic world). The main areas of searching for evidence are: hard drives, removable devices, volatile memory, deleted or hidden files, password protected files, pornographic material etc.

The most important input of a computer forensic investigation is the digital evidence. Digital evidence can be envisaged as the counterpart of fingerprints or DNA in the digital world. Criminals will attempt to cover the traces of their malicious work by using anti-forensic methods to manipulate and tamper the evidence or interfere directly with the process (Harris, 2006).

Anti-forensics is the terminology used to define the activities of hackers or other cyber criminals aiming to undermine or mislead a computer forensic investigation. There are no well-established definitions regarding this discipline since it is quite new and it is yet to be explored. Peron and Legary define it as *"..four categories of evidence destruction, evidence source elimination, evidence hiding and evidence counterfeiting…."*(Harris, 2006), while, Grugq, (Ruxcon, 2004) defines anti-forensics as "*[The attempt] to limit the quantity and quality of forensic evidence.* "

Although Anti-Forensics is a field under development, however, there are already categories of available tools. Grugq seems to be one of the most dedicated anti-forensic researchers so far. With more than five years of anti-forensic studies, he ended up losing his job after publishing "Art of Defiling: Anti-Forensics", (Ruxcon, 2004).

The Metasploit Anti-Forensic project by Vincent Liu is part of the Metasploit project which targets audiences interested in penetration testing. Liu's presentation titled "Bleeding-Edge Anti-Forensics" which was co-presented with Francis Brown for an Infosec World Conference was the most descriptive work of what he did so far about Anti-Forensics.

There are number of techniques that are used to apply anti-forensics. These techniques are not necessarily designed with anti-forensics dimension in mind. For instance, folder shielders have been designed in order to primarily provide a level of security and privacy, but they can be used as an Anti-Forensic tool since they can hide data, (Jahankhani, 2007). The other are;

- **Digital Media Wiping:** A proper wiping of the media that contain the digital evidence, will simply disappear the evidence.
- **Steganography:** Someone can use Steganography to hide a file inside another and make the investigator unable to take advantage of the evidence.
- **Privacy Wipers:** These are tools aim to delete any privacy traces from operating systems, applications or both. If properly used the investigator might find no evidence at all inside the digital media.
- **Rootkits:** Rootkits can subvert the operating system kernel and even react to forensic acquisition processes by hijacking the way the operating system uses areas like process management or memory management to extract the evidence.
- **S.M.A.R.T Anti-Forensics:** This kind of technology can be used by an attacker to suspect if a hard drive has been taken out for a forensic duplication process.
- **Homographic Attacks:** Such an attack can mislead an investigator since some letters that look similar to the human eye can be replaced with others in such a way to make a malicious file look legitimate.
- **File Signature Modification Attacks:** Someone can purposefully change the file signature of a file to make it look something else.
- **Encryption:** This can be used almost in every anti-forensic stage in order to obscure and make unreadable and unusable the evidence.
- **Metadata Anti-Forensics:** Information about data (metadata) can be altered in order to hide user actions.
- **Slack Space Anti-Forensics:** Someone can hide malicious software in areas that operating system might not use, like slack space, because they might be considered as reserved or empty.
- **Secure Digest Functions (MD4, MD5, etc) Collision Generation:** Someone can alter a file and then use Anti-Forensic software to make this file having the same MD4 or MD5 value like before the alteration, thus bypass a forensic integrity check.
- **Digital Memory Anti-Forensics:** There are programs that are able to hide processes or other evidence from memory.
- **Misleading Evidence:** Someone can leave evidence in such a way to mislead the forensic investigation.
- **Packers / Binders:** Someone can use such a program in order to transform a file by changing its structure, thus it can bypass security mechanisms that searches for malicious behaviour patterns inside files.
- **Forensic Tools Vulnerabilities / Exploits:** There are already implementations available to show that some of the computer current Forensic Tools can be bypassed or exploited.
- **Resource Waste:** To purposefully leave traces in a big network in order to make the forensic investigator waste valuable resources and time.

- **Forensic Detection:** Someone can install a mechanism to be triggered after any computer forensic related presence.
- **Anonymous Actions:** It includes every action that can be done by a fake or unknown identity. The result from the investigator is to fail to trace back the malicious activities.
- **Anti-Forensics In Flushable Devices:** Someone can take advantage of devices that can be flashed (like PCI cards or BIOS) and install malicious code inside them, thus they can remain unnoticed.
- **Wireless Anti-Forensics:** Attempts to compromise the availability or usefulness of forensic data related to wireless communications (Harris, 2006).

The aim of this study is to present three of the current anti-forensic approaches and focusing only on memory-based anti-forensics known as anti-forensic live CD's. Memory-based bootable live CD's are specially built Linux operating systems that boot directly from the CD drive into the RAM (Random Access Memory) area. These packages do not load into the hard drive, do not change files or alter other variables in the target system unless specified by the user. Live CD's are used mainly for penetration testing and other security related tasks and they include a variety of software packages used for anti-forensic purposes.

ACTING ANONYMOUSLY

From a forensic scope, anonymity can be considered as a major anti-forensic approach. Below are some of the tools that are used;

Anonymous Mail Accounts

These are accounts that are created using services available on the Internet that facilitate anonymous mailing. This will make the process of e-mail tracking more difficult as the mail headers are altered and no IP address details will be available, (Greene, 2003).

Figure 1, shows that the sender is an anonymous authority – in this case the domain of "bananasplit.

Figure 1. Anonymous mail details

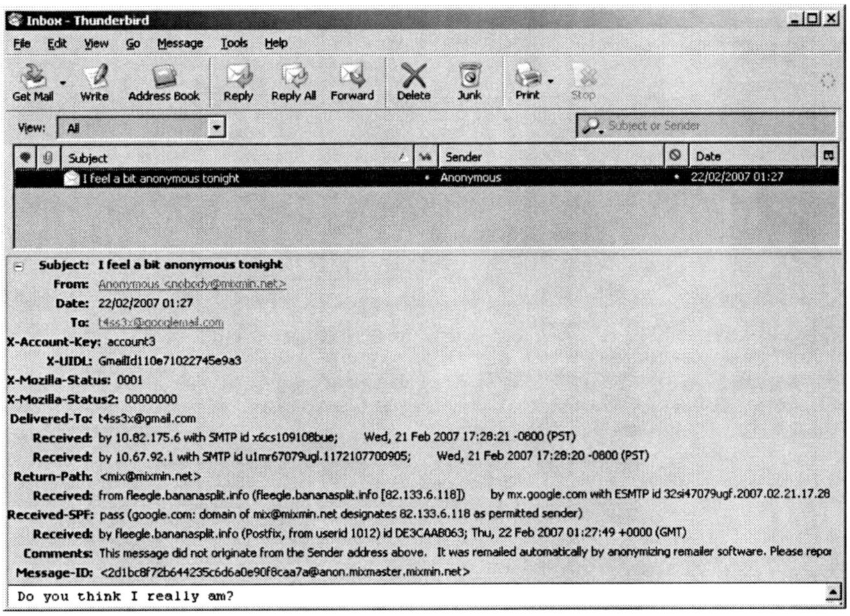

info". There is also an e-mail address (abuse@ bananasplit.info not visible in the picture) where more information about the sender's IP could be requested by the forensics team.

Anonymous Proxies

Nowadays there are plenty of anonymous proxies on the Internet with a significant number of them being free (Free Proxy, 2006), (Anonymous INET, 2006). Although these proxies promise Internet anonymity, they do not always talk about the level of anonymity service they provide.

Below is the result of a test on a high anonymity service in order to find out the amount of information on the user's identity. The anonymity has been checked against a web site (anonymitytest. com) that shows the IP address of the visiting address along with a service ("whois") that aims to trace back the IP.

It is important to note that in special cases even a high anonymity server can reveal all the information regarding its users. All someone has to do is to monitor and analyze the traffic patterns coming to and from that proxy, (Gibson, 2006).

Figure 2, shows that high anonymity proxy puts its own IP number to the visited web page in order to keep the client anonymous.

An attempt to trace back the address will come up with the details of the proxy server and not the user's. (Figure 3)

In order to get more information about the real visitor's identity, the anonymous proxy provider has to be contacted. Here are the problems someone might face:

1. There are cross boarder legal issues. In this case the domain ends with "br" which means that the proxy owner is located somewhere in Brazil.
2. The anonymous proxy provider - no matter the geographical location - might claim that all the logs are deleted and nothing is saved regarding their clients (anonymous visitors). In that case, only a government regulation which enforces IP logging would provide a connection to the client's IPs.

It is important to note that someone can be a part of an anonymous network - like Tor - in order

Figure 2. Anonymity service

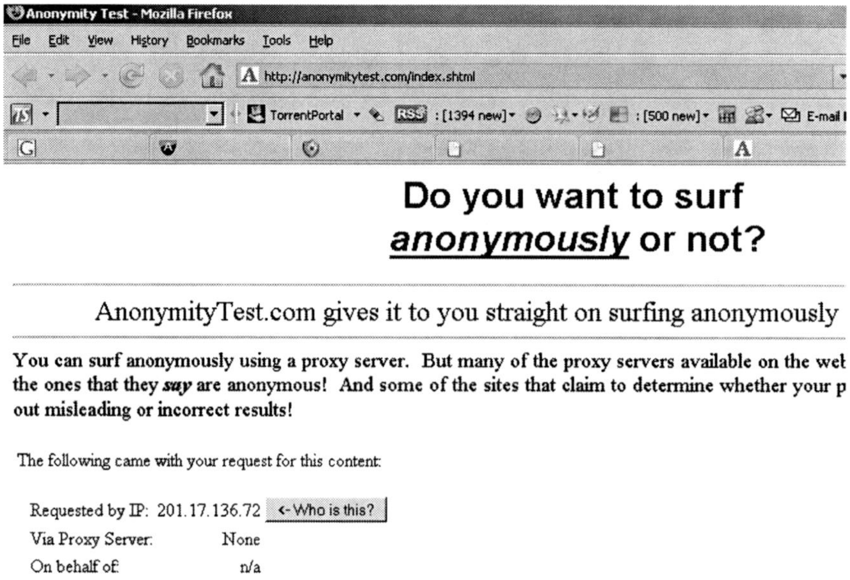

Figure 3. Proxy Server Details

```
nic-hdl-br:   DSS30
person:       Diego Santos Soares
e-mail:       virtua@virtua.com.br
created:      20000424
changed:      20070128

nic-hdl-br:   RII19
person:       Ricardo Ide
e-mail:       ricardoide@globocabo.com.br
created:      20011010
changed:      20011010

remarks:      Security issues should also be addressed to
remarks:      cert@cert.br, http://www.cert.br/
remarks:      Mail abuse issues should also be addressed to
remarks:      mail-abuse@cert.br

% whois.registro.br accepts only direct match queries.
% Types of queries are: domains (.BR), BR POCs, CIDR blocks,
% IP and AS numbers.
```

to achieve anonymity. In this case it is not feasible even for the governments to totally follow an IP or a packet, since the information is going through a variety of interconnected nodes with some encrypted links through several countries. This is a more efficient way to keep the users anonymous.

The only way for someone to monitor an amount of Tor's traffic is to set up a fake Tor server and monitor the traffic of some other servers as well. In a crackdown of a recent crime investigation in Germany, police seized 10 Tor servers for suspicion of a child porn investigation, (Oates, 2006)

WIRELESS ANTI-FORENSICS METHODS

How about if someone launches an attack using multiple access points from roof top of a high building in the middle of a crowded city with the help of a strong directional antenna?

Raul Siles, in his excellent two-part article "Wireless Forensics: Tapping the Air" unveils some "de facto" and some new wireless anti forensics methods. Some of the major approaches are (Siles, 2006 and 2007);

- **The use of covert/illegal channels**:(like channel 14 in US and Europe).Covert channels are channels of communication used illegally for stealth purposes. For example, channel 14 (wireless communications) is not allowed for use; however these frequencies can be used by attackers to exchange data in a covert way. This wireless anti-forensic activity is hard to detect unless the examiner knows what is happening and where to look for them.
- **The use of strong layer-2 encryption**: this is another technique to defeat wireless forensics. Usually the attacker will encrypt his communication channels in order to avoid detection. Because the traffic is encrypted it will be hard or impossible for an examiner to understand the nature of the malicious activity or its content.
- **The modification of the 802.11 specification**: In 2006, a proof of concept approach to wireless communications referred to as "Wi-Fi-Stealth" was presented at the Black Hat US Conference (Butti et al, 2006). Their approach involved modifying the standard 802.11 protocol in such a way that would allow communication between access point with similar configuration.

Different values regarding frame management, control and data frames could be altered easily. Such modifications could make the stealth network invisible most of the wireless device scanners. Even if the frames are detected they will go wrongly mistaken for corrupt frames or communication noise.

- **Wireless MAC spoofing**: MAC address spoofing involves the modification of network cards with the intention of avoiding detection or facilitating attacks. Programs like SMAC can change easily MAC addresses on either wired or wireless communications (SMAC, 2009).By changing wireless MAC addresses, it is possible to achieve high success rates, especially in replay or fake authentication attacks using well-known tools such as aircrack or aireplay. Furthermore, fake wireless MAC addresses will undermine the investigation process.
- **Wireless knocking**: is the counterpart of port knocking in "the wired world" .Port knocking is a form of host to host communication, by generating a connection attempt on a set of specified closed ports. Once a correct sequence of connection attempts is performed connection is established. The concept of wireless knocking was introduced in 2005, at the Black Hat Europe conference in 2005(Oudot, 2005). "Wknock", a software package introduced at the conference, could configure rogue access points in monitor/passive mode, therefore making difficult for examiners to detect their presence since an access point in monitor mode does not sent out packets.
- **802.11n Greenfield mode**: 802.11n is the latest wireless standard developed and released lately to increase the speed of data communication up to 600 Mbit/s. The standard provides two modes: mixed and Greenfield mode. Mixed mode provides support for legacy communications (802.11a/b/g), while the Greenfield mode can not be detected by equipment which is not compatible with the new 802.11n standard. Wireless forensic investigators will not be able to capture data or even detect the presence of 802.11n Greenfield access point unless they have updated their equipment.

While in theory the forensics investigator should monitor every single packet of every channel available around the suspect, in reality the post incident response could end up quite dramatically. This could be due to; ignorance regarding the channels and access points used, legal barriers between the access point and the forensics acquisition, non – cooperative ISP's, etc. The forensic process should be enhanced with security mechanisms which would upgrade the post-incident reaction to real time. The real-time acquisition tools should have capabilities of capturing activity of all the wireless point within a respectable distance.

MEMORY BASED BOOTABLE ENVIRONMENTS

There are plenty of "live CD's" - commercial and freeware – in the market that are made to meet certain user's needs. These are like data recovery (SystemRescueCD, ERD Commander), security (BackTrack, NST), PC benchmarking (StressLinux, Ultimate CD), gaming (LLGP, Freeducgames) or even alternatives for a fully functional operating system (Knoppix, Kanotix). There is also a website called Frozen Tech (www.frozentech.com) that lists a vast majority of the live CD's (if not all of them), (Brand, 2006)

A live CD is nothing more than a compact disk, DVD or USB drive which contain an operating system image file and a boot loader program, used to start or boot a computer system. An image file is a single compressed file that contains the entire operating system programs and files. Bootable CD's, also known as LiveDistros are mostly available freely open source license agreement. According to this agreement "anyone can modify and redistribute the original operating system without asking for permission of retribution from the author" (Open source, 2007).

The concept behind using removable media for storing operating systems is not new. In the early introduction of personal computers, operating systems (such as MS-DOS) were loaded into the memory from removable media (usually floppy disks). With the advent of mainframes (considered the first generation of computer systems), the instructions to hardware components were given by punched cards, which although did not constitute an operating system in per se did introduce the concept of OS. It is worth mentioning though that punched cards were not effective as live CD's since extensive processes required hundreds of them (Gochioco, 2004).

Mainframes were not the only computer systems that used removable media for storing instruction programs or operating systems. For example, diskless computer systems do not have operating systems installed; instead they load from a copy of the operating system located in a network server. Either operating systems such as MINIX are distributed mainly in removable media(CD, floppy etc)because of its extremely small size; MINIX kernel counts only 4000 program lines whereas other operating systems rely on millions of lines of code (Minix, 2007).

Although live CD's are the preferred tools of trade in conducting memory-based anti-forensics, most of them were designed for security testing purposes. Good collections of security testing tools are distributed with these portable media and are usually used by computer security professionals to troubleshoot their computer systems and networks. Unfortunately, even malicious users are making use of these specially build packages to perform illegal activities, amongst all anti-forensics.

The majority of memory-based packages used for anti-forensic purposes, are built on UNIX oriented platforms, although there are similar packages built on platforms such as Windows, Apple MAC, MS-DOS, MINIX etc. There is a reason behind this. Most of UNIX operating system platforms are built on open source code (non proprietary) and for this reason a lot of security focused tools are developed and distributed online. UNIX systems are also the platform of choice for security specialists and hackers because of their reliability, flexibility and amount of accessible security tools. Typically, the graphical user interface (GUI) of a live CD is either a KDE or Gnome interface. These are the most popular user interfaces in UNIX platforms.

Live CD's manage to recognise and work with a variety of hardware components thanks to its device manager called "udev" (Qlogic, 2007), which is the device manger for most of UNIX /LINUX based systems. Device managers are programs designed to handle hardware components of a computer system. Having a device manager that interacts with most of hardware devices promotes the interoperability and portability of livedistros. All the removable media attached to the system are viewable from the livedistro interface.

The OS loaded from the live CD can view, access or copy all the files and programs created by the native operating system (these are stored in the secondary memory area). Linux based operating systems treat all attached devices as files. When the live CD start running it checks for hard drives, removable media and other devices attached to the motherboard. When found, these devices are mounted (connected) to the operating system; at shut down the connection is dropped. Hard disks contents are viewable, can be copied but no alterations can be made to files from the live CD interface. From a malicious user point of view, this is a good opportunity to explore, since information in the native hard disk can be accessed easily without leaving traces. Most of the operating systems run security services which keep records of users or systems that access its files. However, since these systems are not running, no records will be added to the log files of the native OS.

Furthermore, important system files can be copied, scrutinised and important data disclosed. For example, in Windows platforms "SAM" file is the recipient of system passwords. These passwords are encrypted, but if copied and then attacked, systems passwords can be identified. Live CD's provide the necessary tools to unveil the encrypted passwords copied from the hard disk (e.g. Kain, John the Ripper), compromising seriously the security of the system. From the interface of the live CD, the native OS security files such as "SAM" can be copied and saved in ramdisk or in a removable device to be attacked at a later stage. A fully detailed demonstration of this attacking technique will be provided in the next chapter.

Network services are fully accessible from the live CD interface. Connections to internet or local computer systems are easily implemented permitting the user to perform most of the tasks available in operating systems that run from the hard disk.

Memo-based anti-forensics techniques are difficult to beat for a variety of reasons. The most important advantages of using these techniques are:

1. Lack of digital evidence,
2. Compatibility, flexibility and portability,
3. Anonymity,
4. Availability of tools,
5. Freely distributed.

SecurityDistro.com, a site dedicated to memory-based security tools, lists over forty memory-based packages (SecurityDistro, 2007). Most of the packages offer similar tools and interfaces. Among these packages are; Backtrack, Anonym. OS, Helix, Penguin Sleuth and Auditor collection which have a wide range of security tools that might be used to deploy anti-forensic activities.

In this chapter Backtrack anti-forensic tools was used which is freely available. The selection is not casual. Backtrack, does provide the user with the opportunity to use a well-established security focused framework, metasploit (Metasploit, 2007). Metasploit framework is a collection of security tools used to test the security side of computer systems, penetration testing and exploitation.

Furthermore, metasploit contains a special module called anti-forensics, which is a collection of anti-forensic tools (example: timestomp, sam juicer, slacker, transmogrify etc) that can be loaded and used directly from the live CD.

According to Vinnie Liu, a well known anti-forensic researcher, these tools are designed to tamper with or break well recognized industry tools such as Encase, NTFS and PGP desktop, with the final objective of manipulating the digital evidence and compromising the investigation findings (Liu, 2007).

Once Backtrack is downloaded, the Linux based operating systems can be burned in a bootable CD and be ready for use. Instructions on how to accomplish this task are available on various sites online.

Stealing Passwords

Backtrack live CD can be used by malicious users to steal and then crack passwords used to log in computer systems. Windows operating systems store password information locally in the hard drive in a system file called Security Accounts Manager; otherwise known as SAM file. This file is very important from the security point of view since it contains all system user passwords in an encrypted format. Encryption of user passwords is performed by Windows using a proprietary encryption utility called system key which uses "strong encryption techniques to secure account password information that is stored in the SAM file".

In a computer system running Windows OS the system key utility (program) is located at this logical address: C: /Windows/system32/config/

system. The system key program contains also the key used to encrypt the passwords stored in SAM.

Usually access to SAM is restricted since it is a system file. Even if the user manages to copy SAM in a portable media device it will be difficult to unmask the hidden passwords since a key stored in the system utility is needed to decrypt SAM. The key must be extracted first from the system key utility. However with Backtrack, a user can extract quite easily user passwords stored in SAM, without creating digital evidence. By using Backtrack no digital evidence is left since all the operations are performed in RAM and at the same time no digital evidence will be left in the native Windows system since it is not running.

Backtrack can be also used to recover user account passwords of remote computer system running Windows OS. In this case metasploit frame work might be used to achieve the goal. A metasploit 'exploit' called lsass_ms041_011 is used to connect remotely to vulnerable computer systems. Once a connection is established remotely an advanced metasploit package called "meterpreter" is used to fully explore to the target system.

The meterpreter is a sophisticated software package shipped with metasploit that facilitates attack automation, by making it easy to interact with processes, networking, and the file system (Metasploit, 2007). Furthermore, another special module of meterpreter called Sam juicer can be used to copy password hashed from SAM. Sam juicer performs the task without accessing SAM file, the registry or writing any files in the remote computer system hard disk. This is achieved through an advanced connection technique called direct memory injection; no digital evidence is created or left in the target computer system (Liu, 2007).

This technique undermines completely the investigation process because no digital evidence is created in the remote computer system since the communication between computers is conducted in the temporary memory (RAM). Files are not accessed directly, new processes are not created or data added to the log (security monitoring) files. A computer forensic investigator will not be able to gather digital evidence from neither of the computer systems (the attacker's or the attacked system) simply because the evidence does not exist.

Modifying Timestamps

From a computer forensic investigation point of view file timestamps are very important because they provide the necessary evidence to prove if certain anti-forensic activities occurred at a certain moment in time or whether a user was logged in a computer system. For this reason malicious users might attempt to modify timestamps in order to eliminate compromising evidence. Timestamp is the data appended to a file that shows when a file is created, accessed, modified or entry modified. These file attributes are also known as MACE (Modified, Accessed, Created, Entry Modified) attributes. Anti-forensic tools attempt to modify these data parameters in order to mislead computer forensic investigators.

Backtrack again provides the perfect tool to modify timestamps .The tool is called timestomp and is included in the metasploit framework .Timestomp is a program developed by metasploit project which gives the user the opportunity to modify all NTFS timestamp parameters (Metasploit, 2007). NTFS (New File Technology System) is the proprietary file system of modern Windows operating systems including NT, 2000, 2003, XP and Vista (NTFS, 2007).

Timestomp can be used also as a standalone program to modify timestamps; however its potential is fully explored when used within the metasploit framework. Meterpreter module (started from backtrack) permits a user to connect remotely to a target computer system. On the other hand from within this module timestomp can be executed to modify file timestamps. Since all the operations are conducted in temporary memory

(RAM) no digital evidence is left in the systems to indicate traces of anti-forensic activity. Certainly it will almost impossible for a computer forensic investigator to notice timestamp modification since its parameters will look legitimate.

Hiding Data in Slack Space

Slack space is the preferred hard disk area used by malicious user for storing illegal software, documents or pictures because files stored in it are not seen or accessed by windows explorer; "data is hidden in unallocated or unreachable locations that are ignored by the current generation of forensic tools" (Fwiki, 2007). The user is completely unaware of the existence of such files.

A variety of malicious programs might be used to hide data in the slack space; however 'slacker' is one the most proficient tools used to perform such activities. Slacker, which is named after the slack space, is developed by the metasploit team and is released as a module with metasploit framework (Metasploit, 2007). Slacker uses a sophisticated technique to hide programs, files or any other type of data in the slack space. It takes the data, fragments it into thousands of pieces and then distributes it across the slack space in the hard disk. This program mainly stores the data in stable file such as system files (windows/system32 files) which are not examined also buy computer forensic tools .Slacker main features include file splitting and slack space hiding; these features make slacker very hard to trace.

If a computer forensic tool is used to analyze the data in the slack space, no evidence will be discovered since individual fragments of data will not help to construct the true nature of the hidden file; for the forensic tool data is so diffuse that it looks like random noise (Berinato, 2007). Only Slacker can recompose the fragmented pieces of data to create the original file. Slacker have proven to be successful also against PGP desktop, a security tools that includes some tools claming to wipe out completely the slack space. Metasploit researcher Vinnie Liu has proven that data written in the slack space with slacker can not be wiped out even when PGP desktop tool is used (Metasploit, 2007).

Modifying File Extensions and Signatures

In a computer system files are identified by two attributes: file extensions and file signatures. For each file format there is a unique file signature, for example executable files in Windows are identified by file signatures starting with the letters MZ (Liu, 2006). Therefore, to hide a file in a computer system suffices to change its extension and add the letters MZ at the beginning of that file. By using this technique files containing pornographic material for example can be masqueraded as system files and go undetected by computer forensic tools.

Memory-based anti-forensic tools such as Backtrack can be used effectively to manipulate file extension and signatures. Metasploit project has developed and is about to release a tool called Transmogrify that allows a user to masquerade malicious files .Metasploit developers claim that this tool is able to alter file extension and header / signature without being detected by forensic tools like Encase (Metasploit, 2007).

Meterpreter loads remote processes (such as cmd.exe) in memory, therefore no digital evidence will be left in the system .To make things harder for computer forensic investigators, the MACE attributes of wufileuel.sys can be further altered to match the attributes of other system files. Forensic tools will not be able to identify the malicious file since file extension, signature and timestamps look legitimate, and therefore no red flags will be raised.

Live CD Wireless Attacks

Automated wireless attacks are not anymore a thing of the future. Live CDs like Russix are purposely built to automate the process and defeat computer forensics. Russix is a Slax based Wireless Live Linux, designed to be light and dedicated purely to wireless auditing and security (Darknet, 2009). Russix evolved from an internal UK Military Wireless auditing tool based on Debian. It performs WEP cracking in less than thirty minutes and WPA cracking even faster provided that the password of the attacked access point is in its password files. Furthermore this live CD can be used together with rainbow crack tables to increase chances of success against WPA access points. The "Church of the wi-fi" group has released a 33Gb WPA rainbow tables file that can be used with Russix to attack WPA access points (Renderlab, 2009).The table has been generated by combining a very long list of well-known access point names and millions of commonly used or generated passwords. All the operations are performed in RAM; therefore no evidence is left in the attacker's machine once the CDROM is removed from the drive.

Secure Disk Wiping Tools

More tools are now available for download that would support secure data deletion or hard disk formatting. The tools that will be mentioned below do make it really hard or impossible for an investigator to recover even fragments of data.

a) **Darik's Boot And Nuke** ("DBAN") is a self-contained boot disk that securely wipes the hard disks of most computers. DBAN will automatically and completely delete the contents of any hard disk that it can detect, which makes it an appropriate utility for bulk or emergency data destruction (Dban, 2009).
b) **Wipe:** is a tool used to degauss hard disk surfaces, making it virtually impossible to retrieve the data that was stored on it. This tool is designed to make sure that sensitive data is completely erased from magnetic media (Sourceforge, 2009)
c) **Overwrite**: is a UNIX utility that tries to make harder data recovering by overwriting files using random patterns and deterministic patterns (Kyuzz, 2009).
d) **Grind:** overwrites files with certain byte patterns and random data in order to make it impossible to recover the file from magnetic disks(Grind, 2009).
e) **Diskzapper**: automatically begins erasing all the disks as soon as the booting process is completed. No user action is required. (Diskzapper,2009).

Figure 4. Copyright www.securitydistro.com

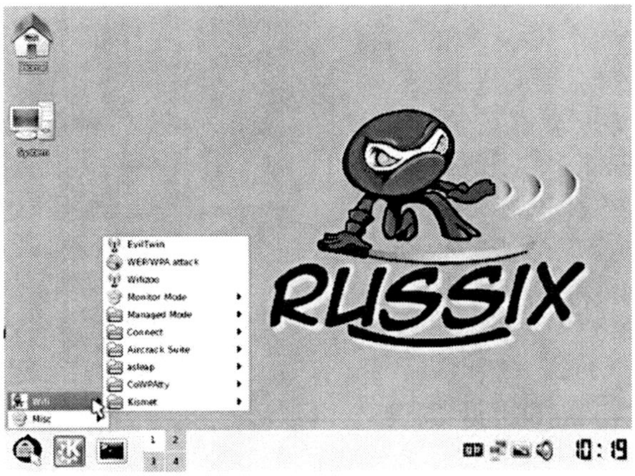

f) **Evidence eliminator**: Evidence Eliminator destroys: Windows SWAP file, Windows Application logs, Windows Temporary Files, Windows Recycle Bin, Windows Registry Backups, Windows Clipboard Data, Start Menu Recent Documents history, Start Menu Run history, Start Menu Find Files (Evidence, 2009).

CRITICAL DISCUSSION AND EVALUATION OF THE MEMORY BASED ANTI-FORENSIC TOOLS AND TECHNIQUES

Memory-based anti-forensic tools and techniques interfere substantially with the investigation process by altering or hiding digital evidence. Because memory-based anti-forensics techniques are deployed directly in temporary memory (RAM), the defence strategies must be focused at this memory area. Slack space must be scrutinised and analysed as well. These defensive strategies must be implemented to achieve some sort of success against memory-based anti-forensic activities:

a) Slack space analysis: Statistical analysis of slack space must be conducted to discover strange or unusual samples of data (Liu, 2006). For example, Slacker splits and distributes file fragments usually across a vast range of system files. For this reason it would be wise to scrutinise carefully the slack space of these files in order to trace potential anti-forensic activity, even though it will be difficult to reconstruct the original hidden data. According to Vinnie Liu, file fragments in slack space and file system information must be analysed together in order to discover elements of slack space anti-forensics (Liu, 2006).

b) Capture data in memory: Memory-based live CDs load, operate and store data in memory, unless specified otherwise by the user. For this reason the main area to look for digital evidence is the memory of the local or remote computer system used for anti-forensic activities. Other data that should be captured is running processes, ports, uploaded or downloaded files which will indicate if such activities occurred.

Memory-based anti-forensics relies mainly on volatile memory, while traditional anti-forensics is deployed in the secondary memory storage area (hard disk). For this reason memory-based live CD activities are hard to detect since volatile memory is unstable and easy to erase. Computer forensic investigators might be able to collect digital evidence only if the perpetrator's computer device is seized and is not shut down; however if the user has removed the live CD and turned off the system, the evidence is lost permanently since memory is volatile. Forensic tools such as Memparser (SourceForge, 2007) or Windows Memory Forensic Toolkit(WFTK, 2007) might be used to collect valuable data in memory.

c) Improving Computer forensic tools: Some memory-based anti-forensic tools (e.g. Timestomp, Transmogrify) tackle computer forensic packages. Timestomp provides a specific option to trick Encase; by using switch–b timestamps are set to blank. Encase and FTK (Forensic Tool Kit), two prestigious computer forensic tools do not recognise Timestomp changes (Liu, 2006). This is a clear indication that these tools must be improved (or re-written) in order to properly detect timestamp alterations, particularly because modified timestamps can compromise the success of a computer forensic case in the court of law. On the other hand Encase and FTK do not detect file extension or signature modification achieved with Transmogrify or Backtrack; even in this case these tools must be improved in order to detect traces of anti-forensic activity. Computer forensic tool designers should examine carefully how memory-based live CDs interact with the system in order to improve their future released tools; the idea behind this is to use the same anti-forensic tools to defeat them.

d) Improving signature analysis: Memory-based anti-forensic tools manage to modify file extensions and signatures. In the meantime some of the most used computer forensic tools (Encase, FTK etc) fail to detect such changes. Encase checks only the first two characters of a file signature which can be easily modified (e.g. MZ for executable files) .Therefore, automated tools will not be able to detect file signature modifications achieved by using memory-based anti-forensic tools. In this case manual investigation should be conducted provided that the investigator identifies the suspicious files. Forensic tools need to be redesigned to tackle file signature modification. A good way forward would be the redesign of the searching process so that files are checked from top to bottom for patterns of data .This method might produce good result since particular patterns of data might be associated to certain files ; if these patterns are not present then further investigation can be conducted.

e) Invest in human and time resources: If digital evidence is well hidden there is a high probability that it will go undetected because of feasibility issues. This is true especially when there are thresholds to meet in terms of time and cost (for e.g. pressure to complete a forensic investigation within a deadline). In the case of memory-based anti-forensics digital evidence is hard to recover because it is well hidden. Therefore more training and time must be provided to investigators so they can succeed. Financial resources must also be committed to develop efficient forensic tools in short periods of space in order to keep up to date with anti-forensic tool development.

Another issue to consider seriously is the way how computer forensic investigators process seized data. Computer forensic cases are usually based on evidence produced by automated tools such as Encase or FTP. Unfortunately these tools do not detect most of anti-forensic activities deployed by live CDs. The solution to this problem is a combination of in depth manual investigation and automated tool searching since manual processing might be able to reveal some of the digital evidence traces.

CONCLUSION

Anti-Forensics is a reality that comes with every serious crime and involves tactics for "safe hacking" and keeps the crime sophistication in a high level. Computer forensic investigators along with the forensic software developers should start paying more attention to anti-forensics tools and approaches.

If we consider the Computer Forensics as the actions of collection, preservation, identification and presentation of evidence, anti-forensics can affect the first three stages. Because these stages can be characterized as "finish to start" between them from a project management point of view, the failure of one of them could end up as a failure of the lot. Thus, there is a high impact of Anti-Forensics to the forensics investigations.

Officially there is no such thing as anti-forensic investigations because the anti-forensic countermeasures are still part of the investigator's skills.

REFERENCES

Anonymous, I. N. E. T. (n.d.). *Fast Proxy Server List*. Retrieved February 28, 2007 from http://www.anonymousinet.com/, [accessed 28 February 2007 Ashroft, J. (2001). *Electronic Crime Scene Investigation: A Guide for First Responders*. Retrieved June 22, 2007 from http://www.iwar.org.uk/ecoespionage/resources/cybercrime/ecrime-scene-investigation.pdf

Berinato, S. (2007). *The Rise of Anti forensics*. Retrieved June 16, 2007 from http://www.whitehatsec.com/home/resources/trade/07tradenews/062607CSO.html

Brand, N. (2007). *Frozen Tech, The LiveCD List*. Retrieved August 11, 2007 from http://www.livecdlist.com/

Butti, L., & Veyssett, F. (2006). *Wi-Fi Advanced Stealth Proc.Black Hat US*. Retrieved April 24, 2009 from http://www.blackhat.com/presentations/bh-usa-06/BH-US-06-Veyssett.pdf

Census Bureau. (2003). *Computer use in 2003*. Retrieved June 21, 2007 from http://www.census.gov/population/pop-profile/dynamic/Computers.pdf

Darknet (2009). Retrieved May 2009 from http://www.darknet.org.uk/2008/02/russix-livecd-linux-distro-for-wireless-penetration-testing-wep-cracking

Dban (2009). Retrieved May 11, 2009 from http://www.dban.org/

Diskzapper (2009). Retrieved March 14, 2009 from http://diskzapper.com/

Evidence (2009). Retrieved May 23, 2009 from http://www.evidence-eliminator.com/product.d2w

Forensicwiki (2007). *Anti-forensic techniques*. Retrieved August 5, 2007 from http://www.forensicswiki.org/wiki/Anti-forensic_techniques

Free Proxy. (n.d.). Retrieved February 16, 2007 from http://www.freeproxy.ru/en/free_proxy/

Frozen Tech. (2006). *Live CD Creation Resources*. Retrieved February 16, 2007 from http://www.livecdlist.com/wiki/index.php/LiveCD_Creation_Resources

Gibson, S. (2006). Gibson Research Corporation. *Security now – Transcript of Episode 70 – Achieving Internet Anonymity*. Retrieved February 23, 2007 from http://www.grc.com/sn/SN-070.pdf

Gichioco, L. (2002). *Computer technology: From punch cards to clustered supercomputers*. Retrieved July 28, 2007 from http://tle.geoscienceworld.org

Greene, T. C. (2003). The Register. *Net Anonymity Service Backdoored*. Retrieved August 21, 2003 from http://www.theregister.co.uk/2003/08/21/net_anonymity_service_backdoored/

Grind (2009). Retrieved January 11, 2009 from http://users.ox.ac.uk/~kebl2949/grind/

Harris, R. (2006). Arriving at an anti-forensics consensus: Examining how to define and control the anti-forensics problem. Retrieved April 12, 2007 from http://www.dfrws.org/2006/proceedings/6-Harris.pdf

Insecure.org. (2006). Top 100 Network Security Tools. *Fyodor*. Retrieved February 3, 2007 from http://www.insecure.org/tools.html

Jahankhani, H., Anastasios, B., & Revett, K. (2007). ECIWS, Digital Anti Forensics: Tools and Approaches. Retrieved June 2007 from http://academic-conferences.org/pdfs/eciw07-booklet.pdf

Johansson, C. (2002). *Forensic and Anti-Forensic Computing*. Retrieved June 24, 2007 from http://www.fukt.bth.se/~uncle/papers/forensics200212.pdf

Kyuzz (2009). Retrieved February 12, 2009 from http://www.kyuzz.org/antirez/overwrite.html

Liu, V., & Stach, P. (2006). *Defeating Forensic Analysis, CEIC 2006 –Technical Lecture1*. Retrieved from http://stachliu.com/files/CEIC2006-Defeating_Forensic_Analysis.pdf

Metasploit.com. (2007). Retrieved June 12, 2007 from http://www.metasploit.com

Minix (2007). Retrieved June 4, 2007 from http://www.minix3.org/

NTFS. (2007). Retrieved August 15, 2007 from http://www.ntfs.com

Oates, J. (2006, September 11). The Register. German Police Seize TOR Servers. Retrieved January 27, 2007 from http://www.theregister.co.uk/2006/09/11/anon_servers_seized/

Opensource.org. (2007). Retrieved July 21, 2007 from http://www.opensource.org/

Oudot, L. (2005). WLAN and stealth issues. Retrieved May 23, 2008 from http://www.blackhat.com/presentations/bh-europe-05/BH_EU_05-Oudot/BH_EU_05_Oudot.pdf

Qlogic.com. (2007). *Persistent Naming Using udev in Linux Environment*. Retrieved August 3, 2007 from http://www.qlogic.com/documents/datasheets/knowledge_data/whitepapers/SN0130979-00.pdf

Renderlab (2009). Church of Wifi WPA-PSK Rainbow Tables. Retrieved February 23, 2009 from:http://www.renderlab.net/projects/WPA-tables/

Ruxcon (2004). The art of defiling. Grugq. Retrieved February 16, 2007 from www.ruxcon.org.au/files/2004/13-grugq.ppt

Schneier, B. (2000). *Secrets and Lies, Digital security in a networked world*. John Wiley and Sons. Inc.

Securitydistro.com. (2007). *Security Distros*. Retrieved August 15, 2007 from http://www.securitydistro.com/index.php?option=com_weblinks&catid=11&Itemid=4

Shinder, D. (2002). *Scene of the Cybercrime. Computer Forensics Handbook*. USA: Syngress Publishing.

Siles, R. (2006, January 16). Security Focus. Sebek 3: Tracking the attackers, part one. Retrieved February 12, 2007 from http://www.securityfocus.com/infocus/1855

Siles, R. (2007). Security Focus. Wireless Forensics: Tapping the Air - Part Two. Retrieved February 16, 2007 from http://www.securityfocus.com/infocus/1885/2

SMAC. (2009). Retrieved January 2009 from http://www.klcconsulting.net/smac/

SourceForge. (2007). *Memparser*. Retrieved August 11, 2007 from http://sourceforge.net/projects/memparser

SourceForge. (2009). Wipe. Retrieved April 5, 2009 from http://wipe.sourceforge.net/

WFTK (Windows Memory Forensic Toolkit). (2007). *Digital Investigations*. Retrieved August 12, 2007 from http://forensic.seccure.net/

Wortley, R., & Smallbone, S. (2004). Child Pornography on the Internet. Retrieved June 21, 2007 from http://www.cops.usdoj.gov/mime/open.pdf?Item=1729

Chapter 14
Life Cycle Pattern Study of Malicious Codes

June Wei
University of West Florida, USA

Randall C. Reid
University of West Florida, USA

Hongmei Zhang
University of South Carolina, USA

ABSTRACT

This chapter investigates the patterns of malicious code attacks based on monthly data of the top 10 virus shares from 1998 to 2005. Three parameters were identified for study, overall pattern of the attack, the number reentries into the top 10 most prevalent attacks, and the maximum percentage share. The dataset was validated by comparing it to an independent dataset that measured the same parameters for a subset of the period of the primary dataset. The effects of malicious code that started before or disappeared outside the collection period were found to not have a significant effect. A multivariate regression analysis showed that the number of entries and the maximum share had a strong relationship with the visible life span. Multivariate cluster analysis was conducted on the reentry parameters and yielded six virus clusters classifications. The high impact viruses, 43 of the 230, are identified and further grouped.

INTRODUCTION

Computer and information security breaches have been a serious threat to the information technology (IT) industry (McClure, 2001; Whitman, 2003). This threat has resulted in dramatic financial losses. Despite the continued efforts of government and industry toward the defense against malicious codes, both the number of attacks and the resulting financial losses continue to increase (Gordon, Loeb, & Richardson, 2006). The CIS/FBI survey estimated the amount of loss due to virus contamination was $15,691,460 (Gordon et al., 2006).

The term "virus" is often used generically to refer to viruses, worms and other forms of malicious code (malware). Sophos defines a virus as "a computer program that copies itself" (www.sophos.com). A virus requires a host program and will not infect a computer until the host program has been run. In this chapter, the generic term virus refers to virus code, worms, Trojans, and all other forms of malware.

While exact numbers are not available, it is estimated that there are over 100,000 viruses in existence in today's computer information systems. Sophos (www.sophos.com) reported that there were 15,907 new malware threats identified during 2005 alone. Virus attacks shows a strong positive correlation with its costs, and denial of service, unauthorized access, and net abuse (Li, Wei, Lai, & Koong, 2004).

Problems and Objectives

Kephardt and White (1991) proposed a theoretical model using an epidemiological model of infectious diseases to study computer viruses. Kephart, Chess and White (1993) defined the epidemiological approach as "characterizing viral invasions at the macro level – has led to some insights and tools that may help society to cope better with the threat (and which may aid the study of biological viruses, too)." This chapter uses a macro level analysis of the life cycle of viruses to help to develop an understanding of how they behave in the environment. An extended discussion of the epidemiological approach to virus analysis can be found in Serazzi and Zanero (2003). A qualitative understanding of the epidemiology of computer viruses has been developed (White, 1998) and a quantitative analysis of the evolving attacks patterns and exploits used by viruses (Coulthard & Vuori, 2002) has been done.

An early approach to defining a virus's life cycle is the Internet Worm Propagation Data Model (IWPDM) (Mcalerney, 1999) which defined the life cycle as having four phases, starting with an activation phase and ending with a death phase. Between these two phases are a series of hibernation and reactivation phases. These would comprise the visible portion of the virus's life cycle. The complete life cycle would include a development phase prior to the activation phase and an epilogue phase where the author(s) are, hopefully, apprehended, tried in criminal court and possibly incarcerated. The development and epilogue phase are not considered part of the visible portion of the virus life cycle due to incomplete and inaccurate data and the lack or relevance to the understanding of the behavior of the virus while it is in the wild. The emergence/disappearance rates and the duration of the active attack period of viruses are not well documented. This emergence, disappearance, and the duration of active infection comprise the visible portion of virus life cycle.

A large number of viruses are developed in a laboratory setting as a proof of concept or as a test of a possible exploit. These are never released into environment so there is no need to directly control them and they often have a very abbreviated and hidden life cycle. In many cases, a patch is developed and released to remove the exploit prior to any attacks taking place. The focus of this chapter is only on those viruses in the general computing environment or "in the wild." The reasons for limiting the focus to those viruses "in the wild" is the abbreviated and often hidden life cycle of the virus while it is in the laboratory. Additionally, it is desirable from an analytical perspective to identify those viruses that passed a real life test as to its virulence and persistence. The ability to compare viruses with varying levels of virulence and persistence should aid in the identification of the characteristics that were the source of its virulence and persistence. Once the mechanism of the attack is understood, hopefully a defense can be developed.

The viruses have been the leading attack sources for the duration of this study and are predicted to remain as the leading attack sources in the future (Gordon et al., 2006). A detailed understanding of virus patterns may provide insights that can be used to develop defenses against these attacks. The objective of this research is to aid information system staffs, executives, directors, and managers or supervisors in business and government in a better understanding of virus threats, enabling them to improve their decisions on how to cope with these threats.

METHODOLOGY

Data on the number of incidents and on different types of viruses are available from Sophos (www.sophos.com), IBM research (www.ibm.com), Vmyths (vmyths.com), McAfee (www.mcafee.com), Kaspersky Labs (www.kaspersky.com), the Wildlist (www.wildlist.com), Symantec (www.symantec.com) and many other antivirus vendors and research organizations. When compared with other vendor's databases, the Sophos database showed a more complete set of information on the viruses for the longest period. This information included reports for every month during the period of analysis with the percentage or share for each virus for each month. The ratings from Sophos have similar patterns as those from the other companies. The chapter includes a detailed comparison of the Sophos data set and the data available from Kaspersky Labs. The Sophos database was selected for the current chapter because of its completeness and duration.

A total of 230 viruses based on monthly top 10 virus data reports from January 1998 to December 2005 were extracted from Sophos' Web site (www.sophos.com) for use in this chapter. The monthly top 10 virus data reports were compiled from the numbers of virus code submissions from Sophos' customers to the Sophos research center. The top 10 virus list encompasses 75% of all of the reported virus incidents during the period (For details regarding a summary of the monthly share data obtained from Sophos for each virus type for the period of January 1998 through December 2005, please contact the authors).

During the period from January 1998 to December 2005, a total of 230 viruses were identified as appearing in the top 10 at least once. Three phases of analysis were used to study the visible portion of the viruses' life cycle. Initially, a pattern analysis was conducted to look at the overall picture in terms of the relationship between the total number of reentry and maximum shares for each virus. Cluster analysis was then performed to investigate similarities in the patterns of each virus entering and leaving the top 10. Further cluster analysis was then done with the high impact viruses identified from the second phase to further refine their classifications.

In the first phase, the study and investigation of the overall virus pattern, three variables were identified: the total time the virus stayed in the top 10 ranks of shares (months) (X1), the total numbers of times reentering top 10 ranks (X2), and the maximum share (%) for all reentries (X3). The total number of reentering top 10 rank (X2) for a particular virus is counted as two entries if that virus appeared for two months, then disappeared for one month, then appeared again.

In the second phase, the investigation of the details of reentering pattern for each virus, 15 variables (X4-X18) were identified in order to conduct grouping based on cluster analysis. X4 refers to the duration from the first (new) entry to the first time it leaves the top 10 ranks (months); X5 refers to the duration from the first leave to the second entry into top 10 ranks (months). This pattern of assignment was repeated for 8 entries and exits. This pattern can be seen graphically in Figure 1.

Even numbered variables, shown above the time line, indicate when a virus was in the top 10 rankings. Odd numbered variables, shown below the time line, measure the number of months between entries into the top 10. Each virus was assigned a unique sequence number that was also used as its identification number upon its initial entry to the top 10.

Figure 1. Variable X4 – X 18 assignments

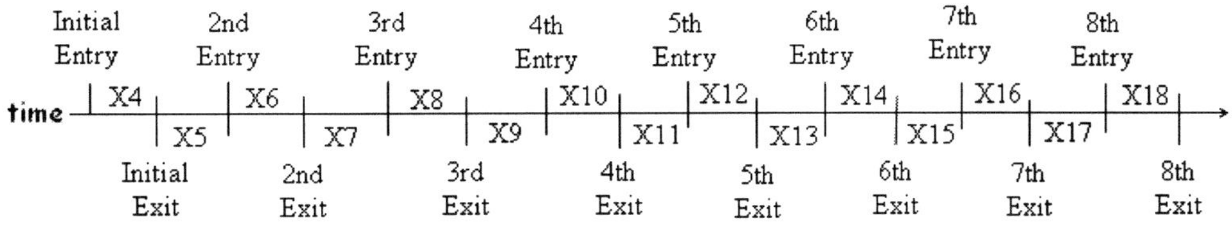

In the third phase, further analyses on the high impact viruses or "bad actors" detected from the second phase were conducted on average shares, total shares, and life spans (X1).

FINDINGS

Analysis of the Visual Portion of Virus Life Span

The analysis in this chapter addresses the relationship between the total number of months of in top 10 ranks (X1) and two factors: total number of reentering top 10 ranks (X2) and the maximum shares (highest percentage) (X3) for each virus. Our relationship study utilized a multiple linear regression model.

The analysis focuses on what kind of virus will stay longer (i.e., the number of months that a virus presents in top 10, X1) at top 10. By doing linear regression analysis, as we expected, each virus tends to experience a peak and then a dying process. In other words, it doesn't matter how large the maximum share a virus is, it will eventually disappear from the top 10. Two contributing used in the classification and understanding of the virus life cycle were the number of re-entries of a virus into top 10 (X2) and the strongest impact (measured by the maximum percentage) (X3) of a virus' whole life. Figure 2 displays the relationships between the numbers of months a virus staying at top 10 (X1) and the numbers of re-entries (X2). Figure 2 displays a clear increasing pattern, which is as expected. On average, the larger the number of reentries of a virus, the longer the virus' visible life span.

Figure 3 represents the relationship between X1 and the strongest impacts (X3) of each virus. The pattern in Figure 3 is not that clear. Points inside the marked area show a strong positive relationship between the life spans and maximum impacts. However, the points outside the marked area can substantially weaken such relationships.

Figure 2. Relationship between visible life spans X1 and numbers of re-entries X2 of a virus

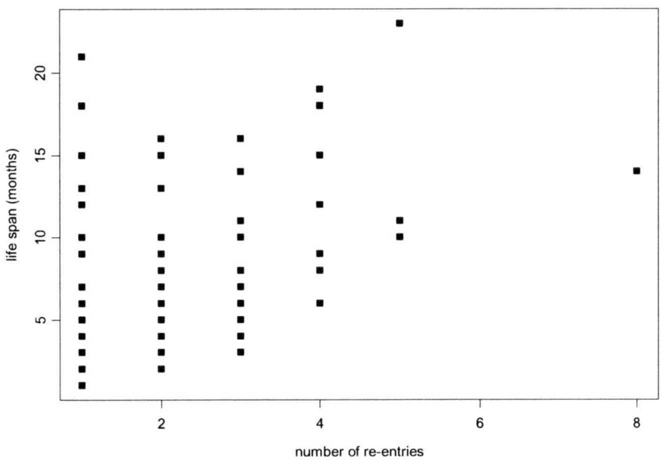

Figure 3. Relationship between life spans X1 and maximum impacts X3

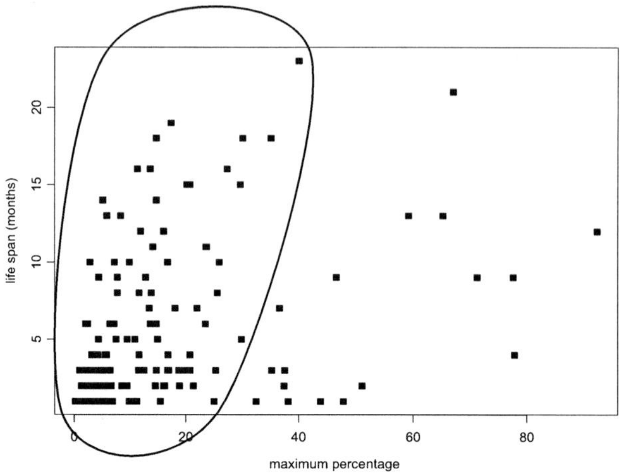

A multiple linear regression line addressing the effects from the number of re-entries (X2), the maximum impacts (X3), and the interaction between the two toward a virus' lifespan (X1) is fitted to the data. To validate the assumptions for a linear regression model (Kutner, Nachtsheim, Neter, & Li, 2004), we used the natural logarithm transformation for the number of months that a virus stays at top 10. The model is:

$$\ln(X_{1i}) = \beta_0 + \beta_1 * X_{2i} + \beta_2 * X_{3i} + \beta_3 * X_{2i} * X_{3i} + \varepsilon_i,$$

where β_m, $m = 0,\ldots,3$, are the coefficients of the linear regression line and ε_i, $i = 1,\ldots,n$, are the errors. Table 1 gives the statistical inferences for the coefficients β_m.

The small P-values indicate that there is a significant relationship between life spans and the three variables, which are the number of entries, the maximum impact, and the interaction of the two.

Table 1. Statistical inference for coefficients

Coefficients	Estimates	P-values
β_0	-0.2791	0.0055
β_1	0.6027	<0.0001
β_2	0.0348	<0.0001
β_3	-0.0063	0.0300

The estimated coefficients clearly indicate that there is a positive relationship between life spans (X1) and the number of re-entries (X2). There is also a positive relationship between life spans (X1) and the maximum impacts (X3). This supports the findings that on average, if a virus enters more frequently, the virus will have a longer life span; and that the higher the maximum impact a virus has the longer the life span of the virus. However, the small value of the estimate of β_2 indicates that the observations outside the marked area in the above figure somehow weakened the relationship between life spans and the maximum impacts.

In order to visualize how the number of entries and the maximum impact interact with each other to affect the life span of a virus, a three-dimensional picture was plotted based on the fitted regression model. Figure 4 shows that the effect from the number of entries gets weaker if the maximum impact is large; the effect from the maximum impact gets weaker when the number of re-entries is large. This observation implies that when the number of re-entries is small, the life span of a virus is highly affected by the value of the maximum impact; when the maximum impact of a virus is low, the life span of a virus is highly affected by the number of entries; when either one of the two variables gets large, the effect from the other gets smaller.

Figure 4. Relationship between life spans X1 and the two factors X2 and X3

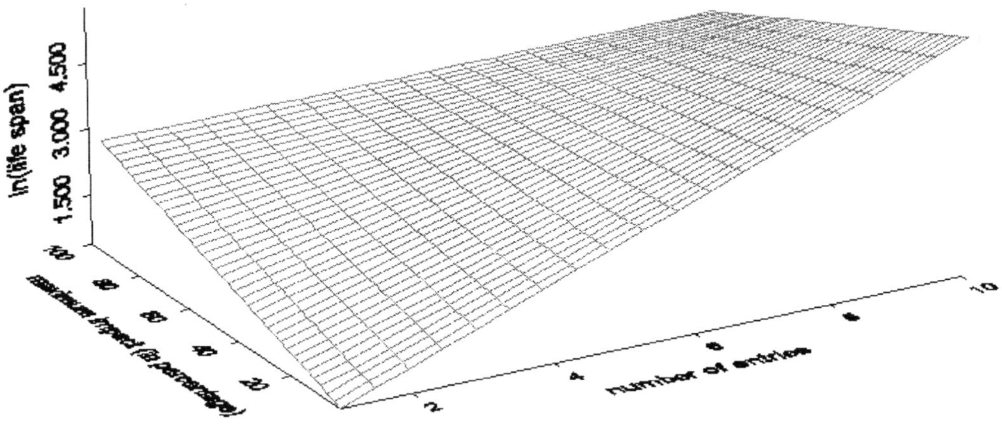

Phase II: Grouping Analysis Based on Cluster Analysis

Cluster analysis is a descriptive technique that allows for the grouping of observations based on the similarity of their characteristics. The goal is to create a number of clusters that are relatively homogeneous in their characteristics and with significant differences between the clusters. If the cluster analysis is done with a large and varied enough data set new observations should fit into an existing cluster. The revealed pattern will help people in computer sciences find a way to keep them from occurring again. However, cluster analysis is unable to predict the individual behavior of any new observations.

In our investigation of the details on virus re-entering and leaving top 10 ranks cluster analysis was performed on the 230 viruses that were members of the top 10 viruses in a month. Several cluster analysis methods exist, including hierarchical and k-means methods. FASTCLUS cluster procedure in SAS is used to find disjoint clusters of observations using a k-means method. The k-means method is especially suitable for large data sets (SAS, 2003).

Cluster analysis was performed on the 230 observations (viruses) that entered the top 10 within a month and 15 entry/exit pattern variables (X4 – X18). A range of 3 to 8 different numbers of clusters was examined. Six clusters provide most reasonable classifications using the cubic clustering criterion. Table 2 summarizes the virus classifications based on cluster analysis.

Further analysis of the data within each of the groups based on the 1-month share of the reported incidents is contained in Table 3.

A seventh cluster was created to assist in the understanding of the impact of each cluster. It is identified as cluster 0 with its members encompassing of all of the reported viruses that did not enter the top 10 at any time. There is very little data available on this seventh cluster other than its total contribution to the overall virus problem.

The most significant findings in Table 4 relate to the dominance of the "one-hit-wonders" and the unexpectedly large impact of viruses in cluster 4. Clusters 0, 1, 2, and 3 basically form a "one-hit-wonders" super cluster that makes up 78.7% (181 of 230) of the viruses in the top 10 and 56.45% of all reported viruses during the period under examination. Technically, clusters 2 and 3 allow two entries, but an examination of the individual members indicated that there is very little difference in the actual impact and behavior of these viruses. The other significant finding is that cluster 4 which had 15 members or 6.7% of the members of the top 10, accounted for 22.08% of all of the reported virus incidents.

Table 2. A summary on virus classification analysis

Cluster Numbers	Virus Patterns	Frequency	Percent %
1	Enter once with the total time in the top 10 less than or equal to 4 months	144	62.61
2	Enter twice with the total time in the top 10 less than or equal to 4 months	27	11.74
3	Enter once or twice with the total time in the top 10 between 5 and 9 months inclusive	16	6.96
4	Enter once or twice with the total time in the top 10 greater than or equal to 10 months	15	6.52
5	Enter the top 10 three or four times	23	10.00
6	Enter the top 10 five or more times	5	2.17

Table 3. Maximum 1-month share of reported incidents

Max Shares	Cluster 1	Cluster 2	Cluster 3	Cluster 4	Cluster 5	Cluster 6
Average	7.3	7.8	20.7	32.9	15.2	18.7
Min	1.0	2.0	4.3	4.0	5.0	6.0
Max	51.1	77.8	77.6	92.4	71.2	40.0
Std Dev	9.9	14.8	19.2	26.5	13.8	13.6

Table 4. Total shares of all reported incidents

Total Shares	Cluster 0	Cluster 1	Cluster 2	Cluster 3	Cluster 4	Cluster 5	Cluster 6
Sum	2457.3	1605.0	465.4	1035.6	2174.8	1501.6	614.4
% of total	24.90	16.30	4.73	10.52	22.08	15.25	6.24
Average	N/A	11.6	17.2	64.7	145.0	65.3	122.9
Min	N/A	2.2	6.2	19.9	26.8	12.1	33.8
Max	N/A	70.5	120.0	234.0	327.8	159.7	356.5
Std Dev	N/A	13.2	21.7	54.9	98.0	43.1	134.6

Phase III: Further Analysis of the High Impact Viruses

The viruses with the highest impact, groups 4, 5 and 6, were combined for further analysis. The names of these viruses and their ID numbers are presented in Appendix A. The 43 members of this group provided for 43.67% of all reported viruses during the 7-year period of the analysis (1998-2005) while comprising only 18.7% (43 of 230) of the viruses that made the top 10 list. The purposes of this additional analysis were to identify the characteristics of these high impact viruses. A cluster analysis was done for 1) the number of months in the top 10, 2) the average share, and 3) the total share. In each case, the group broke into 3 subgroups. Table 5 contains the results of this analysis.

A univariate analysis of each group was performed to identify its basic parameters. Table 6 contains the univariate descriptors of each of the groups.

A coefficient of correlation was calculated to identify any relationships that might exist between the variables. The coefficient of correlation: between the average share per month and the number

Table 5. Clusters formed from the analysis of number of months, average share, and total share

ID Number*	Cluster on Numbers of Months		Clusters on Average Shares		Clusters on Total Shares	
	Month Cluster	Number of Months	Average Cluster	Average Share	Total Share Cluster	Total Share
20	1	4	1	5.63	1	22.5
27	1	6	1	4.85	1	29.1
29	1	3	1	5.6	1	16.8
30	1	6	1	3.33	1	20
32	1	7	1	5.43	1	38
35	1	5	1	5.78	1	28.9
37	1	3	1	4.03	1	12.1
38	1	6	1	3.1	1	18.6
1	2	13	1	5.52	1	71.8
6	2	12	1	5.18	1	62.1
12	2	10	1	2.68	1	26.8
13	2	13	1	3.97	1	51.6
15	2	10	1	3.11	1	31.1
21	2	8	1	5.74	1	45.9
33	2	11	1	5.65	1	62.2
34	2	10	1	5.36	1	53.6
43	2	10	1	3.38	1	33.8
18	1	6	2	6.25	1	37.5
9	3	15	1	5.35	1	80.3
42	3	14	1	2.95	1	41.3
19	2	8	2	6.06	1	48.5
26	2	8	2	9.13	1	73
41	2	11	2	5.98	1	65.8
36	1	6	3	10.63	1	63.8
25	2	12	2	7.12	2	85.4
4	3	16	2	8.79	2	140.6
10	3	18	2	6.63	2	119.4
17	3	16	2	6.31	2	100.9
24	3	18	2	6.35	2	114.3
28	3	16	2	6.38	2	102.1
16	3	15	2	9.21	2	138.2
22	3	14	2	8.02	2	112.3
23	3	19	2	8.41	2	159.7
2	2	10	3	12.32	2	123.2
5	2	13	3	13.91	2	180.8
31	2	9	3	13.13	2	118.2

continues on following page

Table 5. continued

40	2	11	3	10.64	2	117
3	3	15	3	10.31	2	154.6
7	2	12	3	20.72	3	248.6
14	2	13	3	20.39	3	265.1
8	3	21	3	13.86	3	291
11	3	18	3	18.21	3	327.8
39	3	23	3	15.5	3	356.5

*Note: * Id Number and name listing can be found in Appendix A*

Table 6. Univariate descriptors of the groups from the second series cluster analysis

	Number of Months			Average Share			Total Share		
	Cluster 1	Cluster 2	Cluster 3	Cluster 1	Cluster 2	Cluster 3	Cluster 1	Cluster 2	Cluster 3
Count	10	19	14	19	13	11	24	14	5
Mean	5.20	10.74	17.00	4.56	7.28	14.51	43.15	126.19	297.80
Standard Deviation	1.40	1.73	2.66	1.15	1.24	3.78	19.95	25.85	44.39

of months was 0.3997; between the total share and the number of months was 0.7986; and between the average share and total share was 0.8780. All of these results have a reasonable amount of "face validity" because they follow known patterns in the business world. One would expect that as the average share increases so would the total share. The same applies with the relationship between the number of months and the total share. The longer a virus persists, the higher its total reported share would be.

The authors believe that the primary contribution of this additional analysis is the identification of three groups of high impact viruses. Further in-depth analysis of these particular viruses would aid in the development of strategies to counter-act their "cousins" in the future.

Data Validation

Because the data collected by Sophos is a partially self-selected sample of the people who purchased the Sophos product and from the members of the Sophos research team who monitored the Internet, the validity of the data set needs to be established. Founded in 1985, Sophos offers commercial and individual protection products against viruses. Product research and analysis is done in Sophos' high-security research laboratories located worldwide, including in the UK, USA, Canada, and Australia. The information is then posted on its Web site. Over 35 million individuals and organizations of all sizes already used Sophos' products and Web site.

Based on literature review for the top 10 antivirus software companies worldwide, only two companies, Sophos and Kaspersky, were identified as providing monthly updates with market shares on virus threats. Because monthly data from Kaspersky (www.Kaspersky.com) was only available for a shorter period of time, it was used to validate the primary data set from Sophos. Fragmentary data is available from a number of other sources, but they are all incomplete as to the number of months provided or the data available.

The available data from Kaspersky covered the period starting in November 2001 through December 2005. The same period was extracted from the Sophos data and compared with Kaspersky's data set based on Cluster Analysis and Virus Life Span Analysis.

Data Validation Based on Cluster Analysis

The results from a cluster analysis based on two data sets are presented in Table 7. T-Tests were conducted to establish if the distributions were similar. The t-test score for the counts by cluster was 0.1908 indicating that the hypothesis of a common mean and distribution can be accepted at a 99% confidence level. Additional testing was conducted to establish if the percentages of reported virus incidents were similar between the two data sets. For cluster 1, the result of the t-test was 2.2119, which indicated that the hypothesis of the two sources of data having the same proportion of viruses belonging to the first cluster could be accepted but only at a 90% confidence level. The results for clusters 2 through 5 were 1.07, 0.93, 1.06 and 0.89 respectively. These results indicated that the hypothesis of the same proportion of viruses in each cluster could be accepted at a 99% confidence level. Because there were no members of the 6th cluster in the Kaspersky data set, no testing was done.

Data Validation Based on Virus Life Span Analysis

Based on virus life span analysis of two data sets, similar relationships were found between life spans and numbers of re-entries, and between life spans and the maximum impact of a virus. Figure 5 shows the relationship between life spans and numbers of re-entries of a virus, and Figure 6 shows the relationship between life spans and maximum impacts based on the data set from Kaspersky. These two Figures showed similar patterns to the ones given by Figures 2 and 3.

Using the regression model that was developed for the primary data set the reduced data set was evaluated. The original model was:

$$\ln(\text{lifespan}_i)=\beta_0 + \beta_1 *\text{entry}_i + \beta_2 *\text{max\_impact}_i + \beta_3 *\text{entry}_i *\text{max\_impact}_i + \varepsilon_i$$

where β_m, $m = 0,\ldots,3$,, are the coefficients of the linear regression line and ε_i, $i = 1,\ldots,n$, are the errors. Table 8 gives the similar statistical inferences for the coefficients β_m as those in Table 1.

Because the sample size of the second data set (from Kaspersky) is small, there are not many points outside the marked area. Therefore, points

Table 7. A comparison of virus classification between Sophos and Kaspersky

Cluster Numbers	Virus Patterns	Sophos Data 11/01-12/05		Kaspersky Data 11/01-12/05	
		Count	%	Count	%
1	Entry only once with the total time in the top 10 less than or equal to 4 months	88	19.17%	91	10.23%
2	Enter twice with the total time in the top 10 less than or equal to 4 months	8	3.31%	13	1.54%
3	Enter once or twice with the total time in the top 10 between 5 and 9 months inclusive	14	14.25%	11	14.56
4	Enter once or twice with the total time in the top 10 greater than or equal to 10 months	10	28.09%	11	48.66%
5	Enter the top 10 three or four times	5	3.21%	8	3.19%
6	Enters the top 10 five or more times	2	0.19%	0	0%

Figure 5. Relationship between life spans and numbers of re-entries of a virus based on data from Kaspersky

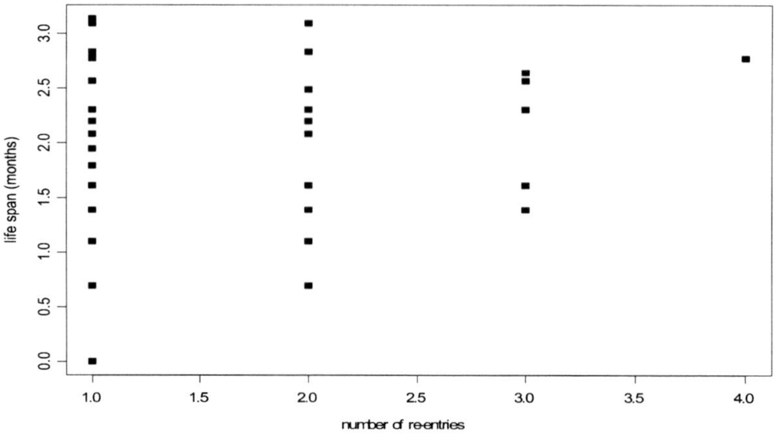

Figure 6. Relationship between life spans and maximum impacts based on data set from Kaspersky

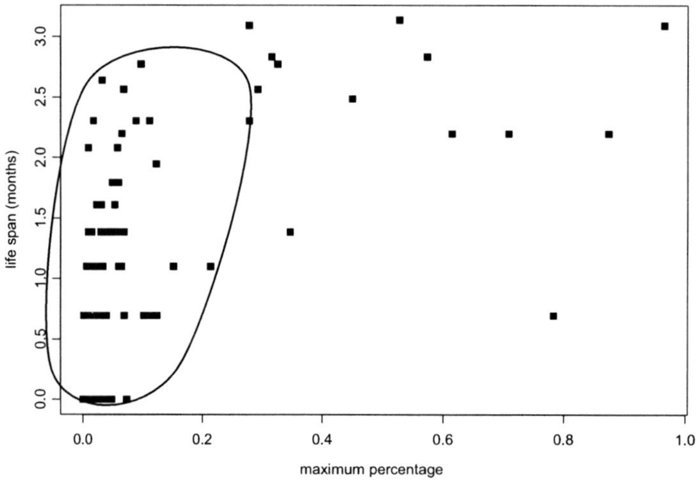

inside the marked area will mostly determine the relationship, which is a strong positive relationship between the life spans and maximum impacts.

The evidence from the analysis of the structure of the two data sets, the pattern analysis and the regression analysis indicates that the Sophos data set is a valid sample that can be further analyzed to establish what the life cycle of malicious code. The Sophos data set was selected for this further analysis because it covers a longer time period than the Kaspersky data set.

Table 8. Statistical inference for coefficients based on Kaspersky's data set

Coefficients	Estimates	P-values
β_0	-0.4426	0.0034
β_1	0.7744	<0.0001
β_2	4.4247	<0.0001
β_3	-1.0435	0.1116

Effects of Truncation of Virus Life Cycle on Life Cycle Classification

Viruses introduced prior to January 1998 or during the latter portion of 2005 would have the possibility of having several values outside the collection period, resulting in the truncation of their life cycle. This truncation had the effect of censoring the additional data that might have been obtained if the viruses' full life cycle were available. To identify the effects on the analysis, a subset of the data set composed of the observations from 1999 through 2004 inclusive was created. The data from 1998 and 2005 was used to identify which clusters a virus belonged to, thus eliminating the effects of truncating the life cycle of an individual virus. This reduced data set has 186 observations compared to the original 230 in the complete data set. Table 9, a summary of the complete data set, and Table 10, a summary of the reduced data set, allow for comparison between the data set on the maximum share of the reported incidents within a given month.

When comparing the univariate descriptors there is very little difference between the data sets. Therefore, this truncation is not assumed to be a significant issue in the analysis.

Table 9. Complete data set (1998-2005)

Max share	1	2	3	4	5	6
Count	144	27	16	15	23	5
Average	7.3	7.8	20.7	32.9	15.2	18.7
Min	1.0	2.0	4.3	4.0	5.0	6.0
Max	51.1	77.8	77.6	92.4	71.2	40.0
Std Dev	9.9	14.8	19.2	26.5	13.8	13.6

Table 10. Reduced data set (1999-2004)

Max share	1	2	3	4	5	6
Count	110	21	15	15	20	5
Average	7.8	8.8	10.6	32.9	16.4	12.7
Min	1.0	2.0	1.0	4.0	5.0	5.0
Max	51.1	77.8	77.8	92.4	71.2	40.0
Std Dev	10.3	16.7	14.4	26.5	14.5	9.6

DISCUSSIONS AND CONCLUSION

By studying the patterns of viruses, this chapter aims to provide a larger picture and understanding of virus attacks. To achieve this objective the 230 most prevalent viruses from January 1998 to December 2005, as reported by Sophos, were analyzed. The analysis was conducted using the virus patterns based on 17 virus pattern variables including life spans, numbers of reentry, maximum shares, and entry/reentry patterns. Specifically, there are a number of major findings that are discussed below.

The investigation of the impact of number of entries (X2) and maximum shares (X3) on lifespan (X1) showed that when the number of entries (X2) is small, the life span (X1) of a virus is highly affected by the value of the maximum impact (X3); when the maximum impact (X3) of a virus is low, the life span (X1) of a virus is highly affected by the number of entries (X2); when one of the two variables (X2 and X3) gets large, the effect from the other variable gets smaller. This supports the finding that on average when a virus enters more frequently, the longer the life span of the virus, and the higher the maximum impact a virus gives the longer the life span of the virus.

Cluster analysis identified five life cycle patterns of reported virus incidents. These can be combined into three "super clusters" to develop a managerial response to the threat. The first super cluster, the "one-hit-wonders," is responsible for 62.61% of all of the reported incidents. The second super cluster is cluster 4, which is composed of a very small number of very persistent and nasty viruses. Cluster 4's 15 members comprised only 6.52% of the total top 10 viruses, but accounted for 22.08% of all of the virus reports. The final super cluster could be called repeat offenders. It is composed of clusters 5 and 6. This super cluster has 28 members and accounted for 22.17% of the incidents. The managerial implications are that an ongoing antivirus program needs to be in place to counter the continuing low-level threat posted by the members on the one-hit-wonder super cluster and a crises response policy to limit the effects of the Cluster 4 and "repeat-offenders."

Clusters 4, 5, and 6 were combined to allow for further analysis of the high impact viruses. These viruses, the really bad actors, comprise a relatively small portion, 18.70% (43 of 230) of all the reported viruses. However, they account for 43.67% of all virus reports for the 7-year period of the study. The "bad actors" can be further subdivided into three groups with a very high impact group of 5 viruses being the worst of the worst.

There are some statistical problems with generalizing the conclusions because the data used in not a random sample of the population of all viruses on the Internet. The sample may have a bias because it is self-selected by those people who purchased Sophos software and then allowed reports of the attacks on their machines to be reported to Sophos for analysis. The data set posted by Sophos is from a very large number of reporting sites because there were more than 10 million copies of their product in use for each year of the reported period. The pattern of the visible portion of the viruses' life cycle and the classification was similar when the data set was available from both Kaspersky and Sophos.

Both of the conclusions about the relationship between maximum share, duration, and number of re-entries to the top 10, and the small number of very significant viruses hold for both the Sophos and Kaspersky data sets. The two independent data sets validate each other but it is difficult to state with certainty that these conclusions hold true for behavior of the entire universe of viruses "in the wild." Anecdotal evidence from practitioners and network administrators support these conclusions, but again they cannot be applied to all viruses "in the wild" with certainty.

Even though the virus data analysis in this chapter covers a large sample and is collected by a leading antivirus company, the results of the research should still be interpreted with several limitations. First, the virus data is collected only from reported virus organizations and individuals. As with all data collected, the nonreported rate might be a problem. Second, there are some factors that might affect the analysis of virus attacks; such as antivirus technologies developed (Wei & Li, 2004) and the changes in legal environments. Third, the data analyzed in this chapter covered the period from January 1998 to December 2005: the entry/reentry of a particular virus outside this time span is not shown in this chapter. However, an analysis of a reduced data set, 1999 to 2004, was conducted and the results indicated that the loss of portions of the visible life cycle on both ends of the period did not significantly affect the results.

The authors believe that this is the first step in a larger study. The study is a post ante analysis of virus life cycle patterns and impacts. As such it is more descriptive than prescriptive. Additional study needs to be done to identify the internal design and mechanisms so that techniques such as AI (pattern recognition (e.g., neural networks, decision trees, etc.) and state-full inspection can be developed to identify new viruses as they are introduced and to block or minimize their effects.

Further research might be conducted to investigate relationships between life span and other factors in order to understand life spans of viruses more deeply. Investigation into other factors may provide critical information enabling users to halt virus re-entry. Distinguishing the patterns between novel pieces of malicious codes and the numerous variants that are released is an additional research direction.

REFERENCES

Coulthard, A., & Vuori, T. A. (2002). Computer viruses: A quantitative analysis. *Logistics Information Management, 15*(5/6), 400-409.

Gordon, L. A., Loeb, W. L., & Richardson, R. (2006). *2006 CSI/FBI computer crime and security survey.* Retrieved November 30, 2007, http://www.gocsi.com

Kephardt, J., & White, S. (1991). Directed-graph epidemiological models of computer viruses. In *Proceedings of the IEEE Computer Society Symposium of Research in Security and Privacy,* Oakland CA, (pp. 343-359).

Kephart, J. O., Chess D. M., & White, S.R. (1993, May). Computers and epidemiology. *IEEE Spectrum.* Retrieved November 30, 2007, from http://www.research.ibm.com/antivirus/SciPapers/Kephart/Spectrum/Spectrum.html

Kutner, M. H., Nachtsheim, C. J., Neter, J., & Li, W. (2004). Applied linear statistical models (5 ed.). McGraw-Hill.

Li, Y., Wei, J., Lai, L., & Koong, K. (2004). Implications of technologies and financial losses on information security. In *Proceedings of the ISOneWorld*, Las Vegas.

McAlerney, J. M. (1999). *An Internet worm propagation model.* Retrieved November 30, 2007, from www.cs.usdavis.edu/research/tech-reports/2005/CSE-2005-10.pdf

McClure, D. (2001). Guarding your gateway. *Association Management*, *53*(8), 60-6.

SAS. (2003). *SAS menu.* SAS Corporation.

Serazzi, G., & Zanero, S. (2003). *Computer virus propagation models. In Tutorials of the 11th IEEE/ACM International Symposium on Modeling, Analysis and Simulation of Computer and Telecom. Systems - MASCOTS 2003.* Retrieved November 30, 2007, from http://citeseer.ist.psu.edu/serazzi03computer.html

Udelson, T. (2000). How secure is your network? *Association Management, 52*(6), 25-6.

Wei, J., & Li, Y. (2004). Forecasting on information systems attack sources. In *Proceedings of the Information Resources Management Association*, New Orleans.

White, S. R. (1998, October 22). Open problems in computer virus research. In *Proceedings of theVirus Bulletin Conference*, Munich Germany.

Whitman, M. (2003). Enemy at the gate: Threats to information security. *Communications of the ACM, 46*(6), 91-95.

Wilcoxon, F., & Wilcox, R.A. (1964). *Some rapid approximate statistical procedures.* Pearl River, NY: Lederle Laboratories.

Yaffee, R. (2000). *Introduction to time series analysis and forecasting with applications of SAS and SPSS.* Academic Press.

APPENDIX A

Listing of the High Impact Viruses from 1998-2005

Id Number	Virus Name	Maximum Share	Months in Top 10	Average Share	Total Share
1	CMOS4	8.30	13	8.30	13.00
2	WM97/Class	26.00	10	26.00	10.00
3	W32/Apology-B	29.80	15	29.80	15.00
4	W32/Magistr-A	27.40	16	27.40	16.00
5	W32/Sircam-A	65.20	13	65.20	13.00
6	W32/Magistr-B	16.10	12	16.10	12.00
7	W32/Badtrans-B	92.40	12	92.40	12.00
8	W32/Klez-H	67.00	21	67.00	21.00
9	W32/Netsky-B	20.20	15	20.20	15.00
10	W32/Netsky-D	30.20	18	30.20	18.00
11	W32/Netsky-P	35.20	18	35.20	18.00
12	W32/Netsky-Q	4.00	10	4.00	10.00
13	W32/Netsky-Z	5.80	13	5.80	13.00
14	W32/Zafi-B	59.20	13	59.20	13.00
15	W32/MyDoom-O	7.20	10	7.20	10.00
16	WM/Cap	20.70	15	20.70	15.00
17	Form	11.40	16	11.40	16.00
18	Parity Boot	6.30	6	6.30	6.00
19	Anticmos	11.70	8	11.70	8.00
20	New Zealand-i	5.00	4	5.00	4.00
21	WM/CopyCap	13.90	8	13.90	8.00
22	W95/CIH-10xx	14.80	14	14.80	14.00
23	WM97/Ethan	17.40	19	17.40	19.00
24	W32/Ska-Happy99	14.80	18	14.80	18.00
25	WM97/Melissa	12.00	12	12.00	12.00
26	W32/Pretty	25.60	8	25.60	8.00
27	W32/Qaz	7.10	6	7.10	6.00
28	W32/Hybris-B	13.70	16	13.70	16.00
29	W32/Verona-B	5.00	3	5.00	3.00
30	W32/Bymer-A	5.00	6	5.00	6.00
31	W32/Nimda-A	71.20	9	71.20	9.00
32	W32/Klez-E	13.50	7	13.50	7.00
33	W32/ElKern-C	23.70	11	23.70	11.00
34	W32/Yaha-E	16.80	10	16.80	10.00
35	JS/NoClose	7.50	5	7.50	5.00
36	W32/Gibe-F	23.50	6	23.50	6.00

continues on following page

APPENDIX A. continued

37	W32/Bagle-Zip	5.00	3	5.00	3.00
38	W32/Netsky-C	5.00	6	5.00	6.00
39	XM/Laroux or XM97/Laroux	40.00	23	40.00	23.00
40	WM97/Marker-A	23.60	11	23.60	11.00
41	WM97/Marker-C	14.20	11	14.20	11.00
42	W32/Fless	6.00	14	6.00	14.00
43	W32/Nimda-D	9.90	10	9.90	10.00

This work was previously published in the International Journal of Information Security and Privacy, Volume 2, Issue 1, edited by H. Nemati, pp. 26-41, copyright 2008 by IGI Publishing (an imprint of IGI Global).

Chapter 15
Trust-Based Usage Control in Collaborative Environment

Li Yang
University of Tennessee at Chattanooga, USA

Chang Phuong
University of Tennessee at Chattanooga, USA

Andy Novobilski
University of Tennessee at Chattanooga, USA

Raimund K. Ege
North Illinois University, USA

ABSTRACT

Most access control models have formal access control rules to govern the authorization of a request from a principal. In pervasive and collaborative environments, the behaviors of a principal are uncertain due to partial information. Moreover, the attributes of a principal, requested objects, and contexts of a request are mutable during the collaboration. A variety of such uncertainty and mutability pose challenges when resources sharing must happen in the collaborative environment. In order to address the above challenges, we propose a framework to integrate trust management into a usage control model in order to support decision making in an ever-changing collaborative environment. First, a trust value of a principal is evaluated based on both observed behaviors and peer recommendations. Second, the usage-based access control rules are checked to make decisions on resource exchanges. Our framework handles uncertainty and mutability by dynamically disenrolling untrusted principals and revoking granted on-going access if access control rules are no longer met. We have applied our trust-based usage control framework to an application of file sharing.

INTRODUCTION

Conventionally registered parties behind firewalls collaborate in well controlled environments. With new virtual communities emerging, parties communicate directly with one another to exchange information or execute transaction in a peer-to-peer (P2P) fashion. The dynamism of the P2P communities means that the principal that offers services will meet requests from unrelated or unknown principals. Peers need to collaborate and obtain services within environments that are unfamiliar or even hostile. Therefore, peers have to manage the risks involved in the collaboration when prior experience and knowledge about each other are incomplete. One way to address this uncertainty is to develop and establish trust among peers. Trust can be built by either a trusted third party (Atif, 2002) or by community-based feedback from past experiences (Resnick, Kuwabara, Zeckhauser, & Friedman, 2000) in a self-regulating system. Trust leads naturally to a decentralized approach to security management that can tolerate partial information.

In such a complex and collaborative world, a peer can protect and benefit itself only if it can respond to new peers and enforce access control by assigning proper privileges to new peers. Access control models (Bertino, 2001a; Jajodia, Samarati, Sapino, & Subrahmanian, 2001) determine authorization based on principals' permission on target objects. Usage of a digital object is temporal and transient in a virtual community, such as online reading, which is beyond an instantaneous access. The usage control (UCON) model (Park & Sandhu, 2004) is proposed to handle continuity of access decisions and mutability of subject and object attributes. Authorization decisions are made before an access and repeatedly checked during the access. The on-going access may be revoked if the security policies are not satisfied due to changes of the subject, object, or system attributes.

The general goal of our work is therefore to investigate the design of a novel approach to addressing both uncertain information and mutable attributes. If successful, this approach will offer significant benefits in emerging applications such as P2P. It will also benefit collaboration over the existing Internet when the identities and intentions of parties are uncertain. We integrate trust evaluation with usage control to handle uncertainty of entities and mutability of attributes. Underlying our framework is a formal computational model of trust and access control that will provide a formal basis to interface authentication with authorization.

Related Works

Most recent research on access control includes task-based authorization controls (Thomas & Sandhu, 1998), team-based access control (Georgiadis, Mavridis, Pangalos, & Thomas, 2001), role-based access control (Gerraiolo, 2001), temporal role-based access control (Bertino, 2001b), and X-GTRBAC (Bhatti, Ghafoor, Bertino, & Joshi, 2005). Recently, UCON (Park & Sandhu, 2004) handles the attribute mutability of a principal or an object when the system makes decision for a request. All of them assume that a principal or an object is defined and represented by its attributes. This means that the identity, role, or group of the subject can be identified through certain authentication mechanisms and that information about behaviors of a principal is certain. However, in a pervasive and collaborative environment, identity may not be identified. Moreover, identity itself can not convey priori information about the likely behavior of a principal. Behaviors of a principal may change between friendly and malicious when privileges are executed. A principal can not make access control decision only based on identity information because identity itself can not ensure friendly behaviors.

Reasoning and building trust for each peer allow peers to make decision when they are interacting with others in a peer-to-peer fashion. Li, Mitchell, and Winsborough (2002) and Yao (2003) use explicit incremental negotiation to establish mutual trust. An overview of trust management is discussed by Grandison, Sloman, and Sloman (2000). Trust management has many applications in e-commerce areas such as works from Atif (2002) and Resnick et al. (2000). Xiong and Liu (2004) handle trust evaluation, especially the community-related context factors and transaction context factor of e-commerce. Zouridaki, Mark, Hejmo, and Thomas (2005) and Yang, Kizza, Cemerlic, and Liu (2007) apply trust evaluation into routing protocols of mobile wireless ad hoc networks (MANETs).

Sandhu and Zhang (2005) apply peer-to-peer access control to trusted computing, enforcing trust and hardware encryption. The SECURE (Cahill, Gray, Seigneur, Jensen, Chen, Shand, et al., 2003) project proposed the seminal ideas to handle trust and secure collaboration in an uncertain environment. Their work can tolerate partial information and overcome initial suspicion to allow secure collaboration to take place by reasoning about trust and risk. Dimmock, Belokosztolszki, Eyers, Bacon, and Moody (2004) incorporate notions of trust into rule inference process of OASIS (Bacon, Moody, & Yao, 2002), a policy-driven access control system. Mutable attributes, obligations, context, and revocation of the authorization are not handled.

Both attribute mutability and uncertain behaviors of a principal are needed to be considered in collaborative resources sharing. In this work we integrate trust management into usage-based access control, which allows collaboration when attributes of a principal are mutable or information on a principal's behaviors is incomplete.

A FRAMEWORK TO INTEGRATION TRUST INTO USAGE CONTROL

Overview of UCON

The UCON model proposed by Park and Sandhu (2004) is a generalization of access control to cover authorization, obligation, conditions, continuity (ongoing controls), and mutability. Authorization handles decisions on user accesses to target resources. Obligations are the mandatory requirements for a subject before or during a usage exercise. Conditions are subject, object, environmental, or system requirements that have to be satisfied before granting of accesses. Subject and object attributes can be mutable. Mutable attributes can be changed because of accesses, whereas immutable attributes can be changed only by administrative actions.

Trust Evaluation

For every request, the owner of resources assigns a trust value within [0, 1] to the requester. The trust is evaluated based on history observations and peer recommendations from referees. The history-based observations are the previous interactions the owner had with the requester. The peer recommendations may include signed trust-assertions from other principals, or a list of referees whom the owner can contact for recommendations.

The owner first computes trust given a sequence of observations from interaction history, then combines the trust with *recommendations* to calculate the *total trust* to the requester. As show in Figure 1, the trust evaluation includes five steps: (1) An owner i calculates the trust value T_{ij} to a requester j based on its observed histories; (2) The owner i receives recommendation ($T_{rj} \ldots T_{zj}$) from multiple peers (reporters) r to z; (3) The owner i does *deviation test* to evaluate trustworthiness of the reporters; (4) The owner

Figure 1. Trust evaluation

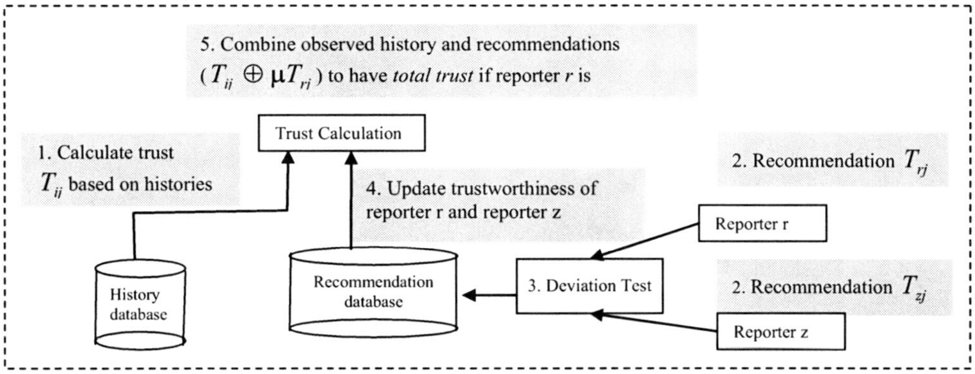

i updates the trustworthiness of the reporter *r* to *z*, for instance, reporter *r* is trustworthy; and (5) The owner *i* merges the observation-based trust T_{ij} with reporter *r*'s recommendation T_{rj}. Each step is detailed as follows.

Step 1: Calculate the Trust of Behaviors Based on Observations of Histories

The owner *i* models the behavior of requester *j* as an actor. Owner *i* thinks that the friendly behaviors of requester *j* follow a probability of T_{ij}, based on outcomes drawn independently from observations. Probability T_{ij} varies for every different requester *j* and every owner *i*. The parameters T_{ij} are unknown, and owner *i* models this uncertainty by assuming that T_{ij} is drawn from a distribution (the prior) that is updated by available new observations. This thinking is under the Bayesian framework. We use the distribution Beta(α,β) for the prior because it is a conjugate distribution. When a conjugate prior is multiplied with= the likelihood function, it gives a posterior probability having the same functional form as the prior, thus allowing the posterior to be used as a prior in further computations. Therefore, T_{ij} ~Beta(α,β) after updating. For a given requester *j* for owner *i*, we define a sequence of variables $T_{ij}^0, T_{ij}^1, T_{ij}^2, \ldots, T_{ij}^{k-1}$. T_{ij}^k characterizes the trust at sampling time points (0, 1, 2, ..., *k*-1, *k*), where T_{ij}^0 ~Beta(α_0, β_0), T_{ij}^1~Beta(α_1, β_1), T_{ij}^2~Beta(α_2, β_2),..., T_{ij}^{k-1}~Beta(α_{k-1}, β_{k-1}), and T_{ij}^k~Beta(α_k, β_k).

Initially, the owner *i* has no knowledge about the requester *j*, therefore, shape parameters are set the same (we use 1 here) as $\alpha_0 = \beta_0 = 1$ at time 0, which means that the trust value has uniform distribution over the interval [0, 1], that is, T_{ij}^0 ~*U*[0, 1] = *Beta*(1, 1). When a new observation is made, α are β are updated. For example, N_k is the number of observed behaviors and G_k is the number of observed friendly behaviors at *k*-th moment. The prior is updated according to $\alpha_k = \alpha_{k-1} + G_k$ and $\beta_k = \beta_{k-1} + N_k - G_k$. In particular, if T_{ij}^{k-1} ~Beta(α_{k-1}, β_{k-1}), we have T_{ij}^k~*Beta*($\alpha_{k-1} + g_k$, $\beta_{k-1} + n_k - g_k$) given that $N_k = n_k$ and $G_k = g_k$. Therefore, T_{ij}^k is characterized by the parameters α_k and β_k defined recursively as follows:

$$\alpha_k = \alpha_{k-1} + g_k \text{ and } \beta_k = \beta_{k-1} + n_k + g_k.$$

We define the trust $\bar{t}_k$ from observation assigned to a requester at a moment *k* to be equal to the expectation value of the Beta(α,β) where

$$\bar{t}_k = \mathrm{E}(T_{ij}^k) = \frac{\alpha_k}{\alpha_k + \beta_k} \qquad (1)$$

Table 1 is a detailed example how an owner *i* updates its trust assignment to the requester *j* based on the interaction owner *i* had with *j* in a given window based on the observed behavior. A window stands for a certain number of observation *i* has to *j*. α and β are the shape parameters for probability density function of the Beta distribution that indicates whether a requester is good

Table 1. Trust calculation based on history observation

Time	Observed behavior # (N_k)	G_k	N_k-G_k
0	0	1	1
1st	10	9	1
2nd	10	8	2
3rd	10	7	3
4th	10	5	5
5th	10	3	7
1 to 5	50	33	19
$\bar{t}_5=E(T_{ij}^5)=\frac{\sum_{k=0}^{5} G_k}{\sum_{k=0}^{5}(N_k - G_k)}$	---	.63	.37
3 to 5	---	15	15
$\bar{t}_5=E(T_{ij}^5)=\frac{\sum_{k=3}^{5} G_k}{\sum_{k=3}^{5}(N_k - G_k)}$	---	.50	.50

or malicious. For example, at first time point, 10 behaviors (n_1 =10) are observed and 9 behaviors (g_1=9) are friendly. The parameters are updated as $\alpha = \alpha_0 + g_1$ and $\beta = \beta + n_1 - g_1$. If owner *i* has observed up to time 5, α and β are updated to 35 and 17, respectively.

There are two alternative ways to update trust values. One is to update trust values based on all the observations and recommendations. The other ways is to update trust values based on recent information only. The advantage of the latter is two folds: reduce the computation complexity and detect the changing of behaviors early. For instance, a requestor is misbehaving in a short time range, and then recent observation together with reports is more reflective to the behavior changing than the overall observation. Table 2 gives the legends in Step 1.

Step 2-3: Deviation Test of Recommendations

If owner *i* also receives the recommendation on requester *j* from its peers (e.g., reporter *r*), it needs to detect and avoid false reports. The owner *i* will do a deviation test to update the trustworthiness of the reporter, thus deciding whether the owner *i* will absorb the recommendation or discard the recommendation. The deviation test is:

$$|E(Beta(\alpha', \beta')) - E(Beta(\alpha, \beta))| \geq d \qquad (2)$$

Where the first term $E(Beta(\alpha', \beta') = T_{ij}$ is the recommendation from reporter *r*, and the second term $Beta(\alpha, \beta)$ is the trust based on owner *i*'s observation, $E(Beta(\alpha, \beta))$ is the expectation of Beta distribution, and *d* is a positive constant as the threshold. If the owner *i* thinks maximum 10% deviation between the recommendation and its own observation is trustworthy, *d* is set to be 0.1.

Step 4: Trustworthiness of Reporters

Trustworthiness of recommendation uses a similar Bayesian approach. Owner *i* thinks trustworthiness of reporter *r*'s recommendation following the probability W_{ir}. $Beta(\phi, \gamma)$ is used to model trustworthiness of reporter *r*'s recommendation and $W_{ir} \sim Beta(\phi, \gamma)$. The parameter ϕ and γ

Table 2. Legends in step 1

i	Owner
j	Requester
r, z	Reporters
k	Index of moments
α	The shape parameter for probability density function of the Beta distribution $Beta(\alpha,\beta)$ that indicates whether a requester is good.
β	The shape parameter for probability density function of the Beta distribution $Beta(\alpha,\beta)$ that indicates whether a requester is malicious.
$Beta(\alpha,\beta)$	The distribution to model the behaviors of requester *j*
T_{ij}	The probability that owner *i* thinks the behaviors of requester j is honest. T_{ij} ~Beta(α,β)
N_k	The variable of observed behaviors at *k-th* moment.
n_k	n_k is the number of observed behaviors at *k-th* moment.
G_k	The variable of observed friendly behaviors at *k-th* moment.
g_k	g_k is the number of observed friendly behaviors at *k-th* moment.
α_k	The shape parameter for probability density function of the Beta distribution $Beta(\alpha_k,\beta_k)$ that indicates whether a requester is good at *k-th* moment. $\alpha_k = \alpha_{k-1} + g_k$
β_k	The shape parameter for probability density function of the Beta distribution $Beta(\alpha_k,\beta_k)$ that indicates whether a requester is malicious at *k-th* moment. $\beta_k = \beta_{k-1} + n_k + g_k$
$Beta(\alpha_k, \beta_k)$	The distribution to model the behaviors of requester *j* at *k-th* moment.
T_{ij}^k	The probability that owner *i* thinks the behaviors of requester *j* is honest at *k-th* moment. T_{ij}^k ~(α_k, β_k)
$E(T_{ij})$	The expectation of Beta(α_k, β_k). $E(T_{ij})=\frac{\alpha_k}{\alpha_k+\beta_k}$
$\overline{t_k}$	The trust value that the owner *i* has to the requester j at *k-th* moment. $\overline{t_k}=E(T_{ij})$

are the shape parameters for probability density function of the Beta distribution $Beta(\phi, \gamma)$ that indicates whether a reporter is trustworthy or not. Initially, owner is ignorant about trustworthiness of reporter, therefore, $\phi_0 = \gamma_0 = 1$ and $Beta(\phi_0, \gamma_0)$ = Beta(1, 1). An update is performed when the results of deviation test are available. A deviation test is used to compare owner *i*'s observation ($T_{ij} \sim Beta(\alpha, \beta)$) with reporter *r*'s recommendation ($T_{ij} \sim Beta(\alpha', \beta')$), as shown in Formula (2). The result of deviation tests is recorded in *s*. Let *s*=1 if the deviation test is positive, otherwise *s*=0 at a moment *k*. The trustworthiness of recommendation if updated by $\gamma_k = \gamma_{k-1} + s$; $\delta_k = \delta_{k-1} + (1 - s)$. If the recommendation of the reporter *r* passes the deviation test, the trustworthiness of reporter *r* is updated, and reporter *r*'s recommendation is merged to owner's history-based trust. If the recommendation of the reporter *r* does not passes the deviation test, the trustworthiness of reporter *r* will also be updated, but its recommendation is ignored. We define the trustworthiness μ_k assigned to a reporter at a moment *k* to be equal to the expectation valued of $Beta(\gamma_k, \delta_k)$, where

$$\mu_k = E(W_{ir}) = \frac{\gamma_k}{\gamma_k + \delta_k} \quad (3)$$

The owner *i* considers reporter *r* is trustworthy if $\mu_k < m$, otherwise the reporter *r* is considered as untrustworthy if $\mu_k \geq m$, where *m* is the threshold of trustworthiness. If the owner *i* trusts a reporter if its ratings deviate no more than in 25% of the cases, the threshold *m* is set to be 0.75.

Step 5: Merge History-based Trust with Recommendations

Meanwhile, recommendations from reporter r bring in new information T_{ij} on the requester's behaviors. The owner combines the new data T_{ij} with its own observation T_{ij} on the condition that the reporter r is trustworthy. If reporter r is trustworthy, its recommendation is merged to owner's history-based trust by $T_{ij} \oplus \mu T_{rj}$ where μ is the trustworthiness that the owner has to the reporter, as shown in Formula (3). After merging the trust based on its own observation with recommendations of the reporter, the owner has the total trust of the requester j.

We define total trust of i towards j as:

$$tt = T_{ij} \oplus \mu T_{rj} = \begin{cases} E(T_{ij}) + \mu E(T_{rj}) & \text{if } E(T_{rj}) > n \\ E(T_{ij}) - \mu E(T_{rj}) & \text{if } E(T_{rj}) \leq n \end{cases} \quad (4)$$

The owner needs to decide if the recommendation T_{ij} from reporter is positive or negative before it merges the recommendation with its own trust. The recommendation is compared against the threshold n that indicates the tolerance to the misbehaviors. If the owner i tolerates a requester j that misbehaves no more than half of the time, it should set the trust threshold n to 0.5. If the recommendation T_{ij} is above the threshold n, the recommendation is positive, which will increase the total trust to the requester, denoted by $tt = T_{ij} \oplus \mu T_{rj} = E(T_{ij}) + \mu E(T_{ij})$. If the recommendation is below the threshold n, the recommendation is negative, which decrease the total trust to the requester, denoted by $tt = T_{ij} \oplus \mu T_{rj} = E(T_{ij}) - \mu E(T_{ij})$. Finally, when an owner decides whether it will trust the behavior of the requester j, it compares the total trust against the threshold n. The owner considers if the requester's behaviors are friendly if $tt < n$ and malicious if $tt \geq n$. Table 3 gives the legends from Step 2 to Step 5.

Table 3. Legends in step 2-5

Beta(α', β')	The distribution to model the friendly behaviors of request j by the reporter r.
T_{ij}	The probability that reporter r thinks the behaviors of requester j is honest. $T_{ij} \sim$ Beta(α', β')
Beta(γ, δ)	The distribution to model trustworthiness of reporter r by owner i.
γ	The shape parameter for probability density function of the Beta distribution $Beta(\gamma, \delta)$ that indicates whether a reporter is honest and trustworthy.
δ	The shape parameter for probability density function of the Beta distribution $Beta(\gamma, \delta)$ that indicates whether a reporter is dishonest and untrustworthy.
W_{ir}	The probability that owner i thinks the reporter r is trustworthy. $W_{ir} \sim Beta(\gamma, \delta)$
Beta(γ_k, δ_k)	The distribution to model trustworthiness of reporter r by owner i at k-th moment.
s	Result of deviation test. s=1 if the deviation test succeeds, and s=0 otherwise.
γ_k	The shape parameter for probability density function of the Beta distribution $Beta(\gamma_k, \delta_k)$ that indicates whether a reporter is honest and trustworthy at a moment k. $\gamma_k = \gamma_{k-1} + s$
δ_k	The shape parameter for probability density function of the Beta distribution $Beta(\gamma_k, \delta_k)$ that indicates whether a reporter is dishonest and untrustworthy at a moment k. $\delta_k = \delta_{k-1} + (1 - s)$
E(W_{ir})	The expectation of $Beta(\gamma_k, \delta_k)$. $E(W_{ir}) = \frac{\gamma_k}{\gamma_k + \delta_k}$
μ	Trustworthiness that the owner has to the reporter. $\mu = E(Beta(\gamma, \delta))$
m	Threshold of trust to behaviors
n	Threshold of trustworthiness to reporters
tt	Total trust. $tt = T_{ij} \oplus \mu T_{rj}$

Trust-Based UCON

A state is an assignment of values to variables which consist of principal attributes, object attributes, and system attributes. The state transition system can be represented by ($\sum$, S, s_0, δ, F) where $\sum$ is input alphabet, S is a set of system states, s_0 is the initial state, δ is the state transition function δ: $S \times \sum \rightarrow S$, and F is the final state. We define a special system state to specify the status of a single request and access process. The system state S includes *initialState*, *preTrust*, *deniedEnroll*, *trusting*, *disEnrolled*, *preAccess*, *deniedAcces*, *accessing*, *revoked*, and *end*. The initialState means the principal has not sent request; preTrust means the principal is waiting for the authentication decision; deniedEnroll means the system denies the enrollment of the principal based on history or recommendations; trusting means the principal is allowed to collaborate and will send access requests; disenrolled means the system revokes the enrollment of a principal based on runtime information; preAccess means the principal is waiting for the authorization decision; deniedAccess means they system denies the authorization request based on access control rules; accessing means the principal is executing granted privilege; revokedAccess means the system denied the privileges of a principal based on runtime mutable attributes; and end means a principal terminates the access. Actions change the state of the system, which is the input alphabet. If the action is performed successfully the action is true; attributes of the principal, object, and system are assigned a new value. A series of actions are defined to change the status of a request. The transition from one state to another is triggered by an action, as shown in Figure 2. These actions include:

1. *requestEnroll*(*p*): Generates a new request (*p*) when a principal tries to join the community.
2. *denyEnroll*(*p*): Rejects a request (*p*) to enroll the community because the requester can not meet the minimum authentication or trust requirement.
3. *enroll*(*p*): Enrolls a principal (*p*) to the community.
4. *revokeEnroll*(*p*): Revokes the allowed enrollment (*p*).
5. *requestAccess*(*p, o, r*): Generates a new access request (*p, o, r*).
6. *denyAccess*(*p, o, r*): Rejects an access request (*p, o, r*).
7. *grantAccess*(*p, o, r*): Grants an access request (*p, o, r*).
8. *revokeAccess*(*p, o, r*): Revokes an on-going and granted access request (*p, o, r*).
9. *endAccess*(*p, o, r*): Terminates an access request (*p, o, r*).

Figure 2. Trust-based UCON model

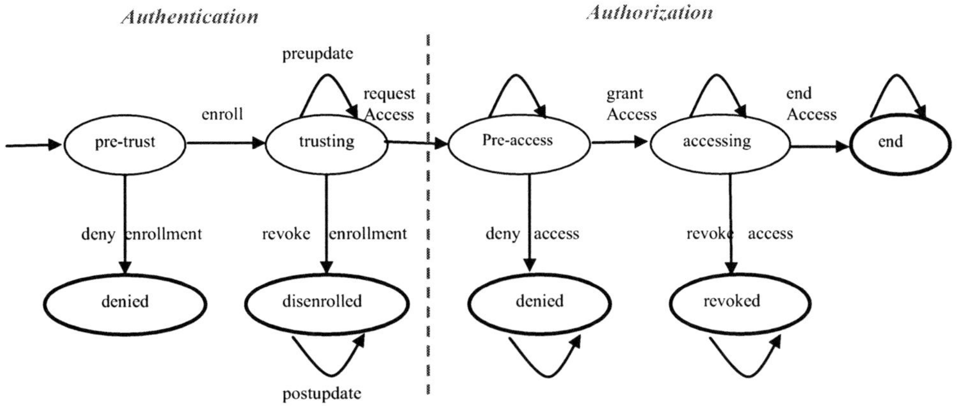

10. *onUpdate(p, o, r)*: Updates the access request *(p, o, r)* when mutable attributes or uncertain behaviors of a principal change.

ARCHITECTURE OF TRUST-BASED USAGE CONTROL IN FILE SHARING

When a principal *p* requests to execute a right *r* on an object *o*, attribute of the principal, permission (right *r* and object *o*), and an optional list credentials are submitted to *secure context handler (SCH)* module. The credentials may include signed trust-assertions (recommendations) from other users or a certificate signed by certificate authority. The SCH looks up the relevant contexts for the requested action and queries the *trust calculator (TC)* module for a trust value about principal *p*. Trust calculator calculates the trust value for a requester based on both observed history and records in recommendation databases. A trust value is passed to *access control manager (ACM)* module for decision. The ACM looks up access control policies that entail several access control constraints. The *constraint service (CS)* module and the *dynamic manager (DAM)* module evaluates access control constraints, for example, time, location, and memberships.

Two categories of trigger events are possible to result in recalculation of trust value and reevaluation of access control policies. Recalculation and reevaluation may cause the revocation of current enrollment or on-going access. The *evidence handler (EH)* module is listening to the peer reports about the misbehaviors of a requester. The negative report can include ignorance of obligation, dishonest behaviors, or the revocation of a requester's certificate. When the trust value of the request drops below a minimum threshold the on-going granted request will be revoked. The result of a trigger event is notified to SCH and execution of request is cancelled. The DAM module is listening to the attribute mutability of the principal, objects, or a context after the permission is granted. For example, the DAM module can be triggered by certain events (e.g., the subject left the group that entails the right). Once the DAM module receives an event, the corresponding access control polices are rechecked by ACM if necessary (e.g., to allow an ongoing usage to continue or revoke it).

Two trigger events may revoke the granted permissions. One is the TC and the other is the DAM. The first one tests whether the behaviors of the requester are too malicious to tolerate and the latter one checks whether the requester is violating the access control rules. Therefore, either one of two trigger events will revoke the in-progress permission.

The update in the TC or DAM may revoke the granted permission. The peers may report the dishonest behavior of the requester or the revocation of the requester's certificate so the trust value of the request is dropped below a scalar. This update will be notified to the *SCH* and cancel the execution of request. After the permission is granted, the DAM will be trigger by certain events such as change of the role. Once the DAM receives an event, the attribute values of the object and subject are retrieved and evaluated and corresponding policies are rechecked by the ACM if necessary (e.g., to allow an ongoing usage to continue or revoke it.)

PROTOTYPE SIMULATION

The architecture outlined in Figure 3 provides the framework for the simulation program of a usage-based access control model. This simulation works under the premise of several users, Tom, Mary, and Lisa, who may request access to files owned by each. Each of these users has been assigned a database to handle the framework described in Figure 3. For every request, a trust value is calculated given past history and current recommendations for the requesting principal. Another factor in consideration is a risk assessment

Figure 3. Trust-based usage control architecture in file sharing

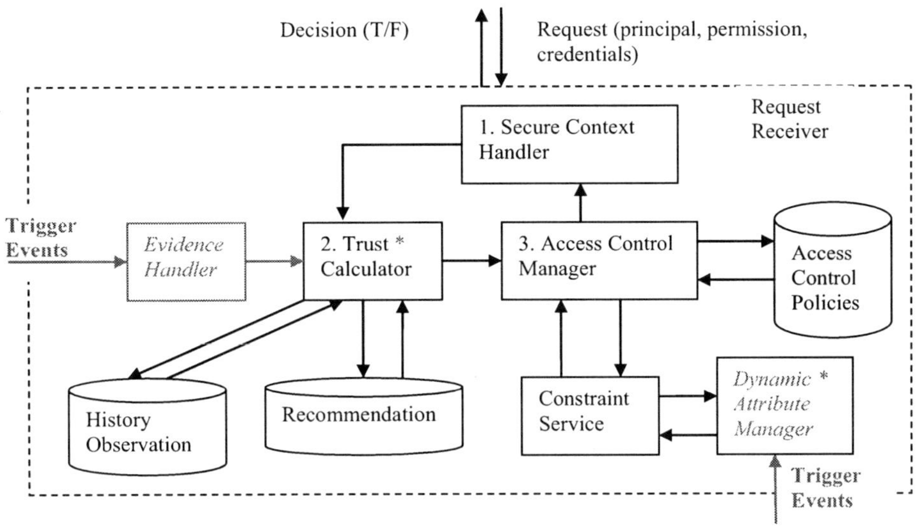

of the action requested based on access control rules assigned to each available file. Each owner assesses the risks based on sensitivity of his/her file. Access to the file may be granted or denied based on trust evaluation, risk assessment, and access control rules. If granted, the continuing usage of this access is contingent on maintaining the trust and risk values within the specified parameters of the rules governing this file. Access is terminated upon completion of the file usage.

Simulation Program

In the context of the simulation, each instantiated user serves as an owner, a requester, or a peer with assigned credentials, trust, and risk values. Additional peers are used and referenced with the purpose of providing recommendations, evidence alerts, and dynamic attribute alerts.

A data model is predefined and populated for use by each instantiated user. The simulation controls the initialization of each instantiated user and communication between instantiated users, configuration of requests, configuration of evidence and dynamic attribute alerts, and display simulation status, and allows for the viewing of the available files, the associated policies, and associated rules. A basic rules engine is implemented to allow simple rule execution.

The simulation program is made up of one single GUI shown in Figure 4 and consists of five sections that provide the means to setup various requests scenarios and alerts. The *user profile* section allows the assignment of the owner and requester. These assignments are required and must be performed to initialize any subsequent actions. The user profile also shows logical and physical credentials associated with the selected user. The *owner* section shows the instantiated user playing the owner role. It also shows the files the owner has exposed for requests. Selecting a file will display any associated policies. Selecting a policy will display any associated rules. The *requester* section shows the instantiated user playing the requester role. Trust and risk are values between 1 and 100. This is the current view the owner has of the requester. The *available files* area displays the list of files available to the requester; these are the same ones shown in the owner's file list. The *requests* are the actions available to the requester. Once a file and a request have been selected, the request button will become available for selection. Clicking on the request button will initialize the request sequence from the requester to the owner and trigger the following events:

Figure 4. Simulation of trust-based usage control

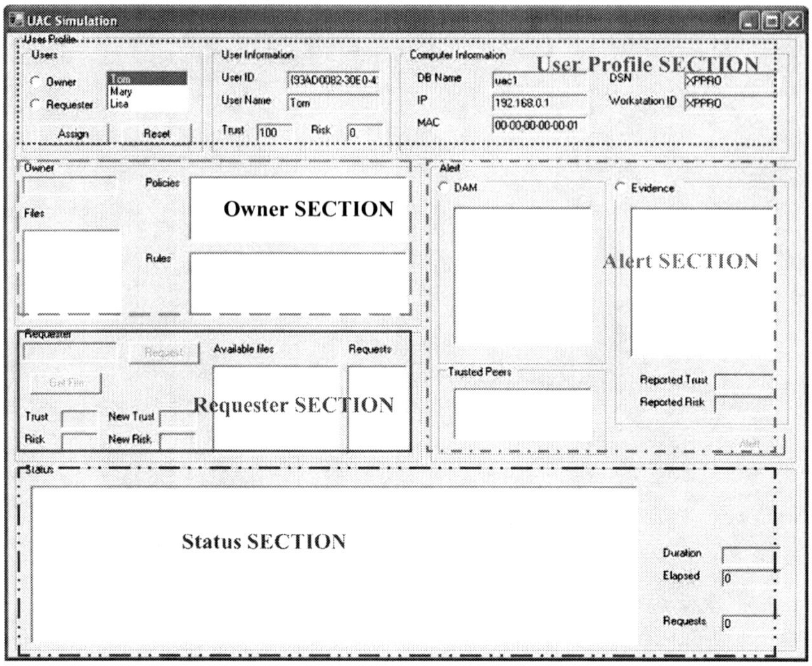

- The requester will receive recommendations for trust and risk values from randomly generated numbers in the range of 30 to 100 and in the range of 1 to 70, respectively.
- The owner will respond to the request by calculating the trust values associated with the requester from history data and recommendations received from peer users.
- Risk associated with the file is calculated from the governing policies and rules.
- The newly calculated trust and risk values for the requester are displayed in new trust and new risk.
- The new trust must be higher than the minimum trust required as defined in the policy.
- The new risk must be lower than the maximum risk allowed as defined and calculated from policy and rules.
- If these parameters are within the acceptable boundaries, the owner releases a token to the requester to be used for the transfer of the requested file.

If the requester is granted access and receives a transfer token then the "get file" button will become available for selection. The alert section allows for two types of in-progress triggers that may halt the current file transfer initiated by the requester. These alerts simulate a trusted peer notification of malicious or negative behaviors. The alerts may be of *evidence* or DAM type. Either will prompt the owner to reevaluate the current trust and risk levels and perform a decision to either allow the transfer to continue or halt it. The status section displays the various messages from the UAC simulation GUI and instantiated users throughout the various states of the transactions. It also displays the number of requests attempted by the requester and the duration and time elapsed for each transfer.

Test Scenario

Several test scenarios are designed to test the ability of our simulation program including how to evaluate trust, evaluate requests against access control rule, and react to evidence alerts and change of mutable attributes. First, a request fails the authentication when the trust value of the request is lower than the minimum trust requirements. Second, a request passes the authentication but fails the authorization when a request does not meet the access control rules although its trust value is higher than the minimum trust requirements. Third, a request passes both authentication and authorization when the trust value of the request is higher than the minimum requirement and the request meets the access control rules. Forth, a request passes both the authentication and authorization; however, the on-going authorized request is revoked by negative evidence reports. Fifth, a request passes both authentication and authorization; however, the on-going authorized request is revoked by mutable attributes such as change of domain or membership, which is triggered by events received from the DAM. All of the above scenarios have passed tests, and we show the update of trust value in Scenario 1 and revocation of granted permission in Scenario 5.

Figure 5 demonstrates a successful request that meets the minimum trust requirements and access control rules and passes authentication and authorization. The requester Lisa requests owner Tom for access to File4 for the purpose of reading the file. Tom's trust in Lisa based on Tom's observation histories is 74 out of 100; the higher Tom's trust in Lisa, the more likely Tom will allow Lisa to read the requested file. Tom will also take into account recommendation from peers (Mary, Peer 1, and Peer 4) with trustworthiness ratings of 65 or higher. Lisa's newly calculated Trust values triggered by her request are 84 and 28 respectively because the total trust value takes into account recommendations provided by Mary, Peerl, and Peer4.

File 4's classification is *public*. Owner Tom has three access control rules for his public file: (1) allow anonymous to read the file classified as public with minimum trust as 30; (2) allow anonymous to write the file classified as public with minimum trust as 80; (3) all the access happens during My Holiday on November 20, 2006. After the successful transfer of File 4, Lisa's trust and risk are increased by 1 point making the final trust equal 85.

Figure 5. Successful request

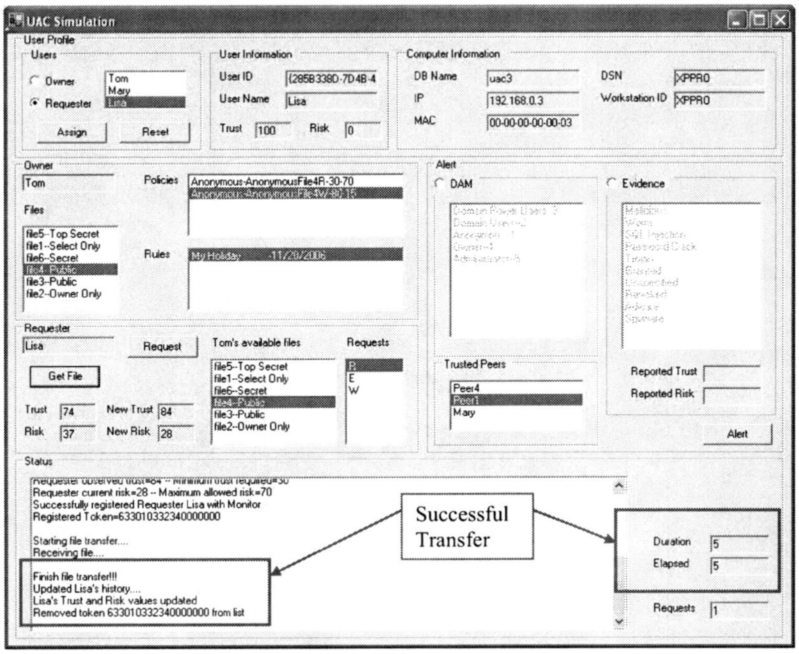

Figure 6 demonstrates the revocation of a granted permission by a DAM alert.

The requester Mary gained the permission to read File 3, but the request was terminated due to the alert generated by Peer4. The alert reflects that Mary's role changed from a *domain power user* to a *domain user* who has no permission to read File 3. Therefore, owner Tom halts the transfer and terminates the connection. Transfer duration was 5. Actual elapsed value is 3 at the point the connection was terminated. The following is the sample output from Scenario 5.

Scenario Output

Owner=Tom
Requester=Mary

Peer4 reported trust value = 51
Peer4 reported risk value = 14
Lisa reported trust value = 78
Lisa reported risk value = 24
Peerl reported trust value = 56
Peerl reported risk value = 47

UserID=d02099fe-8223-4d40-9e08-97567a44e6f4
Requester Name=Mary
Requester IP=192.168.0.2
Requester observed trust=62 -- Minimum trust required=45
Requester current risk=30 -- Maximum allowed risk=79
Successfully registered Requester Mary with Monitor
Registered Token=633010446078906250

Starting file transfer....
Receiving file....
Requester's Security is outside the boundaries -- Aborting connection
Updated Mary' history from Alert...
Mary's Trust and Risk values updated
Removed token 633010446078906250 from list
Requester=Mary

Figure 6. Revoked request triggered by DAM alerts

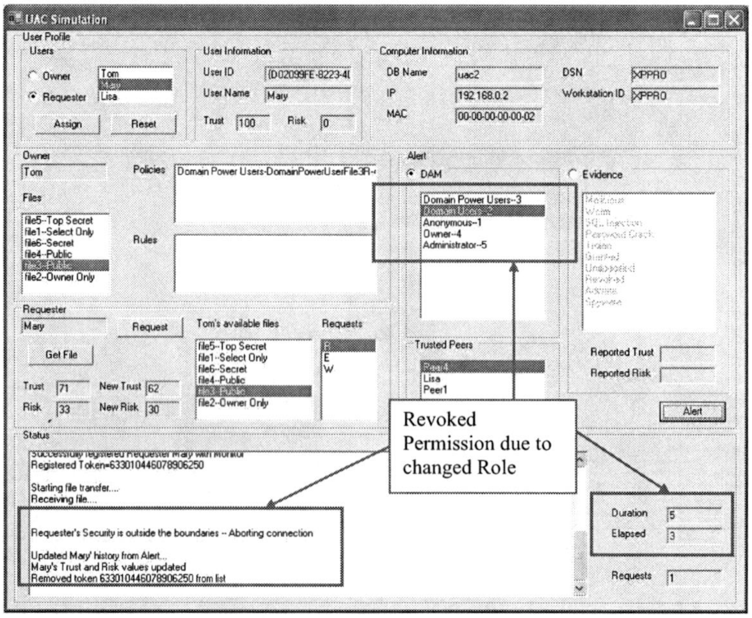

CONCLUSION

We have proposed a framework to integrate trust management into usage-based access control. Our framework is designed to solve uncertainty and attributes mutability in a pervasive and collaborative environment. Our framework was simulated in the application of file sharing in order to demonstrate the feasibility. The authentication and authorization to an on-doing request is checked constantly during the request. The granted request will be terminated if the trust value is lowered due to negative peer reports or when access control rules are not met due to attributes mutability.

REFERENCES

Atif, Y. (2000). Building trust in e-commerce. *IEEE Internet Computing*, *6*(1), 18-24.

Bacon, J., Moody, K., & Yao, W. (2002). A model of OASIS role-based access control and its support for active security. *ACM Transaction Information System Security*, *5*(4), 492-540.

Bertino, E., Bonatti, P. A., & Ferrari, E. (2001). TRBAC: A temporal role-based access control model. *ACM Transactions on Information and System Security*, *4*(3), 191-233.

Bertino, E., Catania, B., Ferrari, E., & Perlasca, P. (2001). A logical framework for reasoning about access control models. In *Proceedings of the Sixth ACM Symposium on Access Control Models and Technologies*, New York, (pp. 41-52). ACM Press.

Bhatti, R., Ghafoor, A., Bertino, E., & Joshi, J. B. D. (2005). X-GTRBAC: An XML-based policy specification framework and architecture for enterprise-wide access control. *ACM Transaction Information System Security*, *8*(2), 187-227.

Cahill, V., Gray, E., Seigneur, J. M., Jensen, C.D., Chen, Y., Shand, B., et al. (2003). Using trust for secure collaboration in uncertain environments. *IEEE Pervasive Computing*, *2*(3), 52-61.

Dimmock, N., Belokosztolszki, A., Eyers, D., Bacon, J., & Moody, K. (2004). Using trust and risk in role-based access control policies. In *Proceedings of the Ninth ACM Symposium on Access Control Models and Technologies*, New York, (pp. 156-162). ACM Press.

Ferraiolo, D. F., Sandhu, R., Gavrila, S., Kuhn, D. R., & Chandramouli, R. (2001). Proposed NIST standard for role-based access control. *ACM Transaction Information System Security*, *4*(3), 224-274.

Georgiadis, C. K., Mavridis, I., Pangalos, G., & Thomas, R. K. (2001). Flexible team-based access control using contexts. In *Proceedings of the Sixth ACM Symposium on Access Control Models and Technologies* (pp. 21-27). ACM Press.

Grandison, T., Sloman, M.: Sloman (2000). A survey of trust in Internet applications. *IEEE Communications Surveys and Tutorials*, *3*(4).

Jajodia, S., Samarati, P., Sapino, M. L., & Subrahmanian, V. S. (2001). Flexible support for multiple access control policies. *ACM Transaction Database System*, *26*(2), 214-260.

Li, N., Mitchell, J. C., & Winsborough, W. H. (2002). Design of a role-based trust management framework. In *Proceedings of the IEEE Symposium on Security and Privacy*, Washington, D.C., (p. 114).

Papoulis, A. (1991). *Probability, random variables, and stochastic processes*. New York: McGraw-Hill.

Park, J., & Sandhu, R. S. (2004) The UCON usage control model. *ACM Transaction Information System Security*, *7*(1), 128-174.

Resnick, P., Kuwabara, K., Zeckhauser, R., & Friedman, E. (2000) Reputation systems. *ACM Communications*, *43*(12), 45-48.

Sandhu, R., & Zhang, X. (2005). Peer-to-peer access control architecture using trusted computing technology. In *Proceedings of the Tenth ACM Symposium on Access Control Models and Technologies*, New York, (pp. 147-158). ACM Press.

Thomas, R. K., & Sandhu, R. S. (1998). Task-based authorization controls (TBAC): A family of models for active and enterprise-oriented authorization management. In *Proceedings of the IFIP TC11 WG11.3 Eleventh International Conference on Database Security* (pp. 166-181). Chapman & Hall, Ltd.

Xiong, L., & Liu, L. (2004). Peertrust: Supporting reputation-based trust for peer-to-peer electronic communities. *IEEE Transaction Knowledge Data Engineering*, *16*(7), 843-857.

Yang, L., Kizza, J. M., Cemerlic, A., & Liu, F. (2007). Fine-grained reputation-based routing in wireless ad hoc networks. In *Proceedings of IEEE Intelligence and Security Informatics Conference* (pp. 75-78). IEEE Computer Society Press.

Yao, W. T. M. (2003). Fidelis: A policy-driven trust management framework. In *Proceedings of the First International Conference on Trust Management* (LNCS 2692, p. 1071).

Zouridaki, C., Mark, B. L., Hejmo, M., & Thomas, R. K. (2005). A quantitative trust establishment framework for reliable data packet delivery in MANETS. In *Proceedings of the Third ACM Workshop on Security of Ad hoc and Sensor Networks*, New York, (pp. 1-10). ACM Press.

This work was previously published in the International Journal of Information Security and Privacy, Volume 2, Issue 2, edited by H. Nemati, pp. 31-45, copyright 2008 by IGI Publishing (an imprint of IGI Global).

Chapter 16
Will it be Disclosure or Fabrication of Personal Information?
An Examination of Persuasion Strategies on Prospective Employees

Xun Li
Nicholls State University, USA

Radhika Santhanam
University of Kentucky, USA

ABSTRACT

Individuals are increasingly reluctant to disclose personal data and sometimes even intentionally fabricate information to avoid the risk of having it compromised. In this context, organizations face an acute dilemma: they must obtain accurate job applicant information in order to make good hiring decisions, but potential employees may be reluctant to provide accurate information because they fear it could be used for other purposes. Building on theoretical foundations from social cognition and persuasion theory, we propose that, depending on levels of privacy concerns, organizations could use appropriate strategies to persuade job applicants to provide accurate information. We conducted a laboratory experiment to examine the effects of two different persuasion strategies on prospective employees' willingness to disclose information, measured as their intentions to disclose or falsify information. Our results show support for our suggestion As part of this study, we propose the term information sensitivity to identify the types of personal information that potential employees are most reluctant to disclose.

INTRODUCTION

High-school seniors have a worry: Will their Facebook or MySpace pages count against them in college admissions? A new survey of 500 top colleges found that 10 percent of admissions officers acknowledged looking at social-networking sites to evaluate applicants. Of those colleges making use of the online information, 38 percent said that what they saw "negatively affected" their views of the applicant.

--Wall Street Journal - September 28, 2008

DOI: 10.4018/978-1-61692-000-5.ch016

Business organizations and other institutions are able to use information systems (IS) to capture and store vast amounts of personal data. Consequently, the public has developed acute anxiety that personal information may be misused, disclosed to unrelated parties, and perhaps even stolen by identity thieves. In early 2000, in a *Wall Street Journal*/NBC poll, "Americans cited loss of personal privacy as their No. 1 concern about the 21st century…" As privacy concerns have been elevated to the forefront of public discourse, organizations are faced with a difficult dilemma: job applicants are reluctant to disclose personal information and sometimes may even provide incomplete or false personal data. Empirical studies show that when people are asked to disclose personal information, they tend to provide false information if they believe their privacy is being compromised (Lwin & Williams, 2003; Fox et al. 2000). But organizations' competitiveness depends on their ability to collect accurate information that will enable them to make critical personnel selection decisions. Therefore, organizations must devise strategies to resolve the conflict between their need for accurate personal information and individuals' wishes to keep their information private.

Because it is partly the increased functionality of IS that has created the heightened awareness of privacy issues, IS researchers and journals are showing keen interest in addressing privacy related issues (e.g., Malhotra et al. 2004; Smith et al. 1996). Although several studies have been conducted, thus far they tend to focus on *consumers'* privacy concerns; research about *employees'* privacy concerns and how organizations must address those issues are relatively scarce (Greenway & Chan, 2005). Current and prospective employees have voiced their information privacy concerns in recent years, and have reported that they have been reluctant to provide accurate personal information when they seek jobs or promotions because they have been afraid that their information will be used in unrelated ways that may impact them adversely (Alge et al. 2006; Stone & Stone, 1990). This is compounded by the fact that privacy practices that protect employee privacy are not standardized across organizations, and employees do not have a clear understanding of policies in their respective organizations (Eddy et al. 1999). Furthermore, in contrast to consumers who can terminate a transaction if they fear their privacy may be compromised, job applicants and employees feel more pressured to disclose personal information to get or hold jobs. In these pressured situations, they tend to fabricate their personal information, as is the case in personality tests (Hough et al., 1990; Hough, 1998; Schmitt & Oswald, 2006). Under the assumption that collecting accurate employee information is desirable, we pose the question: How organizations assuage employees' privacy concerns so that they will be more willing to provide accurate information.

In the persuasion literature, the elaboration likelihood model provides a theoretical perspective on how individuals change their attitudes. Drawing on the persuasion theory, we we tested strategies by which organizations may persuade prospective employees to give accurate personal information. In this study, we focus on prospective employees because job applicants and incumbents respond differently when they are asked to disclose personal information; prior work in psychology suggests that job applicants may be more likely than incumbents to give false information (Hough, 1998; Schmitt & Oswald, 2006). We test whether prospective employees with high levels of privacy concerns are more reluctant to provide personal data and more likely to fabricate what they do provide. Our findings support our arguments that organizations could adopt specific persuasive strategies in privacy policy statements that will persuade prospective employees to increase their intentions to disclose accurate personal information and reduce their intentions to fabricate information.

Researchers have pointed out that employees may be relatively more reluctant to provide specific types of personal information; that is, they may exhibit greater "information sensitivity" to some types of information, but thus far little research has tackled this issue (Thompson & Kaarst-Bown, 2005). To address this gap in information privacy research, we made the first attempt to to identify the types of sensitive information that might be fabricated so that organizations can be more cautious when they ask employees to disclose sensitive information. Our findings indicate that employees are more likely to fabricate such personal information as financial information (e.g. annual income, current debt, annual mortgage payment, checking and saving balances, and other investments), personal medical information (e.g, abortion, drug addiction, allergies, big surgeries and etc.), negative behavior performance (e.g. work mistakes, penalties on driving, fights with colleagues, criminal records and etc.).

RESEARCH FRAMEWORK

IS scholars have noted that privacy is one of the most critical issues that organizations must confront (see review by Greenway and Chan, 2005). According to former Secretary of Commerce Norman Mineta (2000), even the U.S. government regards privacy as one of the most critical issues affecting economic growth. A commonly accepted definition of privacy describes it as the claim of individuals, groups, or institutions to determine for themselves when, how, and to what extent their private information is communicated to others (Westin, 1967).

Specific to disclosure of personal information, it has been found that individuals tend to hold differing levels of concern for information privacy depending on their past experiences, personal characteristics, and the specific contextual conditions (Malhotra et al. 2004). Thus, in a given situation, the extent to which an individual intends to disclose personal information will depend on their general concern for privacy and their beliefs and attitudes toward the entity (i.e., organization) requesting the information. If individuals develop deeper trust toward the entity requiring information instead of harboring fears that disclosure may lead to risks, they are more likely to disclose accurate information (Malhotra et al. 2004). Therefore, organizations that want to obtain accurate information from current and potential employees must take action to alleviate concerns for privacy; organizations must foster the confidence that disclosure of personal information will not lead to undue risks. In a similar vein, IS research has recently started to examine how organizations could alleviate consumers' privacy concerns when transacting through company websites. Findings have indicated that consumers' privacy concerns are more effectively assuaged when organizations provide privacy statements describing their practices rather than the currently used practice of displaying privacy seals such as TRUSTe (Hui et al. 2007). However, research guidelines are not available to explain how organizations may assuage employees' privacy concerns and make them disclose information.

We therefore identify and test one strategy to alleviate employees' concerns for privacy—a strategy that emanates from research on social cognition and specifically uses persuasion theory (Petty & Cacioppo, 1986). *Social cognition* is described as mental processes associated with the ways in which people perceive and react to other individuals, groups, and objects (Wyer, 2004). As per social cognition, individuals' views about particular issues such as their health, their schools, and their privacy make them think, feel, or act positively or negatively toward such issues (Ajzen, 2001; Eagley & Chaiken, 1998; Schwarz & Bohner, 2001). People who have high concerns about privacy may think it is risky to provide information and therefore may react negatively to requests for personal information; that is, they may be unwilling to disclose information. But if their attitudes toward privacy can be changed so that they believe that the organizational practices and procedures will ensure that their information is kept private and not misused, they are more likely to disclose information. Hence, we propose that appropriate persuasion strategies must be used to make employees more willing to disclose personal information, depending on the level of their privacy concerns. Based on the above, we show our research model in Figure 1 and develop our hypotheses. As seen in Figure 1, our research model focuses on the relationship between employees' privacy concerns and their behavioral intentions to disclose or fabricate information. The research model shows that persuasion strategies moderate this relationship.

Employees' Privacy Concerns, and Behavioral Intention

Based on research studies in related disciplines such as management and policy, it is evident that employees believe their privacy has been invaded when organizations use unjust organizational practices or procedures to acquire information from current or prospective employees and use such information in making personnel decisions. Specific organizational practices-related factors have been examined, such as the ability to authorize disclosure, target of release, advanced notice, and purpose of request (Eddy et al. 1999; Stone & Stone, 1990; Stone & Herringshaw, 1991; Stone & Jones, 1992; Stone & Kotch, 1989). The main argument underlying these research studies is that the lesser the perceived invasiveness of organizational practices, the greater the acceptance of such practices and willingness to disclose information.

However, to conduct privacy studies in organizational contexts, currently, no instrument has been developed to measure employees' privacy concerns. Thus, our first step is to develop such an instrument. In the context of consumer privacy, researchers have developed instruments to measure individuals' concerns for information privacy. For example, Smith et al. (1996) developed and validated a multidimensional scale, *concern for information privacy* (CFIP), and Malhotra et al. (2004) developed an instrument measuring *Internet users' information privacy concerns* (IUIPC).

IUIPC is built on *social contract theory* (SC), which has been used to study individuals' perceptions of fairness and justice, particularly in the context of employee-organization relationships. Therefore we believe that by understanding the underlying dimensions and logic in developing this instrument, we can determine the suitability of adapting IUIPC to employee contexts. Organizational justice or fairness has been shown to be an extremely important factor in determining employees' reactions to organizational practices, and employees regard the release of personal information as a risky action because they become vulnerable to the organization's potentially opportunistic behaviors. For this reason, an employee's concerns about information privacy cannot be fully understood without investigating how employees define justice in the context of exchanging personal information with organizations.

Figure 1. Research Model

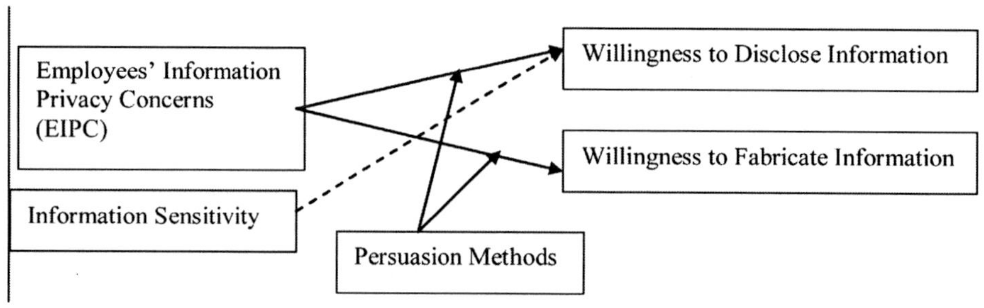

According to Malhotra et al. (2004), SC theory suggests that a firm's collection of personally identifiable data is perceived to be fair only when the individual is granted control over the information, and the consumer is informed about the firm's intended use of the information. As a result, it is possible to characterize IUIPC in terms of three factors: collection, control, and awareness of privacy practices. The collection factor captures the central theme of equitable information exchange based on the agreed social contract. Meanwhile, the control factor represents the freedom to voice an opinion or exit the situation. Finally, the awareness factor indicates understanding about established conditions and actual practices. Following Malhotra et al. (2004), we propose that the employees' level of privacy concerns or their willingness to disclose information center on the same three major dimensions: collection, control, and awareness of privacy practices.

Collection refers to the degree to which an employee is concerned about organizations' collection of personal information. In an equitable exchange, employees give up some information in return for something of value after evaluating the costs and benefits associated with the particular behavior. This collection factor is grounded on SC's principle of *distributive justice*, which relates to "the perceived fairness of outcomes that one receives" (Culnan & Bies, 2003). Thus, employees may have less concern and voluntarily yield accurate information knowing that it will be used fairly to consider their candidacy for a job or a promotion.

Control represents the freedom to voice an opinion (i.e., approval, modification) or exit (i.e., opt-out). This factor is related to SC's principle of procedural justice. According to the principle of procedural justice, individuals view procedures as fair when they are vested with control of the procedures (Thibaut & Walker, 1975; Tyler 1994). In other words, employees will have less concern if they believe they can exercise enough control to decide whether they want to give information and how it will be used.

Awareness of privacy practices refers to policies and procedures by which organizations make employees aware of how the collected information will be used. This awareness factor incorporates two types of justices: interactional and informational. *Interactional justice* includes issues of transparency and propriety of information use stated during enactment of procedures. *Informational justice* relates to the possible disclosure of specified information. When employees are given the specifics of how the information will be used, their perception of fairness increases, and they will have less privacy concerns.

Although the above three dimensions can be used to develop a scale that measures an employee's privacy concerns, the important issue is how these concerns are manifested and measured when employees are asked to provide information. In the privacy literature, a major research stream has emerged that uses intention-based theories such as *theory of reasoned action* (TRA) (e.g., Liu et al., 2005; Malhotra et al., 2004) to predict outcomes of individuals' privacy concerns. According to these theories, the level of individuals' privacy concerns, as an attitude variable, has a negative effect on intentions to disclose information (or willingness to disclose information), and a positive effect on intentions to such actions as fabricating information or terminating transactions. Similarly, it can be expected that the same relationship will apply to the workplace and employees. When asked for personal information, employees with high privacy concerns will be more hesitant and have less intention to disclose information; furthermore, when employees feel compelled to give information to be considered as job candidates, some may develop intentions to fabricate information. Hence, we state our first two hypotheses:

H1a: *employees with high privacy concerns will be less willing to disclose personal information than will prospective employees with low privacy concerns.*

H1b: *employees with high privacy concerns will be more willing to fabricate personal information than will prospective employees with low privacy concerns.*

Behavioral Intentions toward Information Disclosure and Persuasion

As we described earlier, organizations desire to have accurate personal information that will enable them to analyze and make fine-grained hiring and promotion decisions. Employers look for job applicants who show promise of being good citizens and productive members of their organizational work forces (Linowes & Spencer, 1996). But only when employees' privacy concerns are addressed will these prospective employees give accurate information. Therefore, the important question is how to convince employees to recognize the importance of privacy policies so that they develop high intentions to disclose information. This study focuses on two types of behavioral intentions that are related to prospective employees' privacy concerns: intention to disclose personal information and intention to fabricate personal information.

The process of enticing job candidates and incumbents to disclose personal information by soothing their privacy concerns is essentially a process of persuasive communication that influences employees' social cognition. Although many organizations have devised privacy policies and have conveyed these guidelines to their employees, we propose that it is the persuasive communication of organizational policy that can influence concerns for privacy.

Persuasion has been defined as a successful, intentional effort to influence another's mental state or attitude through communication, assuming the person to be persuaded has some measure of freedom (O'Keefe, 1990). Persuasion can help change beliefs about a particular entity by replacing old beliefs with new beliefs (Petty & Cacioppo, 1981).One well-established theory of persuasion is the *elaboration likelihood model* (Petty & Cacioppo, 1986). According to this theory, receivers assess persuasive message differently depending on (among other factors) their involvement with the issue. If the issue is *highly personally relevant* to them, they will focus systematically on the message itself and analyze it according to traditional criteria such as logos, pathos, and ethos. If they find the argument compelling, they will become convinced. This is known as *systematic processing* (Chaiken, 1980), or the *central route to persuasion* (Petty & Cacioppo, 1981). Central route is based on the arguments central to the issue under consideration. It emphasizes the content quality of the persuasive arguments.

However, when the issue is *less personally relevant*, people's attitudes are influenced primarily by heuristic factors such as source qualities and situational elements. This is known as *heuristic processing* (Chaiken, 1980) or the *peripheral route to persuasion* (Petty & Cacioppo, 1981). The peripheral route emphasizes various cues associated with the arguments. The heuristic/peripheral factors that are considered by receivers include whether the source is (1) credible in terms of expertise (Hennigan, Cook, & Gruder, 1982) and lack of self-interested motives (Eagly & Chaiken, 1975); (2) likable in terms of being attractive (Mills & Harvey, 1972) and being similar to the receiver (Mackie & Worth, 1989) and (3) able to engender positive situational factors in terms of, for example, making the receiver feel comfortable (Jorgensen, 1998).

Organizations' signals to job candidates and incumbents about the existence and effectiveness of organizational privacy practices can be viewed as one type of persuasive communication. As Petty & Cacioppo (1981) suggested, personal relevance of an issue is a key factor that determines which persuasion route should be followed. Under high relevance, attitudes are influenced primarily by the quality of the arguments (central route) in the message, whereas under low relevance, attitudes are influenced primarily by peripheral cues.

When employees are presented with a written message about organizational information practices, we expect that the attitudes of employees measured as having *high privacy concerns* are primarily influenced by central route or argument quality, because privacy issue are highly personally relevant to them; whereas employees measured as having *low privacy concerns* will be willing to rely on heuristic factors such as source qualities and situational elements and will take less time to think about the argument quality in the persuasive message. Hence, we propose:

H2a: Prospective employees with high privacy concerns will be more willing to disclose requested information when they are presented with persuasive message of high argument quality than when presented with message with only credible source indicated.
H2b: Prospective employees with low privacy concerns will be more willing to disclose requested information when they are presented with persuasive message with only credible source indicated than when displayed with message of high argument quality.

Laufer & Wolfe (1977) suggested that the exchange of personal information between individuals and an organization is subject to a "calculus of behavior." If the exchange for some economic or social benefits outweighs the risks, people are willing to disclose personal information (Hansell, 2003; Milne & Gordon, 1993). Within an environment where trust between the individual and the organization is lacking, fabrication becomes a means of receiving the benefits without exposure to risk. Lwin & Williams (2003) pointed out that fabrication of information in an online context has emerged as a major consumer response strategy to privacy concerns because fabrication is a less costly and more convenient alternative compared with concerned consumers participating in lobbing efforts of organizations such as the Electronic Privacy Information Center (EPIC).

In the same way, when individuals do not trust an organization's privacy practices they will resort to fabrication to reduce the risk their privacy will be invaded but still maintain the benefits of being considered for a job in the case of job applicants or of being considered a healthy work member in the case of employees. In addition, research indicates that employees are more likely to engage in deviant behaviors directed toward the organization when they perceive lack of organizational support (e.g., Liao et al. 2004). For example, employees may provide inaccurate information to the organization when they think the organization does not have privacy practices adequate to support privacy needs.

Obviously, the persuasiveness of the organization's privacy policy is associated with employees' trust beliefs. Employees who think the message is persuasive establish some trust in the organization. The trust derived from message persuasiveness can reduce the risk beliefs associated with privacy concerns, hence leading to lower intention to adopt measures to protect personal information, such as intention to fabricate information. Thus, we propose:

H2c: Prospective employees with high privacy concerns are less likely to fabricate requested information when they are presented with persuasive message of high argument quality than when presented with message with only credible source indicated.
H2d: Prospective employees with low privacy concerns are less likely to fabricate requested information when they are presented with persuasive message with only credible source indicated than when presented with message of high argument quality.

Information Sensitivity

Prior studies on privacy reveal that to have a complete understanding of individuals' perceptions to information privacy-related issues, researchers must examine not only privacy concerns at a general level, but also must start to address salient beliefs and contextual differences at a specific level (Malhotra et al. 2004; Stewart & Segars, 2002; Smith et al. 1996). Information sensitivity appears to be contextual; that is, what is considered sensitive differs by person and by situation (Cranor et al. 1999; Jones, 1991; Milne, 1997; Weible, 1993). In the situation such as the relation between an employee and organization, some patterns in information type may be identified as being sensitive.

Information sensitivity has been defined as "the level of privacy concern an individual feels for a type of data in a specific situation" (Weibe, 1993, p. 30). Although attempts have been made to differentiate information that is sensitive from information that is not, a consensus has yet to be reached (Hill, 1995). Some guidelines, however, are apparent. According to Wacks (1989), sensitivity of information can be thought of in terms of the value of each piece of information and the ability to link the information to other pieces of information. Individuals appear to be less concerned about the collection and use of information regarding their demographic and lifestyle information and more concerned about the collection and use of medical records, social security numbers, and financial information (Cranor et al. 1999; Nowak & Phelps, 1992; Vidmar & Flaherty, 1985). The FTC (1996) has also noted that financial and medical information are thought, in general, to be sensitive and therefore in need of special protection. In addition, the HIPPA Act of Health specifically addresses the use of sensitive personal medical information.

In workplace settings, studies (e.g., Connetley et al. 1999) have shown that employees are more concerned about the collection and use of information about their negative behaviors such as work mistakes, fights with coworkers, and prior criminal behaviors, and less concerned with the release of their positive behaviors, such as good job performance or volunteer community service.

Hence, based on the above literature review, we expect employees to be highly sensitive to requests asking them to release financial, medical, and prior negative behavior information; hence, we label these classes of information as *high sensitive information*. Employees are less sensitive to demographic, lifestyle, and positive behavior information, so we label it as *low sensitive* information. It is known that individuals' reactions to privacy threats depend on the type of information organizations request (Rohm & Milne, 2004; Malhotra et al., 2004; Woodman et al. 1982). All things being equal, releasing more sensitive information is perceived as more risky than releasing less sensitive information (Milne & Gordan, 1993). Malhotra et al. (2004) have shown that the level of information sensitivity has a negative effect on intentions to disclose requested information. Thus, we propose:

H3: Employees are more willing to disclose low sensitive information than to disclose high sensitive information.

RESEARCH METHOD

To test our hypothesis, we employed an experimental design that used organizational job scenarios (Webster & Trevino, 1995). Specifically, two types of scenarios that described the organizational privacy policies were created as the two different types of persuasion methods, central and peripheral route, on prospective employees.

Development of Scenarios

Scenarios representing persuasive messages were created based on actual privacy policies obtained from several public universities. These are shown in Appendix A. Scenario one presents a privacy policy using the central route to persuasion in which detailed practices and examples are used to address the dimensions of employees' privacy concerns (control, collection, and awareness of privacy policies). For example, described in this scenario are detailed explanations of when employees' information will be disclosed to third parties, who will have access to employees' information, and what the punishments will be if privacy regulations are violated. In scenario two, the peripheral cues are used in the privacy policy to convey its persuasiveness to employees, specifically the source of contacts (credible sources). For example, it is stressed that the president of the university takes the ultimate responsibility for privacy protection.

Measures

Independent Variables. Employees' privacy concerns were measured with an adapted instrument developed from Malhotra et al. (2004). The instrument, which we refer to as EIPC (Employees Information Privacy Concern), was developed based on the arguments presented in section 2.2 and shown in Appendix C. It is directed toward employees in general, including job candidates and incumbents.

Dependent Variables. To measure intentions to disclose information, seven-point semantic scales (MacKenzie & Spreng, 1992) were used. Queries were used such as "Given this hypothetical scenario, specify the extent to which you would reveal (the information)." To measure intention to fabricate information, one question was asked: "When asked for such information, how likely are you going to falsify the information?" (1 = not at all; 2 = under 25%; 3 = 26%–50%; 4 = 51%–75%; 5 = over 75%).

Information sensitivity. To examine the effect of information sensitivity on job candidates' reactions, we asked participants to report their intention to disclose lifestyle information (e.g., hobbies, two favorite magazines, two favorite TV programs), information about positive behavior performance (e.g., rewards for study or work achievement, volunteer community services, blood donation, and charity donation), personal financial information (e.g., annual income, current debt, annual mortgage payment, checking and saving balances, and other investments), personal medical information (e.g., abortion, drug addiction, allergies, and major surgeries), information about negative behavior performance (e.g., work mistakes, driving penalties, fights with colleagues, and criminal records).

Scores on intention to disclose lifestyle information and positive behavior performance were added to construct a variable measuring intention to disclose low sensitive information, whereas scores on intention to disclose the other three types of information were summed to measure intention to disclose high sensitive information.

Control Variables. Factors other than those mentioned may influence employees' reactions to information privacy threats, and therefore we took care to control for these effects by including several control variables. Specifically, we included two demographic variables: sex (Milne & Rohm, 2000) and age (Culnan, 1995; Milne & Rohm, 2000; Wang & Petrison, 1993), and two situation variables: whether the participants' privacy had been invaded in the past (Culnan, 2000) and how much they had been exposed to media reports of incidents of privacy invasion (Smith et al. 1996).

Participants

Participants were 65 students enrolled at a large public university who volunteered for the experiment. In return for volunteering, participants were given course credit. They were informed that the university was developing a human resource information system for people who want to work on campus and that they would be aiding in the development of the system. The system asked participants, as past and potential candidates for campus jobs, to provide various types of information. The persuasive messages delivered through the scenarios explained the school's information privacy policies.

Research Procedures

The sequence of steps in this experiment is shown in Figure 2. Immediately after students volunteered for the study, they were given a pre-experiment questionnaire that was used to collect their background information and measure their general privacy concerns through the instrument we developed, EIPC (Appendix C).

Using participants' EIPC score, the independent variable, we assigned participants with scores above the median of EIPC to the *high privacy concern group*, and those with scores below the median to the *low privacy concern group*. Then participants in each group were randomly assigned to receive either one of the two scenarios. Thus we had four groups to achieve our treatment x block design. Next, participants read the scenario tailored to specific groups. After that, they completed a questionnaire measuring focal variables to identify their behavior intentions (see Appendix C). Finally, they were debriefed and thanked for their participation.

DATA ANALYSIS

The scale of EIPC was validated by conducting confirmatory factor analysis. From the results we found that collection, control, and awareness clearly emerged as independent factors, with items loading on the right designated factors. We also examined discriminant validity of the scale by checking whether a discernable factor loading pattern distinguishes the privacy dimensions from the items measuring intention to disclose. Because each of the items loaded higher than 0.7 on the designated factor and at the same time loaded less than 0.4 on other factors, it can be argued that a reasonable extent of discriminant validity was established.

In addition, to show that intention to disclose information and intention to fabricate information are two distinct constructs, we checked the correlation between the two variables. A correlation value of -0.22 indicates that the two dependent variables are independent constructs.

Figure 2. Experimental Procedures

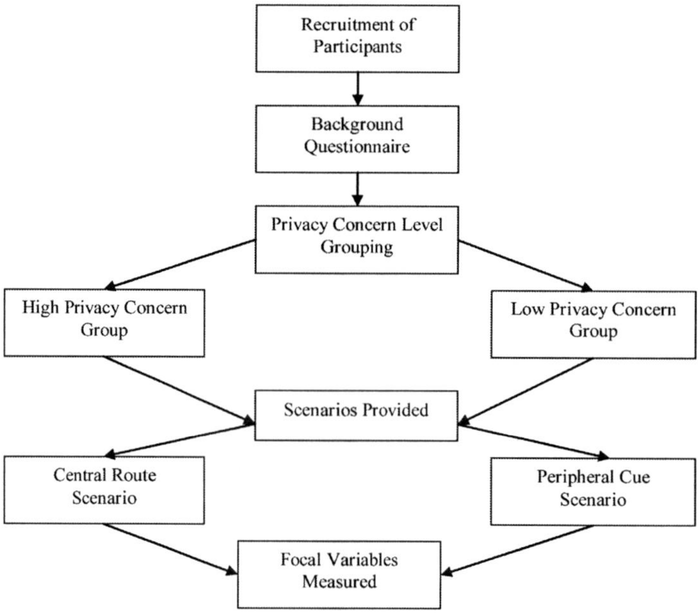

We used the median of EIPC to divide participants into either the *high concern group* or *low concern group*: those with values ranging from the lowest value through median of EIPC were categorized as the *low concern group*, whereas those with values ranging from median through the highest value of EIPC were categorized as the *high concern group*. This resulted in 4 groups of participants (see Table 1).

Manipulation Checks

First, we conducted manipulation checks to see if our treatments via scenarios were successful. We used a questionnaire (see Appendix B) to test whether the participants read the script completely and carefully and whether they perceived differences in the messages. Approximately 90% answered that the scripts were easy to understand and follow. We examined answers to two questions: "What message in the scripts do you think is relevant to you?" and "What items motivated you to respond the way you did in the survey questionnaire?" For those who received scenarios with the peripheral cue treatment, if their answers to the two questions described the credible source of who bears responsibility for privacy protection, their answer was considered *right*. Similarly, for those who received scenarios under the central route treatment, if they described detailed information on how their privacy would be protected, we counted their answers as *right*. About 78% gave right answers. Therefore, we concluded that our treatments were successful in delivering the right type of messages to participants.

Table 1. Participants

	Persuasion Methods	
EIPC	Central Route	Peripheral Cues
Low EIPC	17	14
High EIPC	18	14

We also conducted a check to ensure that our randomization procedures were correct. Among the 65 participants, 44% were females and 56% were males. The average age was 22.36. Control checks were performed on gender, participants' exposure to media reports of privacy invasion incidents, and their experience as victims of privacy invasion. A series analysis of variance (ANOVA) test confirmed that the random assignment of participants to the experimental conditions was successful. No significant differences were seen in gender ($F=1.412$, $p=0.239$), experience as victims ($F=0.969$, $p=0.329$), and exposure to media on privacy ($F=3.33$, $p=0.073$) between the two scenario experiment groups.

Tests of Hypotheses on Intention to Disclose (Fabricate) Information

Table 2 presents the hypothesis testing results for intention to disclose information. Analysis of covariance (ANCOVA) was used in hypothesis testing. The main effect of EIPC on intention to disclose information was significant. Those with high privacy concerns were less likely to disclose personal information than those with low privacy concerns. Therefore Hypothesis 1a was supported.

The interaction effect between EIPC and persuasion methods was not statistically significant. However, a detailed examination of the data pattern (see Figure 3) showed that high concern participants were more likely to disclose information when central route or high quality argument was used as the persuasion method in the privacy policy than when peripheral cues or source credibility were used. On the contrary, low concern participants were more likely to disclose information when peripheral cues were used than when central route was used.

We performed statistical power analysis on this insignificant finding. The effect size of the interaction was 0.002 and power was only 0.062. Although a much larger sample is needed to confirm this finding, no current evidence indicates that the null Hypothesis 2a does not hold.

Table 2. Results on Intention to disclose Information

Dependent Variables	Mean	Standard Deviation	F	P
Intention to disclose Info.				
EIPC	43.45	10.4	7.888	0.007*
Low EIPC	47.146	1.801		
High EIPC	39.632	1.826		
Persuasion Method	43.45	10.4	0.043	0.837
Central route	43.127	1.692		
Peripheral cues	43.651	1.834		
EIPC x Persuasion Method			0.11	0.741
Low EIPC + Central Route	46.467	2.429		
Low EIPC + Peripheral cues	47.825	2.6		
High EIPC + Central Route	39.788	2.513		
High EIPC + Peripheral cues	39.477	2.639		

Notes: We did not report control variables in Table 2 because none have significant effects on intention to disclose information.

Figure 3. Means Level of Intention to Disclose Information

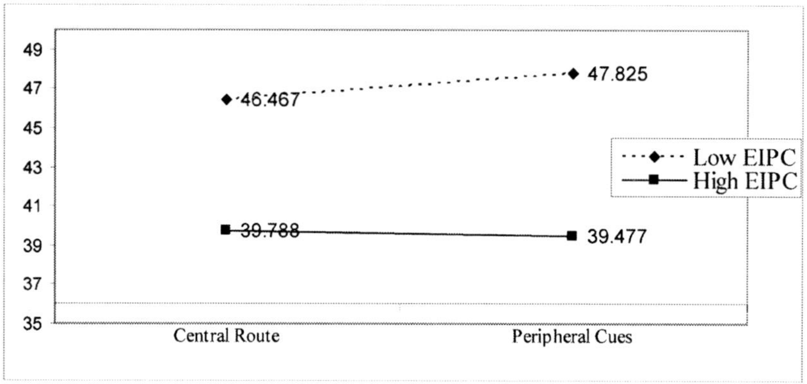

Table 3 presents the hypotheses testing results on intention to fabricate information. The main effect of EIPC on fabrication was not supported. Hypothesis 2b was supported as the interaction between EIPC and persuasion methods was significant.

As shown in Figure 4, participants with low privacy concerns were less likely to fabricate requested information when peripheral cues were used in the persuasive message than when central route was used. In contrast, those with high privacy concerns were more likely to fabricate requested information when only peripheral cues were used than when central route was used.

Tests of Hypotheses on Information Sensitivity

We performed a paired sample t test to compare means of intention to disclose high sensitive information and intention to disclose low sensitive information. As shown in Table 4, participants were more likely to disclose low sensitive information than to disclose high sensitive information,

Table 3. Hypotheses Testing Results on Intention to Fabricate Information

Dependent Variables	Mean	Standard Deviation	F	P
Intention to fabricate Info.				
EIPC	7.18	2.27	0.035	0.851
Low EIPC	7.17	0.4		
High EIPC	7.281	0.407		
Persuasion Method			0.428	0.516
Central route	7.041	0.378		
Peripheral cues	7.411	0.41		
EIPC x Persuasion Method			4.06	0.048
Low EIPC + Central Route	7.553	0.544		
Low EIPC + Peripheral cues	6.788	0.579		
High EIPC + Central Route	6.529	0.553		
High EIPC + Peripheral cues	8.034	0.593		

Notes: We did not report control variables in Table 3, because none have significant effects on intention to fabricate information.

Figure 4. Means Level of Intention to Fabricate Information

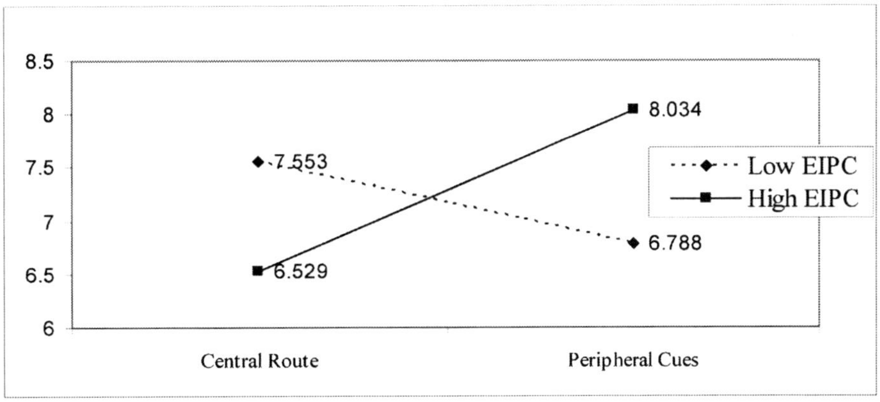

and the difference was statistically significant. Therefore, Hypothesis 3 was supported.

DISCUSSION

Past privacy research has paid little attention to employees' privacy concerns, even though employees are important assets to organizations. Organizations seek accurate personal data from current and prospective employees; hence, we proposed persuasion strategies that could influence prospective employees to willingly disclose truthful personal information. First, we examined

Table 4. Results on Information Sensitivity

	Mean	Standard Deviation
Intention to disclose high sensitive information	20.87	8.54
Intention to disclose low sensitive information	22.60	4.03
Paired difference	1.75	8.38
t =1.686, p=0.0485 (one tail)		

employees' privacy concerns and found that those with higher levels of concern for privacy were indeed less willing to disclose personal information. We then proposed that, depending on the level of privacy concerns of employees, organizations could add appropriate persuasion strategies to their privacy policy statements; by strategically using such persuasions strategies, organizations could increase their ability to collect accurate information. Although our results were not statistically significant, they were in the right direction, suggesting that persuasion strategy is an important factor that affects prospective employees' behavioral intentions to information disclosure. Importantly, results were significant regarding intentions to fabricate information; our results show that inappropriate persuasion strategies could lead prospective employees to have higher intentions to fabricate information, an undesirable outcome from an organizational standpoint. We also found that information sensitivity matters, and employees react to information requests differently based on the type of information requested. Recent research on consumer privacy shows that consumers differ in privacy concerns; appropriate privacy assurances must be provided to persuade consumers to disclose accurate personal information (Smith et al. 1996; Malhotra et al. 2004; Hui et al. 2007). Our study finds complementary results suggesting that employees also may differ in their privacy concerns, and privacy assurance measures must be appropriately designed considering these differences. Thus, organizations must focus their privacy policies and efforts on both consumers of their products or services and on employees who develop these products or provide services.

Implications for Research

This study is one of the first few to address employees' privacy concerns, and we hope this can jump-start more privacy research dealing with employees. Most privacy studies have examined individuals' privacy concerns at a general level, paying little attention to individuals' perceptions in a particular context. Researchers call for privacy studies that consider contextual differences at a specific level (Malhotra et al. 2004; Stewart & Segars, 2002; Smith et al. 1996). This study contributes to the privacy research by providing empirical evidence that persuasion and information sensitivity are important contextual variables that influence employees' evaluation and reactions to organizations' requests to disclose personal information.

Social contract theory suggests that through organizational privacy practices—that is, the making of privacy policy—a more trusting environment for information exchange could be built to significantly ease employees' concerns. We submit that the power of the social contract theories in addressing employees' information privacy concerns may be further enhanced by considering the impact of social cognition, persuasion strategies, and information sensitivity factors. This study is the first that has approached individuals' (employees in this case) information privacy concerns from a social cognition perspective, and it finds that this is a useful foundation to analyze and influence individual privacy concerns. Although privacy can be viewed as a social contract between an individual and an organization as proposed in prior research, we submit that to address and soothe individual privacy concerns we must look at social cognition theories to help us develop strategies. Hence, we believe that future research must pay greater attention to social cognition and its effects on privacy related issues.

This study is among the first few in IS to look closely at employees' intention to fabricate information; thus, more attention should be paid to this issue because organizations are greatly interested in understanding employees' intentions to fabricate. We used a scale that could be tested, enhanced, and improved in future research. Understanding employees' or applicants' intentions will enable organizations to preempt monetary losses stemming from misdirected and ineffective employee recruitment and retention conceived through the use of inaccurate information. Lwin & Williams (2003) present a framework proposing some major situational elements that must be considered to understand an individual's perception of privacy and invasions of privacy. Future research is needed to integrate employees' privacy concerns with Lwin & Williams's framework and to assess the impacts on fabrication of information.

In this study, we focused on the influence of the overall EIPC construct, but more work needs to be done to tease out the individual contributions of its three underlying factors. The research reported here could be extended by examining the impact of collection, control and awareness of privacy practices as individual discrete latent components that, by themselves, have unique impacts on attitude change.

Implication for Practice

From a practical standpoint, this research focuses on a crucial topic in need of attention if organizations' competitiveness depends on their ability to collect accurate employee information. First, we introduce this idea that employees with privacy concern may provide organizations with falsified information. The finding that it is possible to persuade employees to provide accurate personal information is striking and of significant value to policy makers. This suggests that an educational program designed to demonstrate the organization's practices dealing with privacy issues to employees have the potential to increase the collection of accurate employee information significantly. In addition, It is apparent that a one-size –fits-all approach to messaging is unlikely to have universal appeal or success to address private concerns. Much work remains to be done to identify what variables resonate the most with a specific demographic (Angst and Agarwal, 2009).

From an employer's perspective, this research provides insight related to human resource management and IS management. Many organizations have started to use human resource systems to collect employee / prospect employee information. Although all human resource systems have web page for privacy policies, what differentiates one organizational policy from another is whether the policy is properly designed to persuade applicants or employees to provide adequate and accurate personal information. We have shown that effective persuasion strategies will go a long way in presenting policies properly so that current and prospective employees will have a positive understanding of privacy policies and will be willing to disclose, not fabricate, their personal information.

REFERENCES

Ajzen, I. (2001). Nature and operation of attitudes. *Annual Review of Psychology*, *52*, 27–58. doi:10.1146/annurev.psych.52.1.27

Alge, B. J., Ballinger, G. A., Tangirala, S., & Oakley, J. L. (2006). Information Privacy in Organizations: Empowering Creative and Extrarole Performance. *The Journal of Applied Psychology*, *91*(1), 221–232. doi:10.1037/0021-9010.91.1.221

Angst, C. M., & Agarwal, R. (2009). Adoption of electronic health records in the presence of privacy concerns: the elaboration likelihood model and individual persuasion. *Management Information Systems Quarterly*, *33*(2), 339–370.

Chaiken, S. (1980). Heuristic versus systematic information processing and the use of source versus message cues in persuasion. *Journal of Personality and Social Psychology*, *39*, 752–766. doi:10.1037/0022-3514.39.5.752

Connetley, M. L., Mael, F. A., & Morath, R. A. (1999). 'Don't ask—please tell': Selection Privacy from two perspectives. *Journal of Occupational and Organizational Psychology*, *72*, 405–422. doi:10.1348/096317999166752

Cranor, L. F., Reagle, J., & Ackerman, M.S. (1999), Beyond Concern: Understanding Net Users' Attitudes About Online Privacy. *AT&T Labs-Research Technical Report.* TR 99.4.3

Culnan, M. J. (1995). Consumer Awareness of Name Removal Procedures: Implications for Direct Marketing. *Journal of Direct Marketing, 7*, 10–19. doi:10.1002/dir.4000090204

Culnan, M. J. (2000). Protecting Privacy Online: Is Self-Regulation Working? *Journal of Public Policy & Marketing, 19*(1), 20–26. doi:10.1509/jppm.19.1.20.16944

Culnan, M. J., & Bies, R. J. (2003). Consumer privacy: balancing economic and justice considerations. *The Journal of Social Issues, 59*(2), 323–342. doi:10.1111/1540-4560.00067

Eagly, A. H., & Chaiken, S. (1975). An attribution analysis of the effect of communicator characteristics on opinion change: the case of communicator attractiveness. *Journal of Personality and Social Psychology, 32*, 136–144. doi:10.1037/h0076850

Eagly, A. H., & Chaiken, S. (1998). In Gilbert, D., Fiske, S. T., & Lindzey, G. (Eds.), *Handbook of social psychology* (4th ed., *Vol. 1*, pp. 269–322). Boston: McGraw-Hill.

Eddy, E. R., Stone, D. L., & Stone-romero, E. F. (1999). The effects of information management policies on reactions to human resource information systems: and investigation of privacy and procedural justice perspectives. *Personnel Psychology, 52*, 335–358. doi:10.1111/j.1744-6570.1999.tb00164.x

Fox, S., et al. (2000). Trust and Privacy Online: Why Americans want to rewrite the rules? In *The Pew Internet and American Life Project* (pp. 1-9). Available at http://www.pewinternet.org

Greenaway, K. E., & Chan, Y. E. (2005). Theoretical Explanations for Firms' Information Privacy Behaviors. *Journal of the Association for Information Systems, 6*(6), 171–198.

Hansell, S. (2003). Amazon Tries Word of Mouth? Retrieved from http://www.nytimes.com/2003/02/10/business/media/10ADCO.html

Hennnigan, K. M., Cook, T. D., & Gruder, C. I. (1982). Cognitive tuning set, source credibility, and the temporal persistence of attitude change. *Journal of Personality and Social Psychology, 42*, 412–425. doi:10.1037/0022-3514.42.3.412

Hill, R. P. (1995). Researching Sensitive Topics in Marketing, the Special Case of Vulnerable Populations. *Journal of Public Policy & Marketing, 14*(Spring), 143–148.

Hough, L. M. (1998). Effects of intentional distortion in personality measurement and evaluation of suggested palliatives. *Human Performance, 11*, 209–244. doi:10.1207/s15327043hup1102&3_6

Hough, L. M., Eaton, N. R., Dunnette, M. D., Kamp, J. D., & McCloy, R. A. (1990). Criterion-related validities of personality constructs and the effect of response distortion on those validities. *The Journal of Applied Psychology, 75*, 581–595. doi:10.1037/0021-9010.75.5.581

Hui, K. L., Teo, H. H., & Lee, S. Y. T. (2007). The Value of Privacy Assurance: an Exploratory Field Experiment. *Management Information Systems Quarterly, 31*(1), 19–33.

Jones, M. G. (1991). Privacy: A Significant Marketing Issue for the 1990s. *Journal of Public Policy & Marketing, 10*(Spring), 133–148.

Jorgensen, P. F. (1998). Affect, persuasion and communication processes . In Anderson, P. A., & Gurrero, L. K. (Eds.), *Handbook of communication and emotion: Research, theory, application, and contexts* (pp. 403–422). San Diego, CA: Academic Press.

Laufer, R. S., & Wolfe, M. (1977). Privacy as a concept and a social issue: A multidimensional developmental theory. *The Journal of Social Issues*, *33*(3), 22–42. doi:10.1111/j.1540-4560.1977.tb01880.x

Liao, H., Joshi, A., & Chuang, A. (2004). Sticking out like a sore thumb: Employee dissimilarity and deviance at work. *Personnel Psychology*, *57*, 969–1000. doi:10.1111/j.1744-6570.2004.00012.x

Linowes, D. F., & Spencer, R. C. (1996). Privacy in the Workplace in Perspective. *Human Resource Management Review*, *6*(3), 165–181. doi:10.1016/S1053-4822(96)90011-X

Liu, C., Marchewka, J. T., Lu, J., & Yu, C. S. (2005). Beyond concern – a privacy – trust – behavioral intention model of electronic commerce. *Information & Management*, *42*, 289–304. doi:10.1016/j.im.2004.01.003

Lwin, M. O., & Williams, J. D. (2003). A Model Integrating the Multidimensional Developmental Theory of Privacy and Theory of Planned Behavior to Examine Fabrication of Information Online. *Marketing Letters*, *14*(4), 257–272. doi:10.1023/B:MARK.0000012471.31858.e5

MacKenzie, S. B., & Spreng, R. A. (1992). How does motivation moderate the impact of central and peripheral processing on brand attitudes and intentions? *The Journal of Consumer Research*, *18*(March), 519–529. doi:10.1086/209278

Mackie, D. M., & Worth, L. T. (1989). Processing deficits and the mediation of positive affect in persuasion. *Journal of Personality and Social Psychology*, *57*, 27–40. doi:10.1037/0022-3514.57.1.27

Malhotra, N. K., Kim, S. S., & Agarwal, J. (2004). Internet users' information privacy concerns (IUIPC): the construct, the scale and a causal model. *Information Systems Research*, *15*(4), 336–355. doi:10.1287/isre.1040.0032

Mills, J., & Harvey, J. (1972). Opinion change as a function of when information about the communicator is received and whether he is attractive or expert. *Journal of Personality and Social Psychology*, *21*, 52–55. doi:10.1037/h0031939

Milne, G. R. (1997). Consumer Participation in Mailing Lists: A Field Experiment. *Journal of Public Policy & Marketing*, *16*(Fall), 298–309.

Milne, G. R., & Gordon, M. E. (1993). Direct mail privacy-efficiency trade-offs within an implied social contract framework. *Journal of Public Policy & Marketing*, *12*(2), 206–215.

Milne, G. R., & Rohm, A. J. (2000). Consumer Privacy and Name Removal Across Direct Marketing Channels: Exploring Opt-in and Opt-out Alternatives. *Journal of Public Policy & Marketing*, *19*, 238–249. doi:10.1509/jppm.19.2.238.17136

Mineta, N. (2000). *Opening statement at online privacy technology workshop and technology fair.* U.S. Department of Commerce, Washington, D.C. (September 19). Retrieved from http://www.ntia.doc.gov/ntiahome/privacy/900workshop/mineta91900.htm

Nowak, G. J., & Phelps, J. (1992). Understanding Privacy Concerns: An Assessment of Consumers' Information Related Knowledge and Beliefs. *Journal of Direct Marketing*, *6*, 28–39. doi:10.1002/dir.4000060407

O'Keefe, D. J. (1990). *Persuasion: Theory and research* (*Vol. 2*). Newbury Park, CA: Sage.

Petty, R. E., & Cacioppo, J. T. (1981). *Attitudes and persuasion: classic and contemporary approaches*. Dubuque, IA: Brown.

Petty, R. E., & Cacioppo, J. T. (1986). The elaboration-likelihood model of persuasion. In L.Berkowitz (Ed.), *Advances in experimental social psychology,* 19, 123-205.

Rohm, A. J., & Milne, G. R. (2004). Just what the doctor ordered—the role of information sensitivity and trust in reducing medical information privacy concern. *Journal of Business Research*, *57*(9), 1000–1011. doi:10.1016/S0148-2963(02)00345-4

Schmitt, N., & Oswald, F. L. (2006). The Impact of Corrections for Faking on the Validity of Noncognitive Measures in Selection Settings. *The Journal of Applied Psychology*, *91*(3), 613–621. doi:10.1037/0021-9010.91.3.613

Shwartz, N., & Bohner, G. (2001). The construction of attitudes . In Tesser, A., & Shwarts, N. (Eds.), *Blackwell handbook of social psychology: Intraindividual processes* (pp. 436–457). Oxford, UK: Blackwell.

Smith, H. J., Milberg, S. J., & Burk, S. J. (1996). Information privacy: measuring individuals' concerns about organizational practices. *Management Information Systems Quarterly*, 167–196. doi:10.2307/249477

Stewart, K. A., & Segars, A. H. (2002). An empirical examination of the concern for information privacy instrument. *Information Systems Research*, *13*(1), 36–49. doi:10.1287/isre.13.1.36.97

Stone, D. L., & Herringshaw, C. (1991). *Effects of the purpose of the test, perceived relevance, and use of test results on reactions to honesty testing*. Paper presented at the 6th annual conference of the Society for Industrial and Organizational Psychology, St. Louis, MO.

Stone, D. L., & Jones, G. (1992). *Effects of type of test and justification provided on reactions to honesty testing*. Paper presented at the 7th annual conference of the Society for Industrial and Organizational Psychology, Montreal, Quebec, Canada.

Stone, D. L., & Kotch, D. (1989). Individuals' attitudes toward organizational drug testing policies and practices. *The Journal of Applied Psychology*, *3*, 518–521. doi:10.1037/0021-9010.74.3.518

Stone, E. F., & Stone, D. L. (1990). Privacy in Organizations: Theoretical issues, research findings, and protection mechanisms. *Research in Personnel and Human Resource Management*, *8*, 349–411.

Thibaut, J., & Walker, L. (1975). *Procedure justice: a psychological analysis*. Hillsdale, NJ: Erlbaum.

Thompson, E. D., & Kaarst-Brown, M. L. (2005). Sensitive information: A review and research agenda. *Journal of the American Society for Information Science and Technology*, *56*(3), 245–257. doi:10.1002/asi.20121

Tyler, T. R. (1994). Psychological models of the justice motive: antecedents of distributive and procedural justice. *Journal of Personality and Social Psychology*, *67*, 850–863. doi:10.1037/0022-3514.67.5.850

Vidmar, N., & Flaherty, D. H. (1985). Concern for Personal Privacy in an Electronic Age. *The Journal of Communication*, *35*(Spring), 91–103. doi:10.1111/j.1460-2466.1985.tb02236.x

Wacks, R. (1989). *Personal Information, Privacy and the Law*. Oxford: Oxford University Press.

Wang, P., & Petrison, L. A. (1993). Direct Marketing Activities and Personal Privacy: A Consumer Survey. *Journal of Direct Marketing*, *7*, 7–19. doi:10.1002/dir.4000070104

Webster, J., & Trevino, L. K. (1995). Rational and social theories as complementary explanations of communication media choices: Two policy capturing studies. *Academy of Management Journal*, *38*, 1544–1572. doi:10.2307/256843

Weible, R. J. (1993). *Privacy and Data: An Empirical Study of the Influence and Types and Data and Situational Context upon Privacy Perceptions*. Doctoral dissertation, Department of Business Administration, Mississippi State University.

Westin, A. (1967). *Privacy and freedom*. New York: Athenaeum.

Woodman, R. W., Adams, J., Tolchinsky, P. D., & Fromkin, H. (1982). Asurvey of employee perceptions of information privacy in organizations. *Academy of Management Journal*, *25*(3), 647–663. doi:10.2307/256087

Wyer, R. S. Jr. (2004). The cognitive organization and use of general knowledge . In Jost, J. T., & Banaji, M. R. (Eds.), *Perspectivism in social psychology: The yin and yang of scientific progress* (pp. 97–112). Washington, D.C: American Psychological Association. doi:10.1037/10750-008

APPENDIX A. SCENARIO SCRIPTS

Scenario One (Privacy Policy with Central Route)

Welcome to the Human Resource Website of University of KKTT. We are pleased that you have applied for a position in this university. To process your application and select the most suitable candidate for the listed position, we must collect some information from you about your background and your prior job experiences.

Before providing this information, please read the privacy policy of the university below. This policy explains how the university manages and protects the privacy of the information that we collect from you.

Sincerely,
Human Resource Department
University of KKTT

KKTT Information Privacy Policy on Personally Identifiable Information

1. It is the policy of the University of KKTT to ensure that its treatment, custodial practices, and uses of personally identifiable information are in full compliance with all related federal and state statutes and regulations, and demonstrate a rigorous commitment to core values of maximizing trust, integrity, and respect for privacy. In addition to fulfilling all mandatory compliance obligations, the UK is committed to do the utmost to ensure the privacy of personal information.
2. Personally identifiable information may **be released or provided to others only as follows**:
   a. To university employees and/or officers **on an authorized need-to-know basis** who are authorized to use such information as part of their official university duties, and who keep that information confidential, using it only for official university purposes without further disclosing or providing that information to others.

For example, those in supervisory positions may have access to personnel records only of those who report to them, and only for legitimate business purposes; faculty may have access to their students' academic records for authorized purposes such as teaching, advising, grading, and admissions processing; police auditing and supervisors may have access to security and surveillance access information; authorized university computing and network personnel may have access to files, transactions, cookies, profiles, messages, log records call detail, voicemail, and network content containing personally identifiable information; medical professionals may have appropriate access to their patients' records in accordance with medical center policies and all federal and state laws and regulations.

b. As part of the normal directory information, course listings, catalogs, and other official publications of the University or its departments **without providing personal information unless it is explicitly and specifically approved by the individuals.**
c. To external health care, governmental, educational, business, law enforcement, audit, etc., entities to the extent required for the proper and efficient conduct of official University business transactions and processes (e.g., payroll processing and deposits, valid student record maintenance and requests, admissions processing, student loan processing, patient medical record and test results

handling), approved by the responsible University official as specified in sections 3, 4, and 5 below. **Third party handling of such information must meet the strict requirements of keeping that information confidential and using it for authorized official business purposes only.**

d. In accordance with state and federal law concerning personal choice options concerning release of personally identifiable information.

3. **Excluding exceptions granted under sections 4, 5, and 6 below, any other release of personally identifiable information may be made to the minimum extent only** (as determined by specific analysis in each instance of a request for such information) unavoidably required by law. Third party handling of such information (as directed by the University) must meet the strict requirements of keeping that information confidential and using it for authorized official business purposes only.
4. Exceptions to this policy may be made only on specific requests as approved by the authorized university officers responsible for such electronic information as specified in sections 5 and 6 below, and as necessary to achieve the missions and business needs of the University. Any and all exceptions made must be documented with the Executive Vice President and reviewed periodically by the Privacy Officer.
5. Policy Enforcement:
    a. **If any university employee**, student, teaching assistant, research assistant, or other appointee, or any contractor, consultant, or vendor to, or business partner to, the university **discovers evidence of any apparent policy violations, they must notify the Privacy Officer immediately, and preserve the evidence of violation.** Failure to do so is grounds for disciplinary measures up to and including termination of employment or expulsion and other legal actions.
    b. The University should actively create a climate that **encourages all members of the broad university community** (including, but not limited to university personnel, patients, research subjects, vendors, external auditors, volunteers, students, or third parties external to the university); partner institutions and general public to **report to the Privacy Officer instances of credible evidence of possible privacy and security policy violations.**
    c. If anyone discovers apparent evidence of any incidental violation of this policy, he or she should report it to the University Privacy Officer, and the University will take appropriate action to resolve the issue.
    d. If the University discovers violations of this policy substantially caused or perpetuated by careless or deliberate acts or inaction, it will **take disciplinary measures toward those involved up to and including termination of employment and/or expulsion and/or other legal actions.**
    e. On discovery of plausible evidence that this policy or any other privacy or security policy has been or is being violated, the University shall **inform appropriate authorities to investigate and take appropriate action to resolve the issue.**

Scenario Two (Privacy Policy with Peripheral Route)

Welcome to the Human Resource Website of University of KKTT. We are pleased that you have applied for a position in this university. To process your application and select the most suitable candidate for the listed position, we must collect some information from you about your background and prior job experiences.

Before you provide this information, please read the privacy policy of the university below. This policy explains how the university manages and protects the privacy of the information that we collect from you.

Sincerely,
Signed by President,
University of KKTT

KKTT Information Privacy Policy on Personally Identifiable Information

1. The University of KKTT ensures that its treatment, custodial practices, and uses of personally identifiable information are in full compliance with all related federal and state statutes and regulations, and demonstrate a rigorous commitment to core values of maximizing trust, integrity, and respect for privacy. In addition to fulfilling all mandatory compliance obligations, the UK is committed to do the utmost to ensure the privacy of individuals.
2. Responsibilities and authorities for personally identifiable information:
    a. The **Executive Vice President** is the authority who is responsible for **employee personnel information**.
    b. **The Provost** is the authority who is responsible for **faculty personnel information.**
    c. **The Vice President for Student Affairs** is the authority who is **responsible for student or applicant information**.
    d. **The Vice President for Medical Affairs** is the authority who is responsible for **patient information and information about referring physicians**. This information includes medical and dental information in all units across the University.
    e. **Vice President for Computing and Communications** is the authority **who is responsible for personally identifiable information in computer system** authentication, authorization, access, usage, profile, cookie, or other such files or in telecommunications or network records.
    f. **The Executive Vice President** is the authority responsible for **all other personally identifiable information** including but not limited to information relating to alumni, athletic supporters, personally identifiable information in intellectual property, or other technology transfer or royalty, etc., records, personally identifiable information in contract and grant proposals, research records, notes and videos, surveillance tapes, and information on ticket holders, donors, consultants, and contractors, etc.
3. Privacy Officer - As an integral component of the University's commitment to privacy protection and compliance, the **UW has an official University Privacy Officer.** The University Privacy Officer is **appointed by the President of the University and must be a senior member of the University's Administration**. **The Executive Director for the Human Resource Department is presently the University of KKTT's Privacy Officer**. The position is responsible for strategic oversight and coordination of the University's privacy protection and compliance efforts. The Privacy Officer's role also includes
    a. Promptly relaying any credible evidence of or reports concerning violations of University privacy or security policy or law to the appropriate investigative authorities.

b. Providing information as necessary to the University community about existing and emerging legal and compliance requirements with respect to privacy and related best practices. Ensure continuing notification to appropriate parties throughout the University about privacy policy and any revisions to the policy.
c. Supporting security and privacy awareness and education program efforts.
d. Supporting the development, implementation, and maintenance of information systems security and privacy policies and procedures where required in various areas, units, and functions across the University.
e. Supporting monitoring and enforcement of privacy and security policies, and adoption of and adherence to best practices.
f. Working to ensure that vendors, business partners, and others are aware of University privacy and security policies and that University procurement, contracting, and partnering processes not only emphasize adherence to privacy and security policies but, where appropriate, incorporate provisions that punish failures to properly address and comply with the policies.
g. Helping provide direction and oversight concerning risk management practices associated with privacy and security issues and practices.
h. Helping ensure that appropriate audit services and reporting are in place to detect violations and to evaluate the effectiveness of privacy and security policies and of compliance activities.
i. Ensuring and overseeing periodic monitoring of the release of personally identifiable information to ensure continuing compliance with state and federal law.
j. Ensuring review of exceptions to this policy as to their appropriateness and legality.
k. Acting as an advocate for budget and resource requests related to ensuring the maintenance of effective information privacy and security programs.

4. This policy and policy document is to be and **is maintained by the Vice President for Human Resource Department in consultation with the President, the Executive Vice President, the Provost, and the University's Privacy Officer**.

Appendix B. Manipulation Check Questionnaire

A) Did you understand the policy; was it easy to follow?
B) What message in the policy did you feel was most relevant to you?
C) What items in the message made you respond as you did?

Appendix C. Research Constructs and Measures

Table 5.

Control: Seven-point scales anchored with *strongly disagree* and *strongly agree*
(1) Employee privacy is really a matter of employees' right to exercise control and autonomy over decisions about how their information is collected, used, and shared.
(2) Employee control of personal information lies at the heart of employee privacy.
(3) I believe that employee privacy is invaded when control is lost or unwillingly reduced as a result of disclosing information to an organization.
Awareness (of Privacy Practices): Seven-point scales anchored with *strongly disagree* and *strongly agree*
(1) Employers seeking information should disclose the way the data are collected, processed, and used.
(2) A good organizational privacy policy should have a clear and conspicuous disclosure.
(3) It is very important to me that I am aware and knowledgeable about how my personal information will be used.
Collection: Seven-point scales anchored with *strongly disagree* and *strongly agree* (Smith et al. 1996).
(1) It usually bothers me when employers ask me for personal information.
(2) When employers ask me for personal information, I sometimes think twice before providing it.
(3) I'm concerned that employers are collecting too much personal information about me.
Intention to Give Information: Seven-point semantic scales (MacKenzie and Spreng 1992) were used.
Given this hypothetical scenario, specify the extent to which you would reveal: **personal information about lifestyle (e.g. hobbies, two favorite magazines, two favorite TV programs), positive behavior performance (e.g. rewards on study or work achievement, volunteer services in community, blood donation, and charity donation), personal financial information (e.g. annual income, current debt, annual mortgage payment, checking and saving balances, and other investments), your personal medical information (e.g, abortion, drug addiction, allergies, big surgeries and etc.), your negative behavior performance (e.g. work mistakes, penalties on driving, fights with colleagues, criminal records and etc.)**
(1) Unlikely/likely
(2) Not probable/probable
(3) Possible/impossible _r_*
(4) Willing/unwilling _r_
Intention to Fabricate Information:
When asked for such information, how likely are you going to falsify the information? (1 = not at all; 2 = under 25%; 3 = 26%–50%; 4 = 51%–75%; 5 = over 75%)
Covariates:
(1) Sex: (1=male; 2=female).
(2) Age: (1 = 25–34; 2 = 35–44; 3 = 45–54; 4 = 55–64; 5 = over 65 years).
(3) Education: (1 = some school, no degree; 2 = high school graduate; 3=some college, no degree; 4=bachelor's degree; 5 = master's degree; 6 = professional degree; 7 = doctorate degree).
(4) Privacy victim: How frequently have you personally been the victim of what you felt was an improper invasion of privacy? (1=very infrequently; 7=very frequently)
(5) Media exposure: How much have you heard or read during the last year about the use and potential misuse of the information collected from the Internet? (1 = not at all; 7=very much)

Chapter 17
Computer Security Practices and Perceptions of the Next Generation of Corporate Computer Users

S. E. Kruck
James Madison University, USA

Faye P. Teer
James Madison University, USA

ABSTRACT

The purpose of this chapter is to present the results of an empirical study of the computer security practices and perceptions of the next generation of corporate computer users, undergraduate university students. The authors surveyed undergraduate university students who represented 42 different majors. The findings relate to the students' usage of antivirus programs, firewalls, password security, and security patches. Student perceptions of computer security and its importance are also reported. Research in this area is important for two reasons. First, potential employers may find the results useful in assessing their vulnerability to unsafe practices from entry level employees. Secondly, research in this area can give those responsible for providing computer security education a better understanding of students' computer security training needs.

INTRODUCTION

For as long as computers have achieved widespread use in industry, computer security has been critical to the effective functioning of organizations. However, in the mid1990s when the widespread sharing of information over the Internet and the growth of e-commerce became commonplace in organizations, computer security became more important than ever and moved to center stage (Duffy & Walstrom, 2003; Gordon, Loeb, Lucyshyn, & Richardson, 2004). In the first quarter of 2006 alone, e-commerce revenues in the United States totaled $25.2 billion dollars, up 7% from the fourth quarter 2005. Worldwide e-commerce revenues were estimated at $976.1 billion dollars (U.S. Census Bureau, 2006). With extensive commerce taking place online, individuals and companies now face a constant challenge of securing their computers and business transactions from sophisticated cyber criminals.

Society's concerns about the growing threats to computer security are well-founded. For example, when the first virus infected ARPANET in 1987, no one had any idea that within a few years computer viruses would become epidemic. We are now at the point that highly successful new viruses are introduced every week (Schultz, 2004). In addition, the newer, polymorphic viruses are capable of changing their signature every time they replicate and infect a new and different file type in order to keep from being detected.

Clearly, computer security is vital to today's organizations and economy. Industry statistics illustrate the seriousness of threats to computer security. With 137,529 reported security incidents in 2003 alone (CERT/CC, 2006), effective information security has become a necessity rather than an afterthought. Given the substantial number of security incidents in organizations and the growing reliance of corporations on the Internet, there is a need for further research by both practitioners and academicians in the area of computer security. Leach (2003) suggests that the internal threat to computer security is more pressing than external threats and is the "result of poor user security behavior." Goodwin (2005) indicates that IT training is targeted to the CIO, whereas it should be targeted to the "bottom of the pyramid." In light of the known threat caused by the improper computer security practices and perceptions of the users, many researchers bemoan that computer security awareness is just beginning to be addressed in the literature (Collins, Rawlinson, Manwani, & Allen, 2005; Dhillon & Blackhouse, 2001; Goodwin, 2005; Kirkpatrick, 2006; Leach, 2003; Siponen, 2000).

Some authors have issued a call for more computer security awareness, education, and training (Dhillon & Blackhouse, 2001; Kirkpatrick, 2006). Before university professors can effectively respond to this call to action, we must have a clear understanding of students' current computer security practices and perceptions. Therefore, the purpose of this study is to perform empirical research that documents computer security practices and perceptions among undergraduate university students.

Information gleaned through this study should be of interest to both practitioners and academicians. Because the student population of computer users is the next generation of corporate computer users, documenting students' unsafe computing practices and perceptions is important to potential employers as an aid in assessing their vulnerability to unsafe practices from entry level employees.

If universities are to provide organizations with employees who are responsible computer users, the designers of university curriculum must have a better understanding of students' computer security training needs. Computer security training for university students can be more effective if faculty have a profile of students' current computer security practices and perceptions.

RELATED LITERATURE

The related literature that is part of this study examines two aspects of computer security. First, the literature indicating the impact computer crime is having on the bottom line in organizations is summarized. This is followed by an examination of what is found in the literature pertaining to unsafe computing practices by computer users.

Impact of Computer Crime on Organizations

Unfortunately, in spite of attempts to stop or slow the problem, computer crime continues to have a substantial negative impact on the bottom line in organizations. The following recent research cited from practitioner literature warns of the ongoing threat to organizations' computing environment caused by computer crime.

This drain on corporate profits within the USA has been well documented during the last 10 years by the Computer Security Institute (CSI) and the FBI. The 2005 CSI/FBI survey of computer security professionals at 700 organizations documents the growing threat of computer crime. The first, second, and third source of the greatest annual dollar amount of financial loss is virus attacks ($42.8 million); unauthorized access ($31.2); and theft of proprietary information ($30.9), respectively (Gordon, Loeb, Lucyshyn, & Richardson, 2005).

The damage caused by computer viruses can be viewed on a global level. One study reported that the recent Sobig.F virus caused an estimated $29.7 billion of economic damages worldwide (Goldsborough, 2004). Goldsborough (2004) stated that, "It's difficult not to catastrophesize the virus problem."

A further drain on the bottom line in organizations caused by computer crime can be seen from examining the typical size of an organization's information security budget relative to the organization's overall IT budget. In one multi-industry survey, 27% of the respondents indicated that between 6% and 10% of the total IT budget was devoted to security (Gordon et al., 2005). The state government sector had the highest average annual computer security spending per employee at $497 (Gordon et al., 2005). Forrester Research estimates that in Fortune 500 companies, the average spent for IT security is .0025% of revenue or slightly less than what they spend on coffee (Clarke, 2002).

Concerns about issues such as security, privacy, and authentication capabilities are considered by many in the IT industry to be continuing challenges to the implementation of technological advances in use of the Web to conduct business (Becker, 2004; Lim & Wen, 2003). If computer security issues are not adequately addressed, the expansion of Web services will continue to be impeded.

The CSI/FBI computer crime and security survey indicated that Web site computer security incidents exponentially increased in 2004 as compared to the 2003 survey results (Gordon et al., 2005). One example of the security threat to companies utilizing online transactions is that of cyber-extortion. Cyber-extortion, in which cybercriminals threaten to release confidential information unless their pay off demand is met, is growing in frequency and intensity (Gow, 2005.)

Another area of computer crime that is on the rise is identity theft. Becoming a victim of identity theft via an online attack is growing; an estimated 500,000 to 700,000 Americans are victims of identity theft every year (Hinde, 2003). According to the Federal Trade Commission, the dollar loss has grown to over $343 million in 2002 and victims spend over 175 hours attempting to restore their reputation by contacting credit card companies, the credit bureaus, and the local police (Hinde, 2003). Growing at an annual rate of 29%, identify theft is the fastest growing crime in the United Kingdom (Customer Management, 2006.)

Unsafe Practices by Computer Users

Unsafe computing practices by users are documented in the business and academic literature. For example, a recent CNNMoney study that documented unsafe computer practices among users found that too often there is not a match between a user's perception of their computer's security and the actual security of their computer. The study indicated that 77% of the respondents stated their computer was well protected. However, the findings indicated otherwise: four out of five had spyware or adware programs; most computers did not have updated antivirus software and 66% had been infected by a virus; 67% did not have a firewall; and 38% of wireless networks were not encrypted (CNNMoney, 2004).

Cappel (2001/2002) conducted a study among Midwest USA employers. Over half of the respondents held IT/IS leadership positions (e.g., as Chief Information Officers/Vice President, IS/IT Directors, or IS Managers/ Consulting Managers, or Project Leaders) and almost 60% had 13 or more years of professional work experience. The authors found a significant disconnect between expected and actual skills for security.

In another survey, Hinde (2003) documented unsafe computer practices among a sample of 130 IT professionals and chief executive officers. Hinde found that 66% of the respondents had given their password to their colleagues and, of the users with more than 10 passwords, 49% wrote down their passwords or stored them in an unencrypted file on their computer.

In light of those findings, it is not surprising that many researchers are specifically stating that there is a need to require the individual to assume more responsibility when it comes to computer security (Aytes & Connolly, 2004; Goldsborough, 2004; Hinde, 2003; Rhodes, 2001; Schultz, 2004; Tuesday, 2001; Wade, 2004). In a recent editorial, Schultz (2004) reflected the views of many when he issued a call to look more at how to engineer changes in human behavior and develop a wiser user community.

Numerous studies have discussed the importance of reducing computer security risk through giving additional careful attention to computer user security awareness and training (Aytes & Connolly, 2004; Businesses Respond, 2005; Furnell, Gennatou, & Dowland, 2002; Gordon et al., 2005; Kirkpatrick, 2006; Pollitt, 2005; Whitman & Mattord, 2004). Results from one study indicated the majority of computer security practitioners from a cross section of industries viewed security awareness training as important; however, those respondents stated that their organizations did not invest enough in security awareness training (Gordon et al., 2005).

As a part of this study, an exhaustive search of the literature was completed in order to find empirical research that examined university students' practices and perceptions related to computer security. We found one study limited to business students that examined the computer security practices of students at two universities (Aytes & Connolly, 2004). As a part of the Aytes and Connolly study, business students were asked to self-report the frequency of engaging in five security-related practices: share passwords; voluntarily change passwords; open e-mail attachments from an unknown sender without virus checking; open e-mail attachments from a known sender without virus checking; and regular backups. The findings indicated that 49% of the business students engaged in risky computing behavior at least occasionally and 28% did so frequently or all the time (Aytes & Connolly, 2004).

It has been stated that "the more knowledge we have about the causes and consequences of computer security breaches..., the more likely it is that computer security will improve" (Gordon et al., 2004). Yet, the recent study by Aytes and Connolly (2004) reported that there is a lack of research that shows why computer users engage in unsafe behaviors. Those researchers were also unable to find any systematic evidence in the research literature as to which unsafe practices were prevalent.

Stanton, Stam, Mastrangelo, and Jolton (2004) stated that "Because of the importance of end user security-related behaviors, having a systematic viewpoint on the different kinds of behaviors that end users enact could provide helpful benefits for managers..." Thus, the related literature indicates that, if universities are to provide industry with entry-level employees who possess the proper skills and knowledge, those responsible for the university curriculum need to examine the issue of computer security awareness training for students. The authors of this study believe that both practitioners and curriculum designers need to have a clear idea of the knowledge gap that exists among university students regarding safe computing practices. It would help if we better understood the students' perceptions related to numerous computer security issues. This information can be useful in planning an appropriate response to the computer security awareness training needs of students and new hires.

METHODOLOGY

To encourage participation in the survey and increase our response rate, a one-page questionnaire was constructed to obtain data from undergraduate students at one large state university on the east coast. Undergraduate students were selected because the authors felt that this was the first opportunity for young adults to be responsible for their own safe computing practices. The questionnaire was developed based on a previous study of the literature and a previous pilot study directed at three specific majors (Teer, Kruck, & Kruck, 2007). Several questions on the survey were designed to obtain respondents' demographic information. The survey incorporated questions pertaining to the students' behavior concerning: antivirus programs, firewalls, and password security. In addition, the students' perceptions of computer security were obtained.

Students from a variety of courses were surveyed during the spring of 2006. Students were randomly selected around campus, with no more than 10 at any one gathering. Two limitations inherent in this study should be noted. First, the values obtained were participants' stated perceptions and no attempt was made to validate the accuracy of the responses. Second, care should be exercised when generalizing the results to individuals from other universities.

Responses were anonymous and students were not asked to put their names on the questionnaire. Three hundred eighty surveys were completed and 355 students provided usable responses for a response rate of 96%. Some questionnaires were unusable because students did not complete all questions. In addition, students not in an undergraduate program were eliminated. Forty-two different majors, with an average of eight students per major, completed the surveys and those majors listed 10 or more times include over 70% of the participants. Table 1 summarizes the respondent background information.

RESULTS

The findings relative to student behavior are grouped within the topics of antivirus programs, firewalls, number of viruses in the past year, passwords, and operating system security patches. The second section of the findings relate to the students' perceptions of the importance of computer security.

Students' Antivirus Behavioral Practices

As part of this survey, several questions were designed to determine students' general antivirus behavioral practices. Students were asked about their use of antivirus programs, regularity of updating those programs, frequency of antivirus scans, and the number of times their computer has been infected with a virus.

The majority of the students surveyed, 86%, have installed antivirus protection on their own personal computers. The university the students attend provides two different antiviral solutions (Symantec Antivirus Corporate Edition & Norton Antivirus Corporation Edition). In addition, the university used in this survey requires an antivirus solution installed on any computer that is directly attached to the university network (on-campus housing); this may contribute to the high installation rate among respondents. With regard to updating the virus definition file, 67% of the students responded that they "regularly" update. Of those with installed virus protection, only 17% scanned daily, 43% scanned weekly, and 31% scanned monthly, with 9% of the respondents stating they "never" scan.

Table 1. Demographic information

		Number of Responses	Percentage Response
Gender			
	Male	191	54%
	Female	164	46%
Academic Level			
	Freshman	29	8%
	Sophomore	77	22%
	Junior	66	19%
	Senior	183	52%
Major			
	Finance	37	10%
	Management	33	9%
	Marketing	32	9%
	Computer Information Systems	24	7%
	Media Arts and Design	20	6%
	Computer Science	18	5%
	Economics	12	3%
	International Business	12	3%
	Music	12	3%
	Hospitality Tourism Management	11	3%
	English	10	3%
	Integrated Science and Technology	10	3%
	Political science	10	3%
	Psychology	10	3%

The responses to the question regarding the number of viruses students had experienced in the last 12 months indicated that 61% of the students had one or more viruses during the past year. A breakdown of these instances showed that 28% of the students reported encountering 0 viruses in the past 12 months; 38% encountered 1 to 5 viruses; 10% had 6 to 9 viruses; and 13% of the students had 10 or more viruses. The remaining 11% were not sure how many viruses they had in the last 12 months. Table 2 shows the percentage response of antivirus behavior.

Firewalls

When participants were asked about the use of firewalls on their personal computers, 62% of the students report they use a firewall, 22% report they do not have a firewall, and 16% of the students were "unsure" if they have a firewall enabled. The 16% of students who were "unsure" may point to a lack of knowledge about the purpose and use of firewalls by the student population. Table 3 indicates the use of firewall software.

Table 2. Percentage response concerning antivirus behavior

		Percentage Response
Use Antivirus Program		
	Yes	86%
	No/Not sure	14%
Regular updates to Antivirus Program		
	Yes	67%
	No/Not sure	33%
Frequency of Antivirus Scan		
	Daily	17%
	Weekly	43%
	Monthly	31%
	Never	9%
Number of Virus in the Last 12 Months		
	None	28%
	1 to 4	38%
	5 to 9	10%
	10 or more	13%
	Not sure	11%

Passwords

Responses to password security questions were mixed. Sixty-two percent of users reported that they have intentionally given a password to another person. The university attended by the students completing the survey requires an online security training course which covers general computing security along with specific security related to the confidentiality of passwords. This training is required every 6 months, at which time the student's password must be changed.

Seventy-one percent reported that they use passwords on their home computers. This means that more than one-quarter of students' home computers are unprotected. Sixty-two percent of computers are insecure because passwords have been distributed to other people. Table 4 presents the percentage response for each behavior.

Table 3. Percentage response concerning use of firewall

		Percentage Response
Use a Firewall		
	Yes	62%
	No/Not sure	38%

Operating System Security Patches

Sixty-three percent of the respondents stated that they apply patches to their operating system when they become available. Again, this result may be affected by automatic updating recently becoming integrated with operating systems and upgrades. That leaves over one-third of the computers vulnerable to the latest operating system attack. These results are presented in Table 5.

Importance of Security

Students were asked their perceptions of the importance of security as it relates to three aspects: the security of their own personal computer; the importance of computer security to the student; and the importance of computer security to business. The majority of students, 58%, felt that their personal computers are "somewhat secure" and

Table 4. Percentage response concerning password security

		Percentage Response
Intentionally given out password		
	Yes	62%
	No/Not sure	38%
Home computer password protected		
	Yes	71%
	No	29%

Table 5. Percentage response concerning installation of security patches

		Percentage Response
Security patches applied		
	Yes	63%
	No/Not sure	37%

24% indicating they felt their computers are "very secure." Only 14% felt that their computers were "very insecure" or "somewhat insecure."

When asked how important computer security is to a student user, 62% responded that it was "very important" to them and 4% responded that it was "somewhat unimportant" or "not important." Finally, when asked how important computer security is in a business environment, the students were nearly unanimous, 94%, in saying that it is "very important." Table 6 summarizes percentage response of the students' perceptions for each issue.

DISCUSSION AND NEEDED FUTURE RESEARCH DIRECTIONS

Authors of articles in a variety of journals have issued the call for more research in the area of computer security awareness. In response to this increasingly important topic, this empirical research examined the computer security practices and perceptions of undergraduate university students. This chapter indicates a need to develop effective computer security training for university students. University students were surveyed because these young adults are typically away from home and, for the first time, are responsible for their own safe computing practices.

Unfortunately, numerous unsafe computer security practices and perceptions of students were documented in this study.

Fourteen percent of student participants don't use or don't know if they use an antivirus program and 33% reported that they either do not or are unsure if they regularly install updates to their antivirus program. Sixty-one percent of the students stated that they had one or more viruses during the past year. Among the respondents, 37% reported that they do not apply patches to their operating system when they become available or are unsure if patches are applied. Clearly those results indicate students are leaving there personal computers vulnerable to viruses and attacks. Future research is needed to examine the reasons why students fail to protect their personal computer against viruses.

Seventy-one percent reported that they use passwords on their computers. This means over one-quarter of students' home are computers unprotected; and 62% of computers are insecure because passwords have been distributed to other people. Yet 82% of participants state that their own personal computer is "very secure" or "somewhat

Table 6. Perceptions toward computer security

		Percentage Response
Security of Own Personal Computer		
	Very secure	24%
	Somewhat secure	58%
	Not Sure	4%
	Somewhat insecure	9%
	Very insecure	5%
Importance of Computer Security to Student		
	Very Important	62%
	Somewhat Important	31%
	Undecided	3%
	Somewhat Unimportant	2%
	Not Important	2%
Importance of Computer Security to Business		
	Very Important	94%
	Somewhat Important	4%
	Undecided	1%
	Somewhat unimportant	0%
	Not Important	1%

secure." These results are very similar to what was found in the previously cited CNNMoney study that documented a disconnect between a users' perception of their computer security and the actual security of their computer.

This study found that a student's perception of the security on their computer is generally "somewhat secure," and students also felt that it was "very important" for their computer to be secure. When students are asked to respond to the importance of computer security to businesses, they responded almost unanimously that it was important. The combination of these responses may point to a general lack of knowledge among students on both the real need for computer security and how to better protect their computers because more than one-third of the students do not update their virus definition file and almost 40% do not or don't know if they have a firewall. Computer users who update their operating system have their firewall turned on automatically; thus students no longer have to be proactive or knowledgeable about firewalls. But, in so doing they may lack the knowledge and skills to equip themselves with the most effective firewall.

A couple of informal comparisons can be made between the findings of this study and studies of business people. The results of the CSI/FBI 2005 survey indicated 97% had of the respondents had a firewall and 96% of respondents employed a virus protection solution as compared to the 62% of student respondents who stated they used a firewall and 86% employed a virus protection. The results from our survey regarding students' behavior toward passwords can be compared to the password behavior noted among a sample of business people including IT professionals and chief executives. Sixty-six percent of the business respondents had given their password to a colleague (Hinde, 2003), as compared to the 62% of student respondents stating that they shared their passwords.

The present study reported in this manuscript examined the prevalence of unsafe computing practices among one population , undergraduate university students. One possible extension of this research would be to examine unsafe computing practices among students at other universities or graduate students.

University faculty must do more to assure that students sent to the workplace are not bringing unsafe computer practices to the corporate environment and to a university environment (Anderson & Schwager, 2002; Monds & Wang, 2003). The growing vulnerability of corporate computer information systems, as well as threats to the security of personal computers, dictates that those responsible for curriculum in all programs throughout the university must insure that students are obtaining an adequate level of understanding and skill on the fundamentals of computer security. Those who learn and apply effective computer security as students will be better equipped to resolve the computer security challenges they will face in their corporate roles.

REFERENCES

Anderson, J. E., & Schwager, P. H. (2002). Security in the information systems curriculum: Identification and status of relevant issues. *Journal of Computer Information Systems*, *42*(3), 16-24.

Aytes, K., & Connolly, T. (2004). Computer security and risky computing practices: A rational choice perspective. *Journal of Organizational and End User Computing*, *16*(3), 22-40.

Becker, S. A. (2004). A usability study of Internet privacy policies for state and commercial Web sites. *International Journal of Services and Standards*, *1*(1), 52-68.

Businesses respond to cybercrime and security trends. (2005). *The CPA Journal*, *75*(1), 9.

Cappel, J. (2001/2002). Entry-level IT job skills: A survey of employers. *Journal of Computer Information Systems*, *42*(2), 76-83.

CERT/CC Statistics 1988-2006. (2006). Retrieved November 30, 2007, from http://www.cert.org/stats/cert_stats.html

Clarke, R. (2002). *Forum on technology and innovation: Sponsored by Sen. Bill Frist (R-TN), Sen. Jay Rockefeller (D-WV), and the Council on Competitiveness*. Retrieved November 30, 2007, from http://www.techlawjournal.com/security/20020214.asp

CNNMoney. (2004). *Home PCs not so safe*? Retrieved November 30, 2007, from http://money.cnn.com/2004/10/25/technology

Collins, B., Rawlinson, R., Manwani, S., & Allen, K. (2005, September). How can we spread the security message? *Computer Weekly*, *27*, 32.

Customer management: Policing the crisis. (2006, January 27). *Precision Marketing, 22*.

Dhillon, G., & Blackhouse, J. (2001). Current directions in IS security research: Toward socio-organizational perspectives. *Information Systems Journal*, *11*(2), 127-153.

Duffy, T. J., & Walstrom, K. A. (2003). Changes in student computer technology attitudes over time. *Journal of Computer Information Systems*, *43*(3), 27-33.

Furnell, S. M., Bennatou, M., & Dowland, P. S. (2002). A prototype tool for information security awareness and training. *Logistics Information Management*, *15*(5/6), 352-357.

Goldsborough, R. (2004). Arming yourself in the virus war. *Tech Directions*, *63*(8), *14*.

Goodwin, B. (2005, November 29). Investment in security training on the wrong track, say senior staff. *Computer Weekly*, *58*.

Gordon, L. A., Loeb, M. P., Lucyshyn, W., & Richardson, R. (2004). *2004 CSI/FBI computer crime and security survey*. CSI Publications.

Gordon, L. A., Loeb, M. P., Lucyshyn, W., & Richardson, R. (2005). *2005 CSI/FBI computer crime and security survey*. CSI Publications.

Gow, B. (2005). The growing threat of cyber-extortion. *Risk Magazine*, *52*(4), 72.

Hinde, S. (2003). Careless about privacy. *Computers & Security*, *22*(4), 284-288.

Jaques, R. (2004). Demand for IT security pros growing fast. *PC Magazine*. Retrieved November 30, 2007, from http://www.pcmag.co.uk/print/1159247

Kirkpatrick, J. (2006). Protect your business against dangerous information leaks. *Machine Design*, *78*(3), 66.

Leach, J. (2003). Improving user security behavior. *Computers and Security*, *22*(8), 685-692.

Lim, B., & Wen, H. J. (2003). Web services: An analysis of the technology, its benefits, and implementation difficulties. *Information Systems Management*, *20*(2), 49-57.

Monds, K. E., & Wang, C. (2003). CDC's new epidemic: An investigation into awareness, attitudes, actions, and knowledge of computer viruses among students utilizing campus computer labs. *Journal of Computer Information Systems*, *43*(3), 118-126.

Pollitt, D. (2005). Energis trains employees and customers in IT security. *Human Resource Management International Digest*, *13*(2), 25-29.

Rhodes, K. A. (2001). Operations security awareness: The mind has no firewall. *Computer Security Journal*, *17*(3), 1-12.

Schultz, E. (2004). Worms and viruses: Are we losing control? *Computers & Security*, *23*(3), 179-180.

Siponen, M. T. (2000). A conceptual foundation for organizational information security awareness. *Information Management & Computer Security*, *8*(1), 31-41.

Stanton, J. M., Stam, K. R., Mastrangelo, P., & Jolton, J. (2005). Analysis of end user security behaviors. *Computers & Security*, *24*(2), 124-133.

Teer, F. P., Kruck, S. E., & Kruck, G. P. (2007). Empirical study of students' computer security practices/perceptions. *Journal of Computer Information Systems*, *47*(3), 105-110.

Tuesday, V. (2001). Human factor derails best-laid security plans. *Computerworld*, *35*(18), 52.

U.S. Census Bureau. (2006). Retrieved November 30, 2007, from http://www.census.gov/mrts/www/data/html/06Q1.html

Wade, J. (2004). The weak link in IT security. *Risk Management*, *51*(7), 32-37.

Whitman, M. E., & Mattord, H. J. (2004). Making users mindful of IT security. *Security Management*, *48*(11).

This work was previously published in the International Journal of Information Security and Privacy, Volume 2, Issue 1, edited by JH. Nemati, pp. 80-90, copyright 2008 by IGI Publishing (an imprint of IGI Global).

Chapter 18
Design and Implementation of a Framework for Assured Information Sharing Across Organizational Boundaries

Bhavani Thuraisingham
The University of Texas at Dallas, USA

Yashaswini Harsha Kumar
The University of Texas at Dallas, USA

Latifur Khan
The University of Texas at Dallas, USA

ABSTRACT

In this chapter we have designed and developed a framework for sharing data in an assured manner in case of emergencies. We focus especially on a need to share environment. It is often required to divulge information when an emergency is flagged and then take necessary steps to handle the consequences of divulging information. This procedure involves the application of a wide range of policies to determine how much information can be divulges in case of an emergency depending on how trustworthy the requester of the information is.

INTRODUCTION

Organizations including healthcare, military, and financial have to form coalitions and collaborate to solve a particular problem such as the global war on terror or provide joint services to customers. However each organization of a coalition has to enforce policies to ensure that information is shared in a secure manner. This concept as come to be known as assured information sharing (AIS).

In our previous chapter we have described our approach to AIS. Our work was initially based on a need to know paradigm where the organizations share data according to the policies. However in many situations information has to be divulged in case of emergencies. For example, in a high security building, the building plan is not easily accessible and distributable information. But in case of an emergency such as an evacuation in case of fire, people normally unauthorized to such information are permitted to access it. For example consider a case of fire where some employees may be trapped and the building maintenance personnel are unavailable to help out. But using devices that may be able to fetch data, they may be able to use exit routes.

The work described in this chapter discusses the design and implementation of a XML-based framework for a need to share environment. In particular, we have designed and developed a Trusted Computing Base (TCB) which is essentially our framework for Assured Information Sharing that deals with the *"need-to-share"* paradigm. This paradigm shift is one of the recommendations of the 9/11 commission. This deals with dynamic environments with better flow of information. A Trusted Computing base creates a hierarchy for users of the system based on how trustworthy a user is. The level of trust can be as fine grained as possible. Based on the trust level and severity of the emergency, information is disseminated to the user.

The organization of our chapter is as follows. In the Assured Information Sharing section we provide some challenges in assured information sharing including aspects of policy enforcement. The design and implementation of our prototype framework is discussed in the Design and Implementation of the Framework section. The chapter is concluded in the Summary and Directions section. The appendix consists of a set of policies that we have defined for our environment.

ASSURED INFORMATION SHARING

A coalition consists of a set of organizations, which may be agencies, universities and corporations that work together in a peer-to-peer environment to solve problems such as intelligence and military operations as well as healthcare operations. Figure 1 illustrates our architecture for a coalition where three agencies have to share data and information. Coalitions are usually dynamic in nature. That is, members may join and leave the coalitions in accordance with the policies and procedures. A challenge is to ensure the secure operation of a coalition. We assume that the members of a coalition, which are also called its partners, may be trustworthy, untrustworthy or partially (semi) trustworthy.

Security policies include policies for confidentiality, privacy, trust, release, dissemination and integrity. A broader term is dependable systems or trustworthy systems that also include real-time processing and fault tolerance. We will

Figure 1. Architecture for organizational data sharing

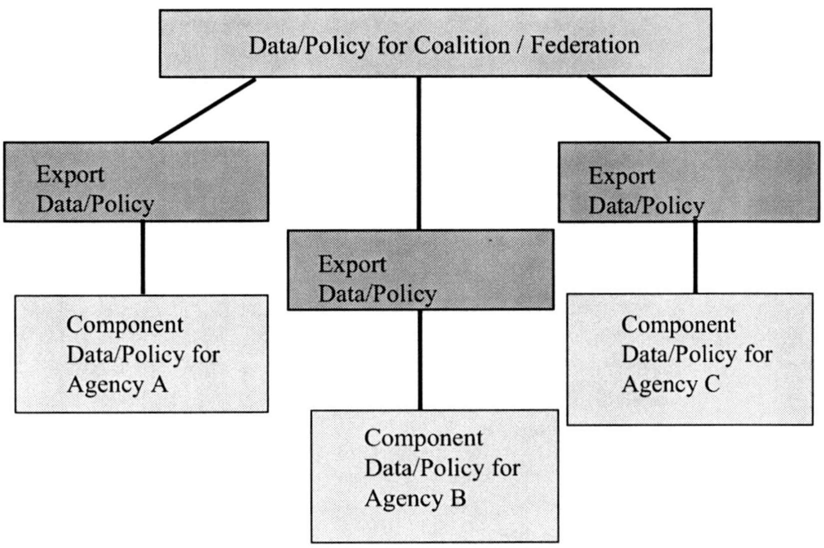

discuss dependability in the next section. By confidentiality we mean that data is only released to individuals who are authorized to get the data. Privacy in general deals with the situation where an individual determines what information should be released about him/her. (Note that different definitions of privacy have been proposed.) Trust policies may add further restriction to privacy and confidentiality policies. For example, a user may be authorized to get the data according to the confidentiality policies, but the system may not trust the individual in which case the data is not released. Similarly a person may give permission to release certain private information about him or her but that person may not trust a particular web site in which case the private information is not released to the web site. Alternatively one could argue that one needs to establish trust first before establishing the confidentiality and privacy policies. For example, a user's (or web site's) trust is established before determining that the user (or web site) can received confidential (or private) information. Release policies specify rules for releasing data while dissemination policies specify rules for disseminating the data. Integrity within the context of security ensures that only authorized individuals can modify the data so that the data is not maliciously corrupted.

There is a critical need for organizations to share data as well process the data in a timely manner, but at the same time enforce various security policies. For example, A and B form a coalition while B and C form a second coalition. A could be California, B could be Texas and C could be Oklahoma. California and Texas could form a coalition as part of the larger states in the US and Texas and Oklahoma could form a coalition as part of the neighboring states in the South of US for emergency management. There is also an urgent need for multiple organizations to share data and at the same time enforce security policies. These policies include policies for confidentiality, privacy, and trust. For example, patient data may be shared by multiple organizations including hospitals, levels of government and agencies. It is important to maintain the privacy of patient data. However it is also important that there are no unnecessary access controls so that information sharing is prohibited. One needs flexible policies so that during emergency situations it is critical that all of the data is shared so that effective decisions can be made. During normal operations, it is important to maintain confidentiality and privacy. In addition, trust policies ensure that data is shared between trusted individuals.

DESIGN AND IMPLEMENTATION OF THE FRAMEWORK

Background

In this section, we will provide some background information including our route finder application, am example as well as some basics in XACML, which is the XML Access Control specification produced by the OASIS standards organization.

Route Finder Application

The application that has been developed is used to demonstrate the concept of TCB is Route Finder. This application is used to find routes between two locations. This application has two parts:

1. Normal Case
2. Emergency Case The application considers some routes to be top secret. Policies are used to specify the conditions in which routes using these locations can be accessed. For example top level Government employees have access to the widest range of locations/ route information. A few policies have been integrated into the implementation and many others are specified using XACML policies.

Example: *Consider a situation where a top level government employee has to access some route information involving the location of a private entity. He can do so based on the level of emergency. Access based on emergency is logged so that this information is available for handling* *the consequences. This involves a certain level of access control implemented through username and password. The Route-Finder application can be accessed anonymously too, in which case the emergency case feature is disabled.*

XACML XACML is an OASIS standard that describes both a policy language and an access control decision request/response language (both encoded in XML) [5]. Using XACML, one can specify access control policies. These policies can be grouped into Policy sets. Based on these policies, it is possible to determine whether a particular access request can be fulfilled. XACML returns the answer to the policy request as one of the four possible values: Permit, Deny, Indeterminate or Not Applicable. XACML specifies two entities: The Policy Enforcement Point (PEP) and the Policy Decision Point (PDP). The PDP returns the answer to a request passed by the PDP in one of the following answer values: Permit, Deny, Intermediate or Not Applicable. This decision is made based on the access control policy specified in the XACML policy files. Though the PDP makes the decision regarding a request, the PEP is responsible for enforcing the decision to grant/deny access.

Policy Files specify rules for "Targets". Each target is composed of three components: Subject, Resource and Action; each target is identified uniquely by its components taken together. The XACML request generated by the PEP contains the target. The PDP's decision-making capability lies in matching the target in the request file with the target in the policy file.

Design of the Framework

The aim of our project is to build a system capable of serving emergency requests on a need-to share basis. The Route-Finder application serves web-based requests. It categorizes the users of the system into a hierarchy in which users at each level have a different level of trust. The users in the topmost layer are the most trusted.

- **The Trusted Computing Base:** The trusted computing base is the wrapper around the Assured Information sharing portal for information dissemination on a need-to-know basis. This wrapper handles the dissemination of information on a need-to-share basis. It contains policies to support emergency requests. The Route Finder application operates in two modes: *Normal:* In this mode, the information is disseminated based on normal access control method of username-password. No information related to secret locations is disseminated. An anonymous user can also use the Route Finder application in the normal mode. Secret locations do not appear on the map. *Emergency:* In this mode, authenticated users can issue requests to obtain route information involving secret locations based on certain policies. For example top government officials have access to many secretive locations. At the same time, the access to such information is also dependent on the level of severity. These operations are logged so that actions can be taken for dealing with the consequences of disseminating confidential information.
- **Database:** The information about routes is stored in database. This database also contains user information for authentication. The route and user information obtained in the databases can be obtained by executing SQL queries. The relational schemas present in the database are as follows:
- **Schemas:** We describe the schemas that we have deigned for the database.

CITY

Name	Id	Private

This schema is used to depict a location. The "Private" field specifies whether this location is a secret location or not

CITY LOC

Cid	X	Y

This schema describes the position of the city or location in terms of X and Y co-ordinates.

DIVISON

Name	DId	Priority

This schema is used to describe a level in the hierarchy. Users belonging to a particular division belong to a level in the hierarchy.

ROUTE

Source	Destination	Path	Private

This schema provides the path information for a different source-destination pairs. Private routes are not available in the normal mode of execution.

USER

Name	Id	Uname	Pwd	DId

This schema provides user information for authentication and for determining privileges.

- **Policy Files:** Each location has a privacy file specifying the access for users in different divisions or levels. Different policies also exist for different users at different severity levels. These policy files can be derived from the trust level of each level of users.

The architecture of the Route finder application is as shown in Figure 2. The Confidential information refers to the Route and location information. The Response Engine refers to the XACML PEP-PDP.

- **Response Engine:** Serves queries passed down from the TCB. This engine translates web interface/web service queries to XACML requests. It also converts the XACML response to a format compliant with the web service/ web interface. The Response Engine contains two sub-components: *Policy Enforcement Point* and the *Policy Decision Point.*

-Policy Enforcement Point (PEP):

- Enforces policies on requests sent by the Web Service.
- Translates this request into an XACML request; sends it to the PDP.

- Policy Decision Point (PDP):

- Makes decisions regarding the request made by the web service.

Figure 2. Architecture

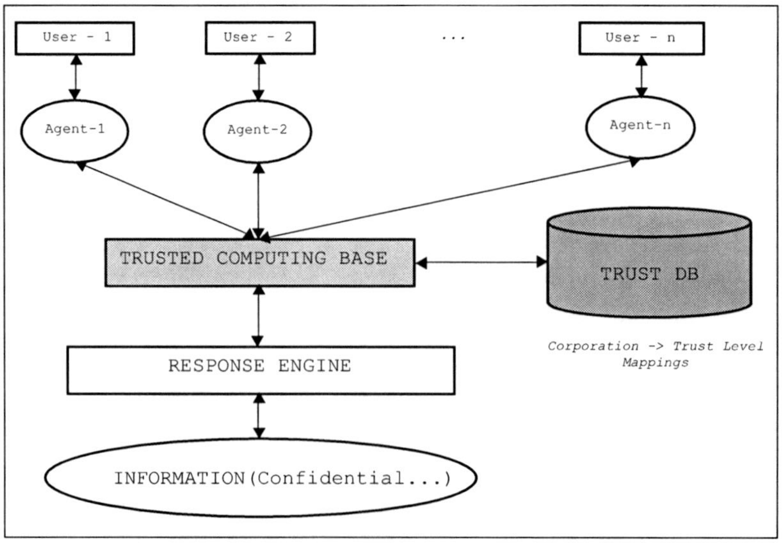

⑩ Conveys the XACML request to the PEP.

Implementation of the Framework

The implementation of the Route-Finder Application has been done in Java using the NetBeans IDE 5.5®. The front end is web based with Java Servlet Pages. The backend consists of a MySQL® database named as TCB and the connection to the database is through JDBC. The business logic is applied thought the use of XACML policies and some integrated into the implementation. The XACML policy files can dynamically change. A policy file exists for each private location containing rules for access in case of emergency situations based on the level of severity. The output (route) of the Route Finder application is a .jpg file pictorially depicting the route obtained. The routes may vary for emergency and normal route situations. If new private locations are added, it is necessary to specify a policy file mentioning the access rules for various levels of severity. The NetBeans IDE® deploys the Tomcat Server. The JSPs, Servlets and XACML policies are all integrated into the NetBeans IDE®.

I. Main screen: This page (illustrated in Figure 3) is accessible to all and is the main page for all users. Any user can anonymously use the system with username "Anonymous" and blank password for limited capabilities. Other users are authenticated based on the username and password that is entered.
II. Normal Mode Execution Screen: Figure 4 illustrates the normal model screen. This mode is accessible by all users. For Anonymous users, the link for 'Emergency Execution mode" does not exists. Secret locations cannot be obtained through the normal execution mode.
III. Normal Execution Route (A -> D): Figure 5 illustrates the normal execution route. The output of normal execution is a .jpg file. The request for normal execution route from Source A to destination D yielded the above route.
IV. Emergency mode Execution Screen: Figure 6 illustrates the emergency mode execution screen. This screen is used for emergency situations and the user is allowed to choose private locations and routes to private locations or those involving them and also the severity to indicate the severity of the emergency. If no special routes are found, the user is directed to the normal execution mode screen.
V. Emergency Execution Route (A -> D): Figure 7 illustrated the emergency execution route; the output of emergency execution is a .jpg file. The request for emergency execution route from Source A to destination D yielded the above route.

The emergency case execution details are logged into a log file. The details include the username, severity level, source and the destinations cities/locations.

Sample Log File Contents

garnet#3#A#D
pepper#3#A#UTD
eutd#1#A#UTD
eutd#3#A#LNT
garnet#3#A#D

- **Setting up the database:** This project makes use of a MySQL database called "TCB". This database is stored in the "Databases" folder. They are all needed for the proper execution of the project. Once MySQL® is installed,

1. Open the DOS command prompt.
2. Change to the "Databases" directory.
3. Execute the following commands:
4. Start MySQL® by executing the command: *mysql*

Sample Rule for an Employee of Division Gov1 to Access a Location "LNT" when the Severity Level is 3

```
<Rule
        RuleId="Emergency:Gov1:3"
        Effect="Permit">
        <Description>
            Government 1 agencies can access this location at severity level 3.
        </Description>
        <Target>
            <Subjects>
                <Subject>
                    <SubjectMatch
                        MatchId="urn:oasis:names:tc:xacml:1.0:function:anyURI-
equal">
                        <AttributeValue

DataType="http://www.w3.org/2001/XMLSchema#anyURI">http://Route-
Finder.com/emergency/Level-3/GOV1</AttributeValue>
                        <SubjectAttributeDesignator

AttributeId="urn:oasis:names:tc:xacml:1.0:subject:subject-id"
                            DataType="http://www.w3.org/2001/XMLSchema#anyURI"/>
                    </SubjectMatch>
                </Subject>
            </Subjects>
            <Resources>
                <Resource>
                    <ResourceMatch
                        MatchId="urn:oasis:names:tc:xacml:1.0:function:anyURI-
equal">
                        <AttributeValue

DataType="http://www.w3.org/2001/XMLSchema#anyURI">http://Route-
Finder/CityUTD</AttributeValue>
                        <ResourceAttributeDesignator

AttributeId="urn:oasis:names:tc:xacml:1.0:resource:resource-id"
                            DataType="http://www.w3.org/2001/XMLSchema#anyURI"/>
                    </ResourceMatch>
                </Resource>
            </Resources>
            <Actions>
                <Action>
                    <ActionMatch
                        MatchId="urn:oasis:names:tc:xacml:1.0:function:string-
equal">
                        <AttributeValue

DataType="http://www.w3.org/2001/XMLSchema#string">access-route</AttributeValue>
                        <ActionAttributeDesignator

AttributeId="urn:oasis:names:tc:xacml:1.0:action:action-id"
                            DataType="http://www.w3.org/2001/XMLSchema#string"/>
                    </ActionMatch>
                </Action>
            </Actions>
        </Target>
    </Rule>
```

- **Running the project:**
  1. Open the NetBeans® IDE.
  2. Click on "Open Project" in the "File" menu.
  3. Browse for the directory "TCB" in the "Source" directory. Once NetBeans® is installed, all project folders are uniquely identified by an icon superimposed on the folder.
  4. Open the TCB project.
  5. Click on "clean and Build Main Project" in the "Build" menu.
  6. Click on "Run Main Project" in the "Run" menu.
  7. A browser opens up with the

Figure 3. Main screen

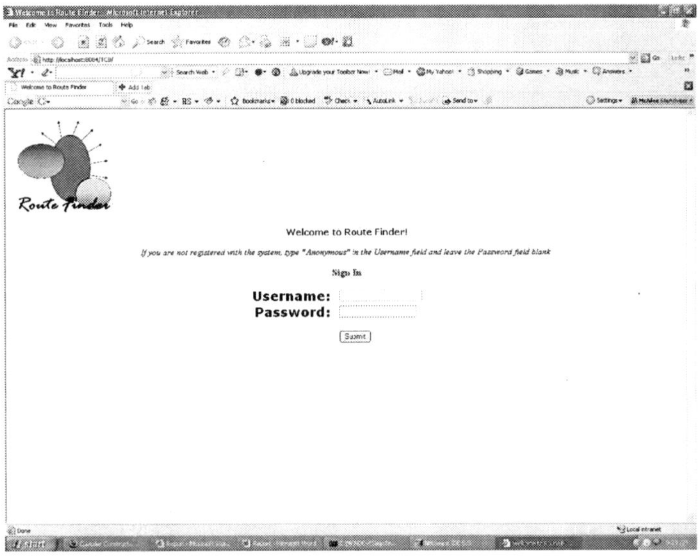

Figure 4. Normal mode execution screen

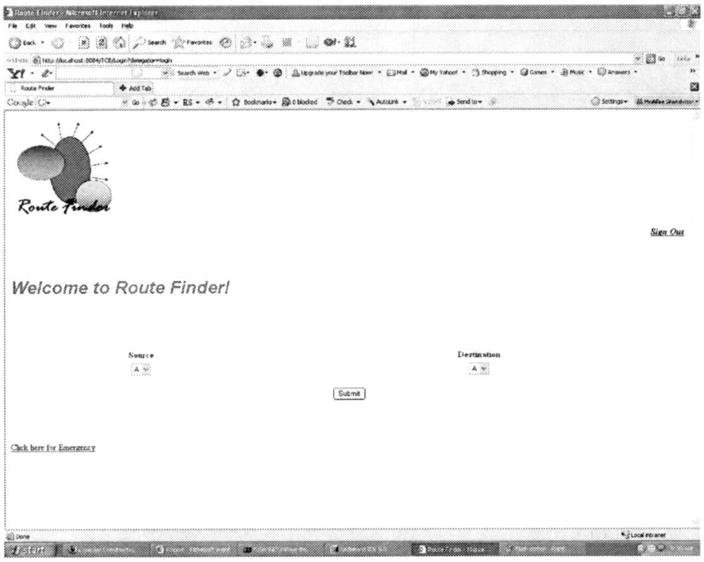

- **Contents of the Database:** *Division:* Currently, 7 divisions exist: GOV1, GOV2, GOV3, UTD, LNT, IMB and XXX for Anonymous. *User:* The database includes one user for each of the divisions: GOV1, GOV2, GOV3, UTD and LNT. Take a look at the databases for username/password. *City:* The city/location currently existing in the database includes: A, B, C, D, UTD and LNT. UTD and LNT are private locations. *Route:* Currently only some select routes

Figure 5. Normal Execution Route (A -> D)

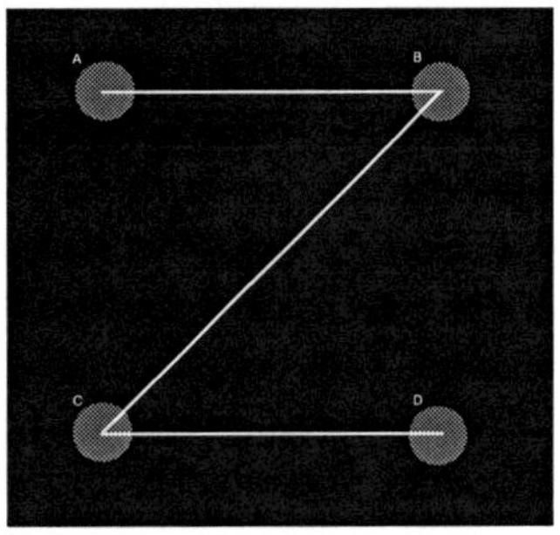

Figure 6. Emergency Mode execution Screen

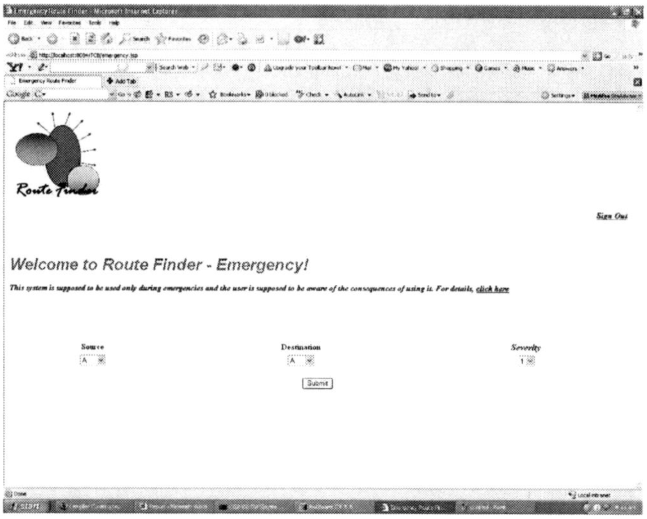

exist. The routes table can be expanded to include more routes.

- **Policy Files:** Policy Files exist for the private locations: UTD and LNT are named CityUTD.xml and CityLNT.xml respectively.
- **Request Files:** Request Files exist for each division to access the private locations.

SUMMARY AND DIRECTIONS

The Trusted Computing Base deals with disseminating information in case of emergencies. The policy of TCB is to disseminate information and then deal with the consequences by auditing, and so forth. This project has been highly influenced by the 9/11 Commission Report Recommendations. The world now recognizes the importance of "need-to-share" information to fight terrorism. Need to share information is also important in cases of medical emergencies and natural calamities.

The next main step would be to integrate this project with the ArcGIS® framework to extend it to various maps. This would involve enriching the dataset that currently exists to include more locations and routes. This would involve considerable effort by persons trained with the ArcGIS® tools. ArcGIS is geospatial and mapping software used to perform spatial analysis, manage large amounts of spatial data and produce maps to aid decision making.

Next would be implementing the interface for web service agents. This would involve exporting an API for webs service agents. This API would allow requests to be sent to the TCB. Web service agents belonging to various organizations can query the TCB without a user having to manually input the query as is the case now. The authentication and authorization mechanisms can work on the identification sent by these agents instead of the current username/password method. This allows the coalition system to be scalable as it can allow for various means of authorization/ authentication; for example, through the use of certificates.

Figure 7. Emergency Execution Route (A -> D)

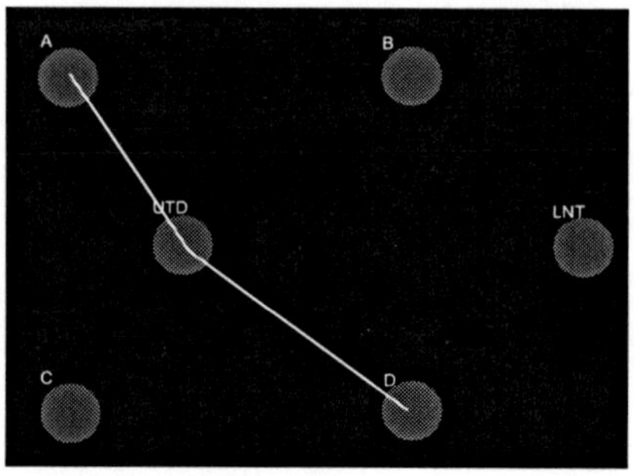

Currently, the requests are hard-coded in the application as request files. This method is not scalable. As the number of divisions and private locations increase in the databases, creating a request file for each access operation manually will be tedious and unmanageable. The request files can be created on-the-fly using APIs provided by Sun's implementation of XACML; it can be included as part of the PEP's business logic.

Policies can be made more granular. This could be done by adding more rules in the policy files or by determining different levels of execution in the implementation. Implementation based could be more to do with the need-to know paradigm and the policy files based decisions could be more need-to share paradigm based.

Auditing of logged data: It is necessary to audit the data that has been logged to monitor possible security breaches. The data is logged into a text file. It is necessary to apply an auditing process to check whether the data that has been disseminated is being used efficiently and that the emergency does really exist with that severity level.

The application can be enhanced to use real location information and depict the locations and routes on a map using the Google Maps® API.

ACKNOWLEDGMENT

The research reported in this chapter was partially funded by grants from the Air Force Office of Scientific Research and from the National Science Foundation under the Cyber Trust program.

REFERENCES

Awad, M., Thuraisingham, B., & Khan, L. "Assured Information Sharing: Volume 2: Experimental Analysis of Data Integration, Mining and Security, Technical Report." The University of Texas at Dallas, UTDCS44-06.

Celikel, E., Kantarcioglu, M., & Thuraisingham, B. (2007). "Assured Information Sharing: Risk-based Data Sharing, Technical Report." The University of Texas at Dallas, 2007 (to appear).

Creating a Trusted Network for Homeland Security, Markle Report. (2003). Editor: Vatis, M.

Finin, T. et al, (May 15, 2007). "Supporting Need to Share with Computational Policies."

Fr´ed´eric Cuppens, et al, "O2O: Managing Security Policy Interoperability with Virtual Private Organizations." retrieved on XXX from, http://www.rennes.enst-bretagne.fr/~fcuppens/articles/hpovua06.pdf

Layfield, R., Kantarcioglu, M., & Thuraisingham, B. (October 2006). "Assured Information Sharing: Volume 3: Using Game Theory to Enforce Honesty Within a Competitive Coalition." Technical Report, The University of Texas at Dallas, UTDCW-46-06.

OASIS, "eXtensible Access Control Markup Language (XACML) Version 1.0." retrieved on XXX from, http://www.oasis-open.org/committees/download.php/2406/oasis-xacml-1.0.pdf

Prepared Statement of James X. Dempsey before the House Committee on Government Reform, "Moving from 'Need-to-Know' to 'Need-to-Share': A review of the 9-11 Commission's Recommendations." (August 3, 2004). Retrieved on XXX from, http://www.cdt.org/testimony/20040803dempsey.shtml

Sun Microsystems, Inc., "Sun's XACML Implementation: Programmer's Guide for Version 1.2." Retrieved on XXX from, http://sunxacml.sourceforge.net/guide.html

Srinivasan, I., & Thuraisingham, B. (October 2006). Design and Simulation of Agent-based Trust Management Techniques for a Coalition Environment. Technical Report, UTDCS45-06 (also published in the Proceedings of IEEE FTDCS, 2007).

W3C, "Web Services Architecture." Retrieved on XXX from, http://www.w3.org/TR/2003/WD-ws-arch-20030808/#id2630981

APPENDIX

Policy File for Location UTD

```
<?xml version=”1.0” encoding=”UTF-8”?>

<!--Document : CityUTD.xmlCreated on : November 13, 2007, 9:30 AMAuthor : yxh068000Description:
    Purpose of the document follows.-->
<Policyxmlns=”urn:oasis:names:tc:xacml:1.0:policy”xmlns:xsi=”http://www.w3.org/2001/XMLSchema-
     instance”xsi:schemaLocation=”urn:oasis:names:tc:xacml:1.0:policycs-xacml-schema-policy-01.
     xsd”PolicyId=”CityUTDPolicy”RuleCombiningAlgId=”urn:oasis:names:tc:xacml:1.0:rule-
     combining-algorithm:deny-overrides”>
   <Description>This policy file specifies the rules to be evaluated for a user accessing the locationUTD
         based on the division the user belongs to and the severity of the emergency.
    </Description>
    <Target>

      <Subjects>
         <AnySubject/>
      </Subjects>
      <Resources>

         <AnyResource/>
      </Resources>
      <Actions>

    <AnyAction/>
</Actions>
</Target>


    <Rule
      RuleId=”Emergency:Gov1:3”
      Effect=”Permit”>
      <Description>

         Government 1 agencies can access this location at severity level 3.
      </Description>
      <Target>
         <Subjects>
              <Subject>

                     <SubjectMatchMatchId=”urn:oasis:names:tc:xacml:1.0:function:anyURI-
                           equal”><AttributeValue
```

```
                    DataType="http://www.w3.org/2001/XMLSchema#anyURI">http:
                    //Route-Finder.com/emergency/Level-3/GOV1</AttributeValue>
                   <SubjectAttributeDesignatorAttributeId="urn:oasis:names:tc:xacml:1.0:subject:sub
                       ject-id"DataType="http://www.w3.org/2001/XMLSchema#anyURI"/>
            </SubjectMatch>
        </Subject>
      </Subjects>
    <Resources>

      <Resource>
        <ResourceMatch
            MatchId="urn:oasis:names:tc:xacml:1.0:function:anyURI-equal"><AttributeValue
                    DataType="http://www.w3.org/2001/XMLSchema#anyURI">http://Route-Finder/
                        CityUTD</AttributeValue>
                   <ResourceAttributeDesignatorAttributeId="urn:oasis:names:tc:xacml:1.0:resource
                        :resource-id" DataType="http://www.w3.org/2001/XMLSchema#anyURI"/>
            </ResourceMatch>

                                  14
        </Resource>
      </Resources>
      <Actions>
        <Action>
              <ActionMatch
                MatchId="urn:oasis:names:tc:xacml:1.0:function:string-equal"><AttributeValue
                    DataType="http://www.w3.org/2001/XMLSchema#string">access-route</Attrib-
                        uteValue>
                   <ActionAttributeDesignatorAttributeId="urn:oasis:names:tc:xacml:1.0:action:acti
                        on-id"
                    DataType="http://www.w3.org/2001/XMLSchema#string"/>
        </ActionMatch>
      </Action>
      </Actions>
      </Target>
      </Rule>


    <Rule
      RuleId="Emergency:Gov1:2"
      Effect="Permit">
    <Description>

        Government 1 agencies can access this location at severity level 2.</Description><Target>
        <Subjects>
```

```
<Subject>

        <SubjectMatchMatchId="urn:oasis:names:tc:xacml:1.0:function:anyURI-
            equal"><AttributeValue
                DataType="http://www.w3.org/2001/XMLSchema#anyURI">http://Route-
                Finder.com/emergency/Level-2/Gov2</AttributeValue>
            <SubjectAttributeDesignatorAttributeId="urn:oasis:names:tc:xacml:1.0:subject:
                subject-id"DataType="http://www.w3.org/2001/XMLSchema#anyURI"/>
    </SubjectMatch>
</Subject>
</Subjects>
<Resources>

    <Resource>
        <ResourceMatch
        MatchId="urn:oasis:names:tc:xacml:1.0:function:anyURI-equal"><AttributeValue
        DataType="http://www.w3.org/2001/XMLSchema#anyURI">http://Route-Finder/
            CityUTD</AttributeValue>
        <ResourceAttributeDesignatorAttributeId="urn:oasis:names:tc:xacml:1.0:resource:res
            ource-id"
        DataType="http://www.w3.org/2001/XMLSchema#anyURI"/>
</ResourceMatch>
</Resource>
</Resources>
<Actions>

<Action>
    <ActionMatch
        MatchId="urn:oasis:names:tc:xacml:1.0:function:string-equal"><AttributeValue
            DataType="http://www.w3.org/2001/XMLSchema#string">access-route</Attribute-
                Value>
        <ActionAttributeDesignatorAttributeId="urn:oasis:names:tc:xacml:1.0:action:action-id"
                DataType="http://www.w3.org/2001/XMLSchema#string"/>
</ActionMatch>
</Action>
</Actions>
</Target>
</Rule>

<Rule
RuleId="Emergency:Gov2:3"
Effect="Permit">
<Description>
```

15

```
Government 2 agencies can access this location at severity level 3.</Description><Target>
<Subjects>
  <Subject>

        <SubjectMatchMatchId="urn:oasis:names:tc:xacml:1.0:function:anyURI-
           equal"><AttributeValue
                DataType="http://www.w3.org/2001/XMLSchema#anyURI">http://Route-
                    Finder.com/emergency/Level-3/GOV2</AttributeValue>
             <SubjectAttributeDesignatorAttributeId="urn:oasis:names:tc:xacml:1.0:subje
                 ct:subject-
             id"DataType="http://www.w3.org/2001/XMLSchema#anyURI"/>
        </SubjectMatch>
     </Subject>
  </Subjects>
  <Resources>

     <Resource>
        <ResourceMatch
           MatchId="urn:oasis:names:tc:xacml:1.0:function:anyURI-equal"><AttributeValue
                DataType="http://www.w3.org/2001/XMLSchema#anyURI">http://Route-
                Finder/CityUTD</AttributeValue>
             <ResourceAttributeDesignatorAttributeId="urn:oasis:names:tc:xacm
                l:1.0:resource:resource-id" DataType="http://www.w3.org/2001/
                XMLSchema#anyURI"/>
        </ResourceMatch>
     </Resource>
  </Resources>
  <Actions>

     <Action>
        <ActionMatch
             MatchId="urn:oasis:names:tc:xacml:1.0:function:string-equal"><AttributeValue
                DataType="http://www.w3.org/2001/XMLSchema#string">access-route</
                AttributeValue>
             <ActionAttributeDesignatorAttributeId="urn:oasis:names:tc:xacml:1.0:action:
                action-id"
              DataType="http://www.w3.org/2001/XMLSchema#string"/>
     </ActionMatch>
  </Action>
  </Actions>
  </Target>
  </Rule>
```

```
<Rule
   RuleId="Emergency:Gov3:3"
   Effect="Permit">
   <Description>

         Government 3 agencies can access this location at severity level 3.</Description><Target>
      <Subjects>
         <Subject>

            <SubjectMatchMatchId="urn:oasis:names:tc:xacml:1.0:function:anyURI-
               equal"><AttributeValue
               DataType="http://www.w3.org/2001/XMLSchema#anyURI">http://Route-
Finder.com/emergency/Level-3/GOV3</AttributeValue>
               <SubjectAttributeDesignatorAttributeId="urn:oasis:names:tc:xacml:1.0:su
                  bject:subject-
               id"DataType="http://www.w3.org/2001/XMLSchema#anyURI"/>
            </SubjectMatch>
         </Subject>
      </Subjects>
      <Resources>

         <Resource>
            <ResourceMatch
               MatchId="urn:oasis:names:tc:xacml:1.0:function:anyURI-
                  equal"><AttributeValue
```


16


```
DataType="http://www.w3.org/2001/XMLSchema#anyURI">http://Route-Finder/CityUTD</Attrib-
   uteValue>
               <ResourceAttributeDesignatorAttributeId="urn:oasis:names:tc:xacm
                  l:1.0:resource:resource-id"  DataType="http://www.w3.org/2001/
                  XMLSchema#anyURI"/>
            </ResourceMatch>
         </Resource>
      </Resources>
      <Actions>

         <Action>
            <ActionMatch
               MatchId="urn:oasis:names:tc:xacml:1.0:function:string-
               equal"><AttributeValue
               DataType="http://www.w3.org/2001/XMLSchema#string">access-route</
AttributeValue>
```

```
                    <ActionAttributeDesignatorAttributeId="urn:oasis:names:tc:xac
                        ml:1.0:action:action-id" DataType="http://www.w3.org/2001/
                        XMLSchema#string"/>
        </ActionMatch>
    </Action>
    </Actions>
    </Target>
    </Rule>

<Rule
    RuleId="Emergency:Utd:3"
    Effect="Permit">
    <Description>

        Users belonging to the UTD division can access this location at Severity level 3.</
Description><Target>
        <Subjects>
            <Subject>

                <SubjectMatchMatchId="urn:oasis:names:tc:xacml:1.0:function:anyURI-
                    equal"><AttributeValue
                        DataType="http://www.w3.org/2001/
                        XMLSchema#anyURI">http://Route-Finder.com/emergency/Level-3/
                        UTD</AttributeValue>
                <SubjectAttributeDesignatorAttributeId="urn:oasis:names:tc:xacml:1.0:subject:
                    subject-id"DataType="http://www.w3.org/2001/XMLSchema#anyURI"/>
            </SubjectMatch>
        </Subject>
    </Subjects>
    <Resources>

        <Resource>
            <ResourceMatch
                MatchId="urn:oasis:names:tc:xacml:1.0:function:anyURI-equal"><AttributeValue
                    DataType="http://www.w3.org/2001/XMLSchema#anyURI">http://Route-
Finder/CityUTD</AttributeValue>
                        <ResourceAttributeDesignatorAttributeId="urn:oasis:names:tc:xacm
                            l:1.0:resource:resource-id" DataType="http://www.w3.org/2001/
                            XMLSchema#anyURI"/>
            </ResourceMatch>
        </Resource>
    </Resources>
    <Actions>
```

```
<Action>
    <ActionMatch
        MatchId="urn:oasis:names:tc:xacml:1.0:function:string-equal"><AttributeValue
            DataType="http://www.w3.org/2001/XMLSchema#string">access-route</At-
tributeValue>
            <ActionAttributeDesignatorAttributeId="urn:oasis:names:tc:xacml:1.0:action
                :action-id" DataType="http://www.w3.org/2001/XMLSchema#string"/>
  </ActionMatch>
</Action>
</Actions>
</Target>
```

17

```
</Rule>

<Rule
RuleId="Emergency:Utd:2"
Effect="Permit">
<Description>

  Users belonging to the UTD division can access this location at Severity level 2.</
Description><Target>
  <Subjects>
      <Subject>

          <SubjectMatchMatchId="urn:oasis:names:tc:xacml:1.0:function:anyURI-
              equal"><AttributeValue
              DataType="http://www.w3.org/2001/XMLSchema#anyURI">http://Route-
Finder.com/emergency/Level-2/UTD</AttributeValue>
            <SubjectAttributeDesignatorAttributeId="urn:oasis:names:tc:xacml:1.0:subject:
                subject-id"DataType="http://www.w3.org/2001/XMLSchema#anyURI"/>
          </SubjectMatch>
      </Subject>
  </Subjects>
  <Resources>

      <Resource>
          <ResourceMatch
              MatchId="urn:oasis:names:tc:xacml:1.0:function:anyURI-equal"><AttributeValue
                  DataType="http://www.w3.org/2001/XMLSchema#anyURI">http://Route-
Finder/CityUTD</AttributeValue>
            <ResourceAttributeDesignatorAttributeId="urn:oasis:names:tc:xacm
                l:1.0:resource:resource-id" DataType="http://www.w3.org/2001/
```

```
                        XMLSchema#anyURI"/>
                </ResourceMatch>
            </Resource>
        </Resources>
        <Actions>

            <Action>
                <ActionMatch
                    MatchId="urn:oasis:names:tc:xacml:1.0:function:string-equal"><AttributeValue
                        DataType="http://www.w3.org/2001/XMLSchema#string">access-route</At-
    tributeValue>
                        <ActionAttributeDesignatorAttributeId="urn:oasis:names:tc:xacml:1.0:action
                            :action-id" DataType="http://www.w3.org/2001/XMLSchema#string"/>
        </ActionMatch>
    </Action>
    </Actions>
    </Target>
  </Rule>

<Rule
  RuleId="Emergency:Utd:1"
  Effect="Permit">
  <Description>

    Users belonging to the UTD division can access this location at Severity level 1.</Description><Target>
      <Subjects>
          <Subject>

                  <SubjectMatchMatchId="urn:oasis:names:tc:xacml:1.0:function:anyURI-
                      equal"><AttributeValue
                      DataType="http://www.w3.org/2001/XMLSchema#anyURI">http://Route-
    Finder.com/emergency/Level-1/UTD</AttributeValue>
                      <SubjectAttributeDesignatorAttributeId="urn:oasis:names:tc:xacml:1.0:subjec
                          t:subject-id"DataType="http://www.w3.org/2001/XMLSchema#anyURI"/>
  </SubjectMatch>
</Subject>
</Subjects>
```


18


```
        <Resources>
            <Resource>

                <ResourceMatch
```

```
                MatchId="urn:oasis:names:tc:xacml:1.0:function:anyURI-equal"><AttributeValue
                    DataType="http://www.w3.org/2001/XMLSchema#anyURI">http://Route-
Finder/CityUTD</AttributeValue>
                    <ResourceAttributeDesignatorAttributeId="urn:oasis:names:tc:xacm
                        l:1.0:resource:resource-id" DataType="http://www.w3.org/2001/
                        XMLSchema#anyURI"/>
            </ResourceMatch>
        </Resource>
    </Resources>
    <Actions>

        <Action>
            <ActionMatch
                MatchId="urn:oasis:names:tc:xacml:1.0:function:string-equal"><AttributeValue
                    DataType="http://www.w3.org/2001/XMLSchema#string">access-route</At-
tributeValue>
                    <ActionAttributeDesignatorAttributeId="urn:oasis:names:tc:xacml:1.0:action
                        :action-id" DataType="http://www.w3.org/2001/XMLSchema#string"/>
        </ActionMatch></Action></Actions></Target></Rule></Policy>
```

Policy File for Location LNT:

```
<?xml version="1.0" encoding="UTF-8"?>

<!--Document : CityLNT.xmlCreated on : November 13, 2007, 9:30 AMAuthor : yxh068000Description:
    Purpose of the document follows.-->

<Policyxmlns="urn:oasis:names:tc:xacml:1.0:policy"xmlns:xsi="http://www.w3.org/2001/XMLSchema-
    instance"xsi:schemaLocation="urn:oasis:names:tc:xacml:1.0:policycs-xacml-schema-policy-01.
    xsd"PolicyId="CityLNTPolicy"RuleCombiningAlgId="urn:oasis:names:tc:xacml:1.0:rule-
    combining-algorithm:deny-overrides">

<Description>This policy file specifies the
    rules to be evaluated for a user
    accessing the locationLNT
    based on the division the user belongs to and the severity of the emergency.
</Description>
<Target>

    <Subjects>
        <AnySubject/>
    </Subjects>
    <Resources>
```

```
        <AnyResource/>
      </Resources>
      <Actions>

    <AnyAction/>
</Actions>
</Target>

<Rule
    RuleId="Emergency:Gov1"
    Effect="Permit">
    <Description>
```

19

```
        Government 1 agencies can access this location at severity level 3.</Description><Target>
        <Subjects>
            <Subject>

                    <SubjectMatchMatchId="urn:oasis:names:tc:xacml:1.0:function:anyURI-
                        equal"><AttributeValue
                         DataType="http://www.w3.org/2001/XMLSchema#anyURI">http://Route-
      Finder.com/emergency/Level-3/GOV1</AttributeValue>
                       <SubjectAttributeDesignatorAttributeId="urn:oasis:names:tc:xacml:1.0:subject:
                            subject-id"DataType="http://www.w3.org/2001/XMLSchema#anyURI"/>
                </SubjectMatch>
            </Subject>
        </Subjects>
        <Resources>

            <Resource>
                <ResourceMatch
                    MatchId="urn:oasis:names:tc:xacml:1.0:function:anyURI-equal"><AttributeValue
                         DataType="http://www.w3.org/2001/XMLSchema#anyURI">http://Route-
      Finder/CityLNT</AttributeValue>
                       <ResourceAttributeDesignatorAttributeId="urn:oasis:names:tc:xacm
                            l:1.0:resource:resource-id" DataType="http://www.w3.org/2001/
                            XMLSchema#anyURI"/>
        </ResourceMatch>
      </Resource>
     </Resources>
     <Actions>

      <Action>
```

```
        <ActionMatch MatchId=”urn:oasis:names:tc:xacml:1.0:function:string-equal”><AttributeValue
            DataType=”http://www.w3.org/2001/XMLSchema#string”>access-route</AttributeValue>
          <ActionAttributeDesignatorAttributeId=”urn:oasis:names:tc:xacml:1.0:action:action-id”
              DataType=”http://www.w3.org/2001/XMLSchema#string”/>
      </ActionMatch>
    </Action>
    </Actions>
    </Target>
    </Rule>

<Rule
    RuleId=”Emergency:Gov1”
    Effect=”Permit”>
    <Description>

      Government 1 agencies can access this location at severity level 2.</Description><Target>
      <Subjects>
        <Subject>

              <SubjectMatchMatchId=”urn:oasis:names:tc:xacml:1.0:function:anyURI-
                  equal”><AttributeValue
                  DataType=”http://www.w3.org/2001/XMLSchema#anyURI”>http://Route-Finder.
      com/emergency/Level-2/GOV1</AttributeValue>
                  <SubjectAttributeDesignatorAttributeId=”urn:oasis:names:tc:xacml:1.0:subject:subj
                      ect-id”DataType=”http://www.w3.org/2001/XMLSchema#anyURI”/>
          </SubjectMatch>
        </Subject>
      </Subjects>
      <Resources>

        <Resource>
          <ResourceMatch
              MatchId=”urn:oasis:names:tc:xacml:1.0:function:anyURI-equal”><AttributeValue
```

20

```
  DataType=”http://www.w3.org/2001/XMLSchema#anyURI”>http://Route-Finder/CityLNT</Attrib-
      uteValue>
                  <ResourceAttributeDesignatorAttributeId=”urn:oasis:names:tc:xacml:1.0:resourc
                      e:resource-id” DataType=”http://www.w3.org/2001/XMLSchema#anyURI”/>
          </ResourceMatch>
        </Resource>
      </Resources>
      <Actions>
```

```
      <Action>
          <ActionMatch
              MatchId="urn:oasis:names:tc:xacml:1.0:function:string-equal"><AttributeValue
                  DataType="http://www.w3.org/2001/XMLSchema#string">access-route</Attrib-
   uteValue>
                 <ActionAttributeDesignatorAttributeId="urn:oasis:names:tc:xacml:1.0:action:acti
                     on-id" DataType="http://www.w3.org/2001/XMLSchema#string"/>
   </ActionMatch>
  </Action>
  </Actions>
  </Target>
  </Rule>

<Rule
  RuleId="Emergency:Gov2"
  Effect="Permit">
  <Description>

  Government 2 agencies can access this location at severity level 3.</Description><Target>
  <Subjects>
   <Subject>

      <SubjectMatchMatchId="urn:oasis:names:tc:xacml:1.0:function:anyURI-
   equal"><AttributeValue
              DataType="http://www.w3.org/2001/XMLSchema#anyURI">http://Route-Finder.com/
   emergency/Level-3/GOV2</AttributeValue>
             <SubjectAttributeDesignatorAttributeId="urn:oasis:names:tc:xacml:1.0:subject:subje
                 ct-id"DataType="http://www.w3.org/2001/XMLSchema#anyURI"/>
      </SubjectMatch>
   </Subject>
  </Subjects>
  <Resources>

   <Resource>
      <ResourceMatch
          MatchId="urn:oasis:names:tc:xacml:1.0:function:anyURI-equal"><AttributeValue
              DataType="http://www.w3.org/2001/XMLSchema#anyURI">http://Route-Finder/
   CityLNT</AttributeValue>
             <ResourceAttributeDesignatorAttributeId="urn:oasis:names:tc:xacml:1.0:resource:reso
                 urce-id" DataType="http://www.w3.org/2001/XMLSchema#anyURI"/>
      </ResourceMatch>
   </Resource>
  </Resources>
  <Actions>
```

```
<Action>
   <ActionMatch
       MatchId="urn:oasis:names:tc:xacml:1.0:function:string-equal"><AttributeValue
           DataType="http://www.w3.org/2001/XMLSchema#string">access-route</Attribute-
Value>
         <ActionAttributeDesignatorAttributeId="urn:oasis:names:tc:xacml:1.0:action:action-id"
              DataType="http://www.w3.org/2001/XMLSchema#string"/>
</ActionMatch>
</Action>
</Actions>
</Target>
```

21

```
  </Rule>
<Rule
    RuleId="Emergency:Gov3"
    Effect="Permit">
    <Description>

  Government 3 agencies can access this location at severity level 3.</Description><Target>
  <Subjects>
     <Subject>

          <SubjectMatchMatchId="urn:oasis:names:tc:xacml:1.0:function:anyURI-
               equal"><AttributeValue
               DataType="http://www.w3.org/2001/XMLSchema#anyURI">http://Route-Finder.
  com/emergency/Level-3/GOV3</AttributeValue>
            <SubjectAttributeDesignatorAttributeId="urn:oasis:names:tc:xacml:1.0:subject:sub
                 ject-id"DataType="http://www.w3.org/2001/XMLSchema#anyURI"/>
        </SubjectMatch>
     </Subject>
  </Subjects>
  <Resources>

     <Resource>
        <ResourceMatch
           MatchId="urn:oasis:names:tc:xacml:1.0:function:anyURI-equal"><AttributeValue
               DataType="http://www.w3.org/2001/XMLSchema#anyURI">http://Route-Finder/
  CityLNT</AttributeValue>
            <ResourceAttributeDesignatorAttributeId="urn:oasis:names:tc:xacml:1.0:resource
                 :resource-id" DataType="http://www.w3.org/2001/XMLSchema#anyURI"/>
        </ResourceMatch>
     </Resource>
```

```
</Resources>
<Actions>

   <Action>
      <ActionMatch MatchId="urn:oasis:names:tc:xacml:1.0:function:string-
   equal"><AttributeValue
         DataType="http://www.w3.org/2001/XMLSchema#string">access-route</Attribute-
   Value>
         <ActionAttributeDesignatorAttributeId="urn:oasis:names:tc:xacml:1.0:action:action-id"
            DataType="http://www.w3.org/2001/XMLSchema#string"/>
   </ActionMatch>
  </Action>
  </Actions>
  </Target>
  </Rule>

<Rule
   RuleId="Emergency:Lnt"
   Effect="Permit">
   <Description>

    Users belonging to the LNT division can access this location at Severity level 3.</Description><Target>
    <Subjects>
      <Subject>

         <SubjectMatchMatchId="urn:oasis:names:tc:xacml:1.0:function:anyURI-
            equal"><AttributeValue
            DataType="http://www.w3.org/2001/XMLSchema#anyURI">http://Route-Finder.
    com/emergency/Level-3/LNT</AttributeValue>
            <SubjectAttributeDesignatorAttributeId="urn:oasis:names:tc:xacml:1.0:subject:subj
               ect-id"DataType="http://www.w3.org/2001/XMLSchema#anyURI"/>
   </SubjectMatch>
</Subject>
</Subjects>
```

22

```
      <Resources>
         <Resource>

            <ResourceMatch
               MatchId="urn:oasis:names:tc:xacml:1.0:function:anyURI-equal"><AttributeValue
                  DataType="http://www.w3.org/2001/XMLSchema#anyURI">http://Route-
   Finder/CityLNT</AttributeValue>
```

```
                    <ResourceAttributeDesignatorAttributeId="urn:oasis:names:tc:xacm
                        l:1.0:resource:resource-id" DataType="http://www.w3.org/2001/
                        XMLSchema#anyURI"/>
                </ResourceMatch>
            </Resource>
        </Resources>
        <Actions>

            <Action>
                <ActionMatch
                    MatchId="urn:oasis:names:tc:xacml:1.0:function:string-equal"><AttributeValue
                        DataType="http://www.w3.org/2001/XMLSchema#string">access-route</At-
    tributeValue>
                    <ActionAttributeDesignatorAttributeId="urn:oasis:names:tc:xacml:1.0:action
                        :action-id" DataType="http://www.w3.org/2001/XMLSchema#string"/>
    </ActionMatch>
    </Action>
    </Actions>
    </Target>
    </Rule>

<Rule
    RuleId="Emergency:Lnt"
    Effect="Permit">
    <Description>


        Users belonging to the LNT division can access this location at Severity level 2.</
    Description><Target>
        <Subjects>
            <Subject>

                <SubjectMatchMatchId="urn:oasis:names:tc:xacml:1.0:function:anyURI-
                    equal"><AttributeValue
                    DataType="http://www.w3.org/2001/XMLSchema#anyURI">http://Route-
    Finder.com/emergency/Level-2/LNT</AttributeValue>
                    <SubjectAttributeDesignatorAttributeId="urn:oasis:names:tc:xacml:1.0:subjec
                        t:subject-id"DataType="http://www.w3.org/2001/XMLSchema#anyURI"/>
            </SubjectMatch>
    </Subject>
    </Subjects>
    <Resources>

    <Resource>
```

```
<ResourceMatch
    MatchId="urn:oasis:names:tc:xacml:1.0:function:anyURI-equal"><AttributeValue
        DataType="http://www.w3.org/2001/XMLSchema#anyURI">http://Route-Finder/
CityLNT</AttributeValue>
        <ResourceAttributeDesignatorAttributeId="urn:oasis:names:tc:xacml:1.0:resource:reso
            urce-id" DataType="http://www.w3.org/2001/XMLSchema#anyURI"/>
</ResourceMatch>
</Resource>
</Resources>
<Actions>

<Action>
<ActionMatch
    MatchId="urn:oasis:names:tc:xacml:1.0:function:string-equal"><AttributeValue
        DataType="http://www.w3.org/2001/XMLSchema#string">access-route</Attribute-
Value> <ActionAttributeDesignatorAttributeId="urn:oasis:names:tc:xacml:1.0:action:action-id"
```

23

```
DataType="http://www.w3.org/2001/XMLSchema#string"/>
</ActionMatch>
</Action>
</Actions>
</Target>
</Rule>

<Rule
RuleId="Emergency:Lnt"
Effect="Permit">
<Description>

Users belonging to the LNT division can access this location at Severity level 1.</
Description><Target>
<Subjects>
    <Subject>

        <SubjectMatchMatchId="urn:oasis:names:tc:xacml:1.0:function:anyURI-
            equal"><AttributeValue
            DataType="http://www.w3.org/2001/XMLSchema#anyURI">http://Route-
Finder.com/emergency/Level-1/LNT</AttributeValue>
            <SubjectAttributeDesignatorAttributeId="urn:oasis:names:tc:xacml:1.0:subject:
                subject-id"DataType="http://www.w3.org/2001/XMLSchema#anyURI"/>
        </SubjectMatch>
    </Subject>
```

```
    </Subjects>
    <Resources>

        <Resource>
                <ResourceMatch MatchId="urn:oasis:names:tc:xacml:1.0:function:anyURI-
                    equal"><AttributeValue
                    DataType="http://www.w3.org/2001/XMLSchema#anyURI">http://Route-
Finder/CityLNT</AttributeValue>
                    <ResourceAttributeDesignatorAttributeId="urn:oasis:names:tc:xacm
                        l:1.0:resource:resource-id" DataType="http://www.w3.org/2001/
                        XMLSchema#anyURI"/>
        </ResourceMatch>
    </Resource>
</Resources>
<Actions>

    <Action>
        <ActionMatch
            MatchId="urn:oasis:names:tc:xacml:1.0:function:string-equal"><AttributeValue
                DataType="http://www.w3.org/2001/XMLSchema#string">access-route</Attrib-
uteValue>
            <ActionAttributeDesignatorAttributeId="urn:oasis:names:tc:xacml:1.0:action:acti
                on-id" DataType="http://www.w3.org/2001/XMLSchema#string"/>
    </ActionMatch></Action></Actions></Target></Rule></Policy>
```


24

This work was previously published in the International Journal of Information Security and Privacy, Volume 2, Issue 4, edited by H. Nemati, pp. 67-90, copyright 2008 by IGI Publishing (an imprint of IGI Global).

Chapter 19
A Secure and Trustful e-Ordering Architecture (TOES) for Small and Medium Size Enterprises (SMEs)

Spyridon Papastergiou
University of Pireaus, Greece

Despina Polemi
University of Pireaus, Greece

ABSTRACT

Although various European Union (EU) directives have established the proper legal framework for the provision of a trustful, legally accepted cross border transaction via electronic means, the consumers are still hesitant to use e-commerce. Lack of confidence with regard to the protection of privacy and security of electronic transactions is among the main reasons. This chapter attempts to form the appropriate confidence framework describing a set of privacy and security requirements that must be taken into account for e-ordering systems. In addition, it presents a trustful e-ordering architecture (TOES) that achieves to address these requirements based on extensible markup language (XML), XML cryptography, public key infrastructure (PKI), Web services policy language (WSPL), and Web services. TOES is an open, secure, interoperable, and affordable e-ordering system that respects the EU legislation.

INTRODUCTION

A significant effort was made during the last decade for the development of the proper infrastructure that would provide most of the appropriate elements that are essential for the international adoption of electronic commerce (e-commerce). The most important requirements that were addressed are the international interconnection that was achieved by the rapid development and spreading of Internet, the domination of extensible markup language (XML) and XML schemas that guaranteed the interoperability among different architectures, different platforms, and different development languages, and finally the development of standards such as the XML common business library (xCBL) that provide a set of XML building blocks and a document framework that allows the creation of robust, reusable XML documents in order to facilitate the global trading.

Nevertheless, a crucial role on the adoption of e-commerce plays the definition of the European Union legal framework. The current legal framework achieves to cover the legal substance of technological possibilities of electronic commerce only at one point. The directive on electronic commerce (2000/31/EC) (European Parliament, 2000) has established a legal framework for suppliers, making it possible for them to do business with customers in other member states without having to apply the laws of those member states.

At the same time, there is a set of European Union directives that clarify the principles that are applied in some of the e-commerce's processes. The Council Directive 2001/115/EC (European Parliament, 2001) of December 20, 2001, amends Directive 77/388/EC with a view to simplifying, modernizing, and harmonizing the conditions laid down for invoicing in respect of value added tax and the Directive 2004/18/EC (European Parliament, 2004) of March 31, 2004, on the coordination of procedures for the award of public works contracts, public supply contracts, and public service contracts constitute representative examples. The chapter is focused on a specific aspect of e-commerce, the electronic ordering (e-ordering) service. The legal framework that concerns the e-ordering service is determined mainly by the EU Directive 2000/31/EC.

The ordering service as a process of e-commerce should allow for true business-to-business (B2B) secure collaboration by giving the possibility to salesmen and purchasers to execute trustful processes of electronic trading for opening new markets. Among the advantages that e-ordering offers include are the generation of more revenue, improvement of sales efficiency, increase of customer retention, accuracy and efficiency of sales, and elimination of costs.

The e-ordering implementations have to satisfy several security and privacy requirements. These requirements arise from the fact that the ordered documents may contain business data (e.g., VAT code, items) or private data that should not be revealed or modified. They should be trustful documents requiring all four dimensions of security (i.e., confidentiality, integrity, authenticity, nonrepudiation) and privacy.

The existing e-ordering systems discriminated in two types. The first are ERP inclusive systems (e.g., SAP) that manage the sources within and beyond an enterprise. These systems are not affordable for small and medium enterprises (SMEs), blocking them from entering B2B profitable applications. In addition, although they satisfy several security requirements, they do not achieve interoperability. The second type is customized solutions offering e-ordering as an autonomous service. Existing systems of this type ignore various security and privacy requirements.

The purpose of this chapter is the presentation of the security and privacy requirements of an e-ordering service, as well as the proposition of an open, affordable, and scalable e-ordering architecture that satisfy these requirements complied with EU regulations and directives. The proposed system, in order to meet these objectives, is built using open technologies, such as XML, XML cryptography, public key infrastructure (PKI), Web services policy language (WSPL), and Web services.

The rest of this chapter is organized as follows. Section 2 provides an overview of the legal framework which supports electronic ordering, it illustrates the fundamental security and privacy requirements of e-ordering and presents the existed e-ordering implementations. Section 3 describes in detail the e-ordering system architecture and its components. Section 4 provides an assessment (technological, organizational, legal, and business) of trustful e-ordering architecture (TOES) architecture and finally Section 5 presents our conclusions and areas for further research.

STATE-OF-THE ART AND REQUIREMENTS

In this section we present the relevant European legislation for secure e-ordering systems, as well as the security and privacy requirements of the e-ordering services (Polemi & Papastergiou, 2006a, 2006c, 2006d) as implied by the W3C working draft "Web Service Architecture (WSA) Requirements" (Austin, 2005) and by the EU legal framework. Current implementations are assessed against these requirements.

Legal Framework

According to the European Union, fundamental objective constitutes the harmonization of national legislative framework for the provision of a uniform framework for the European market. A harmonized legal framework will encourage the adoption of various cross border service as the e-ordering. The community directives that aim in this harmonization include:

i. Directive 2000/31/EC of the European Parliament and of the council of June 8, 2000 (European Parliament, 2000), on certain legal aspects of information society services, in particular electronic commerce, in the internal market.
ii. Directive 1999/93/EC of the European Parliament and of the council of December 13, 1999 (European Parliament, 1999), on a community framework for electronic signatures.
iii. Directive 95/46/EC of the European Parliament and of the council of October 24, 1995 (European Parliament, 1995), on the protection of individuals with regard to the processing of personal data and on the free movement of such data.
iv. Directive 96/9/EC of the European Parliament and of the council of March 11, 1996 (European Parliament, 1996), on the legal protection of databases.
v. Directive 97/66/EC of the European Parliament and of the council of December 15, 1997 (European Parliament, 1997), concerning the processing of personal data and the protection of privacy in the telecommunications sector.
vi. Directive 2002/58/EC of the European Parliament and of the council of July 12, 2002 (European Parliament, 2002), concerning the processing of personal data and the protection of privacy in the electronic communications sector (directive on privacy and electronic communications).
vii. Regulation (EC) 45/2001 of the European Parliament and of the council of December 18, 2000 (European Parliament, 2001a), on the protection of individuals with regard to the processing of personal data by the community institutions and bodies and on the free movement of such data.

Electronic ordering implementations, in order to be harmonized with the legal framework of EU member states, have to comply with the regulations and the directives that were reported previously.

Security Requirements

The e-ordering systems have to satisfy certain fundamental security requirements:

- AR006.2.1: Authentication of origin. The confirmation of the source that sends the orders is a critical issue of the ordering exchange process and its identity needs to be proven.
- AR006.5: Integrity of the content. During transmission or storage time, the orders should be protected from unauthorized (intentionally or accidentally) modification or their replacement.
- AR006.6: Nonrepudiation of origin and receipt. The ordering exchange can not been denied neither from the sender nor from the recipient.

- Long lasting integrity. The electronic signatures of the orders should remain valid over long periods.
- AR006.4: Confidentiality and privacy. The orders should be readable by the designated recipients.
- Integrity of the sequence of the orders. The avoidance of missing orders.
- AR006.1: Availability. The e-ordering service will be able to be used at any time from the enterprises.
- Secure electronic storage. Primary requirements, such as authenticity, integrity, and readability, should be guaranteed throughout the storage period of the e-ordering documents (European Parliament, 1996).
- Legal compliance. All e-ordering implementations have to be compliant to the regulations and directives that described in Section 2.1.

The above requirements impose restrictions that e-ordering systems have to take into account.

Privacy Requirements

Privacy (Polemi & Papastergiou, 2006b) has an important role in the e-ordering service and is discriminated in two directories. The first one concerns the privacy of the information that is published in an untrusted universal description, discovery, and integration (UDDI) protocol. The requirements of this category are the following:

- UDDI Privacy. The e-ordering service can be published to an untrusted directory where the (existed or future) user can invoke it (Carminati, 2005).
- Requestor Privacy. The Web service requestors' query information should be protected (Carminati, 2005).

The second direction has as primary scope to facilitate the interaction and strengthen the interoperability of the participants. These requirements are described as follows:

- AR0020.1: The WSA must enable the expression of a Web service's privacy policy statements.
- AR0020.3: The WSA must give consumer access to a Web service's advertised privacy policy statement.
- AR0020.5: The WSA must enable privacy policy delegation and propagation.
- AR0020.6: Web services must be allowed to support interactions with anonymous parties.

The e-ordering systems have, as obligation, to implement the appropriate privacy countermeasures in order to satisfy all the above mentioned requirements.

Existing E-Orderings Implementations

Existing solutions have not achieved addressing all the above-mentioned requirements.

- Some companies, acting as trusted third parties, offer the service of e-ordering (Accountis). In this solution, the orders are stored and managed centrally, and in certain cases the orders transform from one format to another (E2open). These solutions face interoperability problems, furthermore, in most cases, the orders are not sent or stored in their original format (as created by the sender).
- Most of the solutions are EDI based (Channelinx) revealing deficiencies such as lack of perseverance in the new models of information exchange and they require expensive software. Moreover EDI messages are rigid (e.g., adding new fields cause implementation difficulties).
- Some solutions enable order transaction via e-mail. In certain cases, some of these

solutions give the possibility to the parties to access a secure server in order to administrate their orders (Emergis). The lack of security during the storage time and the weak archiving process (causing the missing orders) are critical issues in these solutions.

- The use of technologies such as message authentication code (MAC) (Message Authentication Codes (MAC), 2002) and secure sockets layer (Secure Sockets Layer [SSL]) is a very common solution in the e-ordering transaction over the Internet in order to ensure the integrity, confidentiality, and authentication during the exchange of orders (Elma). (Accountis; E2open). However, using these technologies, the security requirements can not be satisfied during the storage time.
- Some solutions use PKI and XML signatures in order to achieve nonrepudiation (E2open). Unfortunately, the validity period of the electronic signatures is limited.
- Via a component of a financial suite, the possibility of order management is provided. Moreover, an autonomous Web-based e-ordering facility is used to create e-orders (Accountis; Channelinx).
- Web services and XML are technologies that are used by a number of solutions (Chen & Meixell, 2003). Nevertheless, in these solutions security and privacy remain a crucial issue.
- The XML advanced electronic signatures (XAdES) standard is used from some e-ordering services aiming at the long-term preservation of digitally signed documents. An important drawback of this standard is that only individual signatures are renewed and furthermore the renewed signature does not include all the preceding signatures of the document.

Most of the above solutions satisfy a subset of the security requirements and ignore the privacy requirements as described in Sections 2.2 and 2.3, respectively. Our proposed architecture aims to fill this gap which creates the lack of an e-ordering system that addresses all aforementioned requirements.

TOES: A SECURE AND TRUSTFUL E-ORDERING SERVICE

In this section we will present the standards that are adopted by TOES in order to address the requirements that have been posed. Furthermore, we describe the proposed e-ordering service architecture, the entities that are involved in the service, the procedures, and all the necessary steps that these entities have to follow in order to complete an e-ordering transaction.

Adopted Standards

The proposed system utilizes XML and Web services as the basic technologies for the formulation and transfer of messages within the platform. The choice of these technologies is based on the achievement of security, privacy, and interoperability.

The message format integrated in the system uses the XML common business library version 4.0 (xCBL 4.0) (xCBL.org, 2003), which is a set of XML building blocks and a document framework that allows the creation of robust, reusable XML documents to facilitate global trading. Furthermore, xCBL has been also adopted as the used schema for the e-order documents and for the representing of the buyers', senders', and items' profiles that are stored in a native XML database. The selection of xCBL is based on its maturity level of completeness and clarity that gives the possibility to a system to be parameterized properly in order to be used in different cases. The use of XML for formatting the e-order document allows for the use of XML digital signatures and the integration with a set of standards as has been specified by the ETSI ESI.

The ETSI organization (ETSI work programme, 2001) includes numerous technical bodies that have to, as a main responsibility, support the objectives of the e-Europe initiative. The European Telecommunications Standards Institute, Technical Committee on Security, Electronic Signature and Infrastructure working group (ETSI TCSEC ESI) is the leading body for the electronic signatures and infrastructures standardization within the ETSI program. The ETSI standards that have been taken into account by TOES are the time stamping profile (TS 101 861 v1.2.11), the qualified certificate profile (TS 101 862 v1.2.1), and the XAdES (TS 101 903).

Furthermore, TOES, in order to achieve a secure and trustful e-order transaction, uses an exchange mechanism that relies on the simple object access protocol (SOAP) (Hartman, 2003; Nadalin, 2004) messaging with Web services security extensions. SOAP is a lightweight protocol intended for exchanging structured information in a decentralized, distributed environment such as the proposed architecture.

In such an environment the interoperability has been ensured and strengthened by the use of registries that are based on the UDDI (Bellwood, 2002). UDDI creates a standard interoperable platform that enables companies and applications to publish and retrieve the description of the e-ordering service. This allows for the automatic, dynamic, and quick accomplishment of the e-order transaction among systems that apply similar message formats.

The published services' description is formulated as specified by the Web services description language (WSDL) (Christenssen et al., 2001) and the WSPL (Anderson, 2004). The WSDL has a basic objective to represent the functional aspects of the e-ordering service. Specifically, it is used to describe the interfaces of a service and how to invoke it. TOES uses WSPL (Polemi & Papastergiou, 2006b) in order to represent all nonfunctional attributes of the offered service. For example, the WSPL describes the confidentiality mechanisms, the identity, the role, and the authentication attributes each participant supports and requires.

Service Architecture

Most of the above solutions satisfy a subset of the security requirements and ignore the privacy requirements as described in Sections 2.2 and 2.3, respectively. Our proposed architecture aims to fill this gap which creates the lack of an e-ordering system that addresses all aforementioned requirements.

Figure 1 depicts the four major entities that take part in the e-ordering process. The two major entities that participate in the transaction are the sender that initiates the process and the receiver that receives the order. A detail description of these entities is the following:

a. The Trusted Third Parties (TTPs): The trusted third parties that are required in the proposed architecture are a certification authority (CA), a registration authority (RA), and a time stamping authority (TSA). The CA and the RA offer the PKI services of registration, certification, and revocation status information with online certificate status protocol (OCSP) (Adams, 1999), while the TSA offers standard-based time stamping services (Sklavos, 2001).

Figure 1. Entities of the e-ordering service

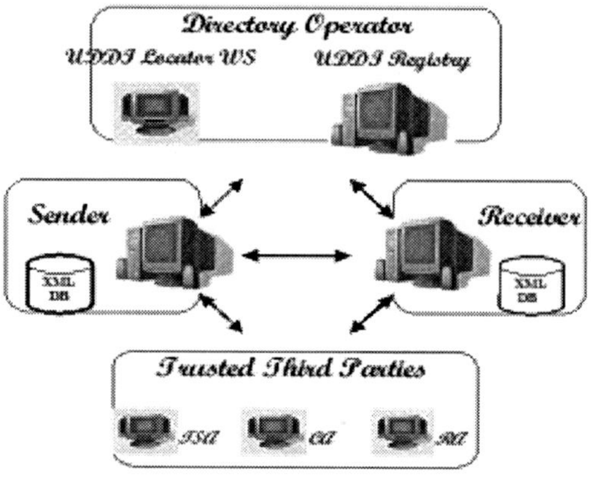

b. Directory Operator: The directory operator entity is composed by the UDDI registry and UDDI locator WS. The UDDI registry is an untrusted directory where Web services can be published, while the UDDI locator WS makes possible the discovery of the UDDI in which a Web service has been published.
c. The Sender: The sender is an organization (e.g., central point of sales, telecomm provider, stores) that hosts TOES architecture. The sender deploys the e-ordering service, publishes it in the UDDI registry, and defines a specific privacy policy that corresponds to the used mechanisms of the e-ordering service. It also communicates with the trusted third parties to get the proper security credentials.
d. The Receiver: The receiver is an organization (e.g., store, private user) that hosts TOES architecture or a similar one. The receiver retrieves the e-ordering service from the UDDI, it is configured to understand the messages, and defines a specific privacy policy that corresponds to the used mechanisms of the e-ordering service. It also communicates with the trusted third parties to get the proper security credentials.

TOES Components

TOES architecture includes a set of components that play an important role in order to satisfy the e-ordering's requirements. These components, as presented in Figure 2, are divided in two distinct tiers, the sender client and the sender server, that are described in details as follows.

Sender Client

The major components of the sender client are the smart card in which the user certificate is stored, the keystore that contains the applet's and the sender WS's certificate, and the client applet that provides the end user interface.

The client applet is running on an Apache HTTP Server 2.0.49 (Apache HTTP Server 2.0.49) which acts as the Web server. The applet gives the possibility to the user to create, manage, and send orders. In order to achieve this, it communicates with seven entities: the smart card to sign the orders, the CA to request certificate status information (Kaliontzoglou, 2003), the TSA to request time stamps in order to produce XAdES signature (Kaliontzoglou, 2003), the applet keystore to receive the appropriate certificates, the XML database to retrieve order's information, the sender WS in which the orders are sent, and the UDDI registry to publish the e-ordering service.

Sender Server

Sender server includes the necessary infrastructure in order to accomplish all the storage's and dispatch's mechanisms in a trustful manner. It is composed of the keystore that contains the sender certificate, the applet certificate, and the receiver certificate, a database in which the orders, the receipts, and the sender WS are stored. The Exist (Meier, 2001) is used as database that constitutes an open source native XML database.

Sender WS has the responsibility to dispatch the e-order to the receiver. The accomplishment of this goal is achieved via the communication with the following five entities: the user interface that sends the orders to the Web service sender, the XML database to store orders and receipts, the sender keystore to receive the sender's and

Figure 2. TOES architecture entities

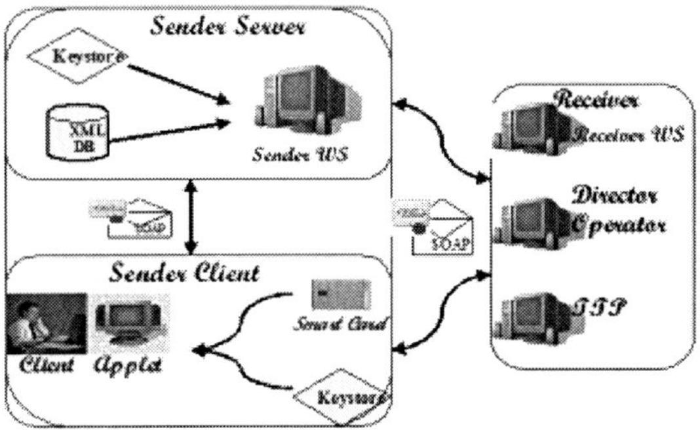

receiver's certificates, the receiver which receives the orders, and the director operator from which retrieves the URL of the Web service receiver. Director operator is composed of:

- UDDI Registry: TOES has integrated the Sun open source implementation of a UDDI, which is part of the Sun Java Web Services Developer's Pack [29]. The Web container on which the UDDI is running is the Tomcat 5.0 for Java WSDP (Tomcat 5.0 for Java WSDP).
- UDDI Locator WS: Offered as a Web service that is running on Axis 1.2 (Axis 1.2).

Sender WS, XML Database, and UDDI Locator WS are integrated into Apache Tomcat 5.5 application server (Jakarta Tomcat 5.5.4).

E-Ordering Process

The sender and the receiver of the order, before initiating the e-ordering process, have to accomplish a set of actions which are divided in three phases.

The first phase includes the definition of the privacy policy of the sender and the receiver. The second phase constitutes the communication with the UDDI registry for the publication and retrieval of the e-ordering service. In the second phase the sender and the receiver have to communicate with the trusted third parties for acquisition of the security credentials.

Phase1-privacy policy: The sender and the receiver have to define the necessary privacy policies using the WSPL. Each of the privacy policy documents is signed using the qualified certificate of the corresponded party. The major objective of the defined privacy policy (Anderson, 2004) is to convey conditions on an e-ordering interaction between two Web service endpoints. All information provided in the privacy policy is aimed to describe the capabilities and requirements of the Web service entities.

Phase2-Service publication and retrieval: The sender produces the WSDL document and uses a standard hash function to a subset of the e-ordering service information that the sender wants to publish to the UDDI registry. The sender's information that remains as clear text corresponds to the contact information, the URL of the WSDL document, and the URL of the privacy policy document. All other information such as Web service properties and private data (e.g., VAT code) are hashed. Then the receiver searching for a service with certain properties generates a query specifying all the conditions on the properties as hashed values. The UDDI registry returns to the receiver the URL of the sender's WSDL document. Now the receiver has the description of the Web service in order to configure its service to be able to receive and send SOAP message according to this description. Furthermore, the receiver produces the WSDL document that corresponds to its service and publishes its service to the UDDI registry following the same way.

Phase3-Set up security credentials: The sender and the receiver take part in the registration and certification procedures as demanded by the certification practice statement of the TTP, and setup the acquired security credentials, in order to achieve a secure communication. Moreover, they have to define the necessary signature policies that will be referenced while producing and validating XAdES signatures, as described in the XAdES standard (Kaliontzoglou, 2006).

When the sender and the receiver have accomplished the aforementioned phases, they are now ready to initiate the ordering process. The necessary steps to complete this process are illustrated below:

Step 1: Access User Interface and Create Order Document (Figure 3: Actions 1-5)

The e-ordering process is initiated by an employee of an organization who authenticates her or himself by means of a smart card and PIN through the user interface. The user accesses a

Figure 3. Sequence diagram for the e-ordering process (Actions 1-15)

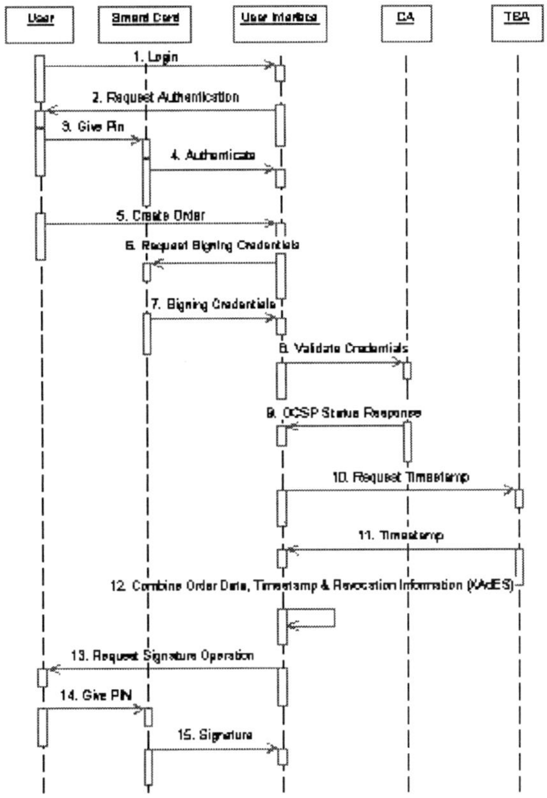

user interface using a browser. The user interface enables the user to complete the necessary data in order to create an order or to manage existing orders. The data input is automatically checked for prevention of errors and is used to create the order document.

Step 2: Sign Order Document (Figure3: Actions 6-15)

When the order document has been created, the user interface transparently gathers the time stamps and revocation status information data from their respective sources. Then, the XAdES signature is formulated based on the cryptographic primitives in the smart card, the user's certificate, and the order data. At the end, the order document is signed using the qualified certificate of the user which is located in the smart card.

Step 3: Dispatching Signed Order Document to Sender WS (Figure 4: Actions 16-30)

After the successful creation of the signed order document, the order document is packaged in a SOAP message employing the WS security extensions and is dispatched to the sender's Web service.

When the sender WS receives the UDDI registry URL, it queries the registry in order to receive the privacy policy's and the WSDL's documents of the deceiver.

All search criteria of the query such as VAT code are hashed. The sender WS retrieves the receiver's and the sender's privacy policy, verifies the digital signatures of the documents, and merges the two policies in order to produce a third one. The sender WS checks the merged policy in order to decide if it is acceptable and thus the e-ordering transaction can be accomplished. Then, the order is packaged in a new SOAP message in which the sender WS applies the acceptable policy, employing the WS security extensions.

The sender WS retrieves the receiver's WSDL document, parses I,t and retrieves the URL of receiver's Web service.

Step 4: Receipt of Order at Receiver WS (Figure4: Actions 31-36)

The protected SOAP message, that has been created, is dispatched over HTTP to the receiver WS. The receiver WS receives the order and follows a fully automated process that requires no human intervention. The SOAP message containing the orders are decrypted with the receiver server's private key and the validity of their WS Security extensions digital signature is verified, so that the point of origin is validated.

Then the e-order document itself is extracted. Validation of the embedded cryptographic information firstly requires communication with a CA for verification of the credentials that were used to sign the e-order as well as verification of any timestamp that was included in the document. Finally the XAdES signature is validated.

Figure 4. Sequence diagram for the e-ordering process (Actions 16-36).

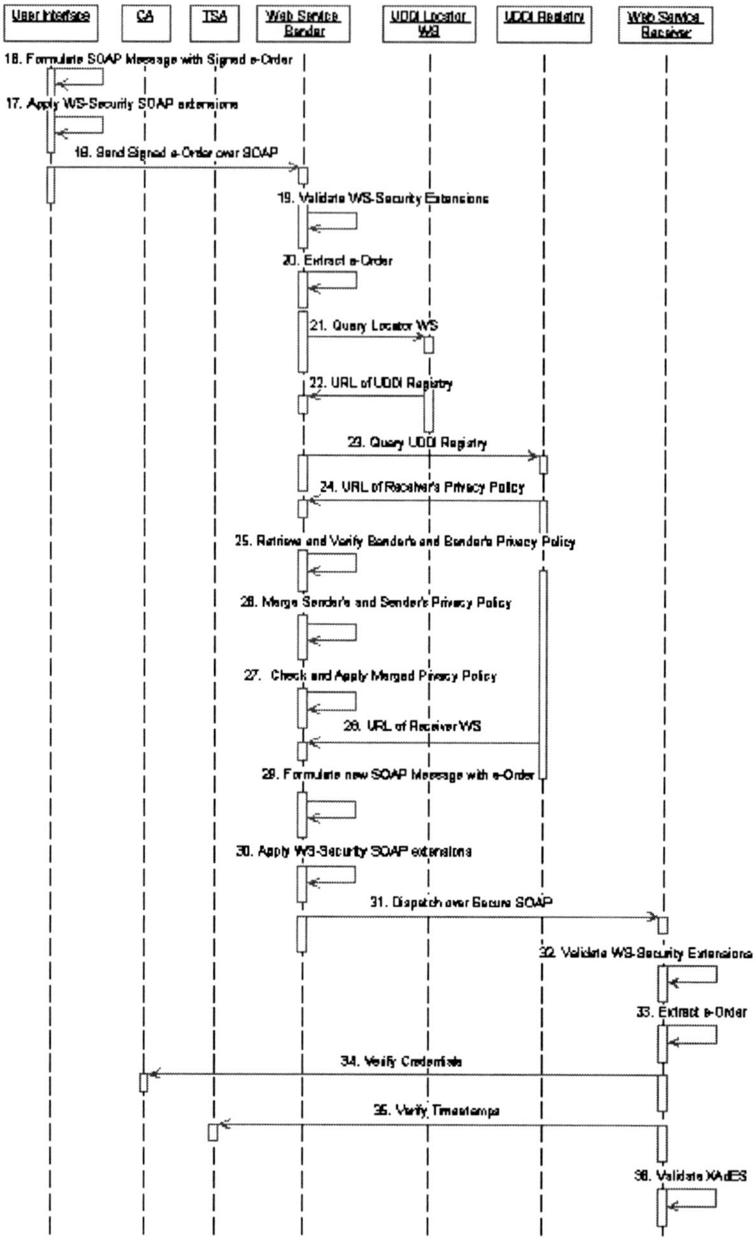

Step 5: Storage of Order at Receiver's XML Database and Dispatching a Receipt (Figure 5: Actions 37-38)

The receiver WS forms a receipt that references to the received order and contains the status of the whole process. Then it stores the e-order and the constructed receipt in a database. From now on, the e-order is available for parsing and further processing by the receiver's users. The receipt is packaged in a SOAP message in which is applied the WS security extensions and is dispatched to issuer WS.

Step 6: Storage of Order at Sender's XML Database (Figure5: Actions 39-40)

Figure 5. Sequence diagram for the e-ordering process (Actions 37-40).

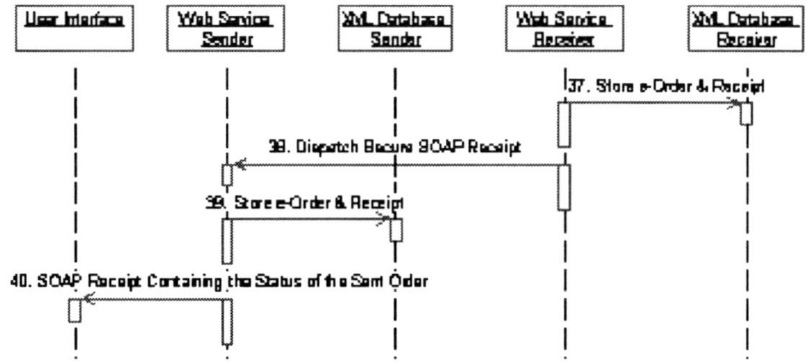

The sender WS receives the SOAP receipt, decrypts it, and verifies the WS security digital signature. Finally it extracts the receipt and stores it in its XML database along with the sent order.

ASSESSMENT

TOES achieves to face a set of issues (security and privacy) in a effective manner. In this section the assessment of the proposed system against specific criteria (i.e., technological, legal, business) is presented.

Security and Trust

Security is the main aspect of each offered e-service in order to meet its business goals. The organizations face serious security problems and they desire to gain the trust of users and to eliminate the lack of confidence.

The proposed system, in order to support a secure production and exchange of e-orders, addresses the security requirements as described in Section 2.2. The countermeasures that are adopted are the following:

- **Authentication of origin:** Using XML digital signatures in combination with tamper resistant cryptographic modules such as smart cards.
- **Integrity of the content:** Using a cryptographic hash function that provides message integrity checks as part of the digital signature process.
- **Non-repudiation of origin and receipt:** Using digital signatures and time stamping.
- **Long lasting integrity:** Using ETSI TS 101 903 that produced ETSI to define XML formats for advanced electronic signatures. A XAdES electronic signature also offers nonrepudiation based on a predefined signature policy.
- **Confidentiality and privacy:** Using XML encryption as specified in the W3C recommendation and the Web services security recommendation for encryption in SOAP messages.
- **Integrity of the sequence of the orders:** Imposing a tight sequence issuance scheme by having a reference number embedded in each order.
- **Availability:** E-ordering Web services are published in registries.
- **Secure electronic storage:** The XML orders are stored with the original format in which they were received. Furthermore, the combination of XAdES and a native XML database guarantees the secure long-term archiving of e-orders.

All the adopted security mechanisms are directly embedded within the proposed system and address the security requirements based on specific policies.

Privacy

All the restrictions that have been imposed by the requirements described in Section 2.3 are faced by the following technical countermeasures:

- **UDDI privacy:** Using hashing techniques to a subset of published data so they can not be revealed.
- **Requestor privacy:** Searching for a service with certain properties, generating a query specifying all the conditions on the properties as hashed values. As a result, the untrusted directory can not infer the search criteria.
- **AR0020.1:** The privacy policy is expressed using WSPL.
- **AR0020.3:** The privacy policy is published in a UDDI registry.
- **AR0020.5:** The privacy policy is used in order to convey conditions under which the e-ordering service is provided.
- **AR0020.6:** Retrieving the privacy policy from UDDI registry anonymously.

The adoption of these countermeasures is crucial for the effectiveness of the proposed architecture.

Legal Compliance

The major objective of each e-ordering implementation is to be compliant with the underlying legal framework described in Section 2.1. The legal framework defines the rules that must be applied in every electronic document transaction. The directive on electronic commerce (2000/31/EC) dictates the main requirements for the e-ordering solutions. The proposed system satisfies these requirements as follows:

- Providing a response to every dispatched e-order.
- Providing the possibility of management of the e-order documents that have not been sent yet.

The European Union directives as described in Section 2.1 impose the need of adoption countermeasures, in addition to the security countermeasures that are presented in Section 4.1. These further countermeasures are the following:

- Integrity of the sequence of the orders: Imposing a tight sequence issuance scheme by having a reference number embedded in each order.
- Secure electronic storage: The XML orders are stored with the original format in which they were received. Furthermore, the combination of XAdES and a native XML database guarantees the secure long-term archiving of e-orders.

The aforementioned countermeasures are integrated in the proposed systems respecting the EU Directives.

Interoperability and Scalability

Interoperability is difficult to accomplish because of the heterogeneity of the different technologies and systems. Interoperability constitutes an objective that requires identifiable and publishable e-services in order to be achieved. In the same manner, scalability is a requirement that have to be satisfied, so that the quality of the offered service remains in an acceptable level.

TOES addresses interoperability and scalability by using widely deployed standards and technologies. Specifically, XML and Web services constitute the most appropriate choices, in combination with open source Java tools. The proposed system utilizes WSDL for the description and UDDI for the publishing of the e-ordering service, while it uses SOAP for the transactions of the e-ordering documents.

Performance

TOES achieves particularly satisfactory performance via the adoption of digital subscriber line (DSL) technology. The use of XML and Web services, as the basic technologies for the formulation and transfer of messages within the platform, expands the size of data several times over. The increase of the data size is translated into more storage, transmission, and processing time. For example, a SOAP message demands extra bandwidth, while in order to format and parse the messages, more processing is needed.

Despite these drawbacks, the advantages of XML and Web services (Hartman, 2003) outweigh the additional storage requirements, transmission time, and processing needed. The flexibility offered by these technologies overcomes differences in platforms, development languages, and architectures, allowing organizations to perform processing tasks cooperatively. Using XML and SOAP, systems from different domains with independent environments, different architectures, and different platforms can engage in a distributed endeavor to address business needs.

Furthermore, a major factor that affects the performance of the TOES is the choice of the proper algorithm for the encryption of the SOAP messages. Secret key encryption is much faster than public key encryption, but secret keys do not scale as well as public keys. So, in order to achieve the higher performance, the adopted process is the following: a one-time generated secret key is used for the encryption and the decryption of the data, using a symmetric algorithm (triple-DES). The generated secret key is encrypted using a public key algorithm (RSA-V1.5) and the encrypted key is embedded in the SOAP message.

Cost

TOES is an affordable solution. SMEs do not have to invest in expensive ERPs in order to offer e-ordering. On the contrary, the costs of TOES adoption are kept low due to the use of open source software (i.e., Exist, Axis 1.2, Apache Tomcat 5.5, and Java Web Services Developer's Pack) and are limited to the operation costs. Furthermore, TOES does not need the existence of dedicated employees that are required to have a specific education background, while the system does not pose additional training and managing requirements.

The tables that follow demonstrate the significant cost and time benefits that the sender and the receiver of e-order receive from the utilizing of TOES platform.

Tables 1 and 2 present a reduction of the time's process and of the cost for the receiver's and sender's organizations, respectively. The time's reduction is reached 75% and 62.5% of the total time of the issuer and the receiver, while the cost is reduced at 70.89% and 59.61%, respectively.

Benefits

The main anticipated benefit by adopting TOES is the significant operating costs reduction, which involves the minimization of the workload required for handling paper orders, as well as the expenses involved in printing and delivering them and without purchasing expensive ERP solutions.

TOES service provides a suitable, friendly interface to the authorized user for inputting the necessary data for the order fields that have been defined in an XML document. A native XML database ensures that the XML orders are stored exactly in the original format in which they were received for any future audit.

TOES is a stand alone, secure e-ordering solution based on open source technologies that can interoperate with existing ERP systems. Service-oriented architecture (SOA) is the key to this possibility. The most widespread ERP systems (Secure Electronic Invoicing System [SELIS], 2006), such as SAP, SSA, and Oracle solutions, have already adopted SOA as the used architecture style. Nowadays, the largest manufacturers, but

Table 1. Receiver organization: Cost and time benefits of utilizing TOES architecture

Handling Stage	Paper Order Time (mins)	e-Order Time (mins)	Paper Order Cost (€)	e-Order Cost (€)
Opening the post – Courier	1		0.9	
Date-stamping	1			
Checking and registering the	2		1.4	
Checking the contents	1	1		
Validation	2	1	5.4	2
Entering the order in the IT system	1.5		1.4	
Archiving	1.5		1.5	
Law litigation			1.8	0.9
Processing errors (10% of Orders)	2	1	1	1
Total	12	3	13.4	3.9
Savings per order		MINS 9		EUR 9.5
Savings, percentage		75 %		70.89 %

Table 2. Sender organization: Cost and time benefits of utilizing TOES architecture

Handling Stage	Paper Order Time (mins)	e-Order Time (mins)	Paper Order Cost (€)	e-Order Cost (€)
Getting the order from the post – Courier				
Marking purchaser number, voucher stamp and voucher number on the order	0.5		0.3	
Accounting	1		0.3	
Entering in order-handling system and checking the contents	1	1	1.2	
Validation	2	1	4.4	2
Archiving	1.5		0.8	
Law litigation			2.4	1.2
Processing errors (10% of Orders)	2	1	1	1
Total	8	3	10.4	4.2
Savings per order		MINS 5		EUR 6.2
Savings, percentage		62.5 %		59.61

also smaller organizations, use SOA in order to design, develop, and implement their solutions.

Finally, TOES satisfies the requirements that have been posed in Section 2.1, respecting the EU legislation.

CONCLUSION, FUTURE WORK AND ACKNOWLEDGMENT

In this chapter, we have presented an architecture for secure e-ordering service. The proposed system is a unique tool for enterprises which desire to send and receive electronic orders via the Internet. The result(s) of these action(s) that use this system permit the enterprise and its employee's to manage the orders in an organized manner with accountability. The user provides the necessary data for the specific order via a browser-accessible form, and when the order is created, it is sent to the supplier, who immediately and automatically sends a confirmation reply of the order in electronic form.

From a technological point of view, the system is compliant and in accordance with state-of-the-art standards that constitute an interoperable, affordable, and scalable solution that address the security and privacy requirements which are described in Sections 2.2 and 2.3. In order to achieve these fundamental objectives, the architecture is based on XML, XML digital signatures and encryption, xCBL, WSPL, and Web services.

Our future plans are to expand our e-ordering service to several directions, with the fundamental objective in mind to enhance the functionality and interoperability features of the proposed architecture. Our forthcoming interests are the mobile submission of e-orders, the automated negotiation of the functional parameters of the e-ordering service, the adoption of a role based authorization system, the verification of the interoperability of the TOES components using two testing tools that have been developed by the WS-I Test Tools Working Group for interoperability assessment, the support of more e-order's XML schemas, the support of a XML key management, the integration with existed ERP systems, and the orchestration of the services using WS-business process execution language (WS-BPEL).

The proposed architecture in combination with the future extensions has a main objective to establish the appropriate confidence framework in order to eliminate every sense of unreliability and mistrust that dominate the electronic transactions, allowing the end user to utilize every aspect of electronic commerce.

The authors would like to thank the GSRT(PENED) programs and the E.C. for its support in funding the SELIS project and all the project partners (Expertnet, UPRC, the Athens Chamber of Commerce and Industry, the Paris Chamber of Commerce and Industry and the Brasov Chamber of Commerce and Industry).

REFERENCES

Accountis. Retrieved December 17, 2007, from http://www.accountis.com

Anderson, A. H. (2004, June 7-9). *An introduction to the Web services policy language (WSPL)*. Paper presented at the 5th IEEE International Workshop on Policies for Distributed Systems and Networks, Yorktown Heights, NY.

Apache HTTP Server 2.0.49. Retrieved December 17, 2007, from http://httpd.apache.org/

Axis 1.2. Retrieved December 17, 2007, from at http://ws.apache.org/axis/

Bajaj, S., et al. (2004, September). *Web services policy framework (WSPolicy).*

Bellwood, T. (Ed). (2002). UDDI version 2.04 API specification. *UDDI committee specification, OASIS standard*. Retrieved December 17, 2007, from www.oasis-open.org/committees/uddi-spec/doc/tcspecs.htm#uddiv2

Channelinx. Retrieved December 17, 2007, from www.channelinx.com

Chen, M., & Meixell, M. J. (2003, November). Web services enabled procurement in the extended enterprise: An architecture design and implementation. *Journal of Electronic Commerce Research, 4*(4), 140-155.

Christenssen, E., et al. (2001). Web services description language (WSDL) 1.1. *W3C note*. Retrieved December 17, 2007, from www.w3.org/TR/wsdl

E2open. Retrieved December 17, 2007, from www.e2open.com

Elma. Retrieved December 17, 2007, from http://www.elma.net/en

Emergis. Retrieved December 17, 2007, from www.emergis.com/en/

ETSI workprogramme. (2001). Retrieved December 17, 2007, from http://www.e-europestandards.org/ETSI_activities.htm

European Parliament. (2001a). *Regulation (EC) 45/2001 of the European Parliament and of the Council of 18 December 2000 on the protection of individuals with regard to the processing of personal data by the community institutions and bodies and on the free movement of such data.*

European Parliament. (2001b). *Council Directive 2001/115/EC of 20 December 2001 amending Directive 77/388/EEC with view to simplifying, modernizing and harmonizing the conditions laid down for invoicing in respect of value added tax.*

European Parliament. (2004). *Council Directive 2004/18/EC of the European Parliament and of the Council of 31 March 2004 on the coordination of procedures for the award of public works contracts, public supply contracts and public service contracts.* Retrieved December 17, 2007, from http://europa.eu.int/eur-lex/en/archive/2004/l_13420040430en.html

Hartman, B. (2003). *Mastering Web services security*. Wiley Publishing.

Jakarta Tomcat 5.5.4. Retrieved December 17, 2007, from at http://tomcat.apache.org/JWSDP. (2005).

Java Web services developer's pack 1.5. Retrieved December 17, 2007, from http://java.sun.com

Message Authentication Codes (MAC). (2002). *Cryptographic message syntax (CMS) algorithms* (IETF RFC 3370).

Polemi, D., & Papastergiou, S. (2006a). *A secure, open and interoperable e-ordering service.* Paper presented at the 2nd International Conference on Web Information Systems and Technologies (Webist, 2006), Setubal, Portugal, (pp. 57-62). INSTICC Press. ISBN: 978-972-8865-46-7.

Polemi, D., & Papastergiou, S. (2006b, March 16-17). *Achievements and future direction in security policies. Electronic democracy challenges of the digital era.* Paper presented at the 2nd National Conference with International Participation, Athens. ACCI.

Polemi, D., & Papastergiou, S. (2006c, August 26-29). *TOES: Trustful and open e-ordering service for SMEs.* Paper presented at the International Conference on Internet Surveillance and Protection (ICISP, 2006), Cap Esterel, Côte d'Azur, France. IEEE Computer Society. ISBN: 0-7695-2649-7.

Polemi, D., & Papastergiou, S. (2006d, October 11-13). *A secure e-ordering Web service.* Paper presented at the Sixth IFIP conference on e-Commerce, e-Business, and e-Government (I3E, 2006) Conference, Turku, Finland, (pp. 352-365). Springer Press. ISBN: 0-387-39226-2.

Secure Electronic Invoicing System (SELIS). (2006). *Deliverable D4.2 interoperability of technology and service across regions/states.* Retrieved December 17, 2007, from http://selis.unipi.gr/

Secure Sockets Layer (SSL). Retrieved December 17, 2007, from http://wp.netscape.com/eng/ssl3/

Tomcat 5.0 for Java WSDP. Retrieved December 17, 2007, from http://java.sun.com/webservices/containers

This work was previously published in the International Journal of Information Security and Privacy, Volume 2, Issue 2, edited by H. Nemati, pp. 14-30, copyright 2008 by IGI Publishing (an imprint of IGI Global).

Chapter 20
SEACON:
An Integrated Approach to the Analysis and Design of Secure Enterprise Architecture-Based Computer Networks

Surya B. Yadav
Texas Tech University, USA

ABSTRACT

The extent methods largely ignore the importance of integrating security requirements with business requirements and providing built-in steps for dealing with these requirements seamlessly. To address this problem, a new approach to secure network analysis and design is presented. The proposed method, called the SEACON method, provides an integrated approach to use existing principles of information systems analysis and design with the unique requirements of distributed secure network systems. We introduce several concepts including security adequacy level, process-location-security matrix, data-location-security matrix, and secure location model to provide built-in mechanisms to capture security needs and use them seamlessly throughout the steps of analyzing and designing secure networks. This method is illustrated and compared to other secure network design methods. The SEACON method is found to be a useful and effective method.

INTRODUCTION

Designing and implementing a secure computer network has become a necessity for companies big or small. Network security is no longer just a technical issue anymore (Sarbanes-Oxley Compliance Journal, 2005). It has also become an economic and legal issue for most companies. According to an IT security management survey, "Two-thirds of those who took part in the survey acknowledged that the wide range of government regulations, such as Sarbanes-Oxley, HIPAA, and GLBA, has affected their company's handling of IT security issues" (Sarbanes-Oxley Compliance Journal, 2005). According to CSI/FBI's Tenth Annual

Computer Crime Security Survey, unauthorized access to information and theft of proprietary information showed significant increases in average loss per respondent (CSI/FBI, 2005). Hackers have also moved to new areas such as identity theft (McMillan, 2005). As a consequence, the cost of information theft has jumped considerably. These surveys indicate that a better computer network design method is needed for designing a more secure computer network.

There has been increased activity in various aspects of security, network system security, and secure network design in the last several years. There are several good articles (Cisco Systems, 2001; Fisch & White, 2001; Ghosh, 2001; Oppenheimer, 2004; Southwick, 2003; Whitman & Mattord, 2005; Whitmore, 2001) that deal with secure network design. For example, Fisch and White (2001) discuss security models and various kinds of security measures in detail. Ghosh (2001) discusses principles of secure network design and an in-depth analysis of ATM networks and their security. Oppenheimer (2004) uses a top-down network design methodology to design an enterprise computer network. The emphasis is on the technical analysis and design of networks. Whitman and Mattord (2005) present a Security Systems Development Life Cycle (SecSDLC) methodology paralleling the basic system development life cycle (SDLC) methodology. There are sophisticated network simulation and performance tools such as OPNET (OPNET, 2005). Most of the existing work on secure network design, however, tends to lean more toward technical details. There is very little research that addresses the issue of security and business requirements of a computer network simultaneously. It is very important to understand an organization's business requirements to design an effective network (Oppenheimer, 2004). It is equally important to understand the organization's security requirements as well. To our knowledge, there is no published design method that integrates secure network requirements with business requirements to develop a secure network. In this chapter, we address the following research questions:

1. How can we identify security and business requirements of a network system seamlessly?
2. How can we identify all possible assets and resources, including business processes and data that need to be protected in a network system?
3. How can we incorporate and document security requirements into conceptual and logical network diagrams?

This chapter follows the DEACON method (Shaw & Yadav, 2001) and presents a new method that provides built-in mechanisms to carry secure network requirements along with business requirements seamlessly throughout the process of analyzing and designing secure network architecture. We have developed, as part of the method, several new concepts such as the security adequacy level, process-location-security matrix, data-location-security matrix, and secure location model to achieve a good interplay between network security requirements and business requirements.

CURRENT WORK ON DEVELOPING SECURE COMPUTER NETWORKS

Computer networking and its security is a vast area of research and study. The topics cover network security concepts, principles, frameworks, techniques, methods, laws, and practices. This chapter draws from research on several of the topics mentioned above; however, it is not practical for this chapter to review even a fraction of the literature covering those topics. Interested readers are kindly referred to Ghosh (2001), Kizza (2005), and Whitman and Mattord (2005) for a good review of topics related to secure computer networks. Here, we limit our literature discussion to research that deals with secure network design methods.

Paul Innella (Innella, 2001) presents a design method based upon the software process model. This is an interesting method but it is, in its current form, too general and too brief to be of any practical use.

Cisco Systems (2001) has developed a secure blueprint for enterprise networks (SAFE) to provide best practice information on designing and implementing secure networks. SAFE is not a design method in the sense of providing specific steps for designing a secure network. Instead, it is a set of design and configuration guidelines that should be followed to design a secure network.

James J. Whitmore presents a method for designing secure solutions. He describes "a systematic approach for defining, modeling, and documenting security functions within a structured design process in order to facilitate greater trust in the operation of resulting IT solutions" (Whitmore, 2001). Using Common Criteria as a basis, he proposes five interrelated security subsystems. These are (Whitmore, 2001):

1. Security audit subsystem
2. Solution integrity subsystem
3. Access control subsystem
4. Information flow control subsystem
5. Identity or credential subsystem

Whitmore's approach develops network security architectures. Once the security requirements have been identified, they can be mapped to the above mentioned security subsystems to develop a security architecture for the system. This is a very important step to designing secure solutions. However, its focus is more on the technical side of the network solution and does not address the identification and determination of security requirements. It also lacks in providing steps for integrating security requirements with business requirements in designing secure network solutions.

Priscilla Oppenheimer presents a top-down network design method consisting of the following major steps (Oppenheimer, 2004):

1. Identifying customer needs and goals
2. Logical network design
3. Physical network design
4. Testing, optimizing, and documenting network design

It is obvious that the top-down network design method parallels the structured systems analysis method for software development. The method provides a detailed discussion of various topics related to computer network design including security. However, the method does not have built-in steps and mechanisms to explicitly address security requirements in addition to business requirements. Also, the method has more focus on technical details of network design. It does not address the issue of network modeling and simulation.

Whitman and Mattord (2005) present a Security Systems Development Life Cycle (SecSDLC) which is based upon the Systems Development Life Cycle (SDLC) waterfall methodology. The SecSDLC methodology consists of investigation, analysis, logical design, physical design, implementation, and maintenance phases (Whitman & Mattord, 2005). The SecSDLC has steps for documenting security policies, analyzing threats, and examining legal issues. However, the SecDLC does not have steps to identify security or business requirements. There is very little support in the form of guidelines and techniques for designing and documenting secure network models and architectures based upon the security and business requirements of an organization.

There is a very limited literature on design methods that provide mechanisms to incorporate security requirements along with business requirements in designing a secure computer network. The next section discusses a new method to analyze and design a secure network and shows how to use these two requirements seamlessly.

AN INTEGRATED APPROACH TO SECURE NETWORK ANALYSIS AND DESIGN

To deal with current security challenges, designing a secure computer network must be an integral part of the overall approach to design a computer network. Security of computer networks cannot be an afterthought anymore. This section discusses the proposed integrated method to analyze and design secure computer networks. The proposed method has been named SEACON (design of Secure Enterprise Architecture-based Computer Networks). The SEACON method has built-in mechanisms to capture a firm's network security needs from the analysis stage and carry them to the implementation stage. Figure 1 shows the detailed steps of the SEACON method. The following paragraphs discuss the SEACON method in detail.

Problem Definition

The first step in the method requires the establishment of goals, objectives, and security policies at three levels, firm, information system (IS), and computer network. The establishment of goals, objectives, and security policies helps determine the context and scope of the problem at hand. Hopefully, the organizational and IS level goals, objectives, and security policies have already been established. If not, then these organizational an IS policies should be defined along with the goals, objectives, and security policies for the network under consideration. Network security policies are determined using the Six-View Perspective of System Security (SVPSS) framework (Yadav, 2006). The SVPSS framework enables an analyst to determine a comprehensive set of security policies by providing a multiview look at system security. Network security policies should be defined under each security view, Threat, Resource, Process, Management, Legal, and Assessment. A firm's security policies act as the bedrock on which secure network and secure information systems are designed and built. Network security policies should be stated as precisely as possible. Access rules and security requirements for internal as well as external entities should be clearly stated. Security risks should be assessed after the initial set of network security policies has been identified. There are several risk assessment methods (GAO, 1999; ASIS, 2003; Verdon & McGraw, 2004) proposed in the literature. All of these methods are quite similar to one another and any of them can be used to assess risks in conjunction with the SVPSS framework.

Requirement Specification

This step involves determining network security needs in addition to traditional business and data modeling activities. We used a process model, data model, network model, and security risk register as major tools to document security requirements and mechanisms that should be included in a network system. We discuss an extended version of these tools below.

The modeling of business activities presented by Shaw and Yadav (2001) and other researchers do not address the integration of security activity modeling with business activity modeling. Security has become too critical to leave it as an afterthought when developing a secure network. Security requirements should be modeled along with business requirements simultaneously. Security requirements modeling can be easily handled by adopting certain conventions as part of existing process and data modeling techniques. For example, a data flow diagram can be easily adapted to capture process security requirements. The entity-relationship data modeling (ERD) can be adapted to capture data security requirements. A business process detailed in a Data Flow Diagram (DFD) must have with it an appropriate security level required to secure the business process. We need to identify and specify security for each process and for each data object-entity. One way

Figure 1. The SEACON method

1. Problem Definition

a. Define organizational goals, objectives, and security policies using SVPSS framework
b. Define IS goals, objectives, and security policies
c. Define network goals, objectives, and security policies

2. Requirement Specification

a. Model business processes and their security requirements (process model)
b. Model organizational data and their security requirements (data model)
c. Identify physical locations to be connected within the network
d. Identify information domains at each location using data-location-security and process-location-security matrices
e. Construct secure location model (secure extended location connectivity diagram) with security annotation
f. Perform the assessment of security risks for each asset such as process, data, and network components and determine appropriate security requirements and mechanisms

3. Secure Network Architecture

a. Identify enclaves and boundary controllers under each information domain
b. Specify security requirements and mechanisms for each enclave based upon the security risk assessment conducted in step 3
c. Assign enclaves and boundary controllers to appropriate nodes
d. Create a secure network architecture diagram
e. Match available technology with specifications on architecture diagram

4. Secure Network Performance Evaluation

a. Simulate secure network operation (e.g. using software such as OPNET)
b. Identify performance bottlenecks and optimize network
c. Identify security holes and correct them
d. Refine secure network architecture

5. Implementation

a. Implement the secure network architecture
b. Prepare a conversion plan
c. Convert to the new secure network system

to state security for processes and entities is to attach a security classification level as a property of an entity or process. We propose a concept of the Security Adequacy Level (SAL) to easily state the nature of security in a data model. SAL refers to the degree of security-strength needed to adequately protect a process or an entity (a data object). The definition of the SAL concept is based on the work on Strength of Mechanism Level (Arber, Cooley, Hirsh, Mahan, & OSterritter, 1999) and four hierarchical divisions of security protection under the trusted computer system evaluation criteria (Department of Defense, 1985). We define six levels of SAL—Low, Basic, Medium, High, and Very High—of security adequacy. The SAL levels are hierarchical in nature, meaning that a given security adequacy level subsumes all the lower level security requirements, Low being the lowest level and Very High being the highest level of security adequacy. These levels are defined in Figure 2.

Defining the security adequacy level of a business process enables one to explicitly pay attention to the security issues of a process and helps a designer to include the appropriate security mechanisms to protect the process in a network system. Figure 3 shows as an example a data flow diagram with two business processes that includes a security adequacy level for each process.

Organizational data modeling involves identifying data objects, their attributes (properties), and relationships among the data objects. The security requirements for data objects are generally ignored when developing a data model. We suggest that security requirements for data objects should be included as part of a data model. We propose to attach an appropriate security adequacy level to each data object in a data model.

Figure 4 shows an illustration of an ER model with a security adequacy level attached to each entity type in the model. The SAL attached to an entity type specifies the level of protection needed for that entity type. The security adequacy level for data and processes should be determined in consultation with the users as well as the managers (owners) of those process and data. We now discuss the idea of a secure location model to identify network components and their security needs.

Secure Location Model

A secure location model shows not only the locations (nodes) and connections, but also the security requirements of those locations and connections. Analysts need a mechanism to include security requirements in various models along

Figure 2. Description of security adequacy levels

Security Adequacy Level (SAL)	Suggested security mechanisms for the level	Correspondence with DOD's Hierarchical divisions (Department of Defense, 1985)	Correspondence with Strength mechanism level (Arber et al., 1999)
Low—defined as minimum protection	Password access	Division D—Minimal Protection	N/A
Basic—defined as basic security practice. It is adequate enough to protect low value data and deter unsophisticated threats	Restrictive granting of rights; enhancement of strict account policies; basic encryption	Division C—discretionary protection (Class C1, Class C2)	Basic
Medium—defined as good security practice. It is adequate enough to protect medium value data and resist sophisticated threats	Deactivation of unnecessary network services; staff training, security update plan; firewall; IPSec VPN	Division B—mandatory protection (Class B1 and Class B2).	Medium
High—defined as high security practice. It is adequate enough to protect high value data and resist high-level threats	Network and host-based Intrusion detection systems (IDS); contingency plans; vulnerability analysis tool; SSL VPN	Division B—mandatory protection (Class B3—security domains)	High
Very High—defined as a formal and very high security practice. It is adequate enough to deal with any kind of threats and protect very high value data.	Application based IDS; formal security protection	Division A—verified design (Class A1 and beyond)	N/A

Figure 3. A list of network security mechanisms

Network Security Mechanisms

Intrusion Detection System (IDS):
1) Network-Based IDS
2) Host-Based IDS
3) Application-Based IDS

Firewall:
1. Packet filtering
2. Proxy server
3. Stateful packet filtering

Virtual Private Network (VPN):
1. IPSec VPN
2. SSL VPN

Secure Configuration of Servers:
1. Deactivation of unnecessary network services
2. Password access protection
3. Restrictive granting of rights
4. Enforcement of strict account policies
5. Audit Logs

Network Policies and Procedures:
1. Staff training
2. Security update plan
3. Contingency plan
4. Vulnerability analysis tools

with the process and data needs of a firm. This mechanism should address the adequacy level of security of various components of a system. The security adequacy level classification discussed in Figure 2 can be used to specify security levels for various network system components.

First, process-location-security and data-location-security matrices should be developed to identify the locational usage of processes and data and to possibly revise their security adequacy levels identified previously in the process and data models. Security adequacy levels for processes and data could be influenced by their locational usage. For example, if a process or data is used from more than one location then it may require a higher level of security. Information domains should be identified next based upon the process-location-security and data-location-security matrices. An information domain is used to group a set of resources with similar characteristics and functionalities. An information domain addresses the issues of data management and data interoperability. Each location can contain one or more information domains.

Secure extended location connectivity diagrams should be developed after the information domains have been determined. A secure extended location connectivity diagram is an extended loca-

Figure 4. An illustration of an ERD with security adequacy levels

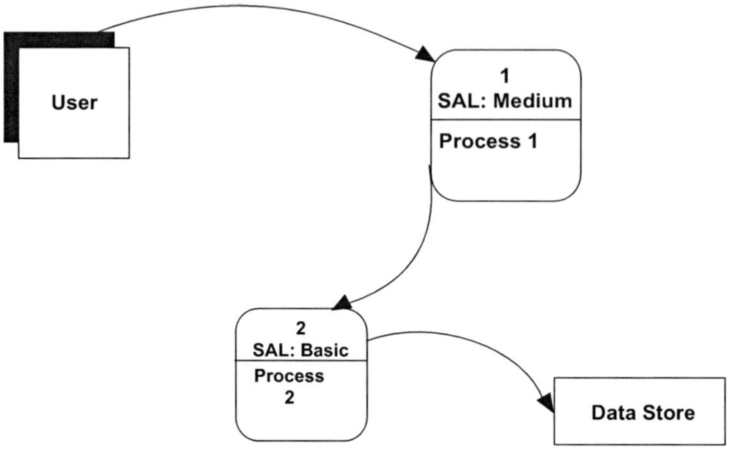

Specification of Security Adequacy Level for Processes

Note: A Security Adequacy Level of Medium for Process 1 means that Process 1 needs a Medium level of security protection.

tion connectivity diagram (Shaw & Yadav, 2001) that incorporates the identification of security adequacy levels for locations and connections. As the process and data are allocated to their appropriate locations, the corresponding process and data security adequacy levels are combined to determine the overall security levels for those locations and connections. The secure extended location diagrams should be drawn using technology independent generic symbols to represent nodes and connections in the diagram. Locations and connectivity links are labeled with security adequacy levels. Figures 8A through 8D show examples of such a diagram.

So far, we have used several types of models—process, data, and network diagrams—to identify various types of assets and resources that need to be protected. Various kinds of threats pose security risks to network assets and resources. Network security requirements should be identified from various perspectives (Yadav, 2006) while keeping in mind the various security risks for the firm. Security risks should be assessed under each view (Yadav, 2006) and prioritized based upon their expected consequences. Users and managers should be consulted in determining network security risks and requirements under each view.

The identification of security risks and their assessment allows a company to compare and evaluate consequences of various types of security risks. It also enables the company to prioritize the security risks and select, given a limited budget, the most consequential security risks for mitigation. A general process for security risk assessment can be stated as follows (Center for Medicare and Medicaid Services, 2002; Yadav, 2006):

1. Identify assets under each security view.
2. Identify sources of risk for each asset.
3. For each source of risk, we estimate its likelihood of occurrence and the consequence (impact) if the risk materializes. Using the levels of likely hood of occurrence and the levels of severity of consequence (impact), determine the risk level for each risk. Also, determine the acceptable risk level for each risk.
4. Evaluate each risk based upon the risk level and the acceptable risk level.

5. Identify security requirements and mechanisms to reduce the risk level to an acceptable level.

The above process is described in detail by CMS (Center for Medicare and Medicaid Services, 2002) and Yadav (2006).

A security risk register can be used to document the outputs of the above risk assessment process. A security risk register is a tabular representation of details about identified risks and security mechanisms for reducing those risks. Figure 5 shows a template for a security risk register.

Security views in the risk register refer to various security perspectives of a network system (Yadav, 2006). Assets are any IT or system resource that needs to be protected. Assets are identified under each security view. For each identified asset, the sources of risks are then determined. For each source of risk, the threat likelihood estimate, the consequence if the threat is realized, and the resultant risk level are computed. An acceptable level of risk is specified for each source of risk. A risk priority level is then computed based upon the resultant risk level and the acceptable risk level. Security requirements and mechanisms are then specified for reducing security threats from the sources of risks having high priority levels. Figure 6 shows a tree-view of the security risk register template shown in Figure 5. It also shows the hierarchical nature of the security risk register. For illustration, only one branch is expanded in Figure 6.

Figure 5. Illustration of an ERD with security adequacy level

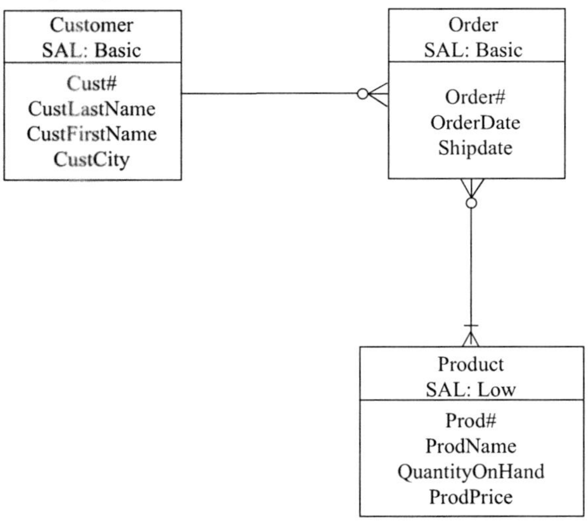

This section has discussed several ways to identify security requirements of a network system. The identified security requirements and mechanisms should be allocated and apportioned among the various components of a network system. A security mechanism is a method, tool, technique, or procedure used to enforce a security policy and to protect resources. Some examples of network security mechanisms are firewalls, intrusion detection systems, virtual private networks (VPN), and network access controls. Figure 7 shows a more complete list of network security mechanisms. For more information about network security mechanisms, please see Irvine and Levine (1999), Fisch and White (2001), Bace and Mell (2001), Rusli (2001), and Warren (2005).

The next section uses the idea of secure network architecture to represent a logical network topology and the security mechanisms apportioned among the various network components.

Secure Network Architecture

Secure network architecture can be represented using a network diagram. A network architecture diagram (NAD) and a secure network architecture diagram (SNAD) are used as a tool to model relationships among network hardware, software, processes, data, and security policies. These diagrams become the basis to simulate and evaluate network architectures. First, information enclaves and boundary controllers (Bionic Buffalo Corporation, 2000; Defense Logistics Agency, 2002) are determined under each information domain. An enclave is a set of resources that are protected at the same level as a group. An

Figure 6. A security risk register template for documenting risks

A Security Risk Register Template									
Security Views	Assets	Sources of Risk	Threat Likelihood Estimate	Consequence, if the threat is realized	Resultant Risk Level	Acceptable Risk Level	Risk Priority Level	Security Requirements	Security Mechanisms

information domain may have several enclaves. Generally, an information domain is physically realized via a set of information enclaves (Bionic Buffalo Corporation, 2000). Enclaves typically contain computing resource components such as switches, servers, printers, and workstations (Defense Logistics Agency, 2002). A boundary controller protects an enclave. For example, a router with a firewall can act as a boundary controller. Appropriate security adequacy levels are assigned to each enclave. A network architecture diagram without security mechanisms could be created as a base network for network analysis and design. Next, the security adequacy level of each enclave is mapped to appropriate network security mechanisms to be included in the secure network architecture.

A secure network architecture diagram is then created. A secure network architecture diagram can be obviously drawn at various levels such as at the logical and physical levels. A logical level diagram uses generic symbols for nodes and links in the diagram. A physical level SNAD, on the other hand, is technology-dependent. Nodes and links in a physical level SNAD are represented by appropriate communication technology available in the market. A SNAD (logical as well as physical) should be designed using network modeling tools such as OPNET (OPNET, 2005) so that they can be simulated and evaluated for performance.

Secure Network Performance Evaluation

A network should be evaluated for various types of performance issues such as security holes, network traffic, response, and throughput. Simulation is a very powerful technique in evaluating computer networks. It is very important to use network modeling software that allows simulation of secure networks under various scenarios. A detailed discussion of network simulation and performance evaluation is beyond the scope of this chapter due to space and other limitations.

Figure 7. A hierarchical (tree) view of a security risk register

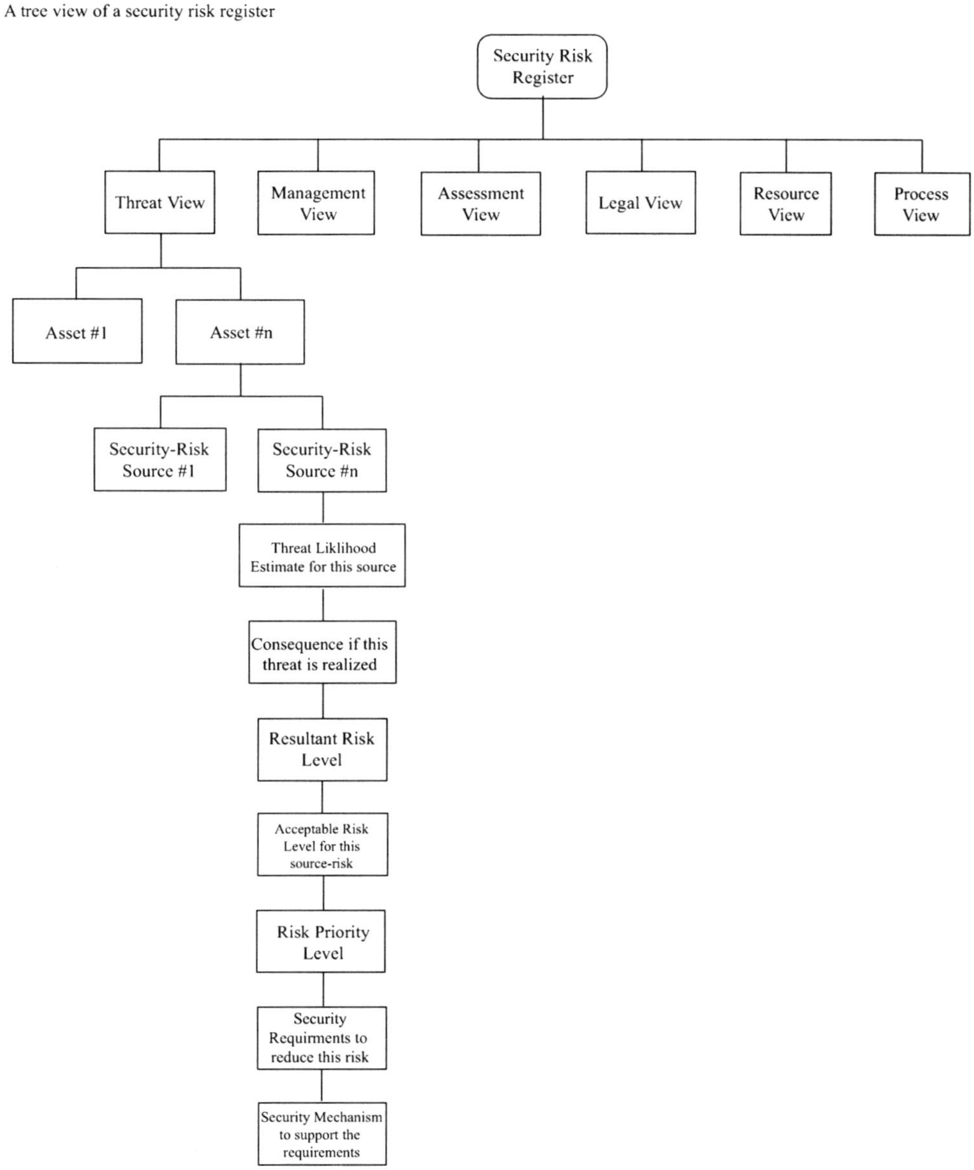

Network Implementation

Implementation will entail buying the required network equipment and other computing resources, and then deploying them onsite. A conversion plan to transition to the new network will have to be prepared. Users should be properly trained in the new network's security and its usage. A detailed discussion of network implementation is beyond the scope of this chapter.

A Comparison of Secure Network Analysis and Design Methods

In order to differentiate the SEACON method further from other methods we compare it with two existing methods, top-down network design (Oppenheimer, 2004) and secure network solutions (Whitmore, 2001). Note that we are including only those methods that address the security of networks.

While there are no standard criteria for comparing secure network design methods, the following criteria are designed to give a representative and objective view of the methods so that an organization may choose the most appropriate secure network design method for a given situation. We extend the criteria developed by Shaw and Yadav (2001) to address the security issues in design. The security-related criteria have been developed using the notion of functional and assurance requirements of the Target of Evaluation (TOE) (Common Criteria Implementation Board, 1999a, 1999b, 1999c) and network security mechanisms. Any secure network design method should have built-in steps to guide an analyst in determining an appropriate set of network security mechanisms to be included in the overall design of a network. A method should also use a network simulation tool for analyzing and evaluating different secure network design scenarios. An extended set of comparison criteria is described below:

- **Capture of multiview security requirements:** (To what extent does the method emphasize a complete and comprehensive security requirements?)—The method should provide guidelines to identify security requirements from multiple perspectives such as legal, privacy, management, assessment, and resource (Yadav, 2006).
- **Mapping of network security mechanisms to firm's security requirements:** (To what extent does the method provide steps to relate security mechanisms to security requirements?)—The method should encourage an analyst to determine the most appropriate set of network security mechanisms to support a given set of security requirements.
- **Interplay between business and security requirements:** (To what extent does the method provide explicit steps in using security and business requirements simultaneously?)—The method should provide built-in steps for using security and business requirements together in creating network architectures.
- **Usability:** (To what extent is the method usable?)—The method should be easy for an organization to apply (Shaw & Yadav, 2001).
- **Integration:** (To what extent is the network integrated with the IS architecture?)—The method should integrate network architecture with the information system(s) in an organization.
- **Documentation:** (What level of documentation does the method provide?)—The method should provide extensive documentation including user requirements, security requirements, and network architecture.
- **Complexity:** (How easy is the method to learn and to apply?)—Ideally, the method should be relatively easy to learn and to apply.
- **Allocation guidelines:** (To what extent does the method help allocate data, processes, and security mechanisms to nodes?)—The method should provide rules and guidelines for determining which processes or data and security mechanisms to assign to each node.
- **Principles:** (What principles does the method emphasize?)—The method should be based on sound principles that have been proven effective instead of relying on intuitive ideas that have a low likelihood of success (Shaw & Yadav, 2001).
- **Outcomes:** (What are the major end products of the methodology and are the products of a high caliber?)—The end products of the method should be relevant to organizational goals and business requirements and should be of high quality.
- **Simulation:** (To what extent does the method emphasize simulation?)—The method should use network simulation tools to evaluate alternative secure network designs.

Table 1 summarizes the results of applying the criteria to each of the existing methods and to the SEACON method. The table shows that each design method has its own strength. However, the SEACON method provides the advantage of a multiview perspective of security, built-in steps for seamless use of security and business requirements in network models, firm level integration of the network, a set of guidelines for allocating business processes and data across network nodes, and emphasis on simulation.

AN ILLUSTRATION OF THE SEACON METHOD

Consider a simple example to illustrate the applicability of the SEACON method. Although the example is not very complex, it does help illustrate the usefulness of the SEACON method. The example problem is described below. It is adapted from Shaw and Yadav (2001). The example problem has been modified to incorporate network security needs.

Table 1. A comparison of secure network design methods

Methods / Criteria	SEACON	Top-Down Network Design (21)	Designing Secure Solutions (28)
Multiview Security Requirements	Yes	No	No
Mapping between Network Security Mechanisms and Security Requirements	High	Medium	High
Interplay between security and business requirements	A seamless use of security and business requirements	No joint use of security and business requirements	No joint use of security and business requirements
Usability	High	High	Medium
Integration	High	Low	Low
Documentation	High	High	Low
Complexity	High	Medium	High
Allocation Guidelines	Good	None	None
Principles	Systems approach; Completeness; Consistency	Technical accuracy; Top-down development	Systems approach; Common criteria
Outcomes	Implementation	Implementation	Network model
Quality	High	High	Medium
Simulation	A secure network architecture that can be used for Simulation model and evaluation	No formal simulation	No formal simulation

Figure 8A. Context level DFD-SHIPIT

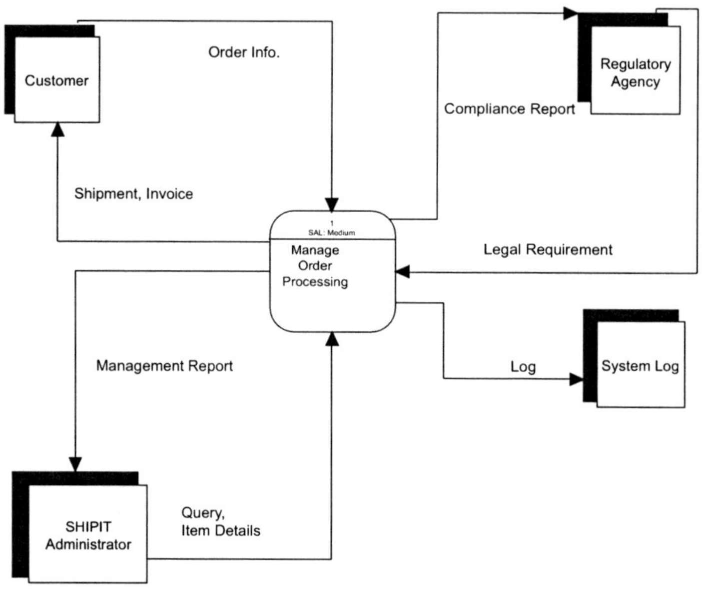

The Context Level DFD for SHIPIT: Manage Order Processing

Figure 8B. Level 0 DFD for SHIPIT

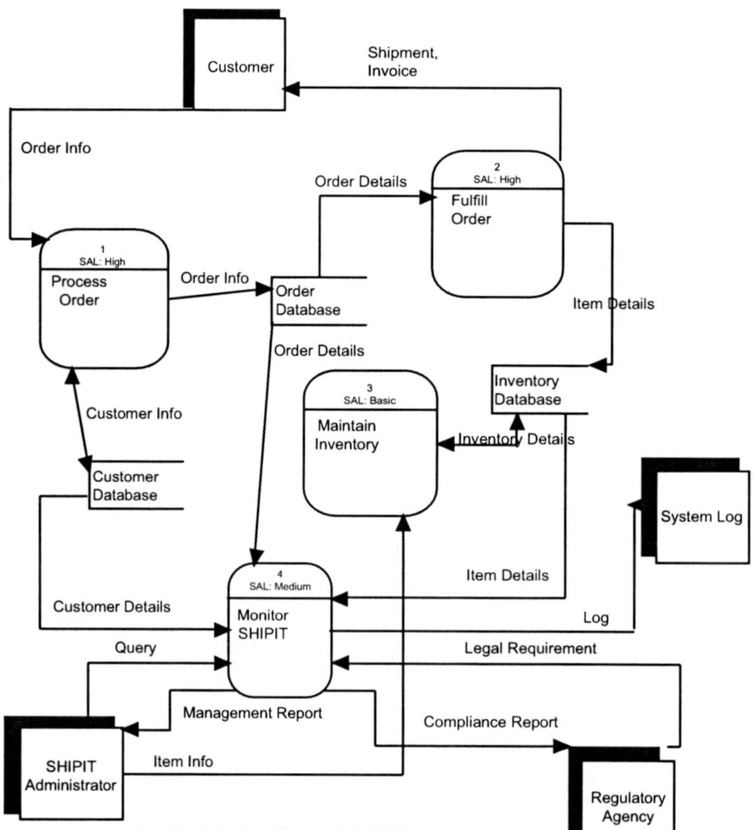

Level 0--Data Flow Diagram for SHIPIT

"SHIPIT is a fictitious firm that provides order processing services for mail order companies. The products are stored in a SHIPIT warehouse, and orders are shipped as they are received. The SHIPIT organization consists of three locations:

- The warehouse in Kansas City, Missouri,
- The office building in Dallas, Texas, and
- The call center in Albuquerque, New Mexico.

Currently, each facility has computers, but they are not integrated, and thus the only mechanism for sharing information is to print reports and physically send them to the other locations. Managers at SHIPIT believe that a computer network allowing them to share information securely over the Internet/intranet would be beneficial, and they decided to develop such a secure network using the SEACON method. The primary business driver for the company in its network design is a desire to reduce business operating costs and expedite access to various data and reports. A secure network may be a little bit more costly, but it will allow the company to run its operation with minimal or no security breaches and avoid costly security fixes down the road. As a general rule, SHIPIT wants to have every computer connected to the SHIPIT network properly administered and secured.

The DEACON method, the forerunner of SEACON method, was applied to the above example (Shaw & Yadav, 2001). In this chapter, we emphasize the discussion of SEACON's security-related steps. Under the SEACON method, the secure network design steps would be similar to the following:

- **Business Problem Definition:** Based upon the brief security description, we infer that SHIPIT should develop a network that provides error-free, reliable, and secure storage; sharing; and transmission of data among the

Figure 9. Entity relationship diagram for SHIPIT

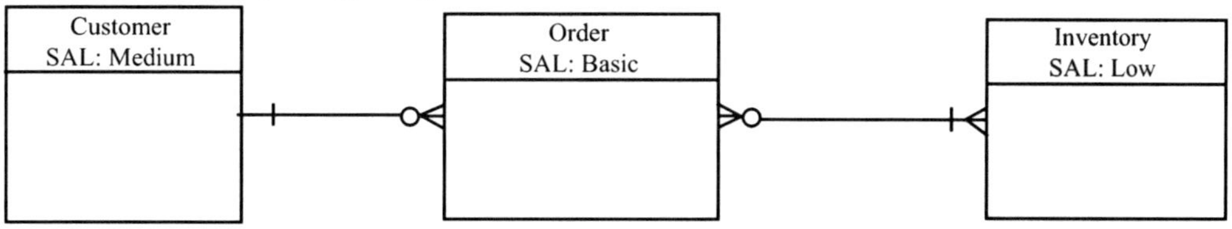

Table 2. Process-location-security matrix

	Office	Warehouse	call center	security adequacy level
Process Order			X	High
Fulfill Order		X		High
Maintain Inventory		X	X	Basic
Monitor SHIPIT	X			Medium
Print Invoice	X			Medium

Table 3. Data-location-security matrix

	Office	Warehouse	call center	security adequacy level
Customer	RUD	R	CRU	Medium
Order	R	RU	CRUD	Basic
Inventory	RU	CRUD	R	Low

C=Create; R=Read; U=Update; D=Delete

facilities at SHIPIT. Appropriate security mechanisms should be built at various levels, those of application, operating system, server, and network levels to safeguard the storage, access, and flow of information on the network. More specifically:

- Network hardware and software assets such as Web servers, database servers, routers, switches, databases, and so forth, should be protected.
- Access to data stored at various locations such as the warehouse, the office building, and the call center should be allowed to only authorized personnel.
- Users should be identified and grouped in various categories and their security responsibilities should be delineated.
- User account administration, user password policy, and privilege review policy should be specified.
- Responsibility for network administration and security should be assigned to a trained and technically competent staff.
- Accounts should be promptly deleted if remained unused for 3 months.
- Accounts belonging to terminated employees should be disposed off immediately.
- Personal equipment should not be connected to the SHIPIT network.

- **Requirement Specification:** We document SHIPIT's security and business needs using process, data, and network models as discussed. We document the final set of security requirements and mechanisms in the form of a security risk register created after the secure location model was developed. The process and data models are shown in Figures 8 and 9, respectively. The security risk register in Table 4 shows security requirements under each security view. Various security views have enabled the SHIPIT firm think about security requirements not only for protecting resources and dealing with various threats but also for legal requirements and continuous security assessment.

Secure Location Model

Figure 10 shows an initial overall location model for SHIPIT. It shows that the three locations, office, warehouse, and call center, are connected to one another. Each location shows, in parenthesis, the number of people who may use SHIPIT network. In some situations, it may be worthwhile to create an initial location model for each location. Next, we need to determine the information domains.

Figure 10. An overall SHIPIT location model (top level)

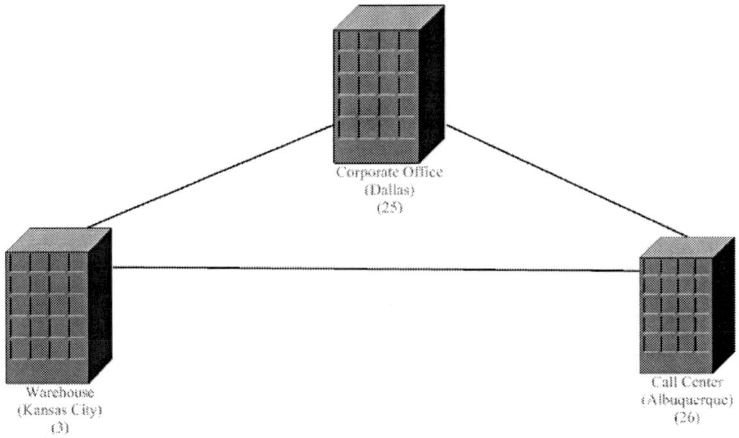

An Initial Overall SHIPIT Location Model (Top Level).

Table 4. Risk register showing risks, requirements, and mechanisms for SHIPIT-under various views

Views	Assets	Sources of Risk	Threat Likelihood Estimate	Consequence, if the threat is realized	Resultant Risk Level	Acceptable Risk Level	Risk Priority Level	Security Requirements	Security Mechanisms
	Physical location (Dallas, Albuquerque, and Kansas City)	Fire	Low	Critical	High	Low	2	1. Protect against fire 2. Backup Data 3. Store backup-data remotely	1. Backup 2. Equipement Security
		Earthquake	Very Low	Damaging	Low	Low	0		
Threat	Inventory Database Server, Customer & Order Database Server	Wrong Configuration	Low	Damaging	Medium	Nil	2	1. Monitor resource integrity 2. Check Resource integrity 3. Adjust resource integrity	1. Audit trails and logs 2. Resource testing 3. Resource maintenance 4. Data back-up 5. Resource monitoring 6. Operational procedures
		Hacker	Medium	Damaging	High	Nil	3	1. Protect against intrusion 2. Protect against denial of service 3. Protect against malicious code	1. Firewall 2. Antivirus software 3. VPN
	General Purpose Server	Wrong Configuration	Medium	Damaging	High	Nil	3	1. Monitor resource integrity 2. Check Resource integrity 3. Adjust resource integrity	1. Audit trails and logs 2. Resource testing 3. Resource maintenance 4. Resource monitoring 5. Operational procedures
		Hacker	High	Damaging	High	Nil	3	1. Protect against intrusion 2. Protect against denial of service 3. Protect against malicious code	1. Firewall 2. Antivirus software 3. VPN
	Web Server	Wrong Configuration	Medium	Damaging	High	Nil	3	1. Monitor resource integrity 2. Check Resource integrity 3. Adjust resource integrity	1. Audit trails and logs 2. Resource testing 3. Resource maintenance 4. Resource monitoring 5. Operational procedures
		Hacker	High	Damaging	High	Nil	3	1. Protect against intrusion 2. Protect against denial of service 3. Protect against malicious code	1. Firewall 2. Antivirus software 3. VPN
	Mail Server	Wrong Configuration	Medium	Damaging	High	Nil	3	1. Monitor resource integrity 2. Check Resource integrity 3. Adjust resource integrity	1. Audit trails and logs 2. Resource testing 3. Resource maintenance 4. Resource monitoring 5. Operational procedures
		Hacker	High	Damaging	High	Nil	3	1. Protect against intrusion 2. Protect against denial of service 3. Protect against malicious code	1. Firewall 2. Antivirus software 3. VPN
	Operation	Technical Failure	Medium	Minor	Low	Low	0		
		Power failure	Low	Minor	Low	Low	0		
		Operational Staff Error	High	Significant	High	Nil	3	1. Verify and confirm input data and commands 2. Save data periodically	1. Staff training 2. Error proof interface
	SIPIT Internet Services	Critical Network Device Failure	Low	Significant	Medium	Low	1	1. Check resource availability 2. Monitor resource availability 3. Maintain redundancy in system	1. Power supplies security 2. Resource testing 3. Resource maintenance 4. Resource monitoring 5. Equipment redundancy
		Denial of Service Attack from the Internet	Very High	Significant	High	Low	2	1. Protect against Denial of Service 2. Protect against intrusion 3. Protect against malicious code	1. Management of outside users' access 2. Resource monitoriing 3. Intrusion detection 4. Audit trails and logs 5. Internet Connection policy
Resource	SHIPIT Data (Customer, Order, Inventory)	Hackers from the Internet	Low	Serious	High	Nil	3	1. Monitor resource integrity 2. Check Resource integrity 3. Adjust resource integrity 4. Monitor resource confidentiality 5. Check resource confidentiality 6. Adjust resource confidentiality	1. Formal standards for establishing user access 2. User ID management 3. Periodic evaluation of access rights 4. Resource monitoring 5. Audit trails and logs 6. Intrusion detection 7. Security scanning Tools
		Insider Attack	Low	Significant	Medium	Nil	2	1. Monitor resource integrity 2. Check Resource integrity 3. Adjust resource integrity 4. Monitor resource confidentiality 5. Check resource confidentiality 6. Adjust resource confidentiality	1. Formal standards for establishing user access 2. User ID management 3. Periodic evaluation of access rights 4. Resource monitoring 5. Audit trails and logs 6. Intrusion detection 7. Security Awareness and Training 8. User ID management
	Technical expertise	Loss of Key Personnel	Medium	Significant	Medium	Low	1	1. Protect from loss of key personnel	1. Succession Planning
Legal	Compliance to the Privacy Act of 1974	Violation of National Privacy Principles	Medium	Damaging	High	Nil	3	1. Establish rules for handling personal data 2. Identify Personal Identifiable Information (PII) 3. Establish proper access control for PII	1. Access policy 2. Transparency and control of access privileges 3. Audit history reporting 4. Centralized authentication and authorization 5. Audit log
	Compliance to the CAN-SPAM Act 2003	Violation of Spam Code of Practice	High	Minor	Medium	Low	1	1. Analyze legal risks for on-line environment 2. Comply with the Spam Act of 2003	1. Spam filtering 2. Spam reporting 3. Spam reduction techniques

continued on following page

Tables 2 and 3 show process-location-security and data-location-security matrices, respectively. These tables reveal that the call center creates customer and order data and the warehouse creates inventory data. The office uses the customer, order, and inventory data. However, the call center does more processing with the customers and orders than any of the other locations. Similarly, the warehouse does more processing with the inventory data. This leads us to propose that customer and order data should reside at the call center and the inventory data should reside at the warehouse. Even though the SHIPIT study case is not very explicit about information on Web sites and e-mails, we assume that it maintains Web site and e-mail services. We propose three

Table 4. continued

Views	Assets	Sources of Risk	Threat Likelihood Estimate	Consequence, if the threat is realized	Resultant Risk Level	Acceptable Risk Level	Risk Priority Level	Security Requirements
Management	Security Policy	Inadequate Policy	Low	Serious	High	Nil	3	1. Identify faulty policy 2. Take corrective action 3. Establish security policy and procedues 4. Review security policy and procedures 5. Monitor security
	Accountability Guidelines	Vague Accountability for Individuals	Medium	Damaging	High	Nil	3	1. Review accountability policy 2. Refine accountability-assignment
		Lack of Accountability Standards	Low	Significant	Medium	Nil	2	1. Review accountability policy 2. Review accountability standards 3. Establish accountability standards
Process	Processes (Process Order, Fulfill Order, Maintain Inventory, Monitor SHIPIT)	Poorly defined Process Steps	Low	Damaging	Medium	Nil	2	1. Evaluate process 2. Rectify weak points/steps of the process 3. Secure each steps of the process 4. Train users in the secure process
	Process Control	Lack of Staff Training	Medium	Damaging	High	Nil	3	1. Train users in the secure process 2. Review secure process
		Lack of Monitoring	High	Minor	Medium	Low	1	1. Review process 2. Establish process ownership 3. Separate duties of actors involved in the process 4. Train users in the secure process 5. Revew process policy
	Process Interface	Lack of Interface Design Standards	Low	Significant	Medium	Low	1	1. Review process policy 2. Review process design standards 3. Evaluate process
Assessment	Assessment Criteria	Inadequate Criteria	Medium	Significant	Medium	Low	1	1. Evaluate assessment criteria 2. Define assessment criteria 3. Define measures for each criterion 4. Collect data on measures 5. Evaluate the measures' effectiveness
		Vague Criteria	Medium	Significant	Medium	Low	1	1. Evaluate assessment criteria 2. Define assessment criteria 3. Define measures for each criterion 4. Collect data on measures 5. Evaluate the measures' effectiveness
	Assessment Method	No Assessment Method	Low	Significant	Medium	Nil	2	1. Define assessment method 2. Train users in assessment
		Inadequate Assessment Training	Low	Significant	Medium	Low	1	1. Train users in assessment 2. Evaluate assessment policy
	Assessment Standards	Inadequate Standards	Low	Significant	Medium	Low	1	1. Evaluate assessment policy 2. Define assessment standards 3. Train users in assessment

information domains, Call Center, Warehouse, and Corporate Office, to segregate and group each set of geographically separated information resources and assets. Tables 2 and 3 also show the security adequacy levels for processes and data. This information makes it easier to determine the security adequacy levels for the location connectivity diagrams. The process-location-security matrix reveals that there are two processes concentrated in the call center. One of the processes, Process Order, requires online interaction with custom-

Figure 11A. SHIPIT secure extended location connectivity diagram (conceptual)

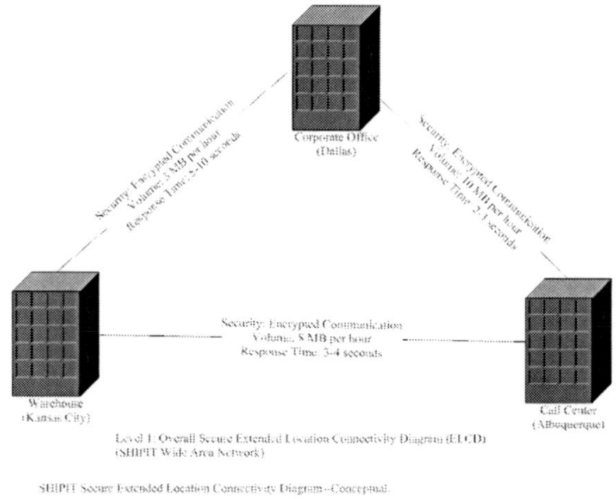

Figure 11B. SHIPIT secure extended location connectivity diagram (conceptual)—corporate office

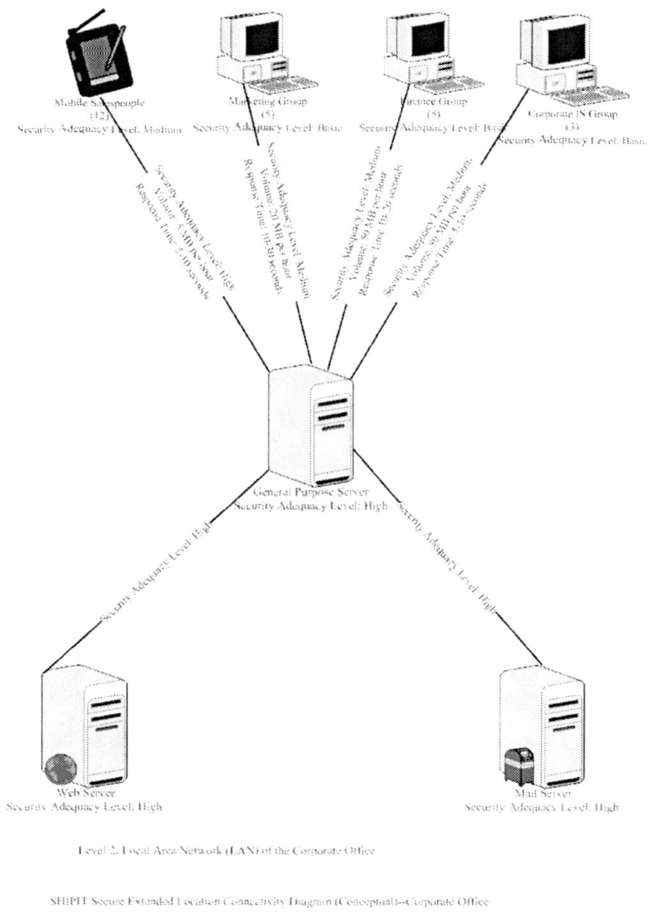

Figure 11C. SHIPIT secure extended location connectivity diagram (conceptual)—warehouse

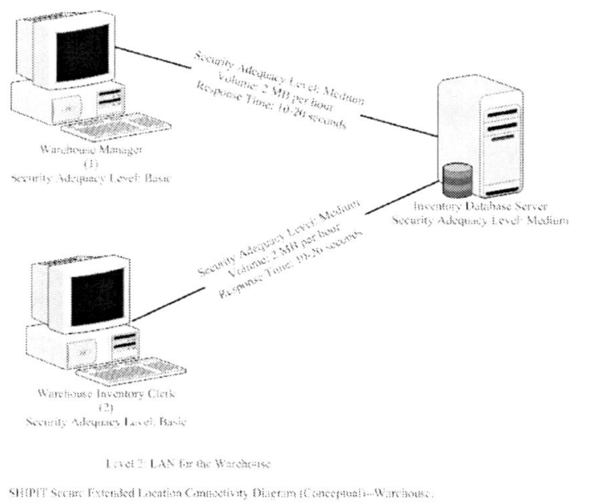

Figure 11D. SHIPIT secure extended location connectivity diagram (conceptual)—call center

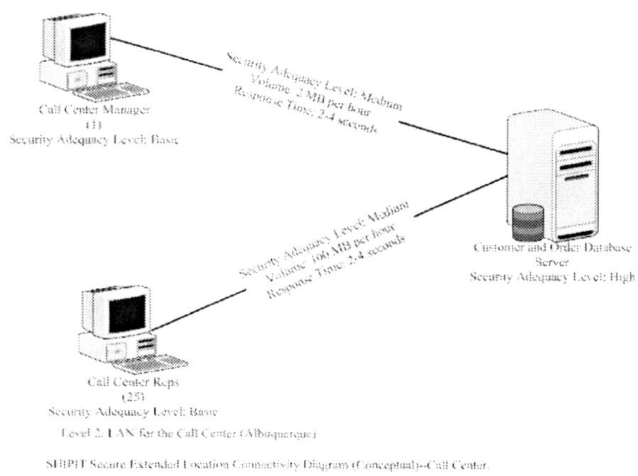

ers. The order processing will require a faster throughput. This suggests that there is a need to have a faster and larger capacity communication requirement for the link between the call center and the rest of the network.

A secure extended location connectivity diagram for the SHIPIT case is shown in Figures 11A, 11B, 11C, and 11D. There are two levels of the secure extended location connectivity diagrams. The first level diagram in Figure 8A shows the overall connection among the three locations, office, warehouse, and call center. Each connection is labeled with security, volume and response time requirements. The second level diagrams in Figures 11B, 11C, and 11D show the network within each location. Each PC and other nodes are labeled with security adequacy level. Connections between servers and client PCs are labeled with security, volume, and response time requirements. Connections between servers are considered local and hence are considered quite fast.

- SHIPIT Secure Network Architecture: There are three enclaves, one under each information domain. Because each information domain has only one enclave, these enclaves are called by the same name as that of their information domain names. We propose to use a specialized router with a firewall as a boundary controller to protect each enclave. The appropriate security mechanisms from the security risk register should be incorporated into the secure network architecture diagram (SNAD). Figure 12 shows a logical SNAD for SHIPIT. The logical SNAD has been drawn using the SmartDraw software. The initial nodes in the diagram were derived from the secure extended location connectivity diagrams as shown in Figures 11A-11D. Figure 12 shows a site-to-site virtual private network (VPN) design to provide a secure network environment for SHIPIT branches. A VPN design is more cost effective than designing a private network using privately leased lines. The concept of enclaves leads us to create a subnet for each enclave. SHIPIT's logical SNAD can be modeled as a hierarchical secure network model consisting of several subnets. The design of hierarchical networks and subnets are better handled and managed by network design and simulation software such as OPNET (OPNET, 2005). Due to space limitation, we do not address the network modeling of SHIPIT using network simulation software.
- Secure Network Performance Evaluation: The logical secure network architecture diagram for SHIPIT can be modeled and simulated using network design and simulation software such as OPNET (2005). OPNET provides tools and techniques to model secure computer networks. Please see OPNET (2005) for more details.

We do not discuss network performance and evaluation here due to space limitations. Discussion of network simulation and performance evaluation is quite involving and will be a subject of another research article.

CONCLUSION AND LIMITATIONS

We have proposed and discussed a new approach to designing secure computer networks for firms. The approach not only emphasizes the importance of using organizational goals and requirements in designing a secure network but also provides built-in mechanisms to capture security needs and use them seamlessly throughout the steps of analyzing and designing secure network architecture. We have proposed and used extended versions of DFD and ERD to not only capture business process and data, but also their security requirements in the same diagrams. Firms can use the SEACON method to design and implement secure computer

Figure 12. A logical secure network architecture diagram for SHIPIT

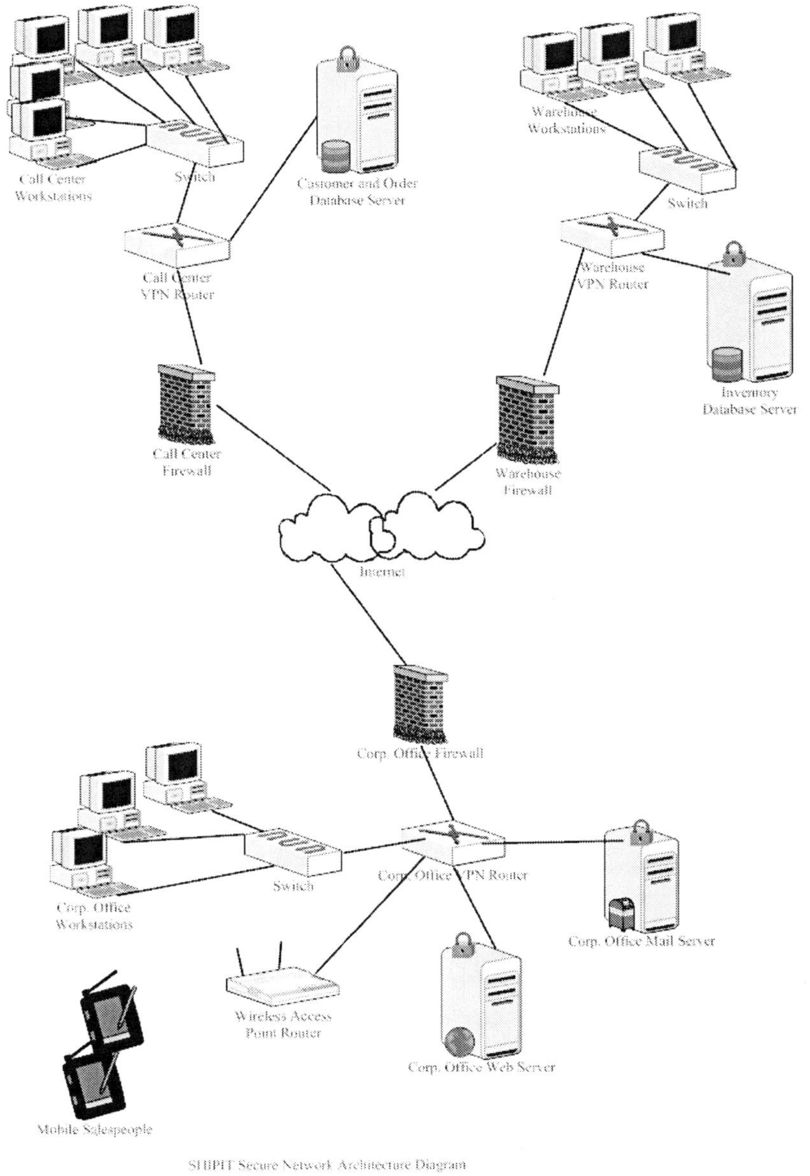

networks that are integrated with the business requirements of that firm. An integrated approach with built-in steps for incorporating security measures right from analysis is superior to pure technical methods because it facilitates seamless support for using business processes, security needs, and the overall IS architecture for a firm.

One of the limitations of the SEACON method is the lack of guidelines for transforming a secure network architecture into a network simulation model that can be easily tested and evaluated using network simulation software. Such guidelines will obviously have to take into account the idiosyncrasies of the target simulation software.

REFERENCES

Arber T., Cooley, D., Hirsch, S., Mahan, M., & Osterritter, J. (1999). *Network security framework: Robustness strategy*. Retrieved November 30, 2007, from http://csrc.nist.gov/nissc/1999/Proceeding/papers/p30.pdf

ASIS. (2003). General security risk assessment guideline. An ASIS International Guideline.

Bace, R., & Mell, P. (2001). *Intrusion detection systems*. NIST Special Publication on Intrusion Detection System. National Institute of Standards and Technology.

Bionic Buffalo Corporation. (2000). *Concept for a secure network computer*. Retrieved November 30, 2007, from http://www.tatanka.com/doc/technote/tn0110.htm

Centers for Medicare & Medicaid Services. (CMS). (2002). *CMS information security risk assessment (RA) methodology* (version #1.1). Baltimore, MD: Department of Health & Human Services.

Cisco Systems. (2001). *SAFE: A security blueprint for enterprise networks*. A white paper. Retrieved November 30, 2007, from http://www.cisco.com/warp/public/cc/so/cuso/epso/sqfr/safe_wp.pdf

Common Criteria Implementation Board. (1999a). *Common criteria for information technology security evaluation, part 1: Introduction and general model* (version 2.1). Retrieved November 30, 2007, from http://csrc.nist.gov/cc

Common Criteria Implementation Board. (1999b). *Common criteria for information technology security evaluation, part 2: Security functional requirements* (version 2.1). Retrieved November 30, 2007, from http://csrc.nist.gov/cc

Common Criteria Implementation Board. (1999c). *Common criteria for information technology security evaluation, part 3: Security assurance requirements* (version 2.1). Retrieved November 30, 2007, from http://csrc.nist.gov/cc/

Defense Logistics Agency. (2002). *Enclave boundary defense*. DLIA 8500.12. Retrieved November 30, 2007, from http://www.dlaps.hq.dla.mil/dlai/i8500.12.htm

Department of Defense Standard. (1985). *Department of defense trusted computer system evaluation criteria*. DoD 5200.28-STD.

Duchessi, P., & Chengalur-Smith, I. (1998). Client/server benefits, problems, best practices. *Communications of the ACM, 41*(5), 87-94.

Fisch, E. A., & White, G. B. (2001). *Secure computers and networks: Analysis, design, and implementation*. CRC Press. Retrieved November 30, 2007, from www.crcpress.com

GAO. (1999, November). *Information security risk assessment—Practices of leading vs. accounting and information management division*. United States General Accounting Office, GAO/AIMD-00-33.

Ghosh, S. (2001). *Principles of secure network systems design*. Springer-Verlag.

Gordon, L. A., Loeb, M. P., Lucyshyn, W., & Richardson, R. (2005). *Tenth annual CSI/FBI computer crime and security survey*. Retrieved November 30, 2007, from www.GoCSI.com

Innella, P. (2001). *Designing secure networks based on the software process model*. A white paper. Retrieved November 30, 2007, from http://www.securityfocus.com/infocus/1191

Irvine, C., & Levin, T. (1999). *A note on mapping user-oriented security policies to complex mechanisms and services* (Tech. Rep.). Retrieved November 30, 2007, from http://cisr.nps.navy.mil/downloads/nps_cs_99_008.pdf

Kizza, J. M. (2005). *Computer network security*. New York: Springer-Verlag.

McMillan, R. (2005). *Computer attacks down, survey says*. Retrieved November 30, 2007, from http://www.networkworld.com/news/2005/072505-security.html

NSW. (2003). *Information security guideline for New South Wales (NSW) government, part 2: Examples of threats and vulnerabilities*. The Office of Information and Communications Technology, Department of Commerce, New South Wales. Retrieved November 30, 2007, from http://www.oict.nsw.gov.au/content/2.3.17-Security-Pt2.asp

NSW. (2003). *Information security guideline for New South Wales (NSW) government, part 3: Information security baseline controls*. The Office of Information and Communications Technology, Department of Commerce, New South Wales. Retrieved November 30, 2007, from http://www.oict.nsw.gov.au/content/2.3.17-Security-Pt2.asp

OPNET Documentation. (2005). Retrieved November 30, 2007, from www.OPNET.com

Oppenheimer, P. (2004). *Top-down network design* (2nd ed.). Indianapolis, IN: Cisco Press.

Rusli, R. (2001). *Secure system architecture and design*. A white paper. SANS GIAC Security Essentials—Practical Assignment, GSEC Web site. Retrieved November 30, 2007, from http://www.giac.org/certified_professionals/practicals/gsec/1422.php

Shaw, N., & Yadav, S. B. (2001). DEACON: An integrated approach to the analysis and design of enterprise architecture-based computer networks. *Communications of the Association for Information Systems, 7*.

Southwick, P. (2003). *Secure network design*. A white paper. Retrieved November 30, 2007, from the Hill Associates Web site: http://www.hill.com/archive/pub/papers/2003/10/paper.pdf#search='secure%20network%20design'

Survey says: Government regulations help secure networks. (2005). Sarbanes-Oxley Compliance Journal. Retrieved November 30, 2007, from http://www.s-ox.com/news/news.cfm?articleID=338

Vaidyanathan, G., & Devaraj, S. (2003). A five framework for analyzing online risks in e-businesses. *Communications of the ACM, 46*(12), 354-361.

Verdon, D., & McGraw, G. (2004). Risk analysis in software design. *IEEE Security & Privacy, 2*(4), 79-84.

Warren, P. (2005). *Ten steps to secure networking*. TechWorld. Retrieved November 30, 2007, from http://www.techworld.com/security/features/index.cfm?FeatureID=1862

Whitman, M. E. (2003). Enemy at the gate: Threats to information security. *Communications of the ACM, 46*(8), 91-95.

Whitman, M. E., & Mattord, H. J. (2005). *Principles of information security* (2nd ed.). Canada: Thomson Course Technology.

Whitmore, J.J. (2001). A method for designing secure solutions. *IBM Systems Journal, 40*(3), 747-768.

Yadav, S. B. (2006). *A six view perspective of system security—issues, risks, requirements, and mechanisms* (Tech. Rep. #ISQSYadav2006-1). Lubbock, TX: Rawls College of Business, Texas Tech University.

This work was previously published in the International Journal of Information Security and Privacy, Volume 2, Issue 1, edited by H. Nemati, pp. 1-25, copyright 2008 by IGI Publishing (an imprint of IGI Global).

Chapter 21
Spam Classification Based on E-Mail Path Analysis

Srikanth Palla
University of North Texas, USA

Ram Dantu
University of North Texas, USA

João W. Cangussu
University of Texas at Dallas, USA

ABSTRACT

Email spam is the most effective form of online advertising. Unlike telephone marketing, email spamming does not require huge human or financial resources investment. Most existing spam filtering techniques concentrate on the emails' content. However, most spammers obfuscate their emails' content to circumvent content-based spam filters. An integrated solution for restricting spam emails is needed as content analyses alone might not provide a solution for filtering unsolicited emails. Here we present a new method for isolating unsolicited emails. Though spammers obfuscate their emails' content, they do not have access to all the fields in the email header. Our classification method is based on the path an email traverses instead of content. Overall, our classifier produced fewer false positives when compared to current filters such as SpamAssassin. We achieved a precision of 98.65% which compares well with the precisions achieved by SPF, DNSRBL blacklists.

INTRODUCTION

E-mail spam is the electronic equivalent of "postage-due" junk mail. Spamming involves sending nearly identical unsolicited messages to thousands or millions of recipients. Out of all the problems pestering the present day e-mail infrastructure, spam has evolved into a nettlesome problem. Unlike its conventional cousins (postal mail and telemarketing) e-mail spam imposes its transactional costs directly on the recipient. Spam can be sent anonymously and turns out to be a highly profitable marketing technique despite some costs incurred in procuring e-mail addresses and dispatching the spam e-mails.

Spam causes more economic damage than hackers or any malware. A London-based computer security firm mi2G (Lemke, 2003) reported that computer outages and lost productivity because of spam led to $10.4 billion in worldwide economic losses whereas viruses and worms cause $8.4 billion in losses, while hackers contribute to $1 billion in financial damage worldwide. Though a wide variety of spam filters and other techniques are currently in use they have not been successful in restricting spam and fraudulent e-mails completely from reaching the recipient.

In recent years there has been a great influx of various spam filtering approaches such as rule-based spam filtering techniques (William, 1996), statistical filtering approaches (Massey, Mick, Raya, & Scott, 2003), machine learning, Bayesian classifiers (Graham, 2003), combinations of rule based and Bayesian approaches (Spa, 2006), and spam filtering using contextual network graphs (Kelleher, 2004). In spite of all these techniques, spam remains an annoying problem because a majority of the spam filtering techniques focus on the content of an e-mail, which is in complete control of the spammers. However, most of the spammers obfuscate their e-mail content in order to circumvent these spam filters. Lowed and Meek (2005) report that a spammer can get through 50% of the blocked spam past the filters by adding 150 or fewer words. If a spammer is able to send test messages to the targeted spam filter, 30 words is sufficient in getting half of the blocked spam through the filters.

Here a new classification method based on the path traversed by an e-mail is proposed. Spammers can forge certain fields in the e-mail header but cannot modify the complete path traversed by an e-mail. Using this information the proposed approach has been able to accurately identify e-mails as legitimate or spam. To achieve this level of accuracy, reputation and trustworthiness of the relays and path traversed by an e-mail are also taken into account. This approach improves upon content analysis as spammers frequently obfuscate their e-mails content. In addition, a social network analysis is also conducted. In this analysis the classifier is fine tuned according to the recipient's social network resulting in a considerable reduction of false positives and false negatives generated by the classifier.

The remainder of this chapter is organized as follow. An overview of several spam filtering techniques including whitelists, blacklists, greylists, machine learning, e-mail authentication, and social networks is presented in Section II. The proposed methodology for the spam identification is delineated in Section III. Section IV presents the validation of the approach using two e-mail corpuses. Also presented in Section IV is a qualitative as well as a quantitative comparison of the proposed approach with relevant existing work. Finally, conclusions as well as potential improvements of the proposed classifier are presented in Section V.

RELATED WORK

A wide variety of filtering techniques and other methods are currently used to restrict spam and phishing e-mails from reaching the user's inbox. Client- and server-side spam filtering using blacklists, whitelists (Hird, 2002) and greylists (Levine, 2005), text classification methods based on predefined rules, Bayesian filtering, pattern recognition, and identity recognition are a few filtering mechanisms currently being used. Also available are e-mail authentication standards such as identified Internet mail (IIM), domain keys identified Internet mail (DKIM), sender policy frame work (SPF), sender ID frame work, and so forth, which identify genuine senders from spammers and fraudsters. In addition to these approaches, new methods based on senders' reputation and accreditation services are emerging rapidly. Next, various spam filtering techniques are briefly discussed. To the best of our knowledge no existing technique applies a multivariate analysis comprising the four analyses described later in Section III.

Spam Filtering using Whitelists, Blacklists and Greylists

Normally client and server side spam filtering is achieved through maintaining a list of trusted and untrusted senders. The list of trusted senders is termed whitelist (Hird, 2002) and the list of untrusted senders is known as blacklist (Hird, 2002). E-mails from the whitelisted senders are not subjected to any filtering mechanisms and delivered to the recipient. The recipient might have sent or replied to the e-mails of the whitelisted senders in the past. The address book of the recipient is normally used as a whitelist. E-mails from the blacklisted senders are always blocked from reaching the recipient.

Most of the newer version mail transmission agents such as procmail, postfix, and qmail support a variation of blacklists know as "domain name system blackhole list" (DNSBL). DNSBL (DNS, 2007) is a real time database of Internet protocol (IP) addresses of the spam sources and relays relaying the spam e-mails. The DNSBL is built on the top of domain name system (DNS) (Allocchio, 1998). When a sender or a relay initiates a connection with the mail server implementing DNSBL filtering, a query on the client's IP addresses is sent to the DNSBL server. If the client's IP address is listed in the DNSBL, the client's connection is dropped. Instead of rejecting the connection, some systems flag the e-mails as spam from that client.

Vixie (MAP, 2005) is credited with the creation of first DNS blacklist known as real time black hole list (RBL). He used RBL for filtering spam in mail abuse prevention system. Levine's (2005) greylisting idea is to temporarily reject the e-mails from unknown sources on the theory that real mail transmission agents will retry sending e-mails where as spamware will not. The inherent problem with greylisting (Levine, 2005) is the delay in the delivery of unknown legitimate e-mails from reaching the recipient. Moreover if legitimate mail transmission agents are not properly configured, the soft failure generated (this is how an e-mail is temporarily rejected) may result in permanent bounce of legitimate e-mails from unknown sources.

Machine Learning

Graham (2003) recommends spam filtering with Bayesian analysis. The content of an e-mail is divided into tokens. Each token is assigned a score based on the number of times it appeared in spam e-mails and legitimate e-mails. A database of such tokens is maintained. For a new incoming e-mail, a list of most effective tokens is computed and the total score is assigned for that e-mail (Spa [2006] uses this approach). Upon subjecting this score to a threshold, a decision is made classifying it as either spam or legitimate.

Soonthornphisaj, Chaikulseriwat, and Piyanan (2006) suggest a technique which works by constructing a centroid vector of an e-mail. Classification is made based on the similarity of the constructed centroid vector with the centroid vector of spam e-mails and the legitimate e-mails. Rigoutsos and Tien's (2004) filtering technique uses a pattern discovery approach for filtering spam e-mails. Their filter is initially trained with a large spam e-mail corpus. Once it is completely trained, it stores certain patterns which are strong indicators of spam. For all incoming e-mail it tries to match these spam patterns; the more patterns it matches, the higher the chances the message being spam.

Authorization and Access Control Mechanisms

E-mail filtering techniques based on access control mechanisms such as authorization and authentication are used to verify the identity of the e-mail's source. To prevent spoofing of the senders e-mail addresses, technology companies and researchers are currently devising protocols for e-mail authentication. The existing proposals for e-mail authentication protocols can be generalized into two categories: (1) IP based authentication such as SPF and sender ID frame work (SIDF) from Microsoft corporation and (2) cryptographic based authentication methods like IIM from CISCO systems and domain keys (DK) from Yahoo! These techniques are described next.

1. **Sender policy frame work (SPF):** Sender policy frame work was originally designed by Wong (Wong & Schlitt, 2005). SPF, formerly known as "sender permitted from," can be considered as an extension to SMTP (Klensin, 2004; Postel, 1982) standard. Here, the envelope sender address in the "MAIL FROM:" SMTP command is verified for forgery. SPF utilizes mail exchanger (MX) (Allocchio, 1998; Klensin, 2004) records for filtering spam. Mail exchange record is a type of resource record present in the DNS, specifying the route information of an e-mail. SPF publishes the reverse MX records identifying the mail servers authorized to send e-mails from that domain. The recipient upon receiving an e-mail can verify these records to make sure that the e-mails were sent by trusted senders having SPF records.
2. **Sender ID frame work (SIDF):** Sender ID framework (Wong & Schlitt, 2006) is based on the caller ID proposal (Atkinson, 2004) and SPF (Wong & Schlitt, 2005). This protocol is proposed to tackle the problem of spoofing the e-mail header and phishing e-mails. When the recipient's in-bound mail server receives new e-mails, it invokes the sender ID frame work, which checks the addresses of the e-mail sources against a registered list of servers that domain has authorized to send the e-mails, that is, it checks the SPF records. Once this verification has been done, the result can be used in conjunction with past behaviors, traffic patterns, sender's reputation information, and so forth. It can also be used with conventional content-based spam filters in making the final judgment, whether to deliver the e-mail to the recipient. Despite verifying the authorization and authentication information, SPF and SIDF alone are not effective in restricting spam and phishing e-mails.
3. **Identified internet mail (IIM):** Fenton and Thomas (2005) propose IIM for e-mail authentication. IIM verifies the integrity of an e-mail using digital signatures and public key cryptography (Elkins, Torto, Levien, & Roessler, 2001). It also validates the senders' e-mail addresses by verifying the authenticity of the associated public key. The digital signature is a hash of the e-mail content signed with the private key. Sender dispatches the e-mail along with its digital signature. When the e-mail reaches the recipient the signature is verified with the corresponding public key. The recipient's machine computes the hash of the message separately using a preagreed hashing algorithm. An exact match between these two hashes indicates that the message has not been modified during the transit.
4. **Domain keys (DK):** Proposed by Delany (2006), it uses public keys advertised in the DNS for e-mail authentication. The objective of domain keys is to verify the integrity of the received e-mail and authenticity of the e-mail source. Domain Keys achieves this by using digital signatures, public key cryptography (Elkins et al., 2001) and DNS. A domain key is completely transparent to the existing e-mail systems with no DK support and is implemented using the optional headers (Resnick, 2001) and DNS records. IIM is integrated with DK to form DKIM standard (Allman, Callas, Delany, Libbey, Fenton, & Thomas,).

Filtering Spam using Social Networks

Recently Boykin and Roychowdhury (2005) proposed a graph theoretic approach for filtering spam e-mails. They constructed an e-mail network solely from the sender and recipient information provided in the e-mail header and used it for identifying the e-mails from the circle of friends. They developed an automatic tool which processes the personal e-mail network and identifies the networks of trusted senders and subnets of spammers. Their tool classified 53% of all the e-mails it verified as spam or nonspam with 100% accuracy with no false negatives and no supervised training.

Detecting Spam at Router Level

Banit, Nitin, and Mart (2005) propose a method for restricting spam at the router level. First, they separate SMTP traffic from other types of traffic at the router. Next, they apply a two step filtering process which identifies bulk e-mail traffic by comparison of the recently received e-mails stored in a repository and then uses a Bayesian classifier to determine whether the bulk e-mail stream is spam. If the bulk e-mail stream is identified as spam then its rate is reduced. Their approach achieved 97% accuracy with negligible amount of false positives when tested on various spam e-mails collected from different sources.

Trust and Reputation Systems

Conventional whitelist filtering has certain advantages and disadvantages. One advantage is that the e-mails routed to the inbox are 100% legitimate e-mails. Where as, the mail in the low priority folders may contain some legitimate e-mails. Moreover there is always the burden of maintaining the whitelists. Golbeck et al. (Jennifer & James, 2004) has developed a new technique by taking the basics from whitelist filtering and social networks. Their technique builds a large reputation network of users, in which users assign a "Reputation" or "Trust" score to the people they know. These ratings affect the score of their e-mails and connect the other users who in turn have their own set of scores. By employing a recursive algorithm the reputation of the senders is inferred and scores are assigned to their e-mails. If there is no direct contact between any two users, their reputation will be inferred through the mutual acquaintances. The assigned scores are displayed next to the e-mails in the recipient's e-mail client. This method can also be used as whitelists by assigning high scores to the senders with whom the recipient is familiar with. Since the reputations are inferred and not taken directly from the lists, this method reduces the number of legitimate e-mails routed into the low priority mailbox.

SPAM CLASSIFICATION METHODOLOGY

As the usage of e-mail grows, the number of mail which are unsolicited or from distributing lists (opt-in e-mails) which reaches our inbox also tends to grow. Even if we are successful in restricting unsolicited e-mails, we are lost in an ocean of unwanted e-mails resulting from mailing or distribution lists. The priority rules defined to classify these e-mails to the designated folders are unable to meet the needs of the users since priorities tend to change significantly over time. Having to sort through the mails to find relevant or important mails might soon become time and resource consuming. We believe that the next generation of spam filters should be capable of more than that. They should be able to sort the mail based on the social proximity of the sender to the user along with preventing any spam and phishing e-mails from reaching one's inbox. This helps the user to get to important/significant messages from co-workers, family, friends, or discussion threads in which the user is an active participant faster.

Though spammers try to counterfeit Web sites or obfuscate the content of the e-mails, they do not have access to all the fields in the e-mail headers. They can forge certain fields like inserting spurious "Received:" header lines before dispatching e-mails but cannot spoof the complete header. Here a novel methodology for isolating spam e-mails is proposed. The classification method is based on the examination of the following items:

1. The header of an e-mail
2. The path traversed by the e-mail
3. The correlations between paths and the relays forwarding the e-mail
4. The social network of the recipient
5. Wantedness and unwantedness of the e-mail's source

In particular, the classifier will analyze the trust and reputation of the contents of the header. Normally, recipients associate their priorities by performing certain actions on the e-mails. For example, reading and deleting, deleting without reading, and archiving. These actions generate an implicit feedback such as storage time of read and unread e-mails of any given sender. In addition to learning from implicit feedback, the classifier also considers the explicit feedback from the recipient when making the decision.

No matter how many authentication techniques we develop, spammers always adapt to them. However, spammers cannot counterfeit or become part of the recipient's social network without the recipient's consent. Considering such issues, the proposed classifier performs the following four analyses: 1) DNS-header analysis; 2) SMTP path and relay analyses; 3) social network analysis; and 4) wantedness analysis. Each analysis complements the others. The individual application of each analysis provides only partial success; however, their combination magnifies the capability of spam detection. For example, DNS-header analyses may not detect a spam e-mail from a legitimate source, but the trust of the traversed path in conjunction with the social network analysis can be used to improve the DNS-header analysis. This can be further improved by means of a wantedness analysis as feedback from the user is taken into account to classify the e-mail. The results presented later in Section IV.C are a clear indication of the effectiveness of the combination of these analyses. The proposed classifier analyzes the "sent" folder of a recipient in social network analyses and uses this information in conjunction with the recipient's feedback in devising static rules for reducing the false positives. One advantage of the proposed technique over existing techniques is that it can be tuned according to the recipient's e-mail activity. It has been tested on a large e-mail corpus as described in Section IV.

Figure 1 illustrates the whole filtering process from Step 1 to Step 4 and the final classification of the e-mail corpus by the classification engine. In Step 1, DNS checks are performed on the host names recorded in the e-mail headers during the SMTP transactions. After that, incoming e-mail traffic is divided into three buckets: 1) Bucket 1[1]: phishing suspects (spoofers, phishers, etc.); 2) Bucket 2: trusted e-mails, and 3) Bucket 3: e-mails not valid for DNS lookups (this is either due to wrongly configured relays or improper recording of hostnames during the SMTP transaction). Buckets 2 and 3 are passed into Step 2.

In Step 2 the legitimacy of the path is verified using Bayesian probability analysis to compute spamminess and legitimateness of the path. The number of spam and nonspam e-mails relayed by the relays recorded in the header of an e-mail is also considered. With this information the spamminess and legitimateness of the relays is computed. Based on the spamminess and legitimateness values of paths and relays, Buckets 2 and 3 are further divided into two buckets: 1) Bucket 4: legitimate e-mails and 2) Bucket 5: spam suspects. Bucket 4 bypasses Step 3, social network analysis, and feeds into wantedness analysis.

Figure 1. The incoming e-mails are filtered through a four step process. The output from the four step process is subjected to the classification engine which further classifies spammers into prospective, recent, suspects, and serial indicating their severity. Nonspam e-mails are classified into socially close, socially distinct, and opt-outs.

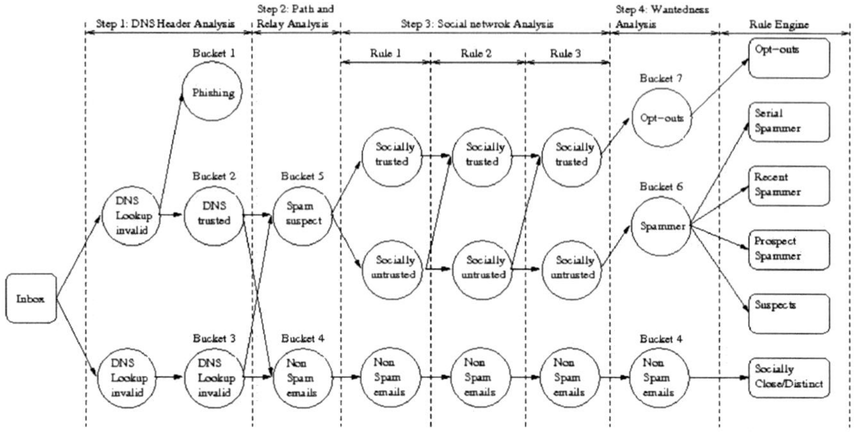

In Step 3 (social network analysis), Bucket 5 from Step 2 is filtered with rules built by analyzing the sent e-mail folder of the recipient. This bucket is further divided into 1) Bucket 6: spammers and 2) Bucket 7: opt-outs (false positives filtered from Bucket 5).

Step 4 (wantedness analysis) calculates 1) the distrust value of the senders, 2) trust value of the senders, 3) wantedness values of the senders e-mails, and 4) unwantedness of the incoming e-mails. Using these parameters the disputed cases in path and relay analyses are resolved. Finally Bucket 4 and 6 along with the computed values of the distrust, trust, wantedness, and so forth are passed into Step 5.

In Step 5 (classification engine) the final classification of spammers and legitimate senders is made. Based on spamminess vs. unwantedness, spammers are further categorized into 1) serial spammers, 2) recent spammers, 3) prospective spammers, and 4) suspects. Legitimateness vs. wantedness is used in classifying legitimate senders into 1) socially close, 2) strangers (socially distinct), 3) opt-ins, and 4) high risk. Next, a detailed description of the steps of the classification methodology is provided.

DNS-Header Analysis

An e-mail passes through at least four relays during its journey towards the recipient. During this entire journey, headers will be added to the message at least three times. The "Received:" header lines are very important to the classification proposed here. The DNS-header analysis is based on these header lines. The header field records the hostnames during the SMTP transactions between the mail relays. DNS-lookups are performed on the recorded hostnames during the mail handoff trying to match the IP addresses specified as the true identity of the machine participating in the handoff.

The purpose of the DNS-header analysis is to identify spoofing in the e-mail headers. In this stage information such as the hostnames of senders, mail servers, and relays recorded in the header during the SMTP authorization process is validated. After passing the e-mail traffic through header validations, the e-mails are divided into two groups: one containing the e-mails which have sufficient information in their headers to perform DNS lookups, and another one containing e-mails which are not suitable for DNS lookups. For instance, improperly configured relays, and purposefully spoofed headers by spammer/phishers. The second group of e-mails is labeled as Bucket 3 and passed into Step 2, SMTP path and relay analyses for further verification.

SMTP Path and Relay Analysis

The SMTP path and relay analysis is designed to filter spam from nonspam e-mails. Two issues need to be resolved: high false positive rate and high false negative rate. Path and relay analysis overcomes these issues by learning from the implicit feedback generated by the recipient's actions on the incoming e-mails, such as whether an e-mail is read and deleted or deleted without being read, and storage time of read and unread e-mails. Mozilla flags (Fla, 2004) are used for this purpose

1. **E-mail paths and path segmentation:** Typically an e-mail passes through at least four relays before reaching the recipient's inbox. During the SMTP transactions the hostname and IP addresses of the relays participating in transmission of the e-mail are recorded in the "Received:" lines in the e-mail header. These lines when read from bottom to top provide the path traversed by an e-mail from the sender to the recipient. Each end-to-end path can be divided into segments of length two relays. For example, let "$S \rightarrow M_1 \rightarrow R_1 \rightarrow M_2 \rightarrow D$" be the path taken by the e-mail, where "S" stands for the sender, "M_1" for the sender's mailserver[2], "R_1" for the intermediate relay, "M_2" for the recipient's mailserver, and "D" for the recipient. This path is reduced into segments of two nodes each: $S \rightarrow M_1$, $M_1 \rightarrow R_1$, and $R_1 \rightarrow M_2$, and $M_2 \rightarrow D$.

The reputation of a relay in the path is dependent on the reputation of their respective predecessor relays. In the above example the reputation of the sender's mailserver "M1" is dependent on the reputation of the sender "S." This is because "S" is the source that generates the e-mails, where as "$M_1$1" just relays these e-mails. Hence if "S" send more spam e-mails than "M_1," the spamminess of "M_1" increases, whereas if "S" send more legitimate e-mails, the legitimateness of "M_1" increases.

2. **Spamminess and legitimateness of the paths:** Let $X = \{X_1, X_2, X_3, \ldots, X_d\}$ denote the path segment vector of an e-mail, where X_1 through X_d are segments of the path traversed by that e-mail. The cumulative spam probability of the path is estimated by constructing a Bayes model as in Equation 1.

$$\Pr(Path = Spam | X_1 = S_{X_1}, \ldots, X_d = S_{X_d}) \tag{1}$$

From Bayes theorem:

$$\Pr(Path = Spam \big| X_1 = S_{X_1}, \cdots, X_d = S_{X_d}) = \left\{ \frac{\Pr(Path = Spam) \times \Pr(X_1 = S_{X_1}, \cdots, X_d = S_{X_d} \big| Path = Spam)}{\Pr(X_1 = S_{X_1}, \cdots, X_d = S_{X_d})} \right\} \tag{2}$$

The right hand side of Equation 2 involves conditional probability of the path segment vector given that the path segment vector's e-mail is spam, which is estimated using Equation 3.

$$\Pr(Path = Spam | X_1 = S_{X_1}, \ldots, X_d = S_{X_d}) = \prod_{i=1}^{d} \Pr(X_i = S_{X_i} \big| Path = Spam) \tag{3}$$

S_{X1} through S_{Xd} in Equations 2 and 3 are spamminess and legitimateness of each segment which are calculated as:

$$Spamminess = \left[\frac{S_n}{S_n + G_n} \right] \tag{4}$$

$$Legitimateness = \left[\frac{G_n}{S_n + G_n} \right] \tag{5}$$

where S_n and G_n are, respectively, the number of spam and nonspam e-mails relayed by each segment.

Next, for all the incoming e-mails, the proposed classifier uses the Bayesian analysis from Equations 2 and 3 to calculate the spam probability of the path traversed by the e-mail. This is a continuous learning mechanism with feedback (both implicit and explicit feedbacks) from the recipient. Using the spamminess and legitimateness calculated from Equation 2 (Bayes model) the paths traveled by the e-mails are classified into three categories: 1) spam paths (very high probability of spam), 2) legitimate paths (very low probability of spam), and 3) disputed paths (unable to classify). Table 1 shows the classification of path segments for e-mails in corpus-I (described later in Section IV).

Then the proposed classifier uses recipients' feedback information and senders' domains wantedness values to classify the e-mails which are relayed over the disputed path segments. In some cases path analysis alone is not sufficient in classifying the e-mails, especially when path segments have IP addresses with less significant segment history information. The classifier uses relay analysis to classify these e-mails.

3. **Spamminess and legitimateness of the relays:** In addition to paths, the transit relays are also classified. For each relay the number of spam e-mails "S_n" and legitimate e-mails "G_n" relayed by that relay are maintained. The classifier uses these values to calculate spamminess and legitimateness using Equations 4 and 5.

Let $R=\{R_1,R_2,\ldots,R_d\}$ denote a relay vector for an e-mail, where "d" is the number of distinct relays that participated in relaying of that particular e-mail. Similarly to path analysis, the cumulative spam probability of these relays is calculated by constructing a Bayes model as in Equations 6 and 7.

$$\Pr(Path = Spam|X_1{=}S_{R_1}, \ldots, R_d = S_{R_d}) \quad (6)$$

$$\Pr(Path = Spam|R_1 = S_{R_1},\cdots,R_d = S_{R_d}) = \left\{\frac{\Pr(Path = Spam)\times \Pr(R_1 = S_{R_1},\cdots,R_d = S_{R_d}|Path = Spam)}{\Pr(R_1 = S_{R_1},\cdots,R_d = S_{R_d})}\right\} \quad (7)$$

The condition probabilities on the right hand side of Equations 7 are estimated using Equation 8.

$$\Pr(Path = Spam|X_1{=}S_{R_1}, \ldots, R_d = S_{R_d}) = \prod_{i=1}^{d}\Pr(R_i = S_{R_i}|Path = Spam) \quad (8)$$

S_{Ri} through S_{Rd} in the above Bayes model represents the spamminess and legitimateness calculated using Equations 4 and 5. For all the incoming e-mails the proposed classifier uses Bayesian analysis to calculate the cumulative spam probability of the relays present in the e-mail. It has been observed that there are some relays forwarding both spam and legitimate traffic. The proposed classifier uses wantedness values (refer to Section III-D) to resolve these disputed relays. Table 1 shows the classification of relays for corpus-I.

4. **Dynamic change in IP addresses:** The dynamic change of IP addresses can be an issue in the path and relay analyses. This is addressed by maintaining the domain level or network level history information.

Table 1. Classification of path segments and relays for corpus-I

Total Path Segments	Spam Path Segments	Nonspam Path Segments	Disputed Path Segments
4844	561	4251	32

Total Relays	Spam Relays	Nonspam Relays	Disputed Relays
9720	1131	8526	63

If the extracted IP address is entirely new, its history is null. In this case the proposed classifier tries to resolve the identity of the new IP address's domain, and checks if there exists any history information for that domain. If the history information is present, it is assigned as the history for the new IP address. The history information of the domain is the average of histories of all the IP addresses belonging to that particular network. Instead of giving 50-50 chance for the new IP address's spamminess and legitimateness, the proposed classifier considers the reputation of the domain to which this IP address belongs.

If the unknown IP address does not have any domain history, the authentication information (spoofed or non spoofed header) obtained from "Step 1: DNS-header analysis" is used. In the worst case scenario if the authentication information is inadequate, the unknown IP address is added as a new member and the learning process is started for that address.

As stated earlier, after performing SMTP path and relay analyses, Buckets 2 and 3 obtained from Step 1, DNS-header analysis are further categorized into legitimate e-mails (Bucket 4) and spam suspects (Bucket 5). Bucket 4 bypasses Step 3 (social network analysis) and heads into Step 4 (wantedness analysis) for further analysis (see Figure 1). Majority of the e-mails in Bucket 5 are spam, but there is a possibility of a few nonspam e-mails being identified as spam (false positives) in path and relay analyses. In order to filter these false positives, Bucket 5 is passed into Step 3 (social network analysis).

Path and relay analyses used the reputation information of the paths and relays for classifying e-mails. Social network analysis analyzes the "sent" e-mail folder of the recipient and builds static rules to filter false positives and false negatives from path and relay analyses. For this purpose, the proposed classifier utilizes the information obtained by analyzing the "sent" e-mail folder of the recipient and the recipient's feedback.

Social Network Analysis

Literature in social science (Granovetter, 1973; Marsden & Campbell, 1984) discusses the social closeness of people based on the amount of time and intensity (frequency) of communication. Granovetter suggests that the time spent in a relationship and the intensity along with the intimacy and reciprocal services form a set of indicators for social tie. The chapter predicts that the strength of an interpersonal tie is a linear combination of amount of time, the emotional intensity, the intimacy (mutual confiding), and the reciprocal services in a relationship. Marsden evaluates the indicators and predicators of strength (tie) described by Granovetter. The chapter concludes that "social closeness" or "intensity" provides the best indicator of strength or tie.

Software companies such as Microsoft (SNARF) (Neustaedter, Brush, Smith, & Fisher, 2005), Lotus (Bifrost) (O.B., & C. L., 2002), among others (Lockerd & Selker, 2003), have developed products that can perform social network analysis and use this information to identify the messages from socially important people. The approach proposed here is more comprehensive as it takes into account many more aspects of the social information than the existing approaches. Here, social proximity of the senders is measured and used to resolve false positives and false negatives as well as to classify the senders into 1) socially trusted and 2) socially untrusted.

This analysis utilizes the information obtained by the recipient's "sent" e-mail folder and the recipient's feedback. The spam suspects are treated with the rules built by analyzing the "sent" folder of the recipient. These rules can be built as per the recipient's e-mail filtering preferences. For example, the following three receiver-centric rules can be used for this analysis. As stated before, these are just an example as rules can be tailored depending on the receiver. The "sent" folder e-mails is analyzed and a list of trusted domains is created from the e-mail IDs of the people to whom the recipient has replied or has ever sent an e-mail. This list is used in formulating the following rules for further analysis.

Example Rule 1: Reciprocity. All the senders in the "sent" folder will be removed from Bucket 2. This rule matches the domain name constructed from the e-mail ID provided at the "Return-path:" field in the e-mails with the domain names in the list extracted from the "sent" folder of the recipient. When Bucket 2 is treated with this rule it results in further categorization of the e-mails into socially trusted e-mails (filtered false positives) and socially untrusted e-mails (spam e-mails).

Example Rule 2: Familiarity with the sender's community. This rule tries to match the domain name constructed from the e-mail ID specified at the "Return-path:" field with the domain names specified in the path of the e-mail. The socially untrusted e-mails obtained by applying Rule 1 on Bucket 2 are treated with this rule. Now, these socially untrusted e-mails can either be confirmed as socially untrusted or further categorized into socially trusted e-mails.

Example Rule 3: Familiarity of the path traveled. This rule filters out the e-mails whose domain names of the relays specified in the path traveled belong to the domains in the trusted list created by analyzing "sent" e-mail folder of the recipient. That is, all the e-mails traversing through trusted relays are forwarded to trusted bucket. This results in a further categorization of the untrusted e-mails (derived from Rule 2) into socially trusted and socially untrusted e-mails.

The approach proposed here provides more thorough analysis than other existing methods. After performing this analysis, all the socially trusted e-mails from Bucket 2 (spam suspects) are classified as opt-outs (identified as Bucket 7) and removed from the bucket since these are false positives. The socially untrusted e-mails from Bucket 2 are classified as spammers (identified as Bucket 6). Next, Bucket 6 (spammers) and 7 (opt-outs) are passed into wantedness analysis for further classification (see Figure 1).

Wantedness Analysis

DNS-header analysis, SMTP path and relay analyses, and social network analysis are performed on the IP addresses and hostnames specified in the e-mail headers. There is a possibility of these analyses being inadequate for classifying some e-mails. This is possible because some senders might be using machines, which are misconfigured in providing the identification information, such as IP address and hostname during the SMTP transactions. In some cases spammers use zombies (commandeered computers) for relaying millions of spam e-mails. Also due to security reasons some organizations deliberately conceal IP addresses of their machines. If the sender's domain employs NAT policy the machines behind the NAT provide private IP addresses, which are not suitable for DNS lookups. Wantedness analysis, described next, is used to address these issues and consequently decrease the number of false positives and false negatives.

1. **Measuring sender's trust (ρ):** The trustworthiness of a sender depends upon the nature of the sender's recent e-mails. The proposed classifier analyze the nature (spam/legitimate) of the recent e-mails sent to the recipient. If the recent e-mails are fraudulent, the new incoming e-mails from the sender are less trustable where as, if the recent e-mails are legitimate, the new incoming e-mails from the sender are more trustable.

Suppose the most recent legitimate e-mail of a sender is sent at time T_j such that $\{T_i<T_j<T_n\}$. The decrease in the trust over the time period $\Delta T=[T_n-T_j]$ can be estimated as $\rho(i,n)^{-\Delta T}$. Hence, the larger the time interval between the sender's legitimate e-mails the larger is the decrease in the trust of a sender.

Figure 2. Decrease in trust of a sender in corpus-I. The average time interval, [$\Delta T_{legitimate\_e\text{-}mails}$] between the most recent e-mail (spam/legitimate) and the last e-mail indicates the periodicity of legitimate e-mails. As time elapses, [$\Delta T_{legitimate\_e\text{-}mails}$] increases, and there is a higher chance of sender's recent e-mails being spam or fraudulent in nature. Hence trust of a sender decreases as [$\Delta T_{legitimate\_e\text{-}mails}$] increases. This decrease in trust may be small but as [$\Delta T_{legitimate\_e\text{-}mails}$] increases, trust decreases exponentially.

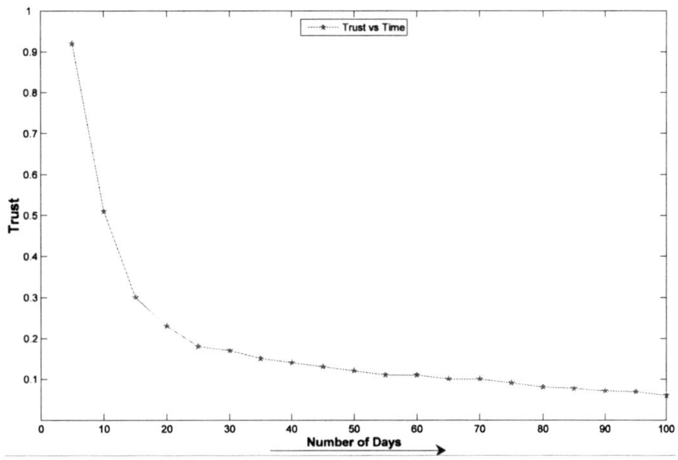

Figure 2 shows a decrease in trust of a phisher whose e-mails are disguised as legitimate e-mails from ebay.com (based on corpus-I). In this group of e-mails, the header of the first e-mail was not spoofed. Due to this, DNS-header analysis classified this e-mail as legitimate, thus initially giving high trust to this e-mail. As time progressed, more and more spoofed e-mails were received from this sender.

The duration between the sender's legitimate e-mail (first e-mail) and the recent e-mail sent by the sender increases [$\Delta T_{legitimate\_e\text{-}mails}$] causing an exponential drop in the sender's trust. Over a period of time, the proposed classifier overcomes these issues by learning from the decrease in trust of the senders computed from their past history and also from the recipient's feedback.

$$Trust(\rho) = \frac{\Delta T_{fraudulent\_emails}}{\Delta T_{legitimate\_emails}} \quad (9)$$

$$\text{Distrust } (\hat{\rho}) = \{1 - \rho\} \quad (10)$$

Equations 9 and 10 are used to calculate trust and distrust of the senders. $\Delta T_{legitimate\_e\text{-}mails}$ in Equation 9 is the average age (relative duration between recent e-mail and legitimate e-mails) of all the legitimate e-mails. $\Delta T_{fraudulent\_e\text{-}mails}$ is the average age of all the fraudulent e-mails from a sender.

Trust for e-mails in Bucket 5 (spam suspects) is computed and used to resolve false positives. Figure 3 is a trust plot for the senders' domains. X-axis is the number of domains of the senders. The star dotted curve represents trust of these domains. The 0.6 threshold value has been derived empirically. A total of 103 domains fall below the threshold except for three domains, marked as O_1, O_2, and O_3 in Figure 3. Here the threshold value selected in Figure 3 is only an example for the corpus-I. During the training phase the optimum threshold value is estimated using receiver operating characteristics analysis (Tom, 2004).

Upon crosschecking with the recipient's feedback, it appears that all the senders from the 1st through 103rd domain are for example, spammers, phishers, and telemarketers. E-mails from these domains are fraudulent in nature. For example, these are Nigerian scam e-mails, and Walmart scam e-mails. Senders' from domains O_2 and O_3 are false negatives. They are opt-ins to the recipient and the wantedness of their e-mails ac-

Figure 3. Trust values of socially untrusted domains. Untrusted domains are classified into high trust domains and low trust domains with one false positive (O_1) and two false negatives (O_2 and O_3).

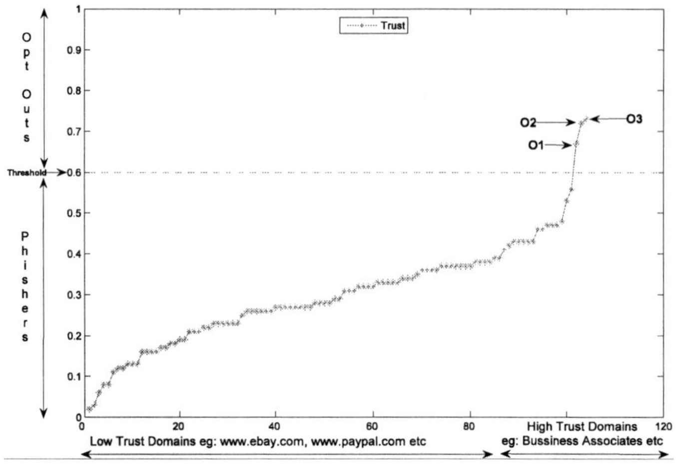

cording to the recipient's feedback is low. E-mail from domain O_1 is a false positive according to the recipient feedback. From the plot in Figure 3 we can observe that the sender's trust value is above the threshold.

On further analysis of untrusted buckets, it was found that mail servers were using private IP addresses where as they provide public IP addresses during SMTP transactions. They were behind NAT. This mismatch in IP addresses during the DNS-header analysis resulted in marking this sender as untrusted. Hence the e-mails from O_1, O_2, and O_3 domains are categorized as opt-outs and removed from the socially untrusted categories.

2. **Measuring the recipient's wantedness (χR):** Two new parameters are introduced here: tolerance and intolerance. They are used to calculate the recipient's wantedness of the senders' e-mails. Normally the storage time of an e-mail in the recipient's inbox depends on its importance to the recipient. The recipient's tolerance towards the sender is directly proportional to the amount of time $\Delta T_{read\_e\text{-}mail}$ e-mails are stored in the inbox.

The more the receiver is interested in a sender, the higher the recipient's tolerance towards that sender (storage time $\Delta T_{read\_e\text{-}mail}$) also increases. The intolerance value is directly proportional to the amount of time $\Delta T_{unread\_e\text{-}mail}$ the sender's unread e-mails are stored. The tolerance value also depends on how frequent the sender is mailing legitimate e-mails. That is, tolerance is inversely proportional to the duration $\Delta T_{legitimate\_e\text{-}mail}$ between the sender's most recent e-mail and the sender's previous legitimate e-mails. If this duration $\Delta T_{legitimate\_e\text{-}mail}$ is small, it indicates that the sender is more frequently sending legitimate e-mails when compared to the sender's fraudulent e-mails (for nonspammers, $\Delta T_{fraudulent\_e\text{-}mail}$ will be zero).

The intolerance value of a sender depends on the frequency of the sender's fraudulent e-mails. The average time interval between the sender's fraudulent e-mails and the most recent e-mail indicates how frequently the sender is mailing fraudulent e-mails. Intolerance is inversely proportional to the $\Delta T_{fraudulent\_e\text{-}mail}$.

$$Tolerance \propto [\Delta T_{read\_email}] \propto \left\{ \frac{1}{\left[\Delta T_{legitimate\_email}\right]} \right\} \quad (11)$$

$$Intolerance \propto \left[\Delta T_{unread\_email}\right] \propto \left\{ \frac{1}{\left[\Delta T_{fraudulent\_email}\right]} \right\} \quad (12)$$

$$Wantedness(\chi_R) = \left\{ \frac{Tolerance}{Intolerance} \right\} = \quad (13)$$

$$\left\{ \left[\frac{\Delta T_{read\_email}}{\Delta T_{unread\_email}} \right] \times \left[\frac{\Delta T_{fraudulent\_email}}{\Delta T_{legitimate\_email}} \right] \right\}$$

$$Unwantedness(\hat{\chi}_R) = \{1 - \chi_R\} \quad (14)$$

The wantedness of the sender's e-mails is computed as the ratio of tolerance to the intolerance as in Equation 13 and used to resolve the disputed cases in path and relay analyses and also in further classifying the e-mails in Bucket 4 (nonspam e-mails) obtained from Step 2 (see Figure 1) into 1) socially close and 2) socially distinct.

Resolving the Disputed Path Segments and Relays

SMTP path and relay analyses classify the paths traveled by an e-mail into three categories: 1) spam paths (very high probability of spam), 2) legitimate paths (very low probability of spam), and 3) disputed paths (unable to classify). The value of "wantedness" is used to resolve the disputed paths.

Our classifier encountered 32 disputed path segments and 63 disputed relays during path and relay analyses in corpus-I. The spamminess and legitimateness of these disputed path segments and relays are more or less similar because the majority of them forward equal proportions of spam and nonspam e-mails. The wantedness values are used for classifying the e-mails relayed over the disputed path segments and relays. Two thresholds are used for classifying these e-mails: 1) an upper threshold of value 0.60 and 2) a lower threshold of value 0.3.

The classification can be observed in Figures 4 and 5. Domains with low wantedness fall below the lower threshold and domains with medium wantedness are confined between the lower and upper thresholds. All the domains above the upper thresholds are highly wanted. E-mails from the senders' whose domains are below the lower threshold are less wanted to the recipient. Mostly this category contains e-mails from spammers, telemarketers, and phishers. E-mails whose domains are between lower and upper thresholds are classified as opt-ins. Majority of the e-mails in this category are, for instance, conference notices, professional news, and club meetings.

Senders' domains labeled as O_1, O_2 in Figure 4 and O_1, $O_{2,}$ and O_3 in Figure 5 are above the threshold, this is because the read e-mails storage time $\Delta T_{read\_e\text{-}mail}$ is more when compared to the unread e-mails. This indicates that a majority of the e-mails from these domains are being read by the recipient. Moreover, the average time interval $\Delta T_{legimate\_e\text{-}mail}$ between the sender's legitimate e-mails and the most recent e-mails from these senders' is small; indicating that the frequency of the legitimate e-mails is greater compared to the frequency of unwanted e-mails (spam or phishing if any in the

Figure 4. Disputed path segments and sender's domains wantedness. All the domains below the lower threshold (0.3) are least wanted domains whereas those above the upper threshold (0.6) are most wanted domains. Domains confined between lower and upper thresholds fall into middle ground.

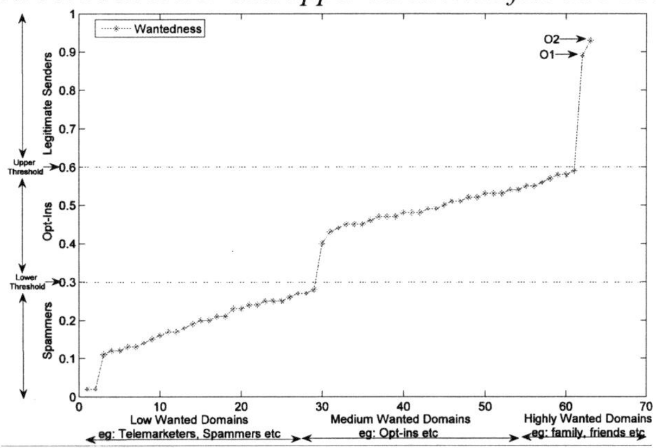

Figure 5. Disputed relays and sender's domain's wantedness. Based on the wantedness values the researched filter classified the e-mails relayed by the disputed relays in low wanted, medium wanted and highly wanted domains.

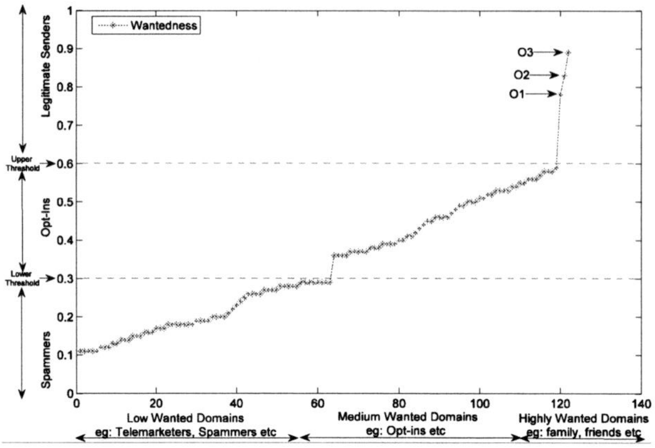

past). Thus all the e-mails from these domains are marked as wanted, enabling them to be labeled as legitimate e-mails. Normally the thresholds are estimated during the training phase by performing ROC (Tom, 2004) analysis. These thresholds can be set as per the user preferences thus giving more importance to the recipients e-mail activity. The wantedness values of the domains in Bucket 4 (nonspam e-mails) from Step 2 are computed and used to further classify nonspam e-mails as socially close or socially distinct.

The e-mails in Bucket 4 contain nonspoofed e-mails. Based on the recipient's social activities these e-mails can be further classified into socially close and socially distinct.

E-mails from family members, friends, and so forth are classified as socially close. E-mails from strangers, opt-ins, and so forth are classified as socially distinct. Figure 6 shows the wantedness for the senders' domains in Bucket 4 (nonspam e-mails); the x-axis represents the number of domains and the y-axis represents their wantedness. Initially, a threshold of 0.5 has been used. All the domains below the threshold are low trust domains. Majority of e-mails from these domains are, for instance, professional news, business announcements, and club meetings. Majority of the senders from the domains above the threshold are considered socially close to the recipient (high

Figure 6. Wantedness values of trusted senders. The analyses identified a total of 157 domains from corpus-I as trusted. By employing threshold (0.5) it classified them as socially close and socially distinct.

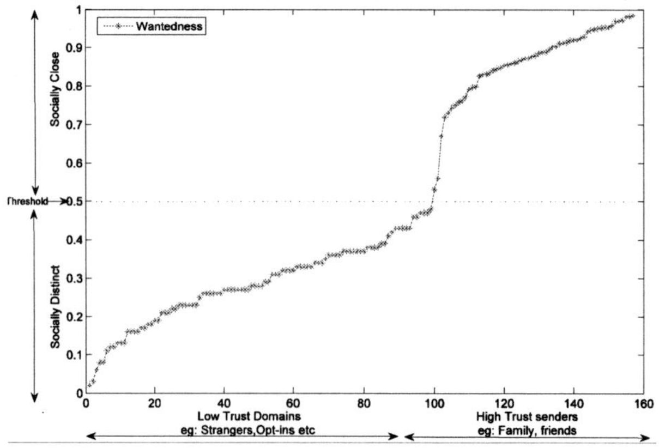

trust domains). All the senders whose domains are below threshold are classified as socially distinct.

Next, Bucket 4 (nonspam e-mails) and Bucket 6 (spammers) along with their values for trust, distrust, wantedness, and unwantedness from Step 4 and spamminess and legitimateness of the paths and relays computed in Step 2 are passed into final Step 5 for further classification.

Classification Engine

Spammers are classified based on the spamminess (from SMTP path and relay analyses) of the paths taken by their e-mails and their unwantedness values computed in Step 4 (wantedness analysis). Figure 7 shows the relation between paths spamminess and their unwantedness. The y-axis is the spamminess of the paths and the x-axis is their respective unwantedness values. Spammers are classified into four categories: 1) prospective spammers, 2) suspects, 3) recent spammers, and 4) serial spammers.

All the senders' who have low spamminess and low unwantedness are classified as "prospective spammers" and those who have high spamminess but low unwantedness are classified as "suspects" because the recent e-mails from these senders are not spam to the recipient but there are occasions in the past when the sender spammed the recipient. Senders' whose spamminess is low and unwantedness is high are categorized as "recent spammers" as the recent e-mails from these senders are spam to the recipient. Senders' with high spamminess and high unwantedness are classified as "serial spammers." Once the final classification of spammer is determined, a blacklist of domains or senders identified as serial spammers is maintained.

Similarly, legitimateness of the paths traversed by legitimate e-mails and their wantedness to the recipient computed in Step 4 (wantedness analysis) are used to classify the legitimate e-mails (See Figure 8). Based on the relationship between the legitimateness of the nonspam paths and wantedness of the senders' domains whose e-mails are relayed over these path segments, they are classified into 1) socially close, 2) socially distinct, 3) opt-ins, and 4) high risk.

E-mails having a low wantedness value and relayed over paths with low legitimateness are labeled as "high risk." Majority of e-mails in this category are telemarketing e-mails. E-mails which have high wantedness but traveled over paths with low legitimateness are marked as "socially distinct." These are the e-mails from senders with whom the recipient previously has none or very low interaction.

E-mails with high legitimateness of their paths and low wantedness are classified as "opt-ins." E-mails which are less significant to the recipient

Figure 7. Classification of spammers based on severity of spamming. Spammers are classified based on spamminess of the paths traveled by their e-mails and their unwantedness values.

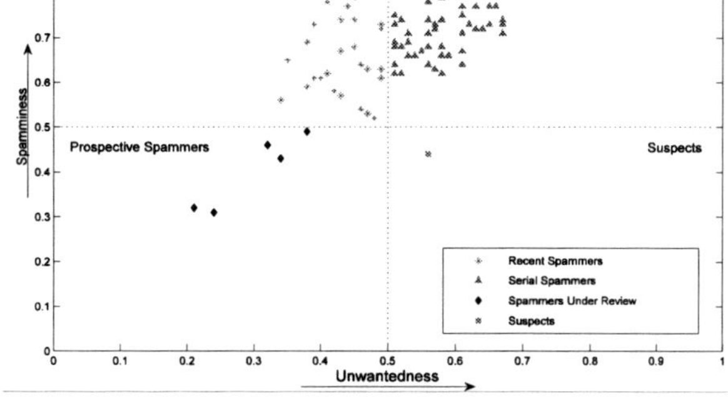

Figure 8. Classification of legitimate e-mails based on the e-mails wantedness values and legitimateness of the paths traveled

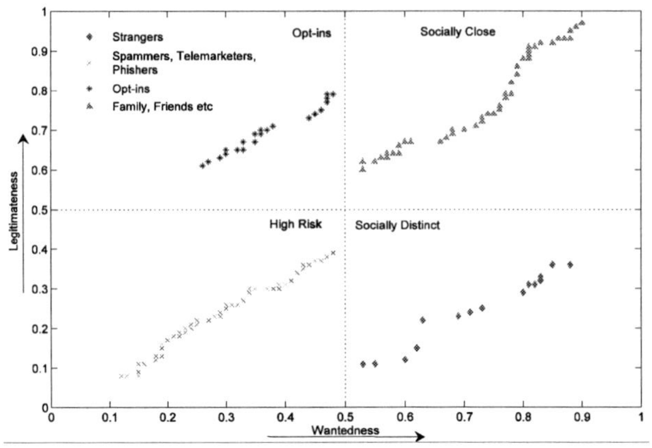

comprise this category; for instance, club announcements, and business news. E-mails with high legitimateness and high wantedness for their paths are classified as "socially close" (e.g., e-mails from family members, and friends). Domains are classified in the same way as legitimate e-mails.

VALIDATION OF THE CLASSIFICATION ARCHITECTURE

The validation of the proposed classifier has been conducted by handlabelling an e-mails corpus, described in Section IV-A, and then comparing the manual classification with the results from the classifier in Section IV-C. A qualitative as well as a quantitative comparison with existing approaches has also been performed and are described, respectively, in Sections IV-D and IV-E.

Corpus Description

Typically the incoming e-mails of a recipient consist of e-mails from various senders; for instance, e-mails from telemarketers, fraudsters, and opt-in senders. In order to verify this, incoming e-mail traffic collected at an enterprise's mailserver has been categorized by asking the recipients to hand-label the different kinds of e-mails. The frequency with which they receive these e-mails has also been computed. It can be noticed that as the number of recipients increase, the type of spam (i.e., telemarketing e-mails, phishing e-mails, opt-in e-mails, etc.) in incoming e-mails (collected at the mailserver) also increases. This is due to an increase in the variation of the recipients' social networks. As the percentage of these unsolicited e-mails increase in the incoming e-mail traffic, annoyance or nuisance increases resulting in loss of productivity.

Most of the available e-mail corpuses do not contain original headers. This is done to anonymize the recipients. Also, the corpuses in existing benchmarks such as TREC (Cormack & Thomas, 2005) contain only spam and nonspam e-mails, and a mixture of different types of spam (i.e., phishing, telemarketing, opt-in e-mails, etc.) is needed for our analyses. These issues have prevented us to use openly available corpuses and a live e-mail corpus has been analyzed with the users' permissions. The corpus, referred as corpus-I, consists of 13,843 e-mails collected over 2.5 years presenting a mix of legitimate, phishing, telemarketing, and opt-in e-mails.

Experimental Setup

Every e-mail passes through three agents (MTA, MDA, and MUA). Spam filtering can be performed on two or even all three agents (MTA, MDA, and MUA). Filters at MTA can be configured with moderate antispam rules rejecting highly suspicious e-mails. This filtered output can be passed to the MDA-based filters where these e-mails are again subjected to the antispam rules built with the recipient's preferences. Users can employ their own spam checking by running the spam filters as proxies on their client machines (SpamAssassin [Spa, 2006] can be configured to run as a proxy on the client).

For filtering spam we have two options. The MDA can examine and discard suspected e-mails, or it can tag suspected spam and hand it along to the MUA. MUAs equipped with the proposed classifier can then use these tags for learning. Employing the knowledge of the user's e-mail preferences gained over a period of time the proposed classifier can discard or move the spam e-mail to the designated special folders.

For the experimental setup the proposed classifier is used in association with procmail. The classifier is deployed on a recipient's local machine running an IMAP proxy and Thunderbird (MUA). All the recipient's e-mails are fed directly into classifier by the proxy. The classifier periodically scans the recipient's inbox for new incoming e-mails. DNS-header analysis, path and relay analyses, social network analysis, and wantedness analysis are performed on the e-mails. The end result is the tagging of e-mails as spam, opt-outs, socially close, and socially distinct. Spammers are further categorized as serial spammers, recent spammers, suspects, or prospective spammers.

Experimental Results

The four analyses (i.e., DNS-header analysis, path and relay analyses, social network analysis and wantedness analysis) are conducted with the final classification of senders by the "classification engine." These are performed sequentially in five steps as described earlier in Section III.

In the experiment, if an instance (here an e-mail) is unwanted (spam) and classified as unwanted then it is counted as true positive (TP). If an instance is wanted (legitimate) and classified incorrectly as unwanted, it is counted as false positive (FP). Let "P" and "N" be the total number of positive and negative instances in a corpus; precision, true positive rate "tp_{rate}", and false positive rate "fp_{rate}" of classifier can be determined as by Equations 15, 16, and 17. Table 2 shows the precision of filtering spam e-mails from the incoming e-mail traffic by performing steps 1, 2, 3, and 4 on corpus-I.

$$Precision = \frac{TP}{TP + FP} \tag{15}$$

$$tp_{rate} = \frac{TP}{P} \tag{16}$$

$$fp_{rate} = \frac{FP}{N} \tag{17}$$

The hit rate also called as true positive rate "tp_{rate}" is calculated for the categorizations of the spammers as the ratio of "TP" instances to the total number of positive instances "P" that originally belongs to their respective categories. Table 3 shows the hit rate for categorizing spammers, legitimate e-mails, and trusted domains. As can be

Table 2. Summary of results for the classification performed on corpus-I

Steps	Analyses	False Positives	False negatives	Precision
1 & 2	DNS-Header + SMTP Path and Relay analysis	31	43	88.57%
2 & 3	SMTP Path and Relay + social network analysis	16	13	94.74%
2 , 3 & 4	SMTP Path and Relay + social network + wantedness analysis	05	07	98.65%

Table 3. Hit rates for classification of spammers, legitimate e-mails, and trusted domains

Category Spammers	Hit Rate
Prospective	91.23%
Suspects	87.45%
Recent	93.21%
Serial	97.76%

Category Legitimate e-mails	Hit Rate
High Risk	90.78%
Socially Distinct	94.35%
Opt-Ins	88.63%
Socially Close	89.03%

Category Trusted Domains	Hit Rate
High Risk	85.83%
Socially Distinct	94.35%
Opt-Ins	93.27%
Socially Close	96.89%

seen, the proposed classifier produces outstanding results. However, there exist scenarios where the proposed classifier presents some limitations. For example, when the path traversed by an e-mail is too short and is coming from an initially trusted sender it may take some time before the trust of the sender decreases and the e-mail is properly filtered as spam.

SMTP Path Analysis vs. Multivariate Analyses

Barry, Ossher, Rajan, Richard, and Mark (2005) describe an algorithm for filtering spam e-mails based on the IP addresses recorded in the e-mail headers. Their analysis is somehow similar to ours as they also classify the incoming e-mails by analyzing the headers rather than the content of the e-mail. However, their algorithm does not combine authentication information with the reputation of the relays and paths traversed by the e-mail. They do not validate the sender/sending domain or detect spoofing in the headers. Instead they rely on the past histories of the IP addresses. Moreover, the method they employ to estimate the spamminess of the unknown IP addresses (IP addresses not present in their database) may not yield justifiable scores (spamminess) as they assume a 50-50 chance for the unknown IP addresses to be either spammy or nonspammy. This assumption holds well if the unknown IP address is a nonspammy address. Since they do not verify the authenticity of the IP addresses in the headers, SMTP path analysis performance in restricting the e-mails with spoofed headers is questionable.

Goodman (2004) describes the problems with IP addresses in the e-mail headers. Since they are highly susceptible to spoofing and cannot be parsed reliably, that is why here their authenticity is verified before using them to filter unsolicited e-mails. The proposed classifier overcomes these issues by establishing the authenticity of the IP addresses during the DNS-header analysis.

The reputation of a relay in a path is dependent on the reputation of its predecessor relays. Bayesian inference is used to learn the reputation of the paths and relays. SMTP path analysis computes weighted averages of the spamminess of the IP addresses in order to estimate the cumulative spamminess of the path traversed by an e-mail. The proposed classifier presents a more comprehensive approach as it detects spoofing in the e-mail headers and uses a multistage feedback among other features. In addition, our classifier presents a better accuracy, 98.65% compared to 70% presented by the SMTP path analysis.

Naive-Bayes vs. Multivariate Analyses

The performance of the proposed classifier with a conventional Naive Bayesian classifier is conducted here using standard ROC curves (see Figure 9). The ROC curves (both Naive Bayes and the proposed classifier) demonstrate various combinations of precision and false positive rate that can be achieved by employing different thresholds for restricting unsolicited e-mails. For instance, if the threshold employed is very high then the false positive rate is almost 0, but

Figure 9. Naive Bayesian classifier vs. multivariate analyses with obfuscated content ROC curve.4

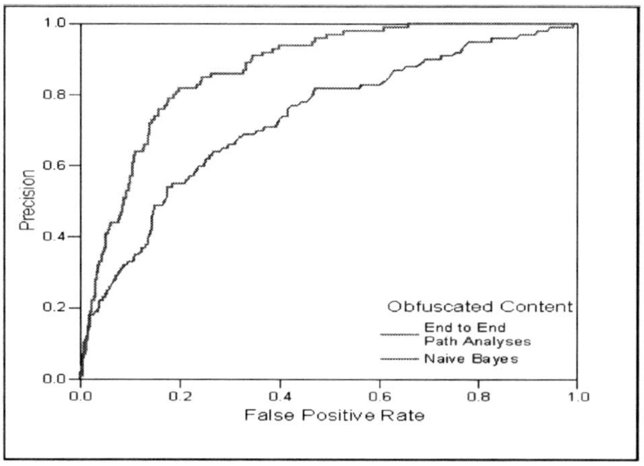

we also detect very few true positives. Hence for high thresholds initially the ROC curve starts at the extreme lower left corner (near to 0).

As the threshold decreases towards a more reasonable value, the classifier detects more true positives resulting in dramatic increase in the precision. So the ROC curve climbs rapidly. Finally, we reach a threshold where there is a noticeable increase in false positive rate. Hence the ROC curve slopes off as we further decrease the threshold.

The Naive Bayes classifier has been initially trained with a set of 6,237 e-mails. This training set is a mixture of different categories of spam and nonspam e-mails (e.g., advertisement e-mails, phishing e-mails, opt-in e-mails, etc.). After the training of the Bayes classifier, both classifiers were tested over corpus- I of 13,843 e-mails. The majority of spam e-mails in corpus-I contain few or no words, e-mails with one or more images, and in some cases only a URL is present.

The idea here is to deny any content to the Naive-Bayes for tokenizing and using them in scoring spam e-mails. One can notice from the ROC curves in Figure 9 that the proposed classifier clearly outperforms Naive-Bayes classifier with a precision of 87.2% where as, naive-Bayes achieves a precision of 73.4% on corpus-I. An interesting observation here is that proposed classification methodology is completely independent of the e-mails content. The ROC curve in Figure 10 shows a remarkable increase in the precision up to 98.65% on corpus-I when the recipient's feedback information is made available to the proposed classifier. This result is an improvement over the precisions achieved by other filters such as SPF and DNSBL blacklists.

CONCLUSION AND FUTURE WORK

The proposed methodology has been applied on a live e-mail corpus consisting of 13,843 e-mails collected over 2.5 years with outstanding precision and accuracy. After analyzing the corpus the proposed classifier successfully separated spam paths from legitimate paths and spam relays from nonspam relays. The concept of wantedness has been introduced and used to measure trust, distrust, wantedness, and unwantedness of senders' domains. These values are further used to classify spammers and legitimate and trusted domains.

During the analyses, 32 disputed path segments and 63 disputed relays were encountered. Spamminess and legitimateness are not decisive in classifying the e-mails relayed by these disputed path segments and relays, since they forward both spam and nonspam e-mails in equal proportions. The proposed classifier successfully resolved these disputed cases during Step 4 (wantedness

Figure 10. Multivariate analyses with recipient's feedback ROC curve

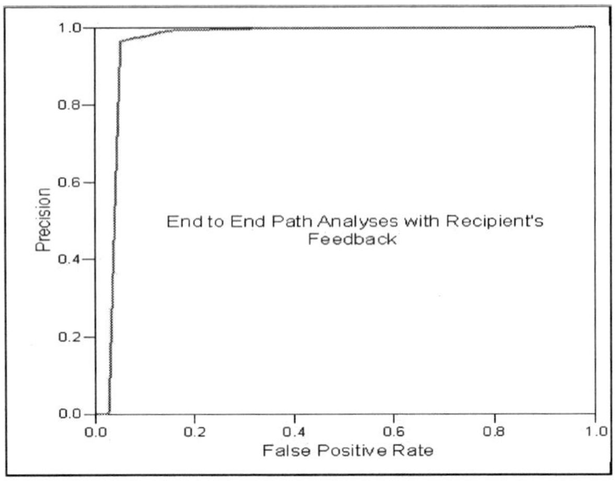

analyses). Therefore, the examination of the information provided in e-mail headers has found to be a valuable addition to spam filtering techniques.

The proposed classifier produced significantly less false positive and false negative rates. However, there are a few cases where the header information is not sufficient to classify e-mails. Under such scenarios, the proposed wantedness analysis based on the nature (any history of spamming or phishing the recipient in the past) of the sender's past interactions with the recipient proved very valuable for the classification scheme. However, some questions could be raised: "What if a relay in the path or sending domain's mail server is compromised?" or "What if the spammers use trusted domains for mailing spam?" Normally it is not possible to compromise the entire path traversed by the e-mails. Only those relays which are in the beginning of the path are susceptible to being commandeered (this is also true in case of zombies or botnets). In such situations more importance can be given to the senders having significant history information. In both cases, initially spam may get through but since the proposed classifier is a continuous learning mechanism, from both implicit and explicit feedbacks, it will overcome these issues by learning from the inferred reputation information. Moreover, if there is any spoofing involved in these headers, they will be detected in the DNS-header analysis.

One may express doubts on the duration of time taken to learn such situations. The proposed classifier performs multivariable analyses on the incoming e-mail traffic. Besides inferring the reputation information, the classifier also measures social proximity of the senders or sending domains. Typically the social proximity measurement is based on senders' actions and recipients' reactions. Senders' actions (such as how frequently they are sending legitimate or spam e-mails, spatial and temporal distribution of their e-mails sent to recipients) along with recipients' reactions (such as how long senders' read e-mails are being stored in the inbox, whether senders' e-mails are being read and deleted or deleted without being read, and percentage of the senders' e-mails that are being read by the recipient) are considered. The learning time is reduced by combining the reputation information with the senders' social association (trust, recipient's tolerance, and wantedness values) measured during the wantedness analysis. Any third party filter can also be used in conjunction with proposed classifier. Another common practice in forging the e-mail header is to add spurious "Received:" header lines before sending the e-mail. Such cases are dealt with the DNS-header analysis. Wantedness analysis (Step 4) is also used for proper classification of such e-mails. The proposed classifier successfully filtered spam e-mails with a precision of 98.65%

on corpus-I. This is superior when compared to the precisions achieved by SPF and DNSRBL lists.

Text, voice, and video are three modes in human communication. Each mode conveys an added level of trust to consumers. With the advent of technologies like voice over IP and video over IP, the differences between online and face-to-face communications are reduced. E-mails have already fallen victim to spam and phishing attacks. It will not be long before phishers pose as a serious threat to VoIP and nideo over IP. In fact, there have already been reports confirming phishing attacks (loosely termed as vishing [Kelly, 2006]) using VoIP. By the time VoIP and video over IP are widely deployed, technology companies must make them as impervious as practicable to spam and phishing attacks. The conjecture is that a similar approach as presented here could be used to identify spam on different media; however this is deferred to future work. Also, spam was the main target here but studies on phishing are already being conducted.

REFERENCES

Allman, E., Callas, J., Delany, M., Libbey, M., Fenton, J., & Thomas, M. Domainkeys identified mail signatures (DKIM). *U.S. Patent No. 6,986,049.*

Allocchio, C. (1998, January). *Using the Internet dns to distribute MIXER conformant global address mapping (MCGAM)* (IETF RFC 2163).

Atkinson, B. (2004, May 20). *Caller id for e-mail.* Microsoft Corp.

Banit, A., Nitin, K., & Mart, M. (2005). *Controlling spam e-mail at the routers*. Paper presented at the EEE International Conference on Communcations (ICC '05), Seoul, Korea.

Barry, L., Ossher, J., Rajan, V. T., Richard, S., & Mark, W. (2005, July 21-22). SMTP path analysis. In *Proceedings of the Second Conference on E-mail and Anti-Spam (CEAS).*

Boykin P. O., & Roychowdhury, V. (2005, April). Personal e-mail networks: An effective anti-spam tool. *IEEE Computer, 38.*

Cormack, G., & Thomas, L. (2005, July 21-22). Spam corpus creation for TREC. In *Proceedings of the Second Conference on E-mail and Anti-Spam (CEAS).*

Delany, M. (2006, January 10). Method and system for authenticating a message sender using domain keys. *U.S. Patent No. 6,986,049.*

DNS. (2007). *Dns-based spam.list of all known dns-based spam databases*. Retrieved December 18, 2007, from http://www.declude.com/JunkMail/Support/ip4r.htm

Elkins, M., Torto, D., Levien, R., & Roessler, T. (2001, August). *Mime security with OPENPGP* (IETF RFC 3156).

Fenton, J., & Thomas, M. (2005, May 13). *Identified Internet mail.* Cisco Systems, Inc.

Fla. (2004). *Message flags*. Retrieved December 18, 2007, from http://lxr.mozilla.org/seamonkey/source/mailnews/base/public/nsMsgMessageFlags.h#45

Goodman, J. (2004). *IP addresses in e-mail clients.* Paper presented at the Conference on E-mail and Anti Spam (CEAS).

Graham, P. (2003). *Better Bayesian filtering.* Paper presented at the MIT Spam Conference.

Granovetter, M. S. (1973). The strength of weak ties. *American Journal of Sociology, 78*, 1360-1380.

Hird, S. (2002). Technical solutions for controlling spam. In *Proceedings of the Annual Meeting of the Australian UNIX and Open Systems User Group (AUUG).*

Jennifer, G., & James, H. (2004). Reputation network analysis for e-mail filtering. In *Proceedings of the First Conference on E-mail and Anti-spam.*

Kelleher, D. (2004). Spam filtering using contextual network graphs. *Final year project in computational linguistics (supervised by Dr. Padraig Cunningham, course 2003–04)*. Department of Computer Science, University of Dublin, Trinity College.

Kelly, J. (2006). *Vishing' attacks use VoIP*. Retrieved September 22, 2006, from http://www.darkreading.com/document.asp?doc id=98787

Klensin, J. (2004, February). *Application techniques for checking and transformation of names* (IETF RFC 3696).

Lemke, T. (2003, November 10). Spam harmed economy more then hackers viruses. *Washington Times*.

Levine, J. (2005, July 21-22). Experiences with greylisting. In *Proceedings of the Second Conference on E-mail and Anti-Spam (CEAS)*.

Lockerd, A., & Selker, T. (2003). Driftcatcher: The implicit social context of e-mail. In *Proceedings of Human-Computer Interaction INTERACT*.

Lowed, D., & Meek, C. (2005). Good word attacks on statistical spam filters. In *Proceedings of the First Conference on E-mail and Anti-Spam (CEAS)*.

MAP. (2005). *Mail abuse prevention system*. Retrieved December 18, 2007, from http://www.mail-abuse.com

Marsden, P. V., & Campbell, K. E. (1984). Measuring tie strength. *Social Forces, 63*(2), 482-501.

Massey, B., Mick, T., Raya, B., & Scott, L. (2003). Learning spam: Simple techniques for freely-available software. In *Proceeding of the 2003 Usenix Annual Technical Conference Freenix Track*.

Neustaedter, C., Brush, A. B., Smith, M. A., & Fisher, D. (2005). Social network relationship finder: Social sorting for e-mail triage. In *Proceedings of the Second Conference on E-mail and Anti-Spam (CEAS)*.

O, B., & C L, S. (2002). Bifrost inbox organizer: Giving users control over the inbox. In *Proceedings of the second Nordic conference on Human-computer interaction Aarhus, Denmark*.

Postel, J. B. (1982, August). *Simple mail transfer protocol* (IETF RFC 821).

Resnick, P. (April, 2001). *Internet message format* (IETF RFC 2822).

Rigoutsos, I., & Tien, H. (2004). Chung-kwei: A pattern-discovery-based system for the automatic identification of unsolicited e-mail messages (spam). In *Proceedings of the First Conference on E-mail and Anti-Spam (CEAS)*.

Soonthornphisaj, N., Chaikulseriwat, K., & Piyanan, T.-O. (2006, August 26-30). *Anti-spam filtering: A centroid-based classification approach*. Paper presented at the Signal Processing, 6th International Conference (Vol. 2, pp. 1096-1099).

Spa. (2006). *Spam assasin*. Retrieved Decemer 18, 2007, from http://www.spamassassin.apache.org

Tom, F. (2004). *Notes and practical considerations for researchers*. The Netherlands: Kluwer Academic Publishers.

William, C. W. (1996). *Learning rules that classify e-mail*. Paper presented at the AAAI Spring Symposium on Machine Learning in Information Access.

Wong, M., & Schlitt, W. (2006). *Sender id: Authenticating e-mail*. Retrieved December 18, 2007, from http://xml.coverpages.org/draft-ietf-marid-core-03.txt

Wong, M., & Schlitt, W. (2005, December 8). *Sender policy framework for authorizing use of domains in e-mail, version 1*. Retrieved December 18, 2007, from http://www.ietf.org/internet-drafts/draft-schlitt-spf-classic-02.txt

ENDNOTES

1 Phishing classification is deferred to future work.

[2] Mailservers forward both spam and nonspam e-mails. Hence their spamminess and legitimateness should be inferred from their predecessor relays.

This work was previously published in the International Journal of Information Security and Privacy, Volume 2, Issue 2, edited by H. Nemati, pp. 46-69, copyright 2008 by IGI Publishing (an imprint of IGI Global).

Chapter 22
A Decentralized Security Framework for Web-Based Social Networks

Barbara Carminati
Università degli Studi dell'Insubria, Italy

Elena Ferrari
Università degli Studi dell'Insubria, Italy

Andrea Perego
Università degli Studi dell'Insubria, Italy

ABSTRACT

The wide diffusion and usage of social networking Web sites in the last years have made publicly available a huge amount of possible sensitive information, which can be used by third-parties with purposes different from the ones of the owners of such information. Currently, this issue has been addressed by enforcing into Web-based Social Networks (WBSNs) very simple protection mechanisms, or by using anonymization techniques, thanks to which it is possible to hide the identity of WBSN members while performing analysis on social network data. However, we believe that further solutions are needed, to allow WBSN members themselves to decide who can access their personal information and resources. To cope with this issue, in this chapter we illustrate a decentralized security framework for WBSNs, which provide both access control and privacy protection mechanisms. In our system, WBSN members can denote who is authorized to access the resources they publish and the relationships they participate in, in terms of the type, depth, and trust level of the relationships existing between members of a WBSN. Cryptographic techniques are then used to provide a controlled sharing of resources while preserving relationship privacy.

INTRODUCTION

The last few years have seen an increasing diffusion of social networking Web sites (Staab, Domingos, Mika, Golbeck, Ding, Finin et al., 2005), where users can establish relationships and share resources, opinions, and contacts for a variety of purposes (recreational, work, dating, etc.). As a result, today Web-based Social Networks (WBSNs) make publicly available a huge amount of possibly sensitive and private information, which may be used with purposes different from what was intended by the users who released it. As a matter of fact, such information is regularly exploited not only by companies for marketing purposes, but also by governments and institutions (such as colleges), in order to track the behaviors/opinions of specific persons, and, in the worst case, by online predators (Barnes, 2006). In addition, Semantic Web technologies, such as FOAF and other RDF-based vocabularies (Brickley & Miller, 2005; Davis & Vitiello Jr., 2005; Golbeck, 2004) are currently widely used in WBSNs for publishing personal profiles, relationships, and trust levels (Ding, Zhou, Finin, & Joshi, 2005), thus making easier to access users' data across multiple WBSNs.

In order to deal with these issues, most of today WBSNs allow a user to specify whether a given information must be public or accessible only by the users with whom he/she has a direct relationship. Such simple access control paradigm has the advantage of being straightforward and easy to be implemented, but it suffers from several drawbacks. On the one hand, it may either grant access to non-authorized users or limit too much information sharing, and, on the other hand, it is not flexible enough to express the heterogeneous access control requirements that different WBSN users may have. For instance, such access control paradigm does not take into account the 'type' of the relationships existing among users. Consequently, it is not possible to state access control policies such as "only my friends or my colleagues can access a given piece of information."

We believe that more sophisticated and flexible mechanisms are therefore required, thanks to which WBSN members can denote which users can access the information they publish. A straightforward solution to this issue is to allow a resource owner to explicitly list the set of WBSN members authorized to access a given resource. This could be expressed by traditional access authorizations, which, in their basic form, are tuples $\langle s, p, o \rangle$, where s is the subject authorized to access object o under privilege p (Bertino & Sandhu, 2005). However, such an approach is not suitable for dynamic and distributed environments as WBSNs are, since a member may be required to update the authorizations applying to his/her resources whenever he/she knows new members or if old relationships he/she has with other members have been revoked. In such a scenario, it is preferable to *intensionally* denote authorized members by specifying the *requirements* they must satisfy to access a given resource. According to this approach, whenever any modification to the state of the WBSN structure occurs, the set of authorized members will dynamically change, without the need of modifying the existing authorizations.

So far, a variety of access control models have been proposed which denote authorized users in terms of their characteristics, and not only by their identities. The role-based model—for example, Ferraiolo, Kuhn, and Chandramouli (2003)—is the most popular one; others are those based on credentials—for example, Agarwal, Sprick, and Wortmann (2004); Winslett, Ching, Jones, and Slepchin (1997)—or certificates—for example, Palomar, Estevez-Tapiador, Hernandez-Castro, and Ribagorda (2006); Thompson, Johnston, Mudumbai, Hoo, Jackson, & Essiari (1999). A similar approach can be applied to WBSNs. In fact, WBSN members usually publish resources having in mind a specific audience consisting, for example, of their friends or colleagues. In other words, in a WBSN context, *relationships* can be used as the basis for specifying authorizations, such as "only my friends can access this resource" or "only WBSN members who are both friends and colleagues of mine can access this resource." Based on such considerations, in (Carminati, Ferrari, & Perego, 2006) we proposed an access control model for WBSNs, where authorizations are specified in terms of the *type*, maximum *depth*, and minimum *trust level* of a relationship. As far as access control enforcement is concerned, we have adopted the *client-side* approach outlined by Weitzner, Hendler, Berners-Lee, and Connolly (2006), where access to resources is granted if the requestor is able to demonstrate that he/she satisfies the requirements stated by the specified access rules, that is, he/she is able to demonstrate to have the relationships required by the authorizations.

However, when client-side access control is adopted, attention should be paid to the fact that relationships themselves that should be protected while enforcing access control. In fact, it may be the case that a WBSN member would like to keep private a relationship he/she has with another member, and thus such relationship should not be exploited by other members for access control purposes. For instance, being aware that a given member participates to a network of consultants of a company *X*, makes one able to infer that he/she is a consultant of company *X*.

To cope with these privacy requirements on WBSN relationships, in (Carminati, Ferrari, & Perego, 2007) we have proposed to specify them through a set of *distribution rules*, which basically state the protection requirements to be enforced on a relationship. By using distribution rules, WBSN members can state which other members can be aware of such relationship (e.g., "only my direct friends and/or colleagues"). However, in that article the focus was only on how to specify relationship privacy requirements, whereas privacy and access control enforcement was just sketched. Moreover, in (Carminati, Ferrari, & Perego, 2007), the sketched enforcement approach was centralized, that is, it relies on the presence of a trusted WBSN management system. Centralized approaches have the disadvantage that the WBSN management system may become a bottleneck for the whole WBSN. This is particularly crucial considering that WBSNs usually consist of a huge number of members. For these reasons, in this chapter we adopt a decentralized architecture that makes use of cryptographic techniques to enforce privacy requirements on WBSN relationships. This decentralized solution grants, on the one hand, the efficiency of the system by avoiding the bottleneck of a central management service, and, on the other hand, prevents sensitive information from being accessed even by the WBSN management system. More precisely, existing relationships and resources are not stored in a central repository, but by WBSN members themselves, who also carry out the tasks related to access control enforcement and privacy protection. In this chapter, we illustrate the protocols we have devised to carry on these tasks, and we analyze their robustness with respect to the most common security attacks. Moreover, in (Carminati, Ferrari, & Perego, 2007) relevant issues were not addressed—for example, the ones related to relationship revocation—which are considered in this chapter.

Although social network security is becoming a more and more relevant research topic (Carminati, Ferrari, & Perego, 2008), to the best of our knowledge, ours is the first proposal of a security framework for WBSNs integrating both privacy protection and access control. Recently, two position articles recognize the relevance of addressing access control issues in WBSNs. In particular, Gates (2007) describes *relationship-based access control* as one of the new security paradigms that addresses the requirements of the so-called Web 2.0. Whereas Hart, Johnson and Stent (2007) identify content-based and relationship-based access control as the key requirements for protecting WBSN resources. Which makes use of relationship information available in WBSNs for denoting authorized subjects. However, those articles do not address access control and privacy requirements enforcement, an issue that we believe is fundamental to make any security solution usable in real-world scenarios. As far as privacy is concerned, research on this issue is currently focusing mainly on privacy preserving data mining techniques that allow social network analysis without disclosing possible sensitive information. Most proposals address this issue by anonymizing WBSN nodes, thus making very difficult to associate them with specific users, and to know which are the members participating in the existing relationships—see, for example,

Wang, Liau, and Hsu (2006); Hay, Miklau, Jensen, Weis, and Srivastava (2007); Backstrom, Dwork, & Kleinberg (2007).

The remainder of this chapter is organized as follows. First, we introduce the main notions on which our approach is based. Then, we illustrate how access control and relationship privacy requirements are enforced. Subsequently, we analyze the robustness of our system to the main security attacks which it may be subject to. We conclude the chapter by outlining future research directions. The algorithms enforcing our protocols are illustrated in the appendix to this chapter.

BACKGROUND

In this section we provide some preliminaries notions used throughout the chapter. In particular, we first introduce how we formally represent a WBSN, and then we illustrate the models we proposed in (Carminati et al., 2006 & 2007) to specify access control and privacy requirements in WBSNs.

Web-Based Social Networks

As depicted in Figure 1, a WBSN can be represented as a graph, where each node corresponds to a WBSN member, whereas edges denote relationships between WBSN members. Since WBSN members can establish relationships of different nature (e.g., "friend of", "colleague", "parent of") and not all relationships are mutual (e.g., "parent of"), we model a WBSN as a directed labeled graph, where the label represents the type of the relationship, whereas the initial node of an edge denotes the WBSN member who established the corresponding relationship, and the terminal node denotes the WBSN member who accepted to establish the relationship. Another relevant characteristic of WBSNs is that relationships may be characterized by a *trust level*, representing how much a given WBSN member considers trustworthy another member. Similarly to the relationship type, the trust level is modeled as a label associated with the corresponding edge.

As an example, referring to the WBSN depicted in Figure 1, each edge is labeled with a pair (*rt*, *t*), where *rt* denotes a relationship type, whereas *t* is a rational number ranging from 0 to 1, denoting the trust level of the corresponding relationship. Thus, we say that *A*(lice) is a friend of *B*(ob), and that she totally trusts him (1 = 100%), Bob is friend of *D*(avid), and he trusts him 50% (0.5), and so forth. Note that the semantics of trust levels depends on the specific purpose of the WBSN. For instance, in WBSNs supporting collaborative

Figure 1. A portion of a WBSN. Labels associated with edges denote, respectively, the type and trust level of the corresponding relationship.

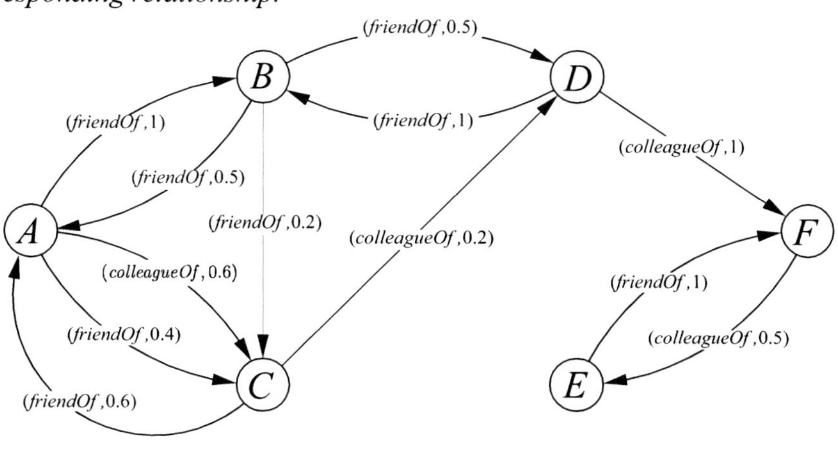

rating, trust levels are used to denote how much a given member trusts the *opinions* of another member, and are independent from the types of relationship existing between them (e.g., whether a member is friend or colleague of mine does not affect how I consider trustworthy his/her opinions about movies). By contrast, in our scenario, trust should convey information about how much confidence I have that a given user does not reveal information to unauthorized users, and thus its purpose has some similarities with the notion of *security level* used in mandatory access control models (see Bertino & Sandhu, 2005 for a survey). Moreover, such confidence may very depending on the relationship I have with that user. As an example, suppose that I have a relationship of type *friendOf* and one of type *colleagueOf* with the same user: in such a case, the confidence I have with him/her as a friend rather than as colleague may be different, since it depends on the context (free time vs. work activities) and on the set of resources to be disclosed (recreative vs. work documents). For these reasons, in our model we use a new type of trust levels, called security trust levels, which are represented by the range of rational numbers between 0 and 1.

Therefore, we formally define a WBSN $\mathcal{SN}$ as a tuple $(V_{\mathcal{SN}}, E_{\mathcal{SN}}, RT_{\mathcal{SN}}, T_{\mathcal{SN}}, \phi_{\mathcal{SN}})$, where $RT_{\mathcal{SN}}$ is the set of supported relationship types, $V_{\mathcal{SN}}$ and $E_{\mathcal{SN}} \subseteq V_{\mathcal{SN}} \times V_{\mathcal{SN}} \times RT_{\mathcal{SN}}$ are, respectively, the nodes and edges of a directed labeled graph $(V_{\mathcal{SN}}, E_{\mathcal{SN}}, RT_{\mathcal{SN}})$, $T_{\mathcal{SN}} = [0, 1]$ is the set of supported trust levels, whereas $\phi_{\mathcal{SN}}: E_{\mathcal{SN}} \rightarrow T_{\mathcal{SN}}$ is a function assigning to each edge $e \in E_{\mathcal{SN}}$ a trust level $t \in T_{\mathcal{SN}}$.

In the following, we say that two WBSN members *A* and *B* *participate* in a relationship of a given type *rt* if there exists a path, consisting only of edges labeled with relationship type *rt*, connecting *A* with *B*. The length of such path corresponds to the *depth d* of the corresponding relationship: if $d = 1$, we say that the relationship is *direct*, whereas, if $d > 1$, we say that the relationship is *indirect*. In case of direct relationship, the trust level is equal to the one associated with the only edge of the corresponding path. By contrast, in case of indirect relationships, the trust level corresponds to the product of the trust levels associated with the edges of which the corresponding path consists. As an example, members *A* and *D* in Figure 1 participate in an indirect relationship of type *friendOf*, depth equal to 2, and trust level equal to $1 \times 0.5 = 0.5$.

Access Control and Privacy Protection in WBSNs

In this section we survey how access control and relationship privacy requirements can be specified in our system. More details can be found in (Carminati et al., 2006 & 2007). Then, in the next sections we show how these requirements are enforced.

As far as access control is concerned, members authorized to access a given resource are denoted in terms of the type, depth, and trust level of existing relationships in the network. In particular, such requirements are specified through one or more *access rules*, each of which consists of a set of *access conditions* denoting *all* the requirements a WBSN member must satisfy to access a given resource.

Formally, an access condition is a tuple AC = (*v*, *rt*, *d*max, *t*min), where $v \in V_{\mathcal{SN}}$ is the member with whom the requestor of a given resource must have a direct or indirect relationship to obtain the access, whereas $rt \in RT_{\mathcal{SN}}$, $d\text{max} \in \mathbf{N}$, and $t\text{min} \in T_{\mathcal{SN}}$ are, respectively, the type, maximum depth, and minimum trust level that the relationship must have. Notice that, since between two nodes there could exist more relationships of type *rt*, we consider as relationship depth the length of the shortest path existing between them. Thus, given a resource *rsc*, owned by a user in $V_{\mathcal{SN}}$, an access rule AR is a pair (*rid*, *ACS*), where *rid* is the identifier (i.e., the URI) of resource *rsc*, whereas *ACS* is the set containing *all* the access conditions to be satisfied to be authorized to access resource *rsc*. Consequently, the semantics of a set of access conditions $ACS = \{\text{AC}_1, \ldots, \text{AC}_n\}$ is a conjunction

$\text{AC}_1 \wedge \ldots \wedge \text{AC}_n$. The same resource can be associated with one or more *alternative* access rules: in other words, the set $ARS = \{\text{AR}_1, \ldots, \text{AR}_n\}$ of access rules applying to a resource corresponds to a disjunction $\text{AR}_1 \vee \ldots \vee \text{AR}_n$.

Example 1

Consider the access rules in Table 1 and the WBSN in Figure 1. The following are examples of access rule sets:

- $ARS_1 = \{\text{AR}_1\}$: *B*(ob), *C*(arol), and *D*(avid) satisfy ARS_1, since the relationship of type *friendOf* they have with *A*(lice) has a depth less than or equal to 2, and a trust level equal to or greater than 0.4.
- $ARS_2 = \{\text{AR}_2\}$: authorized members must be both friends and colleagues of Alice. Both Carol and David participate in relationships of the required depth, but only Carol satisfies the constraint on the minimum trust level.
- $ARS_3 = \{\text{AR}_1, \text{AR}_3\}$: authorized members must be either friends or colleagues of Alice. The policy is satisfied by Bob, Carol, and David.

To protect relationship privacy, we allow members to state whether and how their relationships should be accessible by other WBSN's members. Relationship privacy requirements are established through a set of rules, called *distribution rules*, which basically state how information about existing relationships should be distributed among users in the WBSN.

Distribution rules have a syntax and a semantics analogous to the one of access rules. More precisely, given a relationship of type *rt* corresponding to edge $e_{rt} \in E_{SN}$, a distribution rule DR on e_{rt} consists of a set $DCS = \{\text{DC}_1, \ldots, \text{DC}_n\}$ of *distribution conditions*, denoting *all* the requirements a WBSN member must satisfy in order to be authorized to know the existence of e_{rt}. A distribution condition DC is a tuple (*m*, *rt*, *d*max), where *m* is one of the two members participating in the relationship with which the condition is associated, whereas $rt \in RT_{SN}$ and $d\text{max} \in \mathbf{N}$ express, respectively, constraints on the type and maximum depth of the relationship. Finally, alternative distribution requirements are expressed by associating with the same relationship two or more distribution rules. Thus, a set of distribution rules $DRS = \{\text{DR}_1, \ldots, \text{DR}_n\}$ corresponds to a disjunction $\text{DR}_1 \vee \ldots \vee \text{DR}_n$.

Example 2

Consider the distribution rules in Table 2 and the WBSN in Figure 1. Suppose that Carol establishes a relationship of type friendOf with E(ve), associated with one of the following distribution rule sets:

Table 1. Examples of access rules used throughout the chapter

Access Rule	Access Conditions	Authorized Members
AR_1	{(*A*, *friendOf*, 2, 0.4)}	Members who are in/direct friends of Alice, having a maximum depth equal to 2 and a minimum trust level equal to 0.4.
AR_2	{(*A*, *friendOf*, 2, 0.4), (*A*, *colleagueOf*, 2, 0.2)}	Members who are both in/direct friends and colleagues of Alice, having a maximum depth equal to 2 and, a minimum trust level equal to 0.4 as friend and 0.2 as colleague.
AR_3	{(*A*, *colleagueOf*, 2, 0.2)}	Members who are in/direct colleagues of Alice, having a maximum depth equal to 2 and a minimum trust level equal to 0.2.
AR_4	{(*A*, *friendOf*, *, 0.5)}	Members who are in/direct friends of Alice, with no constraint on the maximum depth, and a minimum trust level equal to 0.5.
AR_5	{(*A*, *colleagueOf*, 4, *)}	Members who are in/direct colleagues of Alice, with no constraint on the minimum trust level, and a maximum depth equal to 4.

Table 2. Examples of distribution rules used throughout the chapter

Distribution Rule	Distribution Conditions	Meaning
DR_1	$\{(C, friendOf, *)\}$	All the in/direct friends of Carol, with no constraint on the maximum depth.
DR_2	$\{(C, friendOf, *), (E, friendOf, 2)\}$	Members who are in/direct friends of Eve, with a maximum depth equal to 2, and, at the same time, satisfy DR_1.
DR_3	$\{(C, colleagueOf, *)\}$	All the in/direct colleagues of Carol, with no constraint on the maximum depth.
DR_4	$\{(D, friendOf, 2)\}$	Members who are in/direct friends of David, with a maximum depth equal to 2.
DR_5	$\{(A, friendOf, 2)\}$	Members who are in/direct friends of Alice, with a maximum depth equal to 2.

- $DRS_1 = \{DR_1\}$: all the members of the WBSN in Figure 1 satisfy DRS_1. More precisely, Alice is a direct friend of Carol, whereas Bob, David, and Fred are indirect friends of Carol, since, respectively, Bob is friend of Alice, David of Bob, and Fred of Eve.
- $DRS_2 = \{DR_2\}$: only Fred satisfies DRS_2.
- $DRS_3 = \{DR_1, DR_3\}$: all the members of the WBSN in Figure 1 satisfy DRS_3, since each member is either friend or colleague of Carol.

Privacy-Aware Access Control Enforcement

In this section, we illustrate the security protocols we have devised to enforce access control while protecting relationships privacy.

We adopt a client-side access control enforcement according to which a requestor is authorized to access a resource only if he/she demonstrates that he/she satisfies the access rules applying to the requested resource. As introduced in the previous section, access rules state which type of relationship should exist between the requestor and another WBSN member, and the maximum depth and minimum trust level required. Adopting a client-side enforcement implies that the requestor has to provide the owner a *proof* showing that there exist the relationships required by the specified access rules, with the required depth and trust level. In order to make a WBSN member able to generate a proof, we assume that with each relationship a certificate is associated, generated by the members participating in it. The relationship certificate is, then, delivered to WBSN members according to the set of distribution rules applying to it (see the section titled "Certificate Protection" for more details).

Therefore, a proof of the existence of a relationship between two members is given by the set of certificates corresponding to each relationship in the path connecting them. When the owner receives this set of certificates, which we refer to as *certificate path*, he/she is able to verify whether the relationship satisfies the constraints on depth and trust level stated in the access rules.

However, enforcing client-side access control without violating privacy requirements of WBSN relationships requires addressing two problems. The first is related to the fact that a certificate path may contain certificates the resource owner is not authorized to access, according to the specified distribution rules, as the following example illustrates.

Example 3

Consider the access rules in Table 1, and suppose that to access Carol's resource rsc, having ID rid, David must satisfy access rule ar1. Referring to

Figure 1, there exists a path ABD satisfying ar1. Suppose that such path is known by David, but that Carol is not authorized to be aware of the relationship corresponding to edge BD, because of the distribution rule specified by Bob. In such a case, if no proper solutions are adopted, when Carol receives the certificate path corresponding to ABD, she will be disclosed information on a relationship (the one corresponding to edge BD) he/she isnot authorized to know.

Another possible privacy breach is due to the possible inferences that a requestor can perform during access control enforcement. Indeed, according to our approach, when a member requests a resource, the owner replies by sending him/her the set of access rules regulating the access to that resource. Based on the information contained in these access rules, some inferences on existing relationships are possible, as illustrated by Example 4.

Example 4

Consider once again access rule AR_1 in Table 1, and suppose that Eve knows only the relationships existing between her and Fred. Now, if she sends an access request to Carol for resource *rsc*, associated with AR_1, when Eve receives such rule she can infer that it is very likely that Alice participates in at least one relationship of type *friendOf*, otherwise no WBSN member will be authorized to access *rsc*.

In order to cope with these issues, our framework implements strategies to protect private information contained in the proof (i.e., the certificate path) and access rules. In the following sections, we first illustrate such strategies and then describe our access control protocol.

However, before doing that, we need to more formally introduce the notion of *relationship certificate*. In our framework, information about relationships is encoded in certificates of the form $\text{RC} = (m, m', rt, t, \text{TS}) \| DSig_{m, m'}$, which denotes a relationship of type *rt* and trust level *t*, established by *m* with *m'* at instant TS. This information is complemented with its digital signature $DSig_{m, m'}$, corresponding to the hash of tuple (m, m', rt, t, TS), encrypted with both *m*'s and *m'*'s private keys. Moreover, each certificate RC has associated an identifier id_{RC}, corresponding to the hash of tuple (m, m', rt, t, TS), and a unique random number rn_{RC}, generated by *m*, whose purpose is to secure certificates when performing the access control protocol, as it will be explained in the next section.

To ensure message authenticity and confidentiality, we assume that each message sent by a member *m* to a member *m'* is encrypted with the private key SK_m of *m* and then with the public key $PK_{m'}$ of *m'*. Moreover, we assume that each WBSN member is equipped with a certificate directory *CD*, storing all the certificates corresponding to the relationships he/she is authorized to know, and that the WBSN service hosts a central certificate revocation list $\mathcal{CRL}$, storing all the certificates revoked by WBSN members. The protocols in charge of delivering certificates to authorized members will be described in the section titled "Certificate Protection".

Protecting Certificate Paths

In order to prevent the owner from using the proof to infer information on private relationships, we introduce the notion of *secured certificate*. More precisely, given a certificate RC, the corresponding secured certificate $\underline{\text{RC}}$ generated by member *m* is a pair $(id_{\text{RC}}, (\text{RC} \| id_m) \oplus rn_{\text{RC}})$, where id_{RC} and rn_{RC} are, respectively, the ID and random number of certificate RC, whereas id_m is the ID of member *m*.[1] Thus, whenever a resource owner *o* receives from a member *m* a set of secured certificates as a proof, in order to validate it, he/she performs, for each of them, the following steps: (1) *o* retrieves from his/her certificate directory the corresponding random number rn_{RC}, if any; (2) *o* computes the following XOR $((\text{RC} \| id_m) \oplus rn_{\text{RC}}) \oplus rn_{\text{RC}}$, which returns $\text{RC} \| id_m$; (3) *o* verifies whether id_m corresponds to the ID of *m*; if they match, $\underline{\text{RC}}$ has been actually sent by *m*.

Secured certificates ensure that the owner accesses only certificates for which he/she satisfies the corresponding distribution rules, that is, certificates for which he/she has been provided with the corresponding random number. In what follows, given a certificate path $CP = \{\mathrm{RC}_1, \ldots, \mathrm{RC}_n\}$ we denote the corresponding *secured certificate path* by $\underline{CP} = \{\underline{\mathrm{RC}}_1, \ldots, \underline{\mathrm{RC}}_n\}$, where $\underline{\mathrm{RC}}_1, \ldots, \underline{\mathrm{RC}}_n$ are the secured certificates corresponding to $\mathrm{RC}_1, \ldots, \mathrm{RC}_n$.

Protecting Access Rules

As already mentioned, whenever a resource owner replies to an access request with a set of access rules, they can be exploited by the requestor to infer information on the relationships existing in a WBSN.

In order to address this issue, our framework makes use of cryptographic techniques. More precisely, we assume that whenever two WBSN members m and m' establish a relationship of a given type rt, they also agree on a symmetric key K, and on the corresponding key identifier id_K, to be associated with relationship type rt. Hereafter, we refer to this key as *relationship key*. Then, when m or m' establish a new relationship of type rt with other WBSN members, they distribute to the new neighbors the key associated with such relationship type. Consequently, all of these members share the same key for relationship type rt. More formally, we can say that a relationship key is a symmetric key shared only by WBSN members who are nodes in a connected subgraph $SNG_{rt} \subseteq \mathcal{SN}$, consisting of edges all labeled with relationship type rt. In what follows, we refer to such subgraphs SNG_{rt} as *social network groups* (SNGs, for short) of type rt, whereas we denote by $K_{SNG\,rt}$ (or simply by K_{rt}) the relationship key shared by the members in SNG_{rt}.

Relationship keys make us able to avoid inferences on private relationship information due to access rules distribution, by simply replacing each access condition in the access rules to be sent to the requestor with the corresponding encrypted version. More precisely, given an access condition AC_{rt} expressing a constraint on a relationship type rt, AC_{rt} is not sent to the requestor in plaintext, but it is first encrypted with the encryption key corresponding to rt, that is, $K_{SNG\,rt}$.

The idea is that, if a requestor r is not able to decrypt AC_{rt}, this means that he/she is not a member of the same SNG of type rt the resource owner o belongs to, and thus no path can exist between r and o.

More precisely, given an access condition AC_{rt}, the corresponding encrypted version, referred to as *secured access condition* $\underline{\mathrm{AC}}$, is a pair $(id_{K\,rt}, E_{K\,rt}(\mathrm{AC}_{rt}))$, where $id_{K\,rt}$ is the ID of key K_{rt}, whereas $E_{K\,rt}(\mathrm{AC}_{rt})$ is the encryption of access condition AC_{rt} with key K_{rt}.

In what follows, we use also the notions of *secured access rule* and *secured access rule set*. More precisely, given an access rule $\mathrm{AR} = (rid, \{\mathrm{AC}_1, \ldots, \mathrm{AC}_n\})$, we denote the corresponding secured access rule by $\underline{\mathrm{AR}} = (rid, \{\underline{\mathrm{AC}}_1, \ldots, \underline{\mathrm{AC}}_n\})$, where $\underline{\mathrm{AC}}_1, \ldots, \underline{\mathrm{AC}}_n$ are the secured access conditions corresponding to $\mathrm{AC}_1, \ldots, \mathrm{AC}_n$. Similarly, given a set of access rules $ARS = \{\mathrm{AR}_1, \ldots, \mathrm{AR}_n\}$, we denote the corresponding secured access rule set by $\underline{ARS} = \{\underline{\mathrm{AR}}_1, \ldots, \underline{\mathrm{AR}}_n\}$, where $\underline{\mathrm{AR}}_1, \ldots, \underline{\mathrm{AR}}_n$ are the secured access rules corresponding to $\mathrm{AR}_1, \ldots, \mathrm{AR}_n$.

Given the relevance of relationship keys, we need to devise a key management scheme ensuring that each relationship key is held only by the members of the corresponding SNG. Moreover, due to the dynamic nature of a WBSN, we need a key management scheme able to securely handle events that could change the SNG structure. As an example, suppose that a member m is disconnected from a SNG because of the revocation of a given relationship. In such a case, m will still hold SNG's relationship key, even though he/she is no longer in the group. By contrast, the establishment of a new relationship between members of two distinct groups SNG_{rt} and SNG'_{rt}, determines the merge of the two SNGs, with the consequence that the members in the new group share two different keys for the same relationship type rt. In order to cope with all these SNG events, we need first to precisely categorize them. More precisely, we can identify two main classes of SNG events, which should be taken into account in devising a management scheme for relationship keys.

The first class of events is fired by the establishment of a new relationship of type *rt*. In particular, a *join* event occurs when a member *m*, not belonging to SNG_{rt}, establishes a relationship of type *rt* with a member of SNG_{rt}. In contrast, a *pseudo-join* occurs if a member of SNG_{rt} establishes a new relationship of type *rt* with another member of SNG_{rt}. In this case, the SNG_{rt} does not enlarge its members since both the members already belong to SNG_{rt}.

Regarding the establishment of new relationships, there could be a further relevant event. Indeed, it could be the case that in a WBSN there exist two or more disjoint SNGs for the same relationship type. In this case, the establishment of a new relationship between members belonging to two disjoint SNGs implies merging the two SNGs. Therefore, a *merge* event occurs when, given two disjoint SNGs for relationship type *rt*, say SNG_{rt} and SNG'_{rt}, a member of SNG_{rt} establishes a new relationship of type *rt* with a member of SNG'_{rt}, or vice versa. After the merge, both SNG_{rt} and SNG'_{rt} cease to exist, whereas a new SNG for relationship type *rt* is created, consisting of all members of SNG_{rt} and SNG'_{rt}.

The latter class of events is fired by the revocation of a relationship of type *rt*. The first event is called *leave*, and occurs when the revocation of a relationship in SNG_{rt} implies that one of its members does not belong any longer to SNG_{rt}. This happens when a member belongs to SNG_{rt} due to a unique relationship of type *rt*, which is being revoked. Similarly to the establishment of a new relationship, relationship revocation does not always imply leaving the group. Indeed, it could be the case that, after the revocation, there still exist further relationships of type *rt* that connect the member to SNG_{rt}. In this case, the revocation fires the so-called *pseudo-leave* event. Finally, the revocation of a relationship may split a SNG into two disjoint subgraphs. This happens when in a SNG_{rt} there exist two subgraphs connected by a unique edge of type *rt*, and the corresponding relationship is revoked. We refer to this event as *partition*.

In order to deal with all these situations, our framework enforces a relationship key generation and distribution protocol, which performs two different procedures depending on whether a relationship is established or revoked. We assume that each WBSN member is equipped with a key directory *KD*, storing, for each relationship key he/she holds, a tuple, $(id_{K\,rt}, K_{rt}, \text{TS}_{K\,rt})$, where $id_{K\,rt}$ and $\text{TS}_{K\,rt}$ are, respectively, the ID and the timestamp of key K_{rt}. Note that the number of relationship keys held by a member is at most equal to $|RT_{SN}|$, where RT_{SN} is the set of relationship types supported by the WBSN. In fact, a member can belong to at most one SNG for each relationship type $rt \in RT_{SN}$, and thus he/she cannot hold more than one key for each of the supported relationship types.

In case a relationship of type *rt* is established by member *m* with another member *m'*, the following procedure is carried out:

1. If both *m* and *m'* already have the same key for relationship type *rt*, this means that *m* and *m'* belong to the same SNG, and thus a pseudo-join event occurred. Consequently, no further operation is performed.
2. Otherwise, we may have three different cases:
    a. if only *m* (*m'*) has a key for relationship type *rt*, a join event occurred;
    b. if *m* and *m'* have two different keys for relationship type *rt*, a merge event occurred;
    c. if *m* and *m'* have no key for relationship type *rt*, this means that they were formerly isolated.

In the above cases, *m* and *m'* generate a new key *K*, with a new timestamp TS_K and ID id_K.

By contrast, in case a relationship of type *rt* between members *m* and *m′* is revoked, the following procedure is carried out by both *m* and *m′*:

1. If *m* and/or *m′* do not participate any longer to at least one relationship of type *rt*, this means that *m*, *m′*, or both, are now isolated from the SNG_{rt}. Thus a leave event occurred. In such a case, if both *m* and *m′* are isolated, they remove the relationship key they hold. By contrast, if *m* (*m′*) still belongs to SNG_{rt} and *m′* (*m*) is isolated, *m′* (*m*) removes the relationship key he/she holds, whereas *m* (*m′*) computes a new key *K* (*K′*), with a new timestamp TS_K ($TS_{K'}$) and ID id_K ($id_{K'}$).
2. If *m* (*m′*) still participates to at least one relationship of type *rt*, this means that he/she is still connected to a SNG of type *rt*. However, *m* (*m′*) cannot verify whether he/she still belongs to the same group of *m′* (*m*) or to a different SNG. Consequently, *m* (*m′*) cannot verify whether a pseudo-leave or a partition event occurred. In such a case, *m* computes a new key *K* (*K′*), having timestamp TS_K ($TS_{K'}$) and ID id_K ($id_{K'}$).

Table 3 summarizes the possible membership events and the corresponding actions performed by WBSN members in order to update relationship keys.

Once the new key has been generated, *m* and/or *m′* distribute it to other SNG members. More precisely, *m* and/or *m′* forward the new relationship key to all their neighbors with which they have established a relationship of type *rt*. In turn, such members forward the new key to their neighbors with which they have established a relationship of type *rt*, and so on. The procedure is iterated by each member receiving the key until it is received by members with no outgoing edges of type *rt*.[2]

Note, however, that key propagation has the drawback that keys do not necessarily reach a member according to the chronological order of their generation. In contrast, the order according to which a member receives new keys is affected by the distance existing between him/her and the members who generated the new keys. A simple solution to this issue is to use the timestamp associated with the key in order to identify the most recent one, which will be then considered as the new relationship key.[3] However, it may be the case that two keys carry the same timestamp.

Table 3. Events and corresponding actions performed by WBSN members on relationship keys

Operation	Event	Event Description	Action
Member *m* establishes a relationship of type *rt* with member *m′*.	*Join*	*m* and/or *m′* did not belong to any SNG of type *rt*.	*m* and *m′* generate a new relationship key.
	Merge	*m* and *m′* belonged to two different SNGs of type *rt*.	
	Pseudo-join	Both *m* and *m′* already belonged to the same SNG of type *rt*.	No action
Member *m* revokes a relationship of type *rt* he/she previously established with member *m′*.	*Leave*	*m* and/or *m′* do not belong any longer to a SNG of type *rt*.	*m* and/or *m′* delete the relationship key. If *m* or *m′* is still in the SNG, he/she computes a new relationship key.
	Pseudo-leave	Both *m* and *m′* still belong to the same SNG of type *rt*.	*m* (*m′*) computes a new relationship key.
	Partition	*m* and *m′* belong to two different SNGs of type *rt*	

Although this is not likely to happen, provided that timestamps have enough precision, such event may occur in case the SNG consists of a high number of members. In order to deal with this issue, we adopt the following policy. Suppose that a member m receives a new relationship key K, with timestamp TS_K, and, later on, another key $K' \neq K$, with timestamp $\text{TS}_{K'} = \text{TS}_K$. In this case, m computes a new relationship key $K'' = K \oplus K'$ and propagates K'. User m will then use K'' to encrypt access conditions expressing constraints on the corresponding relationship type. Indeed, all users receiving K' are able to compute K'' in that they have already received K.

The protocol in charge of generating and distributing relationship keys is illustrated in the appendix to this chapter.[4]

Access Control Protocol

As introduced earlier in this section, our access control protocol adopts a client-side approach according to which the requestor of a resource is in charge of proving to be authorized to access it. The basic idea on which our access control procedure relies is the following. Whenever a requestor submits an access request for a resource, the resource owner sends him/her the corresponding secured access rules. The requestor first verifies whether he/she is able to read *all* the conditions in at least one secured access rule. If this is the case, he/she analyzes the certificates stored in his/her directory in order to check whether he/she satisfies the secured access rules. Then, he/she sends back to the resource owner the set of secured certificate paths, if any, demonstrating that he/she participates in relationships of the required type, depth, and trust level, denoted by at least one of the secured access rules. If the received secured certificate paths are *valid*, the resource owner sends the resource to the requestor. A secured certificate path is considered valid if (a) all the contained secured certificates can be read by the resource owner, and (b) none of such certificates has been revoked.

Our access control protocol is illustrated in the appendix to this chapter. Here, we illustrate it through an example.

Example 5

Consider the WBSN in Figure 1 and the access rules in Table 1. Suppose that Fred requests access to a resource rsc_1, owned by Alice, and that rsc_1 is protected by access rule AR_4. Then, Alice sends Fred a message containing the secured version of AR_4, namely, $\underline{\text{AR}}_4 = (rid_1, \{(id_{K\,friendOf}, E_{K\,friendOf}(A, friendOf, *, 0.5))\})$, where $K_{friendOf}$ is the relationship key of the SNG of type *friendOf* Alice belongs to. According to Figure 1, Fred does not belong to such SNG. Thus, he does not hold $K_{friendOf}$ and therefore he is not able to read $\underline{\text{AR}}_4$. Consequently, Fred is aware that he is not authorized to access resource rsc_1, and he does not perform any further action.

Suppose now that Fred sends another access request to Alice, concerning resource rsc_2, which is associated with the single access rule AR_5. Consequently, Alice sends Fred a message containing the secured version of AR_5, namely, $\underline{\text{AR}}_5 = (rid_2, \{(id_{K\,colleagueOf}, E_{K\,colleagueOf}(A, colleagueOf, 4, *))\})$, where $K_{colleagueOf}$ is the relationship key of the SNG of type *colleagueOf* Alice belongs to. According to Figure 1, Fred belongs to such SNG, and thus he is able to read $\underline{\text{AR}}_5$. Consequently, Fred analyzes his certificate directory in order to build a certificate path demonstrating that the required relationship exists. As we can see from Figure 1, such a path exists, namely, *ACDF*, denoting a relationship of type *colleagueOf* and depth equal to 3. Thus, Fred *might* be authorized to access rsc_2. In such a case, we have two possible scenarios:

1. Fred does not hold all the certificates corresponding to the edges in path *ACDF*. This means that Fred is not aware that such path exists, and he is not able to build a valid certificate path. Consequently, he does not perform any further action. Otherwise,

2. Fred holds all such certificates. Consequently, he is able to build a valid certificate path, denoting a relationship satisfying AR_5. After having sent the secured version of such certificate path to Alice, Fred may receive two possible responses:
   - Alice does not hold one or more of the secured certificates sent by Fred. This means that she will not be able to XOR and read such certificates, because she does not know the corresponding random number. Consequently, Alice sends back to Fred an access denied message.
   - Alice holds all the certificates corresponding to the edges in *ACDF*. Consequently, she is able to read all the certificates by XORing with the correct random number their secured version sent by Fred. Consequently, Alice contacts the central certificate revocation list, and, if none of them has been revoked, she sends Fred resource rsc_2. Otherwise, she sends back to him an access denied message.

In the protocol reported in the appendix to this chapter, the most costly task is the certificate paths generation, which is performed as follows. First, the requestor generates an adjacency list from the certificates stored in his/her directory, by considering only the relationship type specified in the access condition under evaluation. This limits the analysis only to the WBSN view of the requestor with respect to a given relationship type. Then, the adjacency list is analyzed by using a breadth-first search (BFS) algorithm. In addition, the maximum depth and minimum trust level parameters in access conditions are used to limit the search.

The BFS algorithm is performed for each access condition of each access rule in the set. Thus, building certificate paths requires $O(\sum_{\text{AR} \in ARS} \sum_{\text{AC} \in ACS(\text{AR})} (V+E))$ time complexity, where $V \subseteq V_{SN}$ and $E \subseteq E_{SN}$ are the sets of nodes and edges, respectively, corresponding to the WBSN view of the requestor, $ARS = \{\text{AR}_1, \ldots, \text{AR}_n\}$ is the set of access rules, whereas $ACS(\text{AR}) = \{\text{AC}_1, \ldots, \text{AC}_n\}$ is the set of access conditions of an access rule $\text{AR} \in ARS$. However, we believe that, in a real-world scenario, the complexity of the access control protocol is far lower than this upper bound.

First, although, in theory, there is no limit on the number of access rules in *ARS* and of access condition in each rule, in practice we believe that there exists a threshold over which it is very difficult for a resource owner to realize the meaning of a policy. For instance, on one hand, it is very difficult that a WBSN member satisfies an access rule containing a high number of access conditions, each on a different relationship type, and, on the other hand, if the access rules applying to a resource are too numerous, it is very likely that a WBSN member satisfies at least one of them. In such a case, access rules may grant access to unauthorized members or, vice versa, members meant to be authorized may be prevented from accessing resources. Moreover, as far as the cardinality of V and E is concerned, we can assume that the size of the WBSN view of a member is, on the average, far lower than the actual size of the WBSN graph.

It is also important to note that the search is performed for all the access rules in the set only in the worst case. We remind here that a requestor can be granted access to a resource if he/she satisfies at least one of the rules in the set, so the procedure terminates as soon as a rule is satisfied. Finally, since access rules are sent encrypted to the requestor, the search is performed only for those access rules for which the requestor holds the required relationship keys.

CERTIFICATE PROTECTION

As explained in the previous section, relationships are encoded into certificates used to generate the proofs necessary to gain resource access. Certificates are then delivered to WBSN members according to a set of distribution rules. In this section, we illustrate how these rules are enforced, by presenting the overall process carried out once a new relationship is established, or an existing one is revoked or updated. In particular, we can summarize this process into two main tasks: (1) generation, update, or revocation of the corresponding certificate; (2) distribution of the new certificate or of the update/revocation message to WBSN members, according the related distribution rules. It is relevant to note that a further step is needed when a new relationship is established or en existing one is revoked. Indeed, in this case, it could happen that, due to the new relationship, some members may now satisfy distribution rules concerning certificates they currently do not hold, whereas, when a relationship is revoked, some members may no longer satisfy distribution rules concerning certificates they already hold. This requires the re-enforcement of existing distribution rules to deliver existing certificates to neighbors now authorized to hold them, in case of establishment of a new relationship, or to ask neighbors no longer authorized to hold existing certificates to remove them from their directories, in case of relationship revocation.

Certificate Generation, Update, and Revocation

The first task is related to the establishment, update, or revocation of a relationship, and, as a consequence, the generation, update and revocation of the corresponding certificate. Let us start to consider the establishment of a new relationship. In our framework, if a member *m* wishes to establish a relationship with another member *m'*, he/she has to contact *m'* asking whether he/she agrees to establish a relationship of a given type and trust level. Additionally, *m* sends *m'* the corresponding set of distribution rules, denoting the social network members authorized to be aware of such relationship. The relationship is then established only if *m'* agrees on its type, trust level, and the associated distribution rules. In such a case, *m* builds the corresponding certificate, and computes its ID and random number. The protocol for relationship establishment is illustrated in the appendix to this chapter.

By contrast, a relationship can be revoked only by the member who established it, and it is not needed the consent of both members participating in it. Therefore, when a member *m* decides to revoke a relationship there is no need to communicate with the other member to have his/her approval. Consequently, *m* deletes the corresponding certificate from the local directory and notifies of such revocation the certificate revocation list $\mathcal{CRL}$. More precisely, *m* sends $\mathcal{CRL}$ a tuple $(id_{RC}, E_{SK\,m}(DSig_{RC}))$, where id_{RC} is the ID of the revoked certificate, whereas $E_{SK\,m}(DSig_{RC})$ is the digital signature of such certificate, encrypted with *m*'s private key SK_m. Note that denoting a revoked certificate by its identifier and digital signature ensures that the WBSN management system has no access to information concerning the corresponding relationship. Moreover, encrypting the digital signature with the private key of the revoker allows WBSN members holding the certificate to verify whether its revocation is *valid*, that is, it has been done by the member who formerly established the corresponding relationship.

Finally, a relationship update occurs whenever the member who established the relationship wishes to modify its trust level. Note that the only relationship update we are considering is changing the corresponding trust level, in that this operation does not determine any modification of the network graph, and, consequently, it is not necessary to update and distribute relationship keys. By contrast, modifying the members and/or type of a relationship determines the revocation of the existing relationship and the establishment of a completely different one; two events which fire the relationship key generation and distribution protocol.[5] A relationship can be updated only by the member who originally established it, but with the consent of the other member participating in it. This is due to fact that, in order to be valid, the new certificate associated with the updated relationship must be signed by both its members. Since, in practice, updating a relationship implies substituting an old certificate with a new one, the corresponding protocol is a composition of the two protocols described above for relationship establishment and revocation, respectively. First, the new certificate is generated. Then the old one is removed from the certificate directory of the member performing the update, who in addition sends a notification of such event to the certificate revocation list, as explained above. Note that, if the member who receives the update proposal does not agree, the member who sent such proposal has two options: keeping the relationship as it is or revoking it.

Distribution Rule Enforcement

After having established, updated, or revoked a relationship, it is necessary to inform of such event the authorized members—that is, those denoted by the distribution rules associated with the corresponding certificate. For this purpose, the members involved in the new, updated, or revoked relationship start the propagation of a message along the network. More precisely, whenever a WBSN member receives a message concerning the establishment, update, or revocation of a relationship denoted by a given certificate RC, with which is associated a set *DRS* of distribution rules, he/she forwards the message to the members with whom he/she has established a relationship, and satisfying at least one of the rules in *DRS*. Such members carry on the same procedure, until all the authorized WBSN members receive the message. In case of the establishment of a new relationship, the message contains the corresponding certificate and distribution rules. By contrast, in case of a revocation, the message contains the identifier of the certificate to be deleted. Finally, in case of an update, the message contains the new version of the certificate to be updated. Details on the distribution protocol and on the structure and content of the messages propagated along the network are reported in the appendix to this chapter. Here we focus on how distribution rules are enforced.

First of all, it is important to note that, in our approach, WBSN members have only a partial view of the existing relationships, and thus they are not necessarily able to correctly enforce a distribution rule, as the following example shows.

Example 6

Referring to Figure 1 and to the distribution rules in Table 2, suppose that David establishes a *friendOf* relationship with Eve, corresponding to certificate RC, protected by distribution rule DR_4. Thus, the members authorized to receive the corresponding certificate are Alice, Bob, Carol, and Fred. Suppose now that Alice is not aware of the existence of the relationship corresponding to edge *DB*. Thus, when she receives RC, Alice will not be able to enforce DR_4, since, as far as she knows, no path exists between David and her. Moreover, it is important to prevent WBSN members from getting unauthorized information about WBSN relationships from the distribution rules they receive. Suppose that Alice establishes with Carol a relationship, associated with a set of distribution rules stating that only the members who are either friends or colleagues of Alice can receive the corresponding certificate, with no limit about the depth of such relationships. Now, Bob satisfies the former condition (he is a friend of Alice), but he is not one of the colleagues of hers. If both the distribution rules are sent to Bob, he will be aware that Alice most likely participates in at least one relationship of type *colleagueOf*.

In order to address the issues discussed in Example 6, during the propagation of a message concerning the establishment, update, or revocation of a relationship, two modifications are performed on the distribution rules associated with the corresponding certificate. The former aims at avoiding the disclosure of sensitive information to unauthorized members by removing from the set *DRS* of distribution rules those not satisfied by the receiving member. Referring to Example 6, when Alice establish the relationship with Carol, Bob will receive a modified version of the distribution rules associated with the certificate, containing only the rule expressing constraints on the *friendOf* relationship.

Note that this does not affect the correct recursive distribution to be performed by each member receiving a message concerning the establishment, update, or revocation of a relationship, as the following example shows.

Example 7

Referring to Example 6 and to Figure 1, suppose that the *colleagueOf* relationship corresponding to edge *CD* does not exist, and that Bob has established with David a relationship of type *colleagueOf*. Note that, despite this, David does not satisfy the distribution rule on relationship type *colleagueOf* concerning the relationship existing between Alice and Carol, since no path exists between Alice and David consisting of edges all labeled *colleagueOf*. By contrast, David satisfies the distribution rule concerning relationship type *friendOf*, and consequently he will receive from Bob the corresponding distribution rule.

The latter modification of the original set of distribution rules concerns the contained distribution conditions. More precisely, for each condition (*v*, *rt*, *d*max), the original *v* component is replaced with the ID of the receiving member, whereas the *d*max component is decreased of a unit, unless it is set to * (in this case, such value is not modified). Since each member knows all the direct relationships in which he/she participates, he/she is able to correctly propagate the message.

Example 8

Referring to the establishment of the *friendOf* relationship between David and Eve in Example 6, when David sends certificate RC to Eve and Bob, they will receive a modified version of DR_4 = {(*D*, *friendOf*, 2)} corresponding to DR_E = {(*E*, *friendOf*, 1)} and DR_B = {(*B*, *friendOf*, 1)}, respectively. In turn, Alice and Carol will receive from Bob a modified version DR_B = {(*B*, *friendOf*, 1)} corresponding to DR_A = {(*A*, *friendOf*, 0)} and DR_C = {(*C*, *friendOf*, 0)}, respectively, whereas Fred will receive from Eve a modified version of DR_E = {(*E*, *friendOf*, 1)} corresponding to DR_F = {(*F*, *friendOf*, 0)}.

The last issues to be addressed are: (a) avoiding useless waste of network resources because of the propagation of multiple copies of the same message along the network, and (b) being able to verify whether a message is *valid* or not, that is, whether it corresponds to a valid certificate, update, or revocation.

As far as point (a) is concerned, we adopt a policy according to which, if a member *m* receives the same message from more than one neighbor, *m* propagates just the first one, as the following example shows.

Example 9

Consider Figure 1 and the distribution rules in Table 2. Suppose that there exists an edge *CB*, labeled *friendOf*. Suppose also that a relationship of type *friendOf* is established between Carol and Eve, associated with distribution rule DR_1 in Table 2. In this case there exist two paths connecting Carol with Alice, namely *CA* and *CBA*. Consequently, when Carol propagates the corresponding certificate RC along the network, Alice will receive two copies of it, one from Carol, and one from Bob. In such a case Alice does not propagate the later message—that is, if she first receives the message from Carol (Bob), she does not propagate Bob's (Carol's) one.

Regarding the issue at point (b), the validity of a message concerning the establishment, update, or revocation of a relationship can be verified by checking the information contained in the message itself.

Before explaining how this check is performed, let us see how a message concerning the establishment, update, or revocation of a relationship is structured. Independently from the operation they denote, the messages sent to the authorized neighbors have the same structure, consisting of a pair (*oldCertDSig*, *newCert*), where:

- $oldCertDSig = (id_{RC}, E_{SK}(DSig_{RC}))$ is a tuple containing the ID id_{RC} and the signature $DSig_{RC}$ of a given certificate RC to be revoked; in addition, $DSig_{RC}$ is encrypted with the private key *SK* of the member who revoked RC;
- $newCert = (id_{RC'}, rn_{RC'}, RC', DRS)$ is a tuple denoting the new/updated certificate RC′, along with the corresponding ID $id_{RC'}$ and random number $rn_{RC'}$.

In case *oldCertDSig* = NULL, this means that there is no certificate to be revoked/updated (i.e., a new relationship has been established), whereas, in case *newCert* = NULL, this means that no new relationship has been established (i.e., an old relationship has been revoked). When both *oldCertDSig* and *newCert* are not null, the message denotes the update of an existing relationship (which is treated as the revoke of the relationship to be updated and the establishing of a new relationship corresponding to the updated one).

Depending on the specific operation denoted by the message, its validity is determined as follows:

- in case of establishment of a new relationship, the message is valid if the certificate denoted by *newCert* is signed by both the members participating in the corresponding relationship;
- in case of revocation of an existing relationship, the message is valid if the revocation of the certificate denoted by *oldCertDSig* is performed by the member who formerly established the corresponding relationship;
- in case of update of an existing relationship, the message is valid if both the above conditions are satisfied.

Verifying the validity of the new certificate denoted by the *newCert* component of the message, is performed as follows. Suppose that a member *r* receives a message containing a certificate RC = (*m*, *m*′, *rt*, *t*, TS) || *DSig*. Member *r* first computes the hash of (*m*, *m*′, *rt*, *t*, TS), thus obtaining *H*, and then decrypts the digital signature *DSig* of the certificate with the public keys of *m*′ and *m*, respectively, thus obtaining *H*′. If *H* = *H*′ certificate RC is valid.

By contrast, verifying the validity of a revocation is done as follows. Suppose that *r* receives a message where $oldCertDSig = (id_{RC}, E_{SK}(DSig_{RC}))$. As a first step, *r* retrieves from his/her directory the certificate RC denoted by id_{RC}, and fetches from it its digital signature *DSig* and the ID of the member who established the corresponding relationship, say, *m*. Then, *r* decrypts $E_{SK}(DSig_{RC})$ with *m*'s public key. If $DSig_{RC} = DSig$, the revocation is valid.

Finally, verifying whether an update message is valid requires performing the checks illustrated above on both the *newCert* and *oldCertDSig* components of the message.

Distribution Rule Re-Enforcement

The establishment of a new relationship could require to re-enforce some of the existing distribution rules. Indeed, when a new relationship is generated, it could happen that one or both the corresponding members now satisfy distribution rules concerning certificates they formerly were not authorized to hold, thanks to a new path created by the new relationship, as the following example shows.

Example 10

Referring to the WBSN in Figure 1 and the distribution rules in Table 2, consider three members, say, Alice, Bob, and Carol, and suppose that at instant TS_0 there exists only a relationship of type *friendOf* between Alice and Bob, associated with distribution rule DR_5. Since no relationship exists between Bob and Carol, the latter is not authorized to be aware of the relationship between Alice and Bob. However, if Bob, in a subsequent instant $TS_1 > TS_0$, establishes a relationship of type *friendOf* with Carol, the latter will satisfy DR_5. Thus, Carol should receive, besides the certificate corresponding to her relationship with Bob, also the one concerning the relationship between Alice and Bob.

Similarly, the revocation of a relationship could require to re-enforce some of the distribution rules. Indeed, revoking a relationship determines the removal of one or more paths, and thus it may happen that some members are no longer authorized to be aware of relationships corresponding to certificates already stored in their local directory. Referring to Example 10, if Bob, at instant $TS_2 > TS_1$, revokes his relationship with Carol, the latter will be no longer authorized to exploit the existence of the relationship between Alice and Bob, and thus Carol should remove the corresponding certificate from her certificate directory. Of course, if such certificates are not revoked, they may be re-used to obtain unauthorized access to resources.

To cope with these issues, our framework enforces two procedures performed by a WBSN member whenever he/she establishes or revokes a relationship.

Distribution rules re-enforcement due to the establishment of a new relationship. Besides performing the operations described in the previous section, when a member *m* establishes a new relationship with another member *m'*, he/she has also to verify whether *m'* satisfies some of the distribution rules associated with the certificates in his/her directory *CD*. More precisely, *m* considers the set *DRS* of distribution rules applying to each certificate $RC \in CD$, and, if *m'* satisfies one of the rules in *DRS*, *m* sends certificate RC to *m'*. In turn, *m'* performs the same procedure considering the members with whom he/she has established a relationship.

Distribution rules re-enforcement due to the revocation of an existing relationship. When a member *m* revokes a relationship he/she established with another member *m'*, *m* has to verify whether *m'* still satisfies all the distribution rules associated with the certificates he/she sent to *m'*. More precisely, *m* considers the set *DRS* of distribution rules applying to the set *C* of certificates he/she sent to *m'* and, if *m'* does not satisfy any longer any of the rules in *DRS* concerning a given certificate $RC \in C$, *m* notifies the member who established the corresponding relationship to update it by generating and propagating along the network a new version of RC, having a different timestamp. This strategy has the result that RC is revoked, and only the members currently authorized to be aware of the existence of the corresponding relationship receive the new certificate denoting it. By contrast, even if the members who currently are no longer authorized to be aware of the existence of such relationship still maintain RC, for instance because they have not received the update message, they cannot use it during the access control procedure in order to gain access to a resource, since the resource owner can verify that RC has been revoked by contacting the certificate revocation list.

It is, however, true that the presence of revoked certificates in the members' directories determines inconsistencies in their view of the WBSN graph. In order to avoid this, WBSN members can periodically contact the certificate revocation list in order to verify whether they hold revoked certificates.

SECURITY ANALYSIS

The security attacks our framework may be subject to can be grouped into two classes, depending on whether their purpose is gaining unauthorized access to WBSN relationships and resources, or determining a denial of service. In the following sections we discuss the robustness of our system to these two types of attacks.

In the analysis, we assume an honest adversary model, that is, we assume that all WBSN members correctly follow our protocols and do not release private information to unauthorized members. Moreover, we do not consider attacks which exploit network failures. For instance, if one or more WBSN members are temporarily unavailable, some WBSN members may not timely receive the messages propagated along the network concerning relationship keys and (new, updated, or revoked) certificates. However, since a WBSN is a small world network (Watts, 2003), multiple paths exist between its members, and thus it is very difficult, in a WBSN of medium size, that network failures prevent messages from being delivered to online WBSN members. Note, however, that our protocols enforce a fault tolerance mechanism according to which, whenever it is not possible to deliver a message because the receiver is unavailable, such message is repeatedly sent at a given interval.

Attacks for Gaining Unauthorized Access to WBSN Relationships

We remind that the certificate distribution and access control protocols require that resource owner and requestor exchange messages containing sensitive data. These are certificates and access rules that, if accessed by unauthorized members, can determine the disclosure of private information concerning WBSN relationships. To prevent unauthorized access to these messages, in the proposed protocols, all the exchanged messages are encrypted with the private key of the sender and the public key of the receiver. This also avoids identity theft and man-in-the-middle attacks. However, the encryption does not prevent possible inferences by the owner and requestor itself. More precisely, relationship information could be obtained from access rules and certificate paths, as illustrated in the following sections.

Attacks on Access Rules

Whenever an access request is submitted, the resource owner sends the requestor the set of access rules applying to it. We recall that rules may be used by the requestor to infer information about existing relationships he/she is not authorized to know (see the section titled "Privacy-aware Access Control Enforcement"). In order to avoid this privacy breach, access rules are encrypted by using relationship keys (see the section titled "Protecting Access Rules"), as the following example shows.

Example 11

Consider the network graph in Figure 1, and suppose that Fred submits an access request for a resource *rsc* owned by Alice, and that the associated set of access rules is $ARS_1 = \{\{(A, \textit{colleagueOf}, 4, *)\}\}$. Instead of sending ARS_1 in plaintext, Alice encrypts access condition $(A, \textit{colleagueOf}, 4, *)$ with the key she holds for relationship type *colleagueOf*, namely, $K_{colleagueOf}$. As a result, Fred receives the secured access rule set $\underline{ARS}_1 = \{\{(id_{K\,colleagueOf}, E_{K\,colleagueOf}(A, \textit{colleagueOf}, 4, *))\}\}$. Thus, Fred will be able to read the contained condition only if he holds $K_{colleagueOf}$, that is, only if he belongs to the same SNG of type *colleagueOf* Alice belongs to.

According to our key management scheme (see the section titled "Protecting Access Rules"), a relationship key K_{rt} is distributed only to members belonging to the corresponding SNG of type *rt*, that is, to members connected by paths consisting of edges labeled with relationship type *rt*. This is ensured by a neighbor to neighbor propagation according to which, whenever a WBSN member receives K_{rt}, he/she sends it to the members with whom he/she has established a relationship of type *rt* (see the appendix to this chapter for more details). Moreover, whenever a relationship is revoked or a new one is established, a new relationship key is generated. This ensures that if a WBSN member has been disconnected from a SNG after the revocation of a relationship, no path exists between him/her and any member in the SNG, and thus he/she will not receive the new key. As a consequence, it is not possible that WBSN members outside a given SNG hold the corresponding relationship key, even though they previously belonged to it.

Attacks on Certificate Paths

A certificate path is a set of certificates sent by a requestor to a resource owner, in order to demonstrate that he/she satisfies the access rules applying to a given resource. A possible unauthorized disclosure of private information is due to the fact that the certificate paths sent by a requestor might contain one or more certificates that the resource owner is not authorized to access.

Secured certificates are used to prevent this privacy breach (see the section titled "Protecting Certificate Paths"). We remind that secured certificates are obtained by concatenating a certificate with the ID of the member who is sending the certificate paths, and by XORing the resulting string with the certificate random number. Since only WBSN members who are authorized to hold the certificate know the corresponding random number, they are the only ones who can read it, as the following example shows.

Example 12

Referring to Example 11, suppose that Fred holds all the certificates corresponding to path *ACDF*, but that Alice must not be aware of the relationship corresponding to edge *DF*, because of the specified distribution rules. Now, when Alice receives from Fred the secured certificate, she is not able to read it.

Attacks for Gaining Unauthorized Access to WBSN Resources

In our framework, access to a resource is granted only if the requestor can provide to the resource owner a set of certificates demonstrating he/she satisfies at least one of the access rules applying to the requested resource. Therefore, in order to gain unauthorized access to a resource three methods are possible. The first is to forge fake certificates. However, such an attack cannot be performed, since our certificates are digitally signed by both the members participating in the corresponding relationship.

The second method is based on using certificates that the requestor still holds in his/her certificate directory, but that he/she is no longer authorized to have (see the section titled "Distribution Rule Re-enforcement"). As an example, consider the network graph in Figure 1 and the distribution rules in Table 2, and suppose that the *friendOf* relationship existing between Alice and Carol is associated with rule DR_5. Consequently, the members authorized to hold the corresponding certificate are Bob and David. Suppose now that Alice revokes the *friendOf* relationship existing between her and Bob. In such a case, David does not satisfy any longer DR_5, since no path exists connecting Alice to David. However, David still holds the corresponding certificate, and thus he could use it in a certificate path in order to gain access to a resource, even though he is not authorized to do so. This issue is addressed by the policy illustrated in the section titled "Distribution Rule Re-enforcement", according to which, whenever this situation occurs, the old certificate is revoked and substituted with a new version, having a different timestamp. Referring to our example, when the *friendOf* relationship between Alice and Bob is revoked, Bob realizes that Carol is no longer authorized to hold the certificate corresponding to the *friendOf* relationship existing between Alice and Carol. Consequently, he asks Alice to send an update message concerning such relationship. So, if David tries to use the old certificate to gain access to a resource, the resource owner is able to verify that such certificate has been revoked by contacting the certificate revocation list, and access to the resource will be denied.

The third method is to re-use certificates contained in a certificate path, as explained in what follows. Consider the SNG of type *friendOf* depicted in Figure 2, and suppose that *A*(lice) and *E*(ve) hold all the certificates corresponding to the existing relationships, but Alice does not hold the one corresponding to edge *CD*. Suppose now that Eve submits an access request for a resource *rsc* owned by Alice, associated with a single access rule {(*A*, *friendOf*, 5, ∗)}. When Eve sends Alice the set of secured certificates corresponding to path *ABCDE*, Alice is not able to read the secured certificate corresponding to edge *CD*, since she is not authorized to hold it. However, she can keep it in order to re-use such certificate in a later moment. Suppose for instance that, in a subsequent moment, Alice submits an access request for a resource owned by Eve, associated with a single access rule {(*E*, *friendOf*, ∗, ∗)}. Alice is able to read the condition in the secured version of the access rule, since she and Eve belong to the same SNG of type *friendOf*, and thus they share the same key for relationship type *friendOf*. However, she does not satisfy the rule, since she is not authorized to hold the certificate corresponding to edge *CD*, and, consequently, she should not be able to build a valid secured certificate path. However, she could try to re-use the certificate previously obtained by Eve, and gain access to Eve's resource. Secured certificates also prevent this type of attacks. In fact, Alice cannot modify the ID of the sender included into the secured version of a certificate RC without having RC's random number. Thus, if she re-uses the secured certificate previously obtained by Eve, Eve will be able to verify that the certificate path containing it is not valid, and therefore access to the requested resource will not be granted.

Certificates extracted from old certificate paths can be used also to perform timing and replay attacks. Referring to the network graph in Figure 1, suppose that David revokes the relationship he has with Fred, but Fred does not remove the corresponding certificate from his directory. Suppose that Fred sends Alice an access request for resource *rsc*, with which access rule {(*A*, *colleagueOf*, 4, ∗)} is associated. Fred does not satisfy it, since no path exists connecting Fred to Alice. However, Fred could generate a certificate path, since he still holds the revoked certificate corresponding to edge *DF*. Now, if Alice has already received the revocation message concerning such certificate, according to our distribution protocol (see the appendix to this chapter), she will consider the

Figure 2. A portion of WBSN depicting a SNG of type friendOf

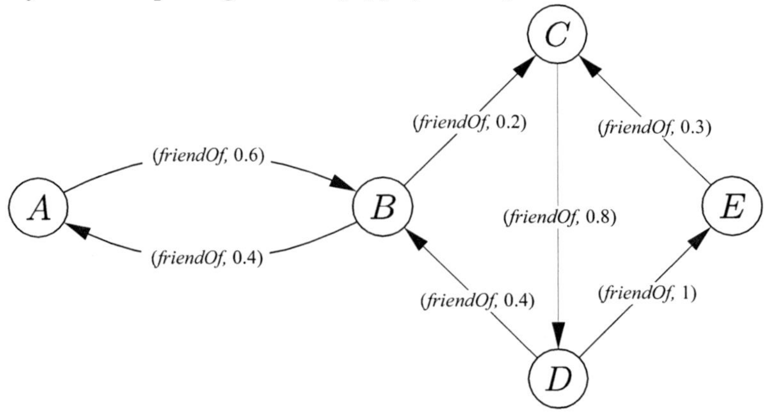

certificate path sent by Fred as invalid, and thus he cannot gain access to the requested resource. However, it is also possible that Fred makes use of the revoked certificate before the corresponding revocation message has been received by Alice, who, in such a case, will consider the certificate path sent by Fred as valid. In order to deal with such a scenario, the WBSN management system is equipped with a certificate revocation list. Thus, the resource owner can contact it in order to know whether a certificate path contains or not revoked certificates.

Denial of Service Attacks

In what follows, we use the term "Denial-of-Service" (DoS) in its broader meaning of an action aimed at making a resource unavailable to its authorized users. By exploiting our protocols, DoS attacks can be performed according to two different methods, depending on whether they modify information on the existing relationships or make unavailable a WBSN service provided by either the WBSN management system or WBSN members.

The former type of attacks is performed by sending fake certificate revocations to the central $\mathcal{CRL}$. In this way, it is possible to prevent authorized members from accessing a resource, by making invalid the certificate paths demonstrating that they satisfy a given access rule. In order to avoid this attack, the revocation message sent to the certificate revocation list consists of the ID of the certificate and its digital signature encrypted with the private key of the sender (see the section titled "Certificate Generation, Update, and Revocation"). This makes a WBSN member authorized to hold this certificate able to verify whether the member who posts the revocation message is authorized to perform the revoke operation, that is, he/she is the one who formerly established the corresponding relationship. Besides avoiding unauthorized revocation of relationship certificates, this also prevents the WBSN service from having access to the information concerning revoked certificates.

The central revocation list, as all other online systems, can be also subject to DoS attacks aiming to make unavailable the service it provides. To address this issue, it is possible to adopt standard strategies used by online systems in order to reduce the risk of denial of service attacks. As an example, we can set an upper bound to the number of requests accepted by the $\mathcal{CRL}$, which may vary depending on the system workload. A similar approach can be adopted in order to avoid denial of service attacks on the side of resource owners, which can be overloaded by access requests or invalid proofs. Such attacks may also be prevented by allowing resource owners to track these kinds of behavior and to maintain a list of malicious members, to be used in order to a priori refuse their requests.

CONCLUSION AND FUTURE WORK

In this chapter, we have illustrated a decentralized security framework for WBSNs, including both access control and privacy protection mechanisms. WBSN members can denote who is authorized to access the resources they publish and the relationships they participate in by specifying access and distribution rules, which express constraints on the type, depth, and trust level of the relationships existing between members of a WBSN. Our framework adopts a decentralized architecture which, on one hand, avoids the bottleneck of a central management service, and, on the other hand, prevents sensitive information from being accessed even by the WBSN service. Existing relationships and resources are not stored in a central repository, but by the WBSN nodes corresponding to the members authorized to hold them, who also carry out the tasks related to access control enforcement and relationship privacy protection through a suite of protocols, exploiting cryptographic techniques in order to prevent security attacks.

We are currently implementing our framework into a prototype, in order to carry out experiments verifying the efficiency and scalability of the overall system. Moreover, we plan to extend our framework by including a negotiation protocol between WBSN members when establishing and updating relationships, and by providing support for existing languages, such as P3P (Cranor, Dobbs, Egelman, Hogben, Humphrey, Langheinrich et al., 2006) and APPEL (Cranor, Langheinrich, & Marchiori, 2002), allowing the specification of privacy preferences and policies. Finally, we will investigate how to provide support to interactions between nodes belonging to different WBSNs.

REFERENCES

Agarwal, S., Sprick, B., & Wortmann, S. (2004). Credential-based access control for Semantic Web services. In *AAAI spring symposium – Semantic Web services*. Retrieved from, http://www.aifb.unikarlsruhe.de/WBS/sag/papers/Agarwal_Sprick_WortmannCredentialBasedAccessControlForSemanticWebServices-AAAI_SS_SWS-04.pdf

Backstrom, L., Dwork, C., & Kleinberg, J. (2007). Wherefore art thou R3579X? Anonymized social networks, hidden patterns, and structural steganography. In *Proceedings of the World Wide Web conference 2007*.

Barnes, S. B. (2006, September). A privacy paradox: Social networking in the United States. *First Monday*, *11*(9). Retrieved from, http://www.firstmonday.org/issues/issue11_9/barnes

Bertino, E., & Sandhu, R. (2005). Database security—Concepts, approaches, and challenges. *IEEE Trans. Dependable Secur. Comput.*, *2*(1), pp. 2–19.

Brickley, D., & Miller, L. (2005, July). *FOAF vocabulary specification* (RDF Vocabulary Specification). Retrieved on XXX from, http://xmlns.com/foaf/0.1

Carminati, B., Ferrari, E., & Perego, A. (2006). Rule-based access control for social networks. In R. Meersman, Z. Tari, P. Herrero, et al. (Eds.), *OTM 2006 workshops proceedings* (pp. 1734–1744). Springer-Verlag.

Carminati, B., Ferrari, E., & Perego, A. (2007). Private relationships in social networks. In *ICDE 2007 workshop proceedings* (pp. 163–171). IEEE CS Press.

Carminati, B., Ferrari, E., & Perego, A. (2008). Security and privacy in social networks. In M. Khosrow-Pour (Ed.), *The encyclopedia of information science and technology, 2nd edition*. (Vol. VII, pp. 3369-3376). Hershey, PA: IGI Publishing.

Cranor, L., Dobbs, B., Egelman, S., Hogben, G., Humphrey, J., Langheinrich, M., et al. (2006, November). *The Platform for Privacy Preferences 1.1 (P3P1.1) specification* (W3C Working Group Note). World Wide Web Consortium. Retrieved from, http://www.w3.org/TR/P3P11/

Cranor, L., Langheinrich, M., & Marchiori, M. (2002, April). *A P3P Preference Exchange Language 1.0 (APPEL1.0)* (W3C Working Draft). World Wide Web Consortium. Retrieved from, http://www.w3.org/TR/P3P-preferences/

Davis, I., & Vitiello Jr., E. (2005, August). *RELATIONSHIP: A vocabulary for describing relationships between people* (RDF Vocabulary Specification). Retrieved from, http://vocab.org/relationship/

Ding, L., Zhou, L., Finin, T., & Joshi, A. (2005). How the Semantic Web is being used: An analysis of FOAF documents. In *Proceedings of the 38th annual Hawaii international conference on system sciences (HICSS'05)*. IEEE CS Press.

Ferraiolo, D. F., Kuhn, D. R., & Chandramouli, R. (Eds.). (2003). *Role-based access control*. Norwood MA: Artech House Publishers.

Fitzpatrick, B. (2005, July). *OpenID 1.1* (Technical Specification). OpenID. Retrieved from, http://www.openid.net/specs.bml

Gates, C. E. (2007). Access control requirements for Web 2.0 security and privacy. In *Proceedings of the Web 2.0 security & privacy 2007 workshop*. Retrieved from, http://seclab.cs.rice.edu/w2sp/2007/papers/paper-205-z_708.pdf

Golbeck, J. A. (2004). *The Trust ontology* (OWL Vocabulary). Retrieved from, http://trust.mindswap.org/ont/trust.owl

Hart, M., Johnson, R., & Stent, A. (2007). More content—less control: Access control in the Web 2.0. In *Proceedings of the Web 2.0 security & privacy 2007 workshop*. Retrieved from, http://seclab.cs.rice.edu/w2sp/2007/papers/paper-193-z_6706.pdf

Hay, M., Miklau, G., Jensen, D., Weis, P., & Srivastava, S. (2007, March). *Anonymizing social networks* (Tech. Rep. No. 07-19). University of Massachusetts Amherst, Computer Science Department. Retrieved from, http://www.cs.umass.edu/~mhay/papers/hay-et-al-tr_0719.pdf

Palomar, E., Estevez-Tapiador, J. M., Hernandez-Castro, J. C., & Ribagorda, A. (2006). Certificate based access control in pure P2P networks. In *Proceedings of the 6th IEEE international conference on peer-to-peer computing (P2P'06)* (pp. 177–184). IEEE CS Press.

Staab, S., Domingos, P., Mika, P., Golbeck, J., Ding, L., Finin, T. W., et al. (2005). Social networks applied. *IEEE Intelligent Systems*, *20*(1), 80–93.

Thompson, M., Johnston, W., Mudumbai, S., Hoo, G., Jackson, K., & Essiari, A. (1999). Certificate based access control for widely distributed resources. In *Proceedings of the 8th USENIX security symposium*. Retrieved from, http://dsd.lbl.gov/~mrt/papers/AkentiUsenixSec.pdf

Wang, D. W., Liau, C. J., & Hsu, T. S. (2006). Privacy protection in social network data disclosure based on granular computing. In *Proceedings of the 2006 IEEE international conference on fuzzy systems (HICSS'05)* (pp. 997–1003). IEEE CS.

Watts, D. J. (2003). Small worlds: *The dynamics of networks between order and randomness*. Princeton, NJ: Princeton University Press.

Weitzner, D. J., Hendler, J., Berners-Lee, T., & Connolly, D. (2006). Creating a policy-aware Web: Discretionary, rule-based access for the World Wide Web. In E. Ferrari & B. M. Thuraisingham (Eds.), *Web and information security* (pp. 1–31). Hershey, PA: IDEA Group Publishing.

Winslett, M., Ching, N., Jones, V. E., & Slepchin, I. (1997). Using digital credentials on the World Wide Web. *Journal of Computer Security*, *5*(3), pp. 255–266.

ENDNOTES

1. Member identifiers are assigned by the WBSN management system upon subscription according to the approach adopted by the OpenID framework (Fitzpatrick, 2005), where each member is identified by a URI consisting of a username appended to the WBSN's URI.

2. Since multiple paths may connect members in the same SNG, the same member may receive the same key more than one time. In such a case, whenever a member receives a key he/she already holds, he/she does not propagate it.

3. We assume that the WBSN management system supports a service that synchronizes the internal clock of each member.

4. For simplicity, the algorithm does not consider the case of multiple keys with the same timestamp.

5. Actually, another information concerning relationships can be modified without modifying the network structure, namely, certificate distribution rules. However, this implies modifying the set of members authorized to be aware of such relationship. In order to manage this operation in a secure and efficient way, updating the distribution

rules of a relationship implies revoking the existing relationship and establishing a new one, different from the former only with respect to the associated distribution rules.

[6] Note that the algorithm describes the most general case. In fact, if m corresponds to the member who established, updated, or revoked the relationship, checking the validity of the message is not needed.

APPENDIX: PROTOCOLS

In this appendix, we present the suite of protocols we have designed for enforcing access control and relationship privacy protection in WBSNs.

Privacy-Aware Access Control

Algorithm 1 illustrates the protocol in charge of relationship key generation and distribution. We recall that this is performed upon the establishment, update, or revocation of a relationship. It has been devised according to the procedure described in the section titled "Protecting Access Rules", to which we refer the reader.

Algorithm 1. Relationship key generation and distribution protocol

Let m and m' be to WBSN members, and let K_m and $K_{m'}$ be the corresponding keys for a given relationship type *rt*.

If m has established a relationship of type *rt* with m'
    If $K_m \neq K_{m'}$
        m and m' generate a new key
        m and m' distribute the new key to the members with whom they have established a relationship of type *rt*, who in turn carry on the same procedure, and so on
    End If
Else
    If m has revoked a relationship of type *rt* he/she had previously established with m'
        If m is isolated
            m removes K_m from his/her directory
        Else
            m computes a new key
            m distributes the new key to the members with whom he/she has established a relationship of type *rt*, who in turn carry on the same procedure, and so on
        End If
        If m' is isolated
            m' removes $K_{m'}$ from his/her directory
        Else
            m' computes a new key
            m' distributes the new key to the members with whom he/she has established a relationship of type *rt*, who in turn carry on the same procedure, and so on
        End If
    End If
End If

Our access control protocol is illustrated by Algorithm 2. In the first step (line 1 of Algorithm 2), a requestor *r* sends an owner *o* an access request for a resource *rsc*. In the next step (lines 2-11), the owner retrieves from his/her rule base the set of access rules applying to *rsc*: if no access rule is associated with *rsc*, the resource cannot be accessed, and consequently *o* sends *r* a denial message. Otherwise, *o* computes the corresponding secured access rule set *ARS*, associates with each contained secured rule a unique nonce value, and sends it to *r*. Then (lines 12-22), *r* starts the evaluation of the first rule AR $\in$ *ARS*. If *r* is able to decrypt all the conditions in AR, he/she builds all the certificate paths satisfying that rule, according to the procedure described in the section titled "Access Control Protocol", he/she computes the corresponding set of secured paths $\underline{\mathcal{CP}}$, and sends it to o along with the nonce value corresponding to $\underline{\text{AR}}$. If o is able to read all the certificates in at least one path $\underline{CP} \in \underline{\mathcal{CP}}$ and they satisfy $\underline{\text{AR}}$, o checks whether any of them is stored into the central certificate revocation list $\mathcal{CRL}$. After being confirmed that none of the certificates in $\underline{CP}$ has been revoked, o sends rsc to r, and r stops evaluating the rules in $\underline{ARS}$ (lines 23-26). Otherwise, o sends r an access denial message (line 29), and r repeats the procedure on the next rule in $\underline{ARS}$.

Algorithm 2. Access control protocol

1: Member r submits to member o an access request for a resource rsc

2: Member o retrieves from his/her rule base the set ARS of access rules associated with resource rsc

3: **If** $ARS = \emptyset$

4: o sends r an access denied message

5: **Else**

6: **For each** $\text{AR} \in ARS$

7: **For each** $\text{AC}_{rt} \in \text{AR}$

8: o computes $\underline{\text{AC}}_{rt} \leftarrow (id_{K\,rt}, E_{K\,rt}(\text{AC}_{rt}))$ and adds it to $\underline{ARS}$

9: **End For**

10: **End For**

11: o sends $E_{PK\,r}(E_{SK\,o}(\{(\underline{\text{AR}}_1, \text{N}_1), \ldots, (\underline{\text{AR}}_n, \text{N}_n)\}))$ to r, where $\underline{\text{AR}}_1, \ldots, \underline{\text{AR}}_n \in \underline{ARS}$, whereas $\text{N}_1, \ldots, \text{N}_n$ are nonce values associated with rules $\underline{\text{AR}}_1, \ldots, \underline{\text{AR}}_n$

12: **For each** $\underline{\text{AR}} \in \underline{ARS}$

13: **If** r is able to decrypt the set of access conditions in rule $\underline{\text{AR}}$

14: Based on the certificates stored into his/her directory CD_r, r builds a set of certificate paths $\mathcal{CP}$ satisfying rule $\underline{\text{AR}}$

15: **If** $\mathcal{CP} \neq \emptyset$

16: **For each** $CP \in \mathcal{CP}$

17: **For each** $\text{RC} \in CP$ having ID id_{RC} and random number rn_{RC}

18: r computes $\underline{\text{RC}} \leftarrow ((id_{\text{RC}}, \text{RC} \,\|\, id_r) \oplus rn_{\text{RC}})$ and adds it to $\underline{CP}$

19: **End For**

20: r adds $\underline{CP}$ to $\underline{\mathcal{CP}}$

21: **End For**

22: r sends $E_{PK\,o}(E_{SK\,r}(\underline{\mathcal{CP}}, \text{N}))$ to o, where N is the nonce value associated with $\underline{\text{AR}}$

23: **If** o can read all the secured certificates in one of the paths $\underline{CP} \in \underline{\mathcal{CP}}$, and $\underline{CP}$ satisfies at least one rule in $\underline{ARS}$

24: **If** no secured certificate in $\underline{CP}$ is stored in the central certificate revocation list $\mathcal{CRL}$

25: o sends rsc to r

26: r terminates the evaluation of the rules in $\underline{ARS}$

27: **End If**

28: **Else**

29: o sends r an access denial message

30: **End If**

31: **End If**

32: **End If**

33: **End For**

34: **End If**

Certificate Protection

Algorithm 3 illustrates the protocol performed when establishing a new relationship. More precisely, suppose that a member m, at a given instant TS, wishes to establish with another member m' a relationship of type rt and trust level t, associated with a set DRS of distribution rules. For this purpose, m sends m' a message consisting of the tuple $((m, m', rt, t, \text{TS}) \,\|\, DSig_m, DRS, \text{N}))$, where $DSig_m$ is the hash of $(m, m', rt, t, \text{TS})$, encrypted with m's private key, whereas N is a nonce value used as session identifier. Member m'

first checks whether $DSig_m$ is valid and, if he/she does not agree to establish the proposed relationship, he/she sends back a reject message consisting of tuple ($DSig_{m'}$, DRS, N). Otherwise, m' encrypts $DSig_m$ with his/her private key, thus obtaining $DSig_{m, m'}$, and then he/she sends back to m a message consisting of tuple ($DSig_{m, m'}$, DRS, N). When m receives such message, he/she builds certificate RC = (m, m', rt, t, TS) || $DSig_{m, m'}$ and generates the corresponding ID id_{RC} and random number rn_{RC}.

Algorithm 3. Protocol for Relationship Establishment

Member m wishes to establish, at instant TS, a relationship of type rt and trust level t with member m', to be distributed according to a given set of distribution rules DRS.

```
m computes DSig_m ← E_{SK m}(MD(m, m', rt, t, TS))
m sends a message E_{PK m'}(E_{SK m}((m, m', rt, t, TS) || DSig_m, DRS, N)) to m', where N is a nonce used as session identifier
If m' checks that DSig_m is valid
    If m' does not agree to establish the relationship proposed by m
        m' rejects the proposal by sending back to m a message E_{PK m}(E_{SK m'}(DSig_{m'}, DRS, N))
    Else
        m' computes DSig_{m, m'} ← E_{SK m}(DSig_m)
        m' sends a message E_{PK m}(E_{SK m'}(DSig_{m, m'}, DRS, N)) to m
        If m checks that D_{PK m'}(DSig_{m, m'}) ← DSig_m
            m builds certificate RC ← (m, m', rt, t, TS) || DSig_{m, m'} and generates the corresponding ID id_RC and random number rn_RC
        End If
    End If
End If
```

By contrast, the revocation of a relationship requires that the member m who formerly established it deletes the corresponding certificate from his/her directory, and notifies of such revocation the central certificate revocation list $\mathcal{CRL}$, by sending the tuple (id_{rc}, $E_{SK\,m}(DSig_{rc})$) (see the section titled "Certificate Generation, Update, and Revocation"). Finally, the protocol performed when updating an existing relationship is a composition of the two protocols for relationship establishment and revocation, since it requires generating a new certificate and revoking an existing one, as illustrated in the section titled "Certificate Generation, Update, and Revocation".

Furthermore, establishing, updating, or revoking a relationship, triggers a procedure where distribution rules are used to propagate along the network a message informing authorized members of such event. For this purpose, our framework implements a protocol consisting of two main steps: (a) the member m who established, updated, or revoked a relationship, denoted by certificate RC and having a set of distribution rules DRS, computes the set $M(m)$ of members with whom he/she established a relationship, and satisfying at least one of the distribution rules in DRS, and (b) member m sends a specific message to the members in $M(m)$, who, in turn, carry on steps (a) and (b) until all the WBSN members denoted by DRS receive the message.

Our distribution protocol is enforced by function DISTRIBUTE(), illustrated by Algorithm 4, which is invoked by a member m and takes as input the two components *oldCertDSig* and *newCert* of the message to be distributed (see the section titled "Distribution Rule Enforcement"). In the first step (lines 2-25), it verifies whether the message to be distributed is valid or not, using the procedure explained in the section titled "Distribution Rule Enforcement".[6] The second part of the algorithm (lines 26-58), which is executed only if the message is valid, considers one by one the members in $M(m)$, that is, the members with whom m has established a relationship, in order to verify which of them is authorized to receive the message. More precisely, for each rule $dr \in DRS$, it first verifies whether the currently considered member $r \in M(m)$ satisfies DR, and adds to *DRS* the corresponding modified version of DR, to be sent to r (lines 30-39). Then, if *DRS* contains at least one rule (line 40), it builds a specific message, depending on the operation to be performed, and sends it to r (lines 41-56). More precisely, in case the message concerns the establishment or update of a relationship, the message consists of the pair (*oldCertDSig*, *newCert*), where variable *newCert* denotes the modified version of *newCert*, containing *DRS* instead of *DRS* (lines 42-51). By contrast, in case of revocation, the message consists of the pair (*oldCertDSig*, null) (lines 52-55).

Algorithm 4. Protocol for distribution rule enforcement

The member invoking function DISTRIBUTE() is denoted by m, whereas r denotes the member to whom m distributes the message concerning the establishment, update, or revocation of a relationship. Moreover, SK_m and PK_r denote, respectively, m's private key and r's public key.

```
1:  Function DISTRIBUTE(oldCertDSig, newCert)
2:      validMsg ← FALSE
3:      Switch
4:          Case oldCertDSig = NULL and newCert ≠ NULL
5:              If the certificate denoted by newCert is valid
6:                  validMsg ← TRUE
7:                  Let DRS be the set of distribution rules applying to the certificate in newCert
8:              End If
9:              Break
10:         End Case
11:         Case oldCertDSig ≠ NULL and newCert ≠ NULL
12:             If the revocation denoted by oldCertDSig and the certificate denoted by newCert are both
                valid
13:                 validMsg ← TRUE
14:                 Let DRS be the set of distribution rules applying to the certificate in newCert
15:             End If
16:             Break
17:         End Case
18:         Case oldCertDSig ≠ NULL and newCert = NULL
19:             If the revocation denoted by oldCertDSig is valid
20:                 validMsg ← TRUE
21:                 Let DRS be the set of distribution rules of the certificate having signature oldCertDSig
22:             End If
23:             Break
24:         End Case
25:     End Switch
26:     If validMsg = TRUE
27:         Let M(m) be the set of members with whom m has established a relationship
28:         For each r ∈ M(m)
29:             DRS ← ∅
30:             For each DR ∈ DRS
31:                 DR ← ∅
32:                 If r satisfies DR
33:                     For each DC ∈ DR
```

```
                    Let DC be a modified version of DC having the dmax component decreased of
                    a unit and m substituted with r
                    Add DC to DR
                End For
                Add DR to DRS
            End If
        End For
        If DRS ≠ ∅
            Switch
                Case oldCertDSig = NULL and newCert ≠ NULL
                    newCert ← (id_RC, id_RC, RC, DRS)
                    Send E_PK r(E_SK m(NULL, newCert)) to r
                    Break
                End Case
                Case oldCertDSig ≠ NULL and newCert ≠ NULL
                    newCert ← (id_RC, id_RC, RC, DRS)
                    Send E_PK r(E_SK m(oldCertDSig, newCert)) to r
                    Break
                End Case
                Case oldCertDSig ≠ NULL and newCert = NULL
                    Send E_PK r(E_SK m(oldCertDSig, NULL)) to r
                    Break
                End Case
            End Switch
        End If
    End For
  End If
EndFunction
```

When r receives the message, he/she first updates his/her certificate directory, and then executes Algorithm 4. Table 4 summarizes the operations performed by r on his/her certificate directory when receiving one of the possible messages.

Table 4. Messages propagated along the network when a new relationship is established, or an existing one is updated or revoked. In the table, SK denotes the private key of the member who sent the message

Operation	Received Message	Action
New relationship	$(\text{NULL}, (id_{RC'}, rn_{RC'}, RC', DRS))$	The member receiving such message stores certificate RC′ in his/her certificate directory.
Update	$((id_{RC}, E_{SK}(DSig_{RC})), (id_{RC'}, rn_{RC'}, RC', DRS))$	The member receiving such message overwrites with RC′ the certificate RC with signature $DSig_{RC}$ stored in his/her certificate directory.
Revoke	$((id_{RC}, E_{SK}(DSig_{RC})), \text{NULL})$	The member receiving such message removes the certificate with signature $DSig_{RC}$ from his/her certificate directory.

Finally, as discussed in the section titled "Distribution Rule Re-enforcement", since the establishment or revocation of a relationship determines a modification of the WBSN graph, it may be necessary to distribute to new neighbors certificates concerning already existing relationships, in case of relationship establishment, or, in case of relationship revocation, to ask the member who established it to update the corresponding certificate with a new one, having a different timestamp. This is enforced by function CheckAndDistribute(), illustrated by Algorithm 5, thanks to which, whenever a member receives a message concerning the establishment, update, or revocation of a relationship, he/she verifies whether there exist any member with whom he/she has established a relationship, and who is now authorized or not authorized to hold already existing certificates.

Algorithm 5. Protocol for distribution rule re-enforcement

Function CHECKANDDISTRIBUTE() takes as arguments the message (*oldCertDSig*, *newCert*), plus the set *C* of certificates held by the member who invokes this function. Such member is denoted by *m*.

```
1:  Function CHECKANDDISTRIBUTE(oldCertDSig, newCert, C)
2:      Switch
3:          Case oldCertDSig = NULL
4:              For each RC ∈ C
5:                  Let DRS be the distribution rules associated with certificate RC
6:                  Let DSig be the digital signature of RC
7:                  Let M(m) be the set of members with whom m has established a relationship, and
                    denoted by DRS before receiving the certificate denoted by newCert
8:                  Let M(m)' be the set of members with whom m has established a relationship, and
                    denoted by DRS after having received the certificate denoted by newCert
9:                  If M(m)'−M(m) ≠ ∅
10:                     For each r ∈ M(m) ∩ M(m)'
11:                         Send E_{PK r}(E_{SK m}(NULL,(id_RC, rn_RC, RC))) to r
12:                     End For
13:                 End If
14:             End For
15:             Break
16:         End Case
17:         Case newCert = NULL
18:             For each RC ∈ C
19:                 Let DRS be the distribution rules associated with certificate RC
20:                 Let DSig be the digital signature of RC
21:                 Let s be the member who established the relationship denoted by RC
22:                 Let M(m) be the set of members with whom m has established a relationship, and
                    denoted by DRS before revoking the certificate denoted by digital signature old-
                    CertDSig
23:                 Let M(m)' be the set of members with whom m has established a relationship, and
                    denoted by DRS after having revoked the certificate denoted by digital signature
                    oldCertDSig
24:                 If M(m)−M(m)' ≠ ∅
25:                     Ask s to update RC with a certificate having a different timestamp
26:                 End If
27:             End For
28:             Break
29:         End Case
30:     End Switch
31: EndFunction
```

Function CHECKANDDISTRIBUTE() takes as input the relationship establishment/revocation message and the set *C* of certificates held by the member *m* invoking such function. In case of relationship establishment (lines 3-16), for each certificate $RC \in C$ and for each member *r* with whom *m* has established a relationship, the function verifies whether *r* is now authorized to hold RC. In such a case, *m* sends *r* the corresponding relationship establishment message. The propagation of such message is then carried out by *r* according to the distribution protocol described by Algorithm 4.

By contrast, in case of relationship revocation (lines 17-29), for each certificate RC $\in C$, the function verifies whether one or more members with whom m has established a relationship, are no longer authorized to hold RC. In such a case, m sends a message to the member s who established the relationship corresponding to RC, asking to update such certificate with a new one, having a different timestamp. The propagation of the corresponding message along the network is then performed according to the distribution protocol described by Algorithm 4.

This work was previously published in the International Journal of Information Security and Privacy, Volume 2, Issue 4, edited by H. Nemati, pp. 22-53, copyright 2008 by IGI Publishing (an imprint of IGI Global).

Compilation of References

Acharya, R., Anand, D., Bhat, S., & Niranjan, U. C. (2001). Compact storage of medical images with patient information. *IEEE Transactions on Information Technology in Biomedicine*, 5, 320–323. doi:10.1109/4233.966107

Adam, N., & Wortmann, J. (1989). Security-control methods for statistical databases: A comparative study. ACM Computing Surveys, 21, 515–556. .doi:10.1145/76894.76895

Adhikari, R. (2008). The Virtualization Challenge, Part 5: Virtualization and Security. TechNewsWorld

Agarwal, S., Sprick, B., & Wortmann, S. (2004). Credential-based access control for Semantic Web services. In AAAI spring symposium – Semantic Web services. Retrieved from, http://www.aifb.unikarlsruhe.de/WBS/sag/papers/Agarwal_Sprick_WortmannCredentialBasedAccessControlForSemanticWebServices-AAAI_SS_SWS-04.pdf

Aggarwal, C. C. (2005). On k-anonymity and the curse of dimensionality. In Proc. of the 31st very large data bases conf. (VLDB'05). Trondheim, Norway.

Aggarwal, G., Feder, T., & etc. (2004). Anonymizing tables for privacy protection. Retrieved from http://theory.standford.edu/ rajeev/privacy.html.

Agrawal, D., & Aggarwal, C. C. (2001). On the design and quantification of privacy preserving data mining algorithms. Paper presented at the 20th ACM SIGMOD SIGACT-SIGART Symposium on Principles of Database Systems, Santa Barbara, CA.

Agrawal, R., Kiernan, J., Srikant, R., & Xu, Y. (2002). Hippocratic databases. In Proc. of the 28th very large data bases conf. (VLDB'02). Hong Kong, China.

Agrawal, R., & Srikant, R. (May 2000). Privacy preserving data mining. Paper presented at the 2000 ACM SIGMOD Conference on Management of Data, Dallas, TX.

Ajzen, I. (2001). Nature and operation of attitudes. Annual Review of Psychology, 52, 27–58. .doi:10.1146/annurev.psych.52.1.27

Alge, B. J., Ballinger, G. A., Tangirala, S., & Oakley, J. L. (2006). Information Privacy in Organizations: Empowering Creative and Extrarole Performance. The Journal of Applied Psychology, 91(1), 221–232. .doi:10.1037/0021-9010.91.1.221

Allman, E., Callas, J., Delany, M., Libbey, M., Fenton, J., & Thomas, M. Domainkeys identified mail signatures (DKIM). U.S. Patent No. 6,986,049.

Allocchio, C. (1998, January). Using the Internet dns to distribute MIXER conformant global address mapping (MCGAM) (IETF RFC 2163).

Altinkemer, K., De, P., & Ozdemir, Z. D. (2006). Toward a consumer-to-healthcare provider (C2H) electronic marketplace. Communications of the Association for Information Systems, 18(19), 413–430.

Alvarez, R. C. (2002). The promise of e-Health - a Canadian perspective. Ehealth International, 1(1), 4. PubMeddoi:10.1186/1476-3591-1-4doi:10.1186/1476-3591-1-4

DOI: 10.4018/978-1-61692-000-5.chcrf

Alvarez, R. C. (2005). In Search of the Holy Grail. One Year Later. Retrieved on July 24, 2009 from http://www.longwoods.com/website/events/docs/BWTCTAlvarez-Mar082005.pdf

Analytics, I. D. (n.d.). Data Breach Harm Analysis from ID Analytics Uncovers New Patterns of Misuse Arising from Breaches of Identity Data. Retrieved November 12, 2007, from http://www.idanalytics.com/news_and_events/20071107.html.

Anderson, A. H. (2004, June 7-9). An introduction to the Web services policy language (WSPL). Paper presented at the 5th IEEE International Workshop on Policies for Distributed Systems and Networks, Yorktown Heights, NY.

Anderson, J. E., & Schwager, P. H. (2002). Security in the information systems curriculum: Identification and status of relevant issues. Journal of Computer Information Systems, 42(3), 16–24.

Andrews, L. W. (2004). Passwords reveal your personality. Retrieved March 13, 2007, from http://cms.psychology-today.com/articles/pto- 20020101-000006.html.

Angst, C. M., & Agarwal, R. (2009). Adoption of electronic health records in the presence of privacy concerns: the elaboration likelihood model and individual persuasion. Management Information Systems Quarterly, 33(2), 339–370.

Anonymous, I. N. E. T. (n.d.). Fast Proxy Server List. Retrieved February 28, 2007 from http://www.anonymousinet.com/, [accessed 28 February 2007 Ashroft, J. (2001). Electronic Crime Scene Investigation: A Guide for First Responders. Retrieved June 22, 2007 from http://www.iwar.org.uk/ecoespionage/resources/cybercrime/ecrime-scene-investigation.pdf

Anton, A. (2007). Is That Vault Really Protecting Your Privacy? Retrieved on July 24, 2009 from The PrivacePlace.org http://theprivacyplace.org/2007/10/is-that-vault-really-protecting-your-privacy/

Anton, A. I., Bochini, D., & He, Q. (2004, May). The use of goals to extract privacy and security requirements from policy statements. In Proc. of the 26th IEEE intl. conf. on software engineering. Scotland, UK.

Antonopoulas, A., (2009, February 17). ATM hack: Organized crime or market forces? Network World Newsletters.

Anzen Consulting Inc. and Sextant for Canada Health Infoway (2007). A 'Conceptual' Privacy Impact Assessment (PIA) on Canada's Electronic Health Record Solution (EHRS) Blueprint Version 2.

Arber, T., Cooley, D., Hirsch, S., Mahan, M., & Osterritter, J. (1999). Network security framework: Robustness strategy. Retrieved November 30, 2007, from http://csrc.nist.gov/nissc/1999/Proceeding/papers/p30.pdf

Atif, Y. (2000). Building trust in e-commerce. IEEE Internet Computing, 6(1), 18–24. doi:10.1109/4236.978365

Atkinson, B. (2004, May 20). Caller id for e-mail. Microsoft Corp.

Awad, M., Thuraisingham, B., & Khan, L. "Assured Information Sharing: Volume 2: Experimental Analysis of Data Integration, Mining and Security, Technical Report." The University of Texas at Dallas, UTDCS44-06.

Aytes, K., & Connolly, T. (2004). Computer security and risky computing practices: A rational choice perspective. Journal of Organizational and End User Computing, 16(3), 22–40.

Bace, R., & Mell, P. (2001). Intrusion detection systems. NIST Special Publication on Intrusion Detection System. National Institute of Standards and Technology.

Backstrom, L., Dwork, C., & Kleinberg, J. (2007). Wherefore art thou R3579X? Anonymized social networks, hidden patterns, and structural steganography. In Proceedings of the World Wide Web conference 2007.

Bacon, J., Moody, K., & Yao, W. (2002). A model of OASIS role-based access control and its support for active security. ACM Transactions on Information and System Security, 5(4), 492–540. doi:10.1145/581271.581276

Bajaj, S., et al. (2004, September). Web services policy framework (WSPolicy).

Bandyopadhyay, S., & Ozdemir, Z. (2009, May 8). On the Potential Role of Independent Personal Health Record Platforms in the Digitization and Sharing of Health Record. Retrieved on July 24, 2009 from SSRN: http://ssrn.com/abstract=1402983.

Banit, A., Nitin, K., & Mart, M. (2005). Controlling spam e-mail at the routers. Paper presented at the EEE International Conference on Communcations (ICC '05), Seoul, Korea.

Barnes, S. B. (2006, September). A privacy paradox: Social networking in the United States. First Monday, 11(9). Retrieved from http://www.firstmonday.org/issues/issue11_9/barnes.

Barry, L., Ossher, J., Rajan, V. T., Richard, S., & Mark, W. (2005, July 21-22). SMTP path analysis. In Proceedings of the Second Conference on E-mail and Anti-Spam (CEAS).

Bas, P., Chassery, J. M., & Macq, B. (2002). Geometrically invariant watermarking using features points. IEEE Transactions on Image Processing, 11, 1014–1028. doi:10.1109/TIP.2002.801587

Bayardo, R. J., & Agrawal, R. (2005). Data Privacy through Optimal k-Anonymization. Paper presented at the ICDE.

Beck, G., & Neufeld, J. (2007). 2007 Indiana Critical Access Hospital Clinical and Technology Needs Survey. Retrieved June 12, 2008, from http://www.in.gov/isdh/publications/partner_rel/pdfs/2007IndianaCriticalAccessHospitalSurvey-p1to17.pdf

Becker, S. A. (2004). A usability study of Internet privacy policies for state and commercial Web sites. International Journal of Services and Standards, 1(1), 52–68. doi:10.1504/IJSS.2004.005688

Bedi, S., Sarma, R. K., & Arya, S. (2002). Outsourcing CT scan: A boon for patients, physicians and hospital administrators in a general hospital. Journal (Academy of Hospital Administration (India)), 14(2), 7–12.

Bellwood, T. (Ed.). (2002). UDDI version 2.04 API specification. UDDI committee specification, OASIS standard. Retrieved December 17, 2007, from www.oasis-open.org/committees/uddi-spec/doc/tcspecs.htm#uddiv2

Bennett, C. (1996). Regulating Privacy in Canada. An Analysis of Oversight and Enforcement in the Private Sector. Advisory Council on Health Infostructure, Paths to Better Health: Final Report. Ottawa: Health Canada Publications.

Berger, E. (2002). Attitudes to privacy, health records and interconnection: Implications for healthcare organizations. Hospital Quarterly, 5(4), 40-5, 2.

Berinato, S. (2007). The Rise of Anti forensics. Retrieved June 16, 2007 from http://www.whitehatsec.com/home/resources/trade/07tradenews/062607CSO.html

Bertino, E., Bonatti, P. A., & Ferrari, E. (2001). TRBAC: A temporal role-based access control model. ACM Transactions on Information and System Security, 4(3), 191–233. doi:10.1145/501978.501979

Bertino, E., Catania, B., Ferrari, E., & Perlasca, P. (2001). A logical framework for reasoning about access control models. In Proceedings of the Sixth ACM Symposium on Access Control Models and Technologies, New York, (pp. 41-52). ACM Press.

Bertino, E., & Sandhu, R. (2005). Database security—Concepts, approaches, and challenges. IEEE Transactions on Dependable and Secure Computing, 2(1), 2–19. doi:10.1109/TDSC.2005.9

Bhatti, R., Ghafoor, A., Bertino, E., & Joshi, J. B. D. (2005). X-GTRBAC: An XML-based policy specification framework and architecture for enterprise-wide access control. ACM Transactions on Information and System Security, 8(2), 187–227. doi:10.1145/1065545.1065547

Bonh, D., & Franklin, M. (2003). Identity-based encryption from weil pairing. SIAM Journal on Computing, 32(3), 586–615. .doi:10.1137/S0097539701398521

Botelho, C. M., & Cazier, J. A. (2008). Guarding Corporate Data from Social Engineering Attacks . In Handbook of Research on Information Security and Assurance. Hershey, PA: IGI Global.

Boucherkha, S., & Benmohamed, M. (2004). A lossless watermarking based authentication system for medical images. Transactions on Engineering. Computing and Technology, 1(1), 240–243.

Boykin P. O., & Roychowdhury, V. (2005, April). Personal e-mail networks: An effective anti-spam tool. IEEE Computer, 38.

Brand, N. (2007). Frozen Tech, The LiveCD List. Retrieved August 11, 2007 from http://www.livecdlist.com/

Brickley, D., & Miller, L. (2005, July). FOAF vocabulary specification (RDF Vocabulary Specification). Retrieved on XXX from, http://xmlns.com/foaf/0.1

Buckovich, S. A., Rippen, H. E., & Rozen, M. J. (1999). Driving toward guiding principles: A goal for privacy, confidentiality, and security of health information. [PubMed]. Journal of the American Medical Informatics Association, 6(2), 122–133.

Burnett, M., & Kleiman, D. (2006). Perfect Passwords. Selection, Protection, Authentication. Syngress.

Bussard, L. (2004). Trust Establishment Protocols for Communicating Devices. PhD Thesis, October 2004

Butti, L., & Veyssett, F. (2006). Wi-Fi Advanced Stealth Proc.Black Hat US. Retrieved April 24, 2009 from http://www.blackhat.com/presentations/bh-usa-06/BH-US-06-Veyssett.pdf

Buyya, R. & Murshed, M. (2002). GridSim: A Toolkit for the Modeling and Simulation of Distributed Resource Management and Scheduling for Grid Computing. The Journal of Concurrency and Computation: Practice and Experience, 1-32.

Byun, J. won, Bertino, E., & Li, N. (2005, Jun.). Purpose-based access control of complex data for privacy protection. In Proc. of the 10th ACM symposium on access control models and technologies (SACMAT'05). Stockholm, Sweden.

Byun, J., Sohn, Y., Bertino, E., & Li, N. (2006) Secure anonymization for incremental datasets. In Proc. of the 3rd VLDB Workshop on Secure Data Management. Korea

Cahill, V., Gray, E., Seigneur, J. M., Jensen, C. D., Chen, Y., & Shand, B. (2003). Using trust for secure collaboration in uncertain environments. IEEE Pervasive Computing / IEEE Computer Society [and]. IEEE Communications Society, 2(3), 52–61.

Cameron, D., Carvajal-Schiaffino, R., Millar, P., Nicholson, C., Stockinger, K., & Zini, F. (2004). OptorSim: A Grid Simulator for Replica Optimisation. In Proceedings of the UK e-Science All Hands Conference. ISBN 1904425216.

Cappel, J. (2001/2002). Entry-level IT job skills: A survey of employers. Journal of Computer Information Systems, 42(2), 76–83.

Carminati, B., Ferrari, E., & Perego, A. (2006). Rule-based access control for social networks . In Meersman, R., Tari, Z., & Herrero, P. (Eds.), OTM 2006 workshops proceedings (pp. 1734–1744). Springer-Verlag.

Carminati, B., Ferrari, E., & Perego, A. (2007). Private relationships in social networks. In ICDE 2007 workshop proceedings (pp. 163–171). IEEE CS Press.

Carminati, B., Ferrari, E., & Perego, A. (2008). Security and privacy in social networks . In Khosrow-Pour, M. (Ed.), The encyclopedia of information science and technology (2nd ed., Vol. VII, pp. 3369–3376). Hershey, PA: IGI Publishing.

Carter, M. (2000). Integrated electronic health records and patient privacy: Possible benefits but real dangers. [PubMed]. The Medical Journal of Australia, 172(1), 28–30.

Cavoukian, A. (2009). Privacy and Government 2.0: The Implications of an Open World Information and Privacy Commisioner of Ontario. Retrieved on July 3, 2009 from http://www.ipc.on.ca/images/Resources/priv-gov-2.0.pdf.

Cazier, J. A., Shao, B. B. M., & St. Louis, R. D. (2007). Sharing information and building trust through value congruence. Information Systems Frontiers, 9, 515–529. .doi:10.1007/s10796-007-9051-6

Cazier, J. A., Wilson, E. V., & Medlin, B. D. (2007). The role of privacy risk in IT acceptance: An empirical study. International Journal of Information Security and Privacy, 1(2), 61–73.

Celikel, E., Kantarcioglu, M., & Thuraisingham, B. (2007to appear). Assured Information Sharing: Risk-based Data Sharing, Technical Report. The University of Texas at Dallas.

Chaiken, S. (1980). Heuristic versus systematic information processing and the use of source versus message cues in persuasion. Journal of Personality and Social Psychology, 39, 752–766. .doi:10.1037/0022-3514.39.5.752

Chellappa, R. K., & Sin, R. G. (2005). Personalization versus privacy: An empirical examination of the online consumer's dilemma. Information Technology Management, 6, 181–202. .doi:10.1007/s10799-005-5879-y

Chen, B., LeFevra, K., & Ramakrishnan, R. (2007) Privacy skyline: Privacy with multidimensional adversarial knowledge. In Proc. of the 3rd Int'l Conf. on Vary Large Data Bases (VLDB'07). Vienna, Austria

Chen, M., & Meixell, M. J. (2003, November). Web services enabled procurement in the extended enterprise: An architecture design and implementation. Journal of Electronic Commerce Research, 4(4), 140–155.

Chow, S., & Holden, R. (1997). Toward an understanding of loyalty: The moderating role of trust. Journal of Managerial Issues, 9(3), 275–299.

Choy, A., Hudson, Z., Pritts, J., & Goldman, J. (2001). Exposed online: Why the new federal health privacy regulation doesn't offer much protection to Internet users. Pew Internet & American Life Project. Retrieved from http://www.pewinternet.org/pdfs/PIP_HPP_HealthPriv_report.pdf

Christenssen, E., et al. (2001). Web services description language (WSDL) 1.1. W3C note. Retrieved December 17, 2007, from www.w3.org/TR/wsdl

Ciampa, M. (2005). Security Guide to Network Fundamentals (2nd ed.). Boston, MA: Course Technology, Cengage Learning.

Clarke, R. (2002). Forum on technology and innovation: Sponsored by Sen. Bill Frist (R-TN), Sen. Jay Rockefeller (D-WV), and the Council on Competitiveness. Retrieved November 30, 2007, from http://www.techlawjournal.com/security/20020214.asp

Clausi, D. A. (2002). An analysis of co-occurrence texture statistics as a function of grey level quantization. Canadian Journal of Remote Sensing, 28(1), 45–62.

Coatrieux, G., Maitre, H., & Sankur, B. (2001). Strict integrity control of biomedical images. Proceedings of Security and Watermarking of Multimedia Contents III, SPIE, 4314, 229–240.

Coatrieux, G., Maitre, H., Sankur, B., Rolland, Y., & Collorec, R. (2001). Relevance of watermarking in medical imaging. In Proceedings of the IEEE CIT: Applications in Biomedicine (pp. 250–252).

Collins, B., Rawlinson, R., Manwani, S., & Allen, K. (2005, September). How can we spread the security message? Computer Weekly, 27, 32.

Collman, J., & Grimes, S. (2005). What healthcare executives should know and do about information security. HIMSS Privacy and Security Toolkit. Retrieved June 12, 2008, from http://www.himss.org/content/files/CEOWhitePaperFinal.pdf

Conca, C., Medlin, D., & Dave, D. (2005). Technology-based security threats: Taxonomy of sources, targets and a process model of alleviation. International Journal of Technology Management, 4(2), 166–177. .doi:10.1504/IJITM.2005.006765

Connetley, M. L., Mael, F. A., & Morath, R. A. (1999). 'Don't ask—please tell': Selection Privacy from two perspectives. Journal of Occupational and Organizational Psychology, 72, 405–422. .doi:10.1348/096317999166752

Cormack, G., & Thomas, L. (2005, July 21-22). Spam corpus creation for TREC. In Proceedings of the Second Conference on E-mail and Anti-Spam (CEAS).

Cornwall, A. (2003). Connecting health: a review of electronic health record projects in Australia, Europe and Canada. Retrieved on September 1, 2006 from http://www.piac.asn.au/publications/pubs/churchill_20030121.html. Sydney, Public Interest Advocacy Centre.

Coulter, A. (2002). After Bristol: Putting patients at the centre. BMJ (Clinical Research Ed.), 324(7338), 648–651. PubMeddoi:10.1136/bmj.324.7338.648doi:10.1136/bmj.324.7338.648

Coulthard, A., & Vuori, T. A. (2002). Computer viruses: A quantitative analysis. Logistics Information Management, 15(5/6), 400–409. doi:10.1108/09576050210447091

Cranor, L., Dobbs, B., Egelman, S., Hogben, G., Humphrey, J., Langheinrich, M., et al. (2006, November). The Platform for Privacy Preferences 1.1 (P3P1.1) specification (W3C Working Group Note). World Wide Web Consortium. Retrieved from, http://www.w3.org/TR/P3P11/

Cranor, L., Langheinrich, M., & Marchiori, M. (2002, April). A P3P Preference Exchange Language 1.0 (APPEL1.0) (W3C Working Draft). World Wide Web Consortium. Retrieved from, http://www.w3.org/TR/P3P-preferences/

Cranor, L. F., Reagle, J., & Ackerman, M.S. (1999), Beyond Concern: Understanding Net Users' Attitudes About Online Privacy. AT&T Labs-Research Technical Report. TR 99.4.3

Croasdell, D. C. (2001). It's role in organizational memory and learning. Information Systems Management, 18(1), 8–11. .doi:10.1201/1078/43194.18.1.20010101/31260.2

Crook, R., Ince, D., & Nuseibeh, B. (2005). On modelling access policies: Relating roles to their organisation context. In Proc. of the 13th IEEE Int'l conf. on requirements engineering (p.157-166). IEEE.

Culnan, M., & Armstrong, P. (1999). Information privacy concerns, procedural fairness, and impersonal trust: an empirical investigation. Organization Science, 10(1), 104–115. .doi:10.1287/orsc.10.1.104

Culnan, M. F., & Armstrong, P. K. (1999). Information privacy concerns, procedural fairness, and impersonal trust: An empirical investigation. Organization Science, 10(1), 104–115. .doi:10.1287/orsc.10.1.104

Culnan, M. J. (1995). Consumer Awareness of Name Removal Procedures: Implications for Direct Marketing. Journal of Direct Marketing, 7, 10–19. .doi:10.1002/dir.4000090204

Culnan, M. J. (2000). Protecting Privacy Online: Is Self-Regulation Working? Journal of Public Policy & Marketing, 19(1), 20–26. .doi:10.1509/jppm.19.1.20.16944

Culnan, M. J., & Bies, R. J. (2003). Consumer privacy: balancing economic and justice considerations. The Journal of Social Issues, 59(2), 323–342. .doi:10.1111/1540-4560.00067

Dalenius, T., & Reiss, S. (1982). Data swapping: A technique for disclosure control. Journal of Statistical Planning and Inference, ▪▪▪, 6.

Damie, P. (2002). Social Engineering: A Tip of the Iceberg. Retrieved January 10, 2008 from http://www.isaca.org/Content/ContentGroups/Journal1/20023/Social_Engineering_A_Tip_of_the_Iceberg.htm.

Davis, I., & Vitiello, E., Jr. (2005, August). RELATIONSHIP: A vocabulary for describing relationships between people (RDF Vocabulary Specification). Retrieved from, http://vocab.org/relationship/

Davis, T., Park, I., Covell, J., Rizzo, L., & Cantor, D. (2006) Health information national trends survey (HINTS): Final report. National Cancer Institute. Retrieved from http://cancercontrol.cancer.gov/hints/docs/HINTS2005Final-Report-0523.pdf

Delany, M. (2006, January 10). Method and system for authenticating a message sender using domain keys. U.S. Patent No. 6,986,049.

DeMuro, P. R., & Grant, A. H. III. (2001). Health privacy standards raise complex implementation issues. Healthcare Financial Management, 55(1), 42–47.

Dennis, A. (2002). Networking in the Internet Age (p. 208). New York: John Wiley and Sons, Inc.

Denton, I. C. (2001). Will patients use electronic personal health records? responses from a real-life experience. [PubMed]. Journal of Healthcare Information Management, 15(3), 251–259.

Dhillon, G., & Blackhouse, J. (2001). Current directions in IS security research: Toward socio-organizational perspectives. Information Systems Journal, 11(2), 127–153. doi:10.1046/j.1365-2575.2001.00099.x

Dignan, L. (2008). Virtualization: What are the security risks? ZDNet Blogs. Retrieved from http://blogs.zdnet.com/security/?p=821

Dimmock, N., Belokosztolszki, A., Eyers, D., Bacon, J., & Moody, K. (2004). Using trust and risk in role-based access control policies. In Proceedings of the Ninth ACM Symposium on Access Control Models and Technologies, New York, (pp. 156-162). ACM Press.

Dinev, T., & Hart, P. (2006). An extended privacy calculus model for e-commerce transactions. Information Systems Research, 17(1), 61–80. .doi:10.1287/isre.1060.0080

Ding, L., Zhou, L., Finin, T., & Joshi, A. (2005). How the Semantic Web is being used: An analysis of FOAF documents. In Proceedings of the 38th annual Hawaii international conference on system sciences (HICSS'05). IEEE CS Press.

Domingo-Ferrer, J., Oganian, A., & Torra, V. (2002). Information-theoretic disclosure risk measures in statistical disclosure control of tabular data. In Proc. of the 14th international conf. on scientist and statistical database management (pp. 227-231). IEEE.

Duchessi, P., & Chengalur-Smith, I. (1998). Client/server benefits, problems, best practices. Communications of the ACM, 41(5), 87–94. doi:10.1145/274946.274961

Duffy, T. J., & Walstrom, K. A. (2003). Changes in student computer technology attitudes over time. Journal of Computer Information Systems, 43(3), 27–33.

Dumitrescu, C., & Foster, I. (2005). Gangsim: A Simulator for Grid Scheduling Studies. Proceedings of the IEEE International Symposium on Cluster Computing and the Grid (CCGrid'05). (pp 1151-1158).

Eagly, A. H., & Chaiken, S. (1975). An attribution analysis of the effect of communicator characteristics on opinion change: the case of communicator attractiveness. Journal of Personality and Social Psychology, 32, 136–144. .doi:10.1037/h0076850

Eagly, A. H., & Chaiken, S. (1998). In Gilbert, D., Fiske, S. T., & Lindzey, G. (Eds.), Handbook of social psychology (4th ed., Vol. 1, pp. 269–322). Boston: McGraw-Hill.

Eastlick, M. A., Lotz, S. L., & Warrington, P. (2006). Understanding online B-to-C relationships: An integrated model of privacy concerns, trust, and commitment. Journal of Business Research, 59, 877–886. .doi:10.1016/j.jbusres.2006.02.006

Eddy, E. R., Stone, D. L., & Stone-romero, E. F. (1999). The effects of information management policies on reactions to human resource information systems: and investigation of privacy and procedural justice perspectives. Personnel Psychology, 52, 335–358. .doi:10.1111/j.1744-6570.1999.tb00164.x

Elkins, M., Torto, D., Levien, R., & Roessler, T. (2001, August). Mime security with OPENPGP (IETF RFC 3156).

Evfimevski, A., Gehrke, J., & Srikant, R. (June 2003). Limiting privacy breaches in privacy preserving data mining. Paper presented at the 22nd ACM SIGMOD-SIGACT-SIGART Symposium on Principles of Database Systems, San Diego, CA.

Evfimevski, A., Srikant, R., Agrawal, R., & Gehrke, J. (July 2002). Privacy preserving mining of association rules. Paper presented at the 8th ACM SIGKDD International Conference on Knowledge Discovery and Data Mining (KDD'02), Edmonton, Alberta, Canada.

Farkas, C., & Jajodia, S. (2002). The inference problem: A survey. SIGKDD Explorations, 4(2), 6–11. .doi:10.1145/772862.772864

Fauzi, M. F. A., & Lewis, P. H. (2006). Automatic texture segmentation for content-based image retrieval application. Pattern Analysis & Applications, 9(4), 307–323. doi:10.1007/s10044-006-0042-x

Fenton, J., & Thomas, M. (2005, May 13). Identified Internet mail. Cisco Systems, Inc.

Ferraiolo, D. F., Kuhn, D. R., & Chandramouli, R. (Eds.). (2003). Role-based access control. Norwood, MA: Artech House Publishers.

Ferraiolo, D. F., Sandhu, R., Gavrila, S., Kuhn, D. R., & Chandramouli, R. (2001). Proposed NIST standard for role-based access control. ACM Transactions on Information and System Security, 4(3), 224–274. doi:10.1145/501978.501980

Finin, T. et al, (May 15, 2007). "Supporting Need to Share with Computational Policies."

Fisch, E. A., & White, G. B. (2001). Secure computers and networks: Analysis, design, and implementation. CRC Press. Retrieved November 30, 2007, from www.crcpress.com

Fitzpatrick, B. (2005, July). OpenID 1.1 (Technical Specification). OpenID. Retrieved from, http://www.openid.net/specs.bml

Forensicwiki (2007). Anti-forensic techniques. Retrieved August 5, 2007 from http://www.forensicswiki.org/wiki/Anti-forensic_techniques

Foster, I., & Kesselman, C. (1998). The Grid: Blueprint for a New Computing Infrastructure. Morgan Kaufman Publishers.

Foster, I., Kesselman, C., Tsudik, G., & Tuecke, S. (1998). A Security Architecture for Computational Grids, Proceedings of the 5th ACM Conference on Computer and Communications Security (pp 83-92).

Fox, S., et al. (2000). Trust and Privacy Online: Why Americans want to rewrite the rules? In The Pew Internet and American Life Project (pp. 1-9). Available at http://www.pewinternet.org

Fox, S., Rainie, L., Horrigan, J., Lenhart, A., Spooner, T., & Carter, C. (2000). Trust and privacy online: Why Americans want to rewrite the rules. Pew Internet & American Life Project. Retrieved from http://www.pewinternet.org/reports/toc.asp?Report=19

Fr'ed'eric, C., et al. "O2O: Managing Security Policy Interoperability with Virtual Private Organizations." retrieved on XXX from, http://www.rennes.enst-bretagne.fr/~fcuppens/articles/hpovua06.pdf

Freeman, L. (2009). Health Law Alert. Retrieved June 30, 2009, from http://www.nixonpeabody.com/publications_detail3.asp?ID=2621.

Fridrich, J., Goljan, M., & Du, R. (2002). Lossless data embedding: New paradigm in digital watermarking. EURASIP Journal on Applied Signal Processing, 2(2), 185–196. doi:10.1155/S1110865702000537

Furnell, S. (2007). An assessment of website password practices. Computers & Security, 26, 445–451.

Furnell, S. M., Bennatou, M., & Dowland, P. S. (2002). A prototype tool for information security awareness and training. Logistics Information Management, 15(5/6), 352–357. doi:10.1108/09576050210447037

Gao, L. Yang, J., Chrobak, M., Zhang, Y., Nguyen, S. & Lee, H.-H. (2006). A low-cost memory remapping scheme for address bus protection. In Proceedings of the 15th International Conference on Parallel Architectures and Compilation Techniques (PACT) (pp. 74-83).

Garfinkel, S. (2001). To know your future . In Database nation: the death of privacy in the 21st century. Sebastopol, CA: O'Reilly & Associates, Inc.

Gartner Inc. (2007). Gartner Says Number of Identity Theft Victims Has Increased More Than 50 Percent Since 2003.

Gassend, B., Suh, G., Clarke, D., Dijk, M., & Devadas, S. (2003). Caches and hash trees for efficient memory integrity verification. In Proceedings of the 9th International Symposium on High-Performance Computer Architecture (HPCA) (pp. 295-306).

Gates, C. E. (2007). Access control requirements for Web 2.0 security and privacy. In Proceedings of the Web 2.0 security & privacy 2007 workshop. Retrieved from, http://seclab.cs.rice.edu/w2sp/2007/papers/paper-205-z_708.pdf

Gaudin, S. (2007, July 3). VA investigator blames IT specialist, lax security for major data loss. InformationWeek. Retrieved from http://www.informationweek.com/news/showArticle.jhtml?articleID=200900263&cid=iwkPrintURL

Georgiadis, C. K., Mavridis, I., Pangalos, G., & Thomas, R. K. (2001). Flexible team-based access control using contexts. In Proceedings of the Sixth ACM Symposium on Access Control Models and Technologies (pp. 21-27). ACM Press.

Ghosh, S. (2001). Principles of secure network systems design. Springer-Verlag.

Gibson, S. (2006). Gibson Research Corporation. Security now – Transcript of Episode 70 – Achieving Internet Anonymity. Retrieved February 23, 2007 from http://www.grc.com/sn/SN-070.pdf

Gichioco, L. (2002). Computer technology: From punch cards to clustered supercomputers. Retrieved July 28, 2007 from http://tle.geoscienceworld.org

Giokas, D. (2004). Privacy Implication for the Interoperable EHR. Retrieved on September 1, 2007 from http://www.rileyis.com/seminars/sept04/GiokasSept04.pdf

Goelman, A. (2006). Work without boundaries: Information technology and the extension of radiology work in time and space. Paper presented at Work Beyond Boundaries, Vancouver, BC.

Goelman, A. (2007, April 26). Teleworks that work: Teleradiology and the emergence of Nighthawk Radiology firms. Presented at Sloan Foundation Industry Studies Annual Conference. Retrieved June 12, 2008, from http://web.mit.edu/sis07/www/goelman.pdf

Golbeck, J. A. (2004). The Trust ontology (OWL Vocabulary). Retrieved from, http://trust.mindswap.org/ont/trust.owl

Goldman, J., & Hudson, Z. (1999). Promoting Health/Protecting Privacy: A Primer. Retrieved on July 24, 2009 from http://www.chcf.org/topics/view.cfm?itemID=12502.

Goldman, J., & Hudson, Z. (2000). Virtually exposed: Privacy and e-health. Health Affairs (Project Hope), 19(6), 140–148. PubMeddoi:10.1377/hlthaff.19.6.140doi:10.1377/hlthaff.19.6.140

Goldsborough, R. (2004). Arming yourself in the virus war. Tech Directions, 63(8), 14.

Goodman, J. (2004). IP addresses in e-mail clients. Paper presented at the Conference on E-mail and Anti Spam (CEAS).

Goodwin, B. (2005, November 29). Investment in security training on the wrong track, say senior staff. Computer Weekly, 58.

Gordis, L., & Gold, E. (1980). Privacy, confidentiality, and the use of medical records in research. Science, 207(4427), 153–156. PubMeddoi:10.1126/science.7350648doi:10.1126/science.7350648

Gordon, L. A., Loeb, M. P., Lucyshyn, W., & Richardson, R. (2004). 2004 CSI/FBI computer crime and security survey. CSI Publications.

Gordon, L. A., Loeb, M. P., Lucyshyn, W., & Richardson, R. (2005). 2005 CSI/FBI computer crime and security survey. CSI Publications.

Gordon, L. A., Loeb, W. L., & Richardson, R. (2006). 2006 CSI/FBI computer crime and security survey. Retrieved November 30, 2007, http://www.gocsi.com

Gostin, L. (1997). Health care information and the protection of personal privacy: Ethical and legal considerations. [PubMed]. Annals of Internal Medicine, 127(8 Pt 2), 683–690.

Gow, B. (2005). The growing threat of cyber-extortion. Risk Magazine, 52(4), 72.

Gragg, D. (2007). A Multi-Level Defense Against Social Engineering. SANS. Retrieved July 1, 2007, from http://www.sans.org/reading_room/whitepapers/engineering/920.php.

Graham, P. (2003). Better Bayesian filtering. Paper presented at the MIT Spam Conference.

Grandison, T., Sloman, M.: Sloman (2000). A survey of trust in Internet applications. IEEE Communications Surveys and Tutorials, 3(4).

Granovetter, M. S. (1973). The strength of weak ties. American Journal of Sociology, 78, 1360–1380. doi:10.1086/225469

Greenaway, K. E., & Chan, Y. E. (2005). Theoretical Explanations for Firms' Information Privacy Behaviors. Journal of the Association for Information Systems, 6(6), 171–198.

Greene, T. C. (2003). The Register. Net Anonymity Service Backdoored. Retrieved August 21, 2003 from http://www.theregister.co.uk/2003/08/21/net_anonymity_service_backdoored/

Hair, J. E., Black, W. C., Babin, B. J., Anderson, R. E., & Tatham, R. L. (2006). Multivariate Data Analysis (6th ed.). Upper Saddle River, N.J: Pearson, Prentice Hall.

Hammond, C. (2001). The intelligent enterprise. InfoWorld, 23(6), 45–46.

Han, J. (2001). Data mining concepts and techniques. Magan Kanfmann Publishers.

Hansell, S. (2003). Amazon Tries Word of Mouth? Retrieved from http://www.nytimes.com/2003/02/10/business/media/10ADCO.html

Haralick, R. (1979). Statistical and structural approaches to textures. Proceedings of the IEEE, 67(5), 786–804. doi:10.1109/PROC.1979.11328

Harris, R. (2006). Arriving at an anti-forensics consensus: Examining how to define and control the anti-forensics problem. Retrieved April 12, 2007 from http://www.dfrws.org/2006/proceedings/6-Harris.pdf

Hart, M., Johnson, R., & Stent, A. (2007). More content—less control: Access control in the Web 2.0. In Proceedings of the Web 2.0 security & privacy 2007 workshop. Retrieved from, http://seclab.cs.rice.edu/w2sp/2007/papers/paper-193-z_6706.pdf

Hartman, B. (2003). Mastering Web services security. Wiley Publishing.

Hassol, A., Walker, J. M., et al. (2004). Patient Experiences and Attitudes about Access to a Patient Electronic Health Care Record and Linked Web Messaging. Journal of the American Medical Informatics Association, 11, 505–513. PubMeddoi:10.1197/jamia.M1593doi:10.1197/jamia.M1593

Hay, M., Miklau, G., Jensen, D., Weis, P., & Srivastava, S. (2007, March). Anonymizing social networks (Tech. Rep. No. 07-19). University of Massachusetts Amherst, Computer Science Department. Retrieved from, http://www.cs.umass.edu/~mhay/papers/hay-et-al-tr_0719.pdf

Henderson, S., & Snyder, C. (1999). Personal information privacy: implications for MIS managers. Information & Management, 36, 213–220. .doi:10.1016/S0378-7206(99)00019-1

Hennnigan, K. M., Cook, T. D., & Gruder, C. I. (1982). Cognitive tuning set, source credibility, and the temporal persistence of attitude change. Journal of Personality and Social Psychology, 42, 412–425. .doi:10.1037/0022-3514.42.3.412

Hesse, B. W., Nelson, D. E., Kreps, G. L., Croyle, R. T., Arora, N. K., Rimer, B. K., & Viswanath, K. (2005). Trust and sources of health information. Archives of Internal Medicine, 165, 2618–2624. .doi:10.1001/archinte.165.22.2618

Hill, R., Bartrum, T., & Guy, B. (2006). Hospital outsourcing trends in clinical services. Retrieved June 12, 2008, from http://www.wallerlaw.com.sitemason.com/files/WallerSurvey.pdf

Hill, R. P. (1995). Researching Sensitive Topics in Marketing, the Special Case of Vulnerable Populations. Journal of Public Policy & Marketing, 14(Spring), 143–148.

Hinde, S. (2003). Careless about privacy. Computers & Security, 22(4), 284–288. doi:10.1016/S0167-4048(03)00403-6

Hintoglu, A. A., Inan, A., Saygin, Y., & Keskinoz, M. (2005). Suppressing Data Sets to Prevent Discovery of Association Rules. Paper presented at the ICDM.

Hird, S. (2002). Technical solutions for controlling spam. In Proceedings of the Annual Meeting of the Australian UNIX and Open Systems User Group (AUUG).

Hoffman, D. L., & Novak, T. P. (1996). Marketing in hypermedia computer-mediated environments: Conceptual foundations. Journal of Marketing, 60(3), 50–68. .doi:10.2307/1251841

Hollaman, L. (2002). Practical realities: Potential impact of the draft privacy of personal information act on healthcare providers. [PubMed]. Hospital Quarterly, 5(4), 46–48.

Holt, G. D., Love, P. E. D., & Li, H. (2000). The learning organization: toward a paradigm for mutually beneficial strategic construction alliances. International Journal of Project Management, 18(6), 415–421. .doi:10.1016/S0263-7863(99)00066-6

Homan, Q. (2003) Healthcare satisfaction study final report. Harris Interactive/ARiA Marketing. Retrieved from http://www.harrisinteractive.com/news/downloads/harrisariahcsatrpt.pdf

Hong, T. (2006). Contributing factors to the use of health-related websites. Journal of Health Communication, 11(2), 149–165. .doi:10.1080/10810730500526679

Hoofnagle, C. J. (2005). Privacy Self Regulation: A Decade of Disappointment. Retrieved on March 13, 2007 from http://www.epic.org/reports/decadedisappoint.html.

Hough, L. M. (1998). Effects of intentional distortion in personality measurement and evaluation of suggested palliatives. Human Performance, 11, 209–244. .doi:10.1207/s15327043hup1102&3_6

Hough, L. M., Eaton, N. R., Dunnette, M. D., Kamp, J. D., & McCloy, R. A. (1990). Criterion-related validities of personality constructs and the effect of response distortion on those validities. The Journal of Applied Psychology, 75, 581–595. .doi:10.1037/0021-9010.75.5.581

Howarth, P., & Rüger, S. (2004). Evaluation of texture features for content-based image retrieval (Book chapter: Image and video retrieval). [Berlin/Heidelberg: Springer.]. Lecture Notes in Computer Science, 3115, 978–1039.

Huh, J., DeLorme, D., & Reid, L. N. (2005). Factors affecting trust in online prescription drug information and impact of trust on behavior following exposure to DTC advertising. Journal of Health Communication, 10(8), 711–731. .doi:10.1080/10810730500326716

Hui, K. L., Teo, H. H., & Lee, S. Y. T. (2007). The Value of Privacy Assurance: an Exploratory Field Experiment. Management Information Systems Quarterly, 31(1), 19–33.

Human Memory. (n.d.). Intelegen, Inc. Retrieved December 12, 2007, from http://brain.web-us.com/memory/human_memory.htm.

Hupp, M. (2007). Protecting patient medical records from the nosy. Retrieved November 30, 2007, from http://www.bizjournals.com/milwaukee/stories/2007/11/12/focus3.html?t=printable.

Innella, P. (2001). Designing secure networks based on the software process model. A white paper. Retrieved November 30, 2007, from http://www.securityfocus.com/infocus/1191

Irvine, C., & Levin, T. (1999). A note on mapping user-oriented security policies to complex mechanisms and services (Tech. Rep.). Retrieved November 30, 2007, from http://cisr.nps.navy.mil/downloads/nps_cs_99_008.pdf

Itagaki, M. W., Berlin, R. B., & Schatz, B. R. (2002) The rise and fall of e-health: Lessons from the first generation of Internet healthcare. Medscape General Medicine, 4(2). Retrieved from http://www.medscape.com/viewarticle/431144_Print

Ives, B., Walsh, K. R., & Schneider, H. (2007). The domino effect of password reuse. Communications of the ACM, 47(4), 75–78. .doi:10.1145/975817.975820

Jackman, M. W. A digital imaging transformation in radiology departments. Healthcare Technology: The Digital Healthcare System. Retrieved June 12, 2008, from http://www.hctproject.com/content/PDF/HCT2_wp_jackman.pdf

Jahankhani, H., Anastasios, B., & Revett, K. (2007). ECIWS, Digital Anti Forensics: Tools and Approaches. Retrieved June 2007 from http://academic-conferences.org/pdfs/eciw07-booklet.pdf

Jajodia, S., Samarati, P., Sapino, M. L., & Subrahmanian, V. S. (2001). Flexible support for multiple access control policies. ACM Transactions on Database Systems, 26(2), 214–260. doi:10.1145/383891.383894

Jaques, R. (2004). Demand for IT security pros growing fast. PC Magazine. Retrieved November 30, 2007, from http://www.pcmag.co.uk/print/1159247

Jarvenpaa, S. L., & Tractinsky, N. (1999). Consumer trust in an Internet store: A cross-cultural validation. Journal of Computer-Mediated Communication, 5(2). Retrieved from http://www.ascusc.org/jcmc/vol5/issue2/jarvenpaa.html.

Jarvenpaa, S. L., Tractinsky, N., & Vitale, M. (2000). Consumer trust in an Internet store. Information Technology Management, 1, 45–71. .doi:10.1023/A:1019104520776

Järvinen, O. P. (2009). Privacy management of patient-centered e-health . In Wilson, E. V. (Ed.), Patient-Centered E-Health. Hershey, PA: IGI Global.

Jennifer, G., & James, H. (2004). Reputation network analysis for e-mail filtering. In Proceedings of the First Conference on E-mail and Anti-spam.

Johansson, C. (2002). Forensic and Anti-Forensic Computing. Retrieved June 24, 2007 from http://www.fukt.bth.se/~uncle/papers/forensics200212.pdf

Jones, M. G. (1991). Privacy: A Significant Marketing Issue for the 1990s. Journal of Public Policy & Marketing, 10(Spring), 133–148.

Jorgensen, P. F. (1998). Affect, persuasion and communication processes . In Anderson, P. A., & Gurrero, L. K. (Eds.), Handbook of communication and emotion: Research, theory, application, and contexts (pp. 403–422). San Diego, CA: Academic Press.

Kanaley, R. (2001). Login error trouble keeping track of all your sign-ons? Here's a place to keep your electronic keys, but you better remember the password. San Jose Mercury News, 3G.

Kargupta, H., Datta, S., Wang, Q., & Sivakumar, K. (2003). On the Privacy Preserving Properties of Random Data Perturbation Techniques. Paper presented at the ICDM.

Kelleher, D. (2004). Spam filtering using contextual network graphs. Final year project in computational linguistics (supervised by Dr. Padraig Cunningham, course 2003–04). Department of Computer Science, University of Dublin, Trinity College.

Kelley, M. (2006, April 9). Web handling radiology night work. The Charleston Gazette. Retrieved June 12, 2008, from http://redorbit.com/news/health/463659/web_handling_radiology_night_work/index.html

Kelly, J. (2006). Vishing' attacks use VoIP. Retrieved September 22, 2006, from http://www.darkreading.com/document.asp?doc id=98787

Kephardt, J., & White, S. (1991). Directed-graph epidemiological models of computer viruses. In Proceedings of the IEEE Computer Society Symposium of Research in Security and Privacy, Oakland CA, (pp. 343-359).

Kephart, J. O., Chess, D. M., & White, S. R. (1993, May). Computers and epidemiology. IEEE Spectrum. Retrieved November 30, 2007, from http://www.research.ibm.com/antivirus/SciPapers/Kephart/Spectrum/Spectrum.html

Kieke, R. L. (2003). What's next, now that the April deadline has passed? Journal of Health Care Compliance, 5(4), 41–43.

Kirkpatrick, J. (2006). Protect your business against dangerous information leaks. Machine Design, 78(3), 66.

Kishimoto, H., Savva, A., & Snelling, D. (2003). OGSA Fundamental Services: Requirements for Commercial GRID Systems. Technical Report, Open Grid Services Architecture Working Group. OGSA-WG.

Kita, C. I. (2003)... IEEE Annals of the History of Computing, 25(3), 62–67. .doi:10.1109/MAHC.2003.1226656

Kizza, J. M. (2005). Computer network security. New York: Springer-Verlag.

Klensin, J. (2004, February). Application techniques for checking and transformation of names (IETF RFC 3696).

Klop, R., van Wijmen, F. C., & Philipsen, H. (1991). Patients' rights and the admission and discharge process. Journal of Advanced Nursing, 16(4), 408–412. PubMeddoi:10.1111/j.1365-2648.1991.tb03430.xdoi:10.1111/j.1365-2648.1991.tb03430.x

Kong, J., Acricmez, O., Seifert, J., & Zhou, H. (2009). Hardware-Software Integrated Approaches to Defend Against Software Cache-based Side Channel Attacks. In Proceeding of the 15th IEEE International Symposium on High-Performance Computer Architecture (HPCA) (pp. 393-404).

Kong, X., & Feng, R. (2001). Watermarking medical signals for telemedicine. IEEE Transactions on Information Technology in Biomedicine, 5(3), 195–201. doi:10.1109/4233.945290

Kosseim, P. (2003). The landscape of rules governing access to personal information for health research: A view from afar. [PubMed]. Health Law Journal, 11, 199–215.

Krane, D. (2005) Number of 'cyberchondriacs'—U.S. adults who go online for health information—increases to estimated 117 million. Healthcare News, 8(5). Retrieved from http://www.harrisinteractive.com/news/newsletters_healthcare.asp

Kutner, M. H., Nachtsheim, C. J., Neter, J., & Li, W. (2004). Applied linear statistical models (5 ed.). McGraw-Hill.

Kyuzz (2009). Retrieved February 12, 2009 from http://www.kyuzz.org/antirez/overwrite.html

Laufer, R. S., & Wolfe, M. (1977). Privacy as a concept and a social issue: A multidimensional developmental theory. The Journal of Social Issues, 33(3), 22–42. .doi:10.1111/j.1540-4560.1977.tb01880.x

Laughlin Memorial Hospital Discovers a Treatment for Password Problems. (n.d.). Retrieved December 12, 2007, from http://www.imprivataonesign.com/documents/casestudies/memorial_healthcare.pdf

Layfield, R., Kantarcioglu, M., & Thuraisingham, B. (October 2006). "Assured Information Sharing: Volume 3: Using Game Theory to Enforce Honesty Within a Competitive Coalition." Technical Report, The University of Texas at Dallas, UTDCW-46-06.

Leach, J. (2003). Improving user security behavior. Computers & Security, 22(8), 685–692. doi:10.1016/S0167-4048(03)00007-5

Lee, R. B., Kwan, P. C. S., McGregor, J. P., Dwoskin, J., & Wang, Z. (2005). Architecture for protecting critical secrets in microprocessors. In Proceedings of the 32nd International Symposium on Computer Architecture (ISCA) (pp. 2-13).

LeFevre, K., DeWitt, D. J., & Ramakrishnan, R. (2006). Mondrian Multidimensional K-Anonymity. Paper presented at the ICDE.

Lefevre, K., DeWitt, J. D., & Ramakrishnan, R. (2005, June). Incognito: Efficient full-domain k-anonymity. In Proc. Of the 24th ACM Intl. Conf. on Management of Data (SIGMOD'05). Maryland, USA.

Legrand, A., Marchal, L., & Casanova, H. (2003). Scheduling Distributed Applications: The SimGrid Simulation Framework. In Proceedings of the 3rd IEEE/ACM International Symposium on Cluster Computing and the Grid 2003 (CCGrid2003) (pp 138-145).

Lemke, T. (2003, November 10). Spam harmed economy more then hackers viruses. Washington Times.

Lengnick-Hall, C. A., & Sanders, M. M. (1997). Designing effective learning systems for management education: Student roles, requisite variety, and practicing what we teach. Academy of Management Journal, 40(6), 1334–1368. .doi:10.2307/257036

Levine, J. (2005, July 21-22). Experiences with greylisting. In Proceedings of the Second Conference on E-mail and Anti-Spam (CEAS).

Levy, F., & Goelman, A. (2005). Offshoring and radiology . In Brainard, L., & Collins, S. M. (Eds.), Offshoring white-collar work: The issues and implications. The Brookings Trade Forum.

Lewis, N. (2008). Virtualization Security. IT Security Magazine. http://www.itsecurity.com/features/virtualization-security-061708/

Li, C.-T., Lou, D.-C., & Chen, T.-H. (2000). Image authentication via content-based watermarks and a public key cryptosystem. In. Proceedings of IEEE International Conference on Image Processing, III, 694–697.

Li, J., & Shaw, J. (2004). Protection of health information in data mining. International Journal of Healthcare Technology and Management, 6(2), 210–222. .doi:10.1504/IJHTM.2004.004977

Li, N., Mitchell, J. C., & Winsborough, W. H. (2002). Design of a role-based trust management framework. In Proceedings of the IEEE Symposium on Security and Privacy, Washington, D.C., (p. 114).

Li, X.-B., & Sarkar, S. (2006). A Tree-Based Data Perturbation Approach for Privacy-Preserving Data Mining. IEEE Transactions on Knowledge and Data Engineering, 18(9), 1278–1283. .doi:10.1109/TKDE.2006.136

Li, Y., Wei, J., Lai, L., & Koong, K. (2004). Implications of technologies and financial losses on information security. In Proceedings of the ISOneWorld, Las Vegas.

Li, Z., & Ye, X. (2008). Privacy Protection on Multiple Sensitive Attributes (LNCS 4861, pp. 141-152).

Li, Z., Zhan, G., & Ye, X. (2006a). Towards an anti-inference (k, l)-anonymity model with value association rules. In Proc. Of the 17th Intl Conf. on Database and Expert System Applications (DEXA'06). Krakow, Poland.

Li, Z., Zhan, G., & Ye, X. (2006b). Towards a more reasonable generalization cost metric for k-anonymity. In Proc. Of the 23rd British National Conf. on Database (BNCOD'06). Northern, Ireland.

Liao, H., Joshi, A., & Chuang, A. (2004). Sticking out like a sore thumb: Employee dissimilarity and deviance at work. Personnel Psychology, 57, 969–1000. .doi:10.1111/j.1744-6570.2004.00012.x

Lie, D., Thekkath, C., Mitchell, M., Lincoln, P., Boneh, D., Mitchell, J., & Horowitz, M. (2000). Architectural support for copy and tamper resistant software. In Proceedings of the 9th International Conference on Architectural Support for Programming Languages and Operating Systems (ASPLOS) (pp. 168-177).

Lim, B., & Wen, H. J. (2003). Web services: An analysis of the technology, its benefits, and implementation difficulties. Information Systems Management, 20(2), 49–57. doi:10.1201/1078/43204.20.2.20030301/41470.8

Lin, H.-Y., & Harn, L. (1991). A generalized secret sharing scheme with cheater detection. In Proceedings of the International Conference on the Theory and Applications of Cryptology: Advances in Cryptology (LNCS 739, pp. 149-158).

Linowes, D. F., & Spencer, R. C. (1996). Privacy in the Workplace in Perspective. Human Resource Management Review, 6(3), 165–181. .doi:10.1016/S1053-4822(96)90011-X

Liu, C., Marchewka, J. T., Lu, J., & Yu, C. (2005). Beyond concern–a privacy-trust-behavioral intention model of electronic commerce. Information & Management, (42): 289–304. .doi:10.1016/j.im.2004.01.003

Liu, V., & Stach, P. (2006). Defeating Forensic Analysis, CEIC 2006 –Technical Lecture1. Retrieved from http://stachliu.com/files/CEIC2006-Defeating_Forensic_Analysis.pdf

Loch, K., Carr, H., & Warkentin, M. (1992). Threats to information systems: today's reality, yesterday's understanding. Management Information Systems Quarterly, 16(2), 173–186. .doi:10.2307/249574

Lockerd, A., & Selker, T. (2003). The implicit social context of e-mail . In Proceedings of Human-Computer Interaction INTERACT. Driftcatcher.

Lohr, S. (2007, August 14) Google and Microsoft look to change health care. New York Times. Retrieved from http://www.nytimes.com/2007/08/14/technology/14healthnet.html

Lombardi, R. (2008). Consumer control over personal medical data is coming to Canada, Retrieved on July 24, 2009 from InterGovWorld.com, 08 September 2008, http://www.itworldcanada.com/a/search/42adc562-c0a8-0006-001d-88f64bbf545d.html

Lowed, D., & Meek, C. (2005). Good word attacks on statistical spam filters. In Proceedings of the First Conference on E-mail and Anti-Spam (CEAS).

Lu, C.-S., & Liao, H.-Y. (2003). Structural digital signature for image authentication: An incidental distortion resistant scheme. IEEE Transactions on Multimedia, 5, 161–173. doi:10.1109/TMM.2003.811621

Luo, X. (2002). Trust production and privacy concerns on the Internet: A framework based on relationship marketing and social exchange theory. Industrial Marketing Management, 31(2), 111–118. .doi:10.1016/S0019-8501(01)00182-1

Lwin, M. O., & Williams, J. D. (2003). A Model Integrating the Multidimensional Developmental Theory of Privacy and Theory of Planned Behavior to Examine Fabrication of Information Online. Marketing Letters, 14(4), 257–272. .doi:10.1023/B:MARK.0000012471.31858.e5

Machanavajjhala, A., Gehrke, J., Kifer, D., & Venkitasubramaniam, M. (April 2006). L-Diversity: Privacy Beyond k-Anonymity. Paper presented at the 22nd IEEE International Conference on Data Engineering (ICDE 2006), Atlanta, Georgia.

MacKenzie, S. B., & Spreng, R. A. (1992). How does motivation moderate the impact of central and peripheral processing on brand attitudes and intentions? The Journal of Consumer Research, 18(March), 519–529. .doi:10.1086/209278

Mackie, D. M., & Worth, L. T. (1989). Processing deficits and the mediation of positive affect in persuasion. Journal of Personality and Social Psychology, 57, 27–40. .doi:10.1037/0022-3514.57.1.27

Macq, B., & Dewey, F. (1999). Trusted headers for medical images. In Proceedings of the DFG VIII-D II Watermarking Workshop.

MacVittie, D. (2007, Aug. 31). Javelin 2006 Identity Fraud Report. Network Computing.

Malhotra, N. K., Kim, S. S., & Agarwal, J. (2004). Internet users' privacy concerns (IUIPC): The construct, the scale, and a causal model. Information Systems Research, 15(4), 336–355. .doi:10.1287/isre.1040.0032

Manjunath, B., Wu, P., Newsam, S., & Shin, H. (2000). A texture descriptor for browsing and similarity retrieval. Journal of Signal Processing: Image Communication, 16, 33–43. doi:10.1016/S0923-5965(00)00016-3

Manos, D. (2007) Microsoft joins lawmakers, activists to demand patient privacy rights. (Healthcare IT News, 18 October 2007).

Marsden, P. V., & Campbell, K. E. (1984). Measuring tie strength. Social Forces, 63(2), 482–501. doi:10.2307/2579058

Martin, D., Kifer, D., Machanavajjhala, A., Gehrke, J., & Halpern, J. (2007). Worst-case background knowledge for privacy-preserving data publishing. In Proc. of the 23nd international conference on data engineering (ICDE'07).

Martin, K. M. (1993). Untrustworthy participants in perfect secret sharing schemes . In Ganley, M. J. (Ed.), Cryptography and Coding III (pp. 255–264). Oxford University Press.

Martins, C. S., & Martins, S. J. (2005). The Impact of the USA PATRIOT Act on Records Management. Information Management Journal, 39(3), 52–58.

Mason, J. K. McCall, & Smith, R.A. (1999) Law and Medical Ethics (5th Ed.). London: Butterworths.

Mason, R. (1986). Four ethical issues of the information age. Management Information Systems Quarterly, 10(1), 5–12. .doi:10.2307/248873

Massey, B., Mick, T., Raya, B., & Scott, L. (2003). Learning spam: Simple techniques for freely-available software. In Proceeding of the 2003 Usenix Annual Technical Conference Freenix Track.

Masys, D., Baker, D., Butros, A., & Cowles, K. E. (2002). Giving patients access to their medical records via the Internet: The PCASSO experience. Journal of the American Medical Informatics Association, 9(2), 181. .doi:10.1197/jamia.M1005

Matt, S. (2002). Auditing and reporting for HIPAA compliance. Health Management Technology, 23(12), 12–13.

Matthew, I. K., & Johnson, K. B. (2002). Personal Health Records: evaluation of functionality and utility. Journal of the American Medical Informatics Association, 9(2), 171–180. PubMeddoi:10.1197/jamia.M0978doi:10.1197/jamia.M0978

Mayer, R. C., Davis, J. H., & Schoorman, F. D. (1995). An integrative model of organizational trust. Academy of Management Review, 20(3), 709–734. .doi:10.2307/258792

McAlerney, J. M. (1999). An Internet worm propagation model. Retrieved November 30, 2007, from www.cs.usdavis.edu/research/tech-reports/2005/CSE-2005-10.pdf

McCarthy, D. B., Shatin, D., Drinkard, C. R., Kleinman, J. H., & Gardner, J. S. (1999). Medical records and privacy: Empirical effects of legislation. [PubMed]. Health Services Research, 34(1 Pt 2), 417–425.

McClure, D. (2001). Guarding your gateway. Association Management, 53(8), 60–66.

McGaw, D. (2008). Comprehensive Privacy and Security: Critical for Health Information Technology. Centre for Democracy and Technology Version 1.0. Washington, DC: CDT Health Privacy Project.

Mcguckin, R. H., & Nguyen, S. V. (1990). Public use microdata: Disclosure and usefulness. Journal of Economic and Social Measurement, 16, 19–39.

McMillan, R. (2005). Computer attacks down, survey says. Retrieved November 30, 2007, from http://www.networkworld.com/news/2005/072505-security.html

McQuade, S. (2006). Understanding and Managing Cybercrime. Upper Saddle River, New Jersey: Prentice Hall.

Medlin, B. D., & Cazier, J. A. (2005). An Investigative Study: Consumer Password Choices on an E-Commerce Site. Journal of Information Privacy and Security, 1(4), 33–52.

Medlin, B. D., & Cazier, J. A. (2007). An Empirical Investigation: Health Care Employee Passwords and Their Crack Times in Relationship to HIPAA Security Standards. International Journal of Healthcare Information Systems and Informatics, 2(3), 39–48.

Melville, N. (2009). Today's interconnectivity can also put patient data in jeopardy. Retrieved June 1, 2009, from http://compliance.utimaco.com/na/tag/hitech-act/

Menezes, A. J., van Oorschot, P. C., & Vanstone, S. A. (1997). Handbook of applied cryptography. CRC Press. LLC.

Menon, A. M., Deshpande, A. D., Perri, M., & Zinkhan, G. M. (2002). Trust in online prescription drug information among Internet users: The impact on information search behavior after exposure to direct-to-consumer advertising. Health Marketing Quarterly, 20(1), 17–35. .doi:10.1300/J026v20n01_03

Menon, S., & Sarkar, S. (2006). Exploiting Problem Structure to Efficiently Sanitize Very Large Transactional Databases. Paper presented at the WITS.

Menon, S., Sarkar, S., & Mukherjee, S. (2005, September). Maximizing Accuracy of Shared Databases when Concealing Sensitive Patterns. Information Systems Research, 16(3), 256–270. .doi:10.1287/isre.1050.0056

Messmer, E. (2003). HIPAA deadline ups healthcare anxiety. New World (New Orleans, La.), 20(10), 30.

Meyerson, A., & Williams, R. (2004). On the complexity of optimal k-anonymity. In Proc. of the 23th ACM SIGMOD principles of database systems (PODS'04). France.

Milberg, S., Smith, H., & Burke, S. (2000). Information privacy: corporate management and national regulation. Organization Science, 11(1), 35–57. .doi:10.1287/orsc.11.1.35.12567

Miller, A., & Tucker, C. (2009). Privacy protection and technology diffusion: the case of electronic medical records. Management Science, 55(7), 1077–1093. .doi:10.1287/mnsc.1090.1014

Mills, J., & Harvey, J. (1972). Opinion change as a function of when information about the communicator is received and whether he is attractive or expert. Journal of Personality and Social Psychology, 21, 52–55. .doi:10.1037/h0031939

Milne, G. R. (1997). Consumer Participation in Mailing Lists: A Field Experiment. Journal of Public Policy & Marketing, 16(Fall), 298–309.

Milne, G. R., & Gordon, M. E. (1993). Direct mail privacy-efficiency trade-offs within an implied social contract framework. Journal of Public Policy & Marketing, 12(2), 206–215.

Milne, G. R., & Rohm, A. J. (2000). Consumer Privacy and Name Removal Across Direct Marketing Channels: Exploring Opt-in and Opt-out Alternatives. Journal of Public Policy & Marketing, 19, 238–249. .doi:10.1509/jppm.19.2.238.17136

Mineta, N. (2000). Opening statement at online privacy technology workshop and technology fair. U.S. Department of Commerce, Washington, D.C. (September 19). Retrieved from http://www.ntia.doc.gov/ntiahome/privacy/900workshop/mineta91900.htm

Mitnick, K., & Simon, W. L. (2002). The Art of Deception. Hoboken, New Jersey: John Wiley & Sons.

Monds, K. E., & Wang, C. (2003). CDC's new epidemic: An investigation into awareness, attitudes, actions, and knowledge of computer viruses among students utilizing campus computer labs. Journal of Computer Information Systems, 43(3), 118–126.

Monegain, B. CCHIT receives slew of applications just under deadline. Retrieved June 12, 2008, from http://www.healthcareitnews.com/story.cms?id=9060

Moore, S., & Shepherd, J. P. (2006). The cost of fear: shadow pricing the intangible costs of crime. Applied Economics, 38(3), 293–300. .doi:10.1080/00036840500367781

Morgenstern, M. (1987, December). Security and inference in multilevel database and knowledge-based systems. [ACM.]. SIGMOD Record, 16(3), 357–373. .doi:10.1145/38714.38752

Morizot, J., & Le Blanc, M. (2005). Searching for a developmental typology of personality and its relations to antisocial behavior: A longitudinal study of a representative sample of men. Journal of Personality, 73(1), 139–182. .doi:10.1111/j.1467-6494.2004.00307.x

Morrissey, J. (2006). A Day in the Life of a Medical Record - Lifting the veil on the security of today's paper-based environment. The National Alliance for Health Information Technology. Retrieved on September 1, 2007 from http://www.nahit.org/dl/A_Day_in_the_Life.pdf

Moye, S. (2006). Congress Assesses Data Security Proposals. Information Management Journal. 40(1), 20-23.

Myers, J., Frieden, T., Bherwanj, K., & Henning, K. (2008). Ethics in public health research: privacy and public health at risk: public health confidentiality in the digital age. American Journal of Public Health, 98(5), 793–801. PubMeddoi:10.2105/AJPH.2006.107706doi:10.2105/AJPH.2006.107706

Naqvi, S., & Riguidel, M. (2004). Security Architecture for Heterogeneous Distributed Computing Systems. Proceedings of the IEEE International Carnahan Conference on Security Technology 2004 (IEEE ICCST2004). (pp. 34-41).

Neustaedter, C., Brush, A. B., Smith, M. A., & Fisher, D. (2005). Social network relationship finder: Social sorting for e-mail triage. In Proceedings of the Second Conference on E-mail and Anti-Spam (CEAS).

Norris, P. (2001). Digital divide: Civic engagement, information poverty, and the Internet worldwide. Cambridge, UK: Cambridge University Press.

Nowak, G. J., & Phelps, J. (1992). Understanding Privacy Concerns: An Assessment of Consumers' Information Related Knowledge and Beliefs. Journal of Direct Marketing, 6, 28–39. .doi:10.1002/dir.4000060407

O, B., & C L, S. (2002). Bifrost inbox organizer: Giving users control over the inbox. In Proceedings of the second Nordic conference on Human-computer interaction Aarhus, Denmark.

O'Connor Hospital Privacy Policy. (n.d.). Retrieved December 3, 2007, from http://www.oconnorhospital.org/contact/Pages/HospitalPrivacyPolicy.aspx.

O'Keefe, D. J. (1990). Persuasion: Theory and research (Vol. 2). Newbury Park, CA: Sage.

Oates, J. (2006, September 11). The Register. German Police Seize TOR Servers. Retrieved January 27, 2007 from http://www.theregister.co.uk/2006/09/11/anon_servers_seized/

Oliveira, S., & Zaiane, O. R. (2002). Privacy Preserving Frequent Itemset Mining. Paper presented at the IEEE International Conference on Privacy, Security and Data Mining, Maebashi City, Japan.

Oppenheimer, P. (2004). Top-down network design (2nd ed.). Indianapolis, IN: Cisco Press.

Oudot, L. (2005). WLAN and stealth issues. Retrieved May 23, 2008 from http://www.blackhat.com/presentations/bh-europe-05/BH_EU_05-Oudot/BH_EU_05_Oudot.pdf

Pai, H., & Lau, F. (2005). Web-based electronic health information systems for prostate cancer patients. [PubMed]. The Canadian Journal of Urology, 12(3), 2582–2591.

Palomar, E., Estevez-Tapiador, J. M., Hernandez-Castro, J. C., & Ribagorda, A. (2006). Certificate based access control in pure P2P networks. In Proceedings of the 6th IEEE international conference on peer-to-peer computing (P2P'06) (pp. 177–184). IEEE CS Press.

Papoulis, A. (1991). Probability, random variables, and stochastic processes. New York: McGraw-Hill.

Park, J., & Sandhu, R. S. (2004). The UCON usage control model. ACM Transactions on Information and System Security, 7(1), 128–174. doi:10.1145/984334.984339

Payton, F. C., & Brennan, P. F. (1999). How a community health information network is really used. Communications of the ACM, 42(12), 85–89. .doi:10.1145/322796.322814

Pearlman, L., Welch, V., Foster, I., Kesselman, C., & Tuecke, S. (2002). A Community Authorization Service for Group Collaboration. In Proceedings of the 3rd International Workshop on Policies for Distributed Systems and Networks (POLICY '02) (pp 50-59).

Pedersen, T. P. (1991). A threshold cryptosystem without a trusted party. In Proceedings of EUROCRYPT (pp. 522-526).

Pederson, T. P. (1992). Non-interactive and information-theoretic secure verifiable secret sharing. In Proceedings of the 11th Annual International Cryptology Conference on Advances in Cryptology (LNCS 576, pp. 129-140).

Pellegrini, A. (1988). Rough-and-tumble play and social competence. Developmental Psychology, 24, 802–806. .doi:10.1037/0012-1649.24.6.802

Petty, R. E., & Cacioppo, J. T. (1981). Attitudes and persuasion: classic and contemporary approaches. Dubuque, IA: Brown.

Petty, R. E., & Cacioppo, J. T. (1986). The elaboration-likelihood model of persuasion. In L.Berkowitz (Ed.), Advances in experimental social psychology, 19, 123-205.

Pfleeger, C. P., & Pfleeger, S. L. (2007). Security in Computing (4th ed.). Prentice Hall Protecting the Privacy of Patients' Health Information. Retrieved November 13, 2009, from http://www.hhs.gov/news/facts/privacy.html

Plater, S., Seeley, E., & Dixon, L. A. (1998). Two routes to privacy protection: A comparison of health information legislation in Canada and the United States. Journal of Women's Health / the Official Publication of the Society for the Advancement of Women's Health Research, 7(6), 665-672.

Polemi, D., & Papastergiou, S. (2006a). A secure, open and interoperable e-ordering service. Paper presented at the 2nd International Conference on Web Information Systems and Technologies (Webist, 2006), Setubal, Portugal, (pp. 57-62). INSTICC Press. ISBN: 978-972-8865-46-7.

Polemi, D., & Papastergiou, S. (2006b, March 16-17). Achievements and future direction in security policies. Electronic democracy challenges of the digital era. Paper presented at the 2nd National Conference with International Participation, Athens. ACCI.

Polemi, D., & Papastergiou, S. (2006c, August 26-29). TOES: Trustful and open e-ordering service for SMEs. Paper presented at the International Conference on Internet Surveillance and Protection (ICISP, 2006), Cap Esterel, Côte d'Azur, France. IEEE Computer Society. ISBN: 0-7695-2649-7.

Polemi, D., & Papastergiou, S. (2006d, October 11-13). A secure e-ordering Web service. Paper presented at the Sixth IFIP conference on e-Commerce, e-Business, and e-Government (I3E, 2006) Conference, Turku, Finland, (pp. 352-365). Springer Press. ISBN: 0-387-39226-2.

Pollack, A. (2003, November 16). Who's reading your x-ray? The New York Times. Retrieved June 12, 2008, from http://query.nytimes.com/gst/fullpage.html?sec=health&res=9503E5D91238F935A25752C1A9659C8B63&fta=y

Pollitt, D. (2005). Energis trains employees and customers in IT security. Human Resource Management International Digest, 13(2), 25–29.

Postel, J. B. (1982, August). Simple mail transfer protocol (IETF RFC 821).

Qibin, S., & Chang, S.-F. (2005). A secure and robust digital signature scheme for JPEG2000 image authentication. IEEE Transactions on Multimedia, 7(3), 480–494. doi:10.1109/TMM.2005.846776

Quelch, J. A., & Klein, L. R. (1996). The Internet and international marketing. Sloan Management Review, 37(3), 60–76.

Rastogi, V., Hong, S., & Suciu, D. (2007). The boundary between privacy and utility in data publishing. In Proc of the 33rd Intl. Conf. on Very Large Data Bases (VLDB'07). Vienna, Austria.

Reis, N., & Moysa, G. (2005). Legal Protections of ElectronicHealth Records: Issues of Consent and Security. [PubMed]. Health Law Review, 14(1), 18–25.

Renderlab (2009). Church of Wifi WPA-PSK Rainbow Tables. Retrieved February 23, 2009 from:http://www.renderlab.net/projects/WPA-tables/

Resnick, P. (April, 2001). Internet message format (IETF RFC 2822).

Resnick, P., Kuwabara, K., Zeckhauser, R., & Friedman, E. (2000). Reputation systems. ACM Communications, 43(12), 45–48. doi:10.1145/355112.355122

Rhodes, K. A. (2001). Operations security awareness: The mind has no firewall. Computer Security Journal, 17(3), 1–12.

Richardson, R. (2008). 2008 CSI Computer Crime & Security Survey. Retrieved August, 2009 from the CSI Web site: http://www.gocsi.com/

Rigoutsos, I., & Tien, H. (2004). Chung-kwei: A pattern-discovery-based system for the automatic identification of unsolicited e-mail messages (spam). In Proceedings of the First Conference on E-mail and Anti-Spam (CEAS).

Robinson, D. M. (1994). Health information privacy: Without confidentiality. [PubMed]. International Journal of Bio-Medical Computing, 35(Suppl), 97–104.

Rogers, M. P., & Solihin, Y. (2006). Efficient data protection for distributed shared memory multiprocessors. In Proceedings of the 15th International Conference on Parallel Architectures and Compilation Techniques (PACT) (pp. 84-94).

Rohm, A. J., & Milne, G. R. (2004). Just what the doctor ordered—the role of information sensitivity and trust in reducing medical information privacy concern. Journal of Business Research, 57(9), 1000–1011. .doi:10.1016/S0148-2963(02)00345-4

Rountree, P. W., & Land, K. C. (1996). Perceived risk versus fear of crime: empirical evidence of conceptually distinct reactions in survey data. Social Forces, 74(4), 1353–1376. .doi:10.2307/2580354

Rozovsky, L. E., & Rozovsky, F. A. (1985). The computer challenge: More data, less privacy. [PubMed]. Canadian Doctor, 51(3), 46–48.

Runy, L. A. (2008). The best defense. Hospitals & Health Networks, 7(3), 22–26.

Rusli, R. (2001). Secure system architecture and design. A white paper. SANS GIAC Security Essentials—Practical Assignment, GSEC Web site. Retrieved November 30, 2007, from http://www.giac.org/certified_professionals/practicals/gsec/1422.php

Sacco, V. F. (1993). Social support and the fear of crime. Canadian Journal of Criminology, 35(2), 187–196.

Samarati, P. (2001). Protecting Respondents' Identities in Microdata Release. TKDE, 13(6), 1010–1027.

Samarati, P., & Sweeney, L. (1998). Protecting privacy when disclosing information: K-anonymity and its enforcement through generalization and suppression. Technical Report. SRI Computer Science Lab.

Sandhu, R., & Chen, F. (1998). The multilevel relational (MLR) data model. ACM Transactions on Information and System Security, 1(1), 93–132. .doi:10.1145/290163.290171

Sandhu, R., & Zhang, X. (2005). Peer-to-peer access control architecture using trusted computing technology. In Proceedings of the Tenth ACM Symposium on Access Control Models and Technologies, New York, (pp. 147-158). ACM Press.

Saygin, Y., Verykios, V. S., & Clifton, C. (2001). Using unknowns to prevent discovery of association rules. SIGMOD Record, 30(4), 45–54. .doi:10.1145/604264.604271

Schmitt, N., & Oswald, F. L. (2006). The Impact of Corrections for Faking on the Validity of Noncognitive Measures in Selection Settings. The Journal of Applied Psychology, 91(3), 613–621. .doi:10.1037/0021-9010.91.3.613

Schneck, P. A., & Schwan, K. (1998). Dynamic Authentication for High-Performance Networked Applications. (Technical Report GIT-CC-98-08). Georgia Institute of Technology, College of Computing.

Schneier, B. (2000). Secrets and Lies, Digital security in a networked world. John Wiley and Sons. Inc.

Schultz, E. (2004). Worms and viruses: Are we losing control? Computers & Security, 23(3), 179–180. doi:10.1016/j.cose.2004.04.001

Schwartz, M. (2007, December 6). Application developlment: Buy, build or cobble? Retrieved June 12, 2008, from http://www.bmighty.com/hardware_software/showArticle.jhtml?articleID=204701872

Segars, A. (1997). Assessing the unidimensionality of measurement: a paradigm and illustration within the context of information systems research. Omega, 25(1), 107–121. .doi:10.1016/S0305-0483(96)00051-5

Serazzi, G., & Zanero, S. (2003). Computer virus propagation models. In Tutorials of the 11th IEEE/ACM International Symposium on Modeling, Analysis and Simulation of Computer and Telecom. Systems - MASCOTS 2003. Retrieved November 30, 2007, from http://citeseer.ist.psu.edu/serazzi03computer.html

Shaw, A. (2000). Provincial Health Networks Slowed Down by the Lack of New Privacy Legislation. Canadian Health Care Technology (March 2000).

Shaw, N., & Yadav, S. B. (2001). DEACON: An integrated approach to the analysis and design of enterprise architecture-based computer networks. Communications of the Association for Information Systems, 7.

Shi, W., Lee, H.-H., Falk, L., & Ghosh, M. (2006). An integrated framework for dependable and revivable architectures using multicore processors. In Proceedings of the 33rd International Symposium on Computer Architecture (ISCA) (pp. 102-113).

Shi, W., Lee, H.-H., Ghosh, M., & Lu, C. (2004). Architectural support for high speed protection of memory integrity and confidentiality in multiprocessor systems. In Proceedings of the 13th International Conference on Parallel Architectures and Compilation Techniques (PACT) (pp.123-134).

Shi, W., Lee, H.-H., Ghosh, M., Lu, C., & Boldyreva, A. (2005). High efficiency counter mode security architecture via prediction and precomputation. In Proceedings of the 32nd International Symposium on Computer Architecture (ISCA) (pp. 14-24).

Shinder, D. (2002). Scene of the Cybercrime. Computer Forensics Handbook. USA: Syngress Publishing.

Shwartz, N., & Bohner, G. (2001). The construction of attitudes . In Tesser, A., & Shwarts, N. (Eds.), Blackwell handbook of social psychology: Intraindividual processes (pp. 436–457). Oxford, UK: Blackwell.

Siles, R. (2006, January 16). Security Focus. Sebek 3: Tracking the attackers, part one. Retrieved February 12, 2007 from http://www.securityfocus.com/infocus/1855

Siles, R. (2007). Security Focus. Wireless Forensics: Tapping the Air - Part Two. Retrieved February 16, 2007 from http://www.securityfocus.com/infocus/1885/2

Sillence, E., Briggs, P., Harris, P., & Fishwick, L. (2006). A Framework for understanding trust factors in Web-based health advice. International Journal of Human-Computer Studies, 64, 697–713. .doi:10.1016/j.ijhcs.2006.02.007

Sillence, E., Briggs, P., Harris, P., & Fishwick, L. (2007). Going online for health advice: Changes in usage and trust practices over the last five years. Interacting with Computers, 19, 397–406. .doi:10.1016/j.intcom.2006.10.002

Simpson, R. L. (2002)... Nursing Management. Chicago, 33(12), 46–48.

Siponen, M. T. (2000). A conceptual foundation for organizational information security awareness. Information Management & Computer Security, 8(1), 31–41. doi:10.1108/09685220010371394

Smith, H. J., Milberg, S. J., & Burk, S. J. (1996). Information privacy: measuring individuals' concerns about organizational practices. Management Information Systems Quarterly, 20(2), 167–196. .doi:10.2307/249477

Smith, J. R., & Protti, D. J. (2001). Personal Health Records: part II: a potential model for Canada?Healthcare Information Management and Communications.

Soonthornphisaj, N., Chaikulseriwat, K., & Piyanan, T.-O. (2006, August 26-30). Anti-spam filtering: A centroid-based classification approach. Paper presented at the Signal Processing, 6th International Conference (Vol. 2, pp. 1096-1099).

Southwick, P. (2003). Secure network design. A white paper. Retrieved November 30, 2007, from the Hill Associates Web site: http://www.hill.com/archive/pub/papers/2003/10/paper.pdf#search='secure%20network%20design'

Srinivasan, I., & Thuraisingham, B. (October 2006). Design and Simulation of Agent-based Trust Management Techniques for a Coalition Environment. Technical Report, UTDCS45-06 (also published in the Proceedings of IEEE FTDCS, 2007).

Srivastava, R. P., & Mock, T. J. (1999). Evidential reasoning for web trust assurance services. Journal of Management Information Systems, 16(3), 11–32.

Staab, S., Domingos, P., Mika, P., Golbeck, J., Ding, L., & Finin, T. W. (2005). Social networks applied. IEEE Intelligent Systems, 20(1), 80–93. doi:10.1109/MIS.2005.16

Staddon, J. (2003, June). Dynamic inference control. In Proc. of the 8th ACM SIGMOD workshop on research issues in data mining and knowledge discovery (DMKD'03) (p. 94-100). San Diego, CA, USA: ACM.

Stallings, W. (2003). Cryptography and network security: Principles and practices (3rd ed.). Prentice Hall.

Stanton, J. M., Stam, K. R., Mastrangelo, P., & Jolton, J. (2005). Analysis of end user security behaviors. Computers & Security, 24(2), 124–133. doi:10.1016/j.cose.2004.07.001

Stewart, K. A., & Segars, A. H. (2002). An empirical examination of the concern for information privacy instrument. Information Systems Research, 13(1), 36–49. .doi:10.1287/isre.13.1.36.97

Stone, D. L., & Herringshaw, C. (1991). Effects of the purpose of the test, perceived relevance, and use of test results on reactions to honesty testing. Paper presented at the 6th annual conference of the Society for Industrial and Organizational Psychology, St. Louis, MO.

Stone, D. L., & Jones, G. (1992). Effects of type of test and justification provided on reactions to honesty testing. Paper presented at the 7th annual conference of the Society for Industrial and Organizational Psychology, Montreal, Quebec, Canada.

Stone, D. L., & Kotch, D. (1989). Individuals' attitudes toward organizational drug testing policies and practices. The Journal of Applied Psychology, 3, 518–521. .doi:10.1037/0021-9010.74.3.518

Stone, E. F., & Stone, D. L. (1990). Privacy in Organizations: Theoretical issues, research findings, and protection mechanisms. Research in Personnel and Human Resource Management, 8, 349–411.

Straub, D. (1990). Effective IS security: an empirical study. Information Systems Research, 1(3), 255–276. .doi:10.1287/isre.1.3.255

Suh, G. E., Clarke, D., Gassend, B., van Dijk, M., & Devadas, S. (2003). AEGIS: architecture for tamper-evident and tamper-resistant processing. In Proceedings of the 17th International Conference on Supercomputing (ICS) (pp. 160-171).

Swartz, N. (2004). A prescription for electronic health records. Information Management Journal, 38(4), 20–26.

Sweeney, L. (1997). Guaranteeing anonymity when sharing medical data, the datafly system. Journal of the American Medical Informatics Association.

Sweeney, L. (2002a). Achieving k-anonymity privacy protection using generalization and suppression. International Journal on Uncertainty, Fuzziness and Knowledge-based Systems, 10(5), 571–588. .doi:10.1142/S021848850200165X

Sweeney, L. (2002b). K-anonymity: A model for protecting privacy. Intl. Journal on Uncertainty, Fuzziness and Knowledge-based Systems, 10(5), 557–570. .doi:10.1142/S0218488502001648

Takefusa, A., Matsuoka, S., Aida, K., Nakada, H., & Nagashima, U. (1999). Overview of a Performance Evaluation System for Global Computing Scheduling Algorithms. In Proceedings of the Eighth IEEE International Symposium on High Performance Distributed Computing (HPDC'99) (pp 97-104).

Tapp, A. (2004). CNPS: Respecting your privacy rights. [PubMed]. The Canadian Nurse, 100(4), 36–37.

Taylor, H. (2003). Most people are 'privacy pragmatists' who, while concerned about privacy, will sometimes trade it off for other benefits. Harris Interactive. Retrieved from http://www.harrisinteractive.com/harris_poll/printer-friend/index.asp?PID=365

Taylor, H., & Leitman, R. (2001). U.S. trails other English speaking countries in use of electronic medical records and electronic prescribing. Harris Interactive, 1(28), 1–3.

Taylor, H., & Leitman, R. (2002). The future use of the Internet in four countries in relation to prescriptions, physician communication and health information. Health Care News, 2(13). Retrieved from http://www.harrisinteractive.com/news/newsletters_healthcare.asp

Taylor, I. (1995). Fear of crime: interpreting victimization risk. Urban Studies (Edinburgh, Scotland), 32(9), 1567–1568.

Teer, F. P., Kruck, S. E., & Kruck, G. P. (2007). Empirical study of students' computer security practices/perceptions. Journal of Computer Information Systems, 47(3), 105–110.

Thibaut, J., & Walker, L. (1975). Procedure justice: a psychological analysis. Hillsdale, NJ: Erlbaum.

Thiemert, S., Sahbi, H., & Steinebach, M. (2005). Applying interest operators in semifragile video watermarking. IS&T SPIE International Symposium on Electronic Imaging Security, Steganography and Watermarking of Multimedia Contents VII, 5681, 353–362.

Thomas, R. K., & Sandhu, R. S. (1998). Task-based authorization controls (TBAC): A family of models for active and enterprise-oriented authorization management. In Proceedings of the IFIP TC11 WG11.3 Eleventh International Conference on Database Security (pp. 166-181). Chapman & Hall, Ltd.

Thompson, E. D., & Kaarst-Brown, M. L. (2005). Sensitive information: A review and research agenda. Journal of the American Society for Information Science and Technology, 56(3), 245–257. .doi:10.1002/asi.20121

Thompson, M., Johnston, W., Mudumbai, S., Hoo, G., Jackson, K., & Essiari, A. (1999). Certificate based access control for widely distributed resources. In Proceedings of the 8th USENIX security symposium. Retrieved from, http://dsd.lbl.gov/~mrt/papers/AkentiUsenixSec.pdf

Thompson, S.T. Helping the Hacker? (2006). Library Information, Security, and Social Engineering. Information Technology and Libraries. Chicago, 25(4), 222–226.

Tian, J. (2002). Reversible watermarking by difference expansion. In Proceedings of the Workshop on Multimedia and Security (pp. 19–22).

Tom, F. (2004). Notes and practical considerations for researchers. The Netherlands: Kluwer Academic Publishers.

Tomcat 5.0 for Java WSDP. Retrieved December 17, 2007, from http://java.sun.com/webservices/containers

Townsend, A., & Bennett, J. (2003). Privacy, technology, and conflict: emerging issues and action in workplace privacy. Journal of Labor Research, XXIV(2), 195–205. .doi:10.1007/BF02701789

Truta, T. M., Fotouhi, F., & Barth-Jones, D. (2003). Disclosure risk measures for microdata. Proc. of the 15th Intl. Conf. on Scientific and Statistical Database Management (pp. 15-22).

Tuesday, V. (2001). Human factor derails best-laid security plans. Computerworld, 35(18), 52.

Tyler, T. R. (1994). Psychological models of the justice motive: antecedents of distributive and procedural justice. Journal of Personality and Social Psychology, 67, 850–863. .doi:10.1037/0022-3514.67.5.850

Udelson, T. (2000). How secure is your network? Association Management, 52(6), 25–26.

Upshur, R. E., Morin, B., & Goel, V. (2001). The privacy paradox: Laying Orwell's ghost to rest. [PubMed]. Canadian Medical Association Journal, 165(3), 307–309.

Vaidyanathan, G., & Devaraj, S. (2003). A five framework for analyzing online risks in e-businesses. Communications of the ACM, 46(12), 354–361. doi:10.1145/953460.953522

Vatis, M. (Ed.). (2003). Creating a Trusted Network for Homeland Security, Markle Report.

Verdon, D., & McGraw, G. (2004). Risk analysis in software design. IEEE Security & Privacy, 2(4), 79–84. doi:10.1109/MSP.2004.55

Verykios, V. S., Bertino, E., Fovino, I. N., Provenza, L. P., Saygin, Y., & Theodoridis, Y. (2004, March). State-of-the-art in Privacy Preserving Data Mining. SIGMOD Record, 33(1), 50–57. .doi:10.1145/974121.974131

Verykios, V. S., Elmagarmid, A. K., Elisa, B., Saygin, Y., & Elena, D. (2004, April). Association Rule Hiding. IEEE Transactions on Knowledge and Data Engineering, 16(4), 434–447. .doi:10.1109/TKDE.2004.1269668

Vidmar, N., & Flaherty, D. H. (1985). Concern for Personal Privacy in an Electronic Age. The Journal of Communication, 35(Spring), 91–103. .doi:10.1111/j.1460-2466.1985.tb02236.x

Volonino, L., & Robinson, S. R. (2004). Principles and Practice of Information Security: Protecting Computers From Hackers and Lawyers. Upper Saddle River, NJ: Pearson Prentice Hall.

Vu, K. L., Proctor, R. W., Bhargav-Spantzel, A., Tai, B. L., Cook, J., & Schultz, E. (2007). Improving password security and memorability to protect personal and organizational information. International Journal of Human-Computer Studies, 65, 744–757. .doi:10.1016/j.ijhcs.2007.03.007

Wacks, R. (1989). Personal Information, Privacy and the Law. Oxford: Oxford University Press.

Wade, J. (2004). The weak link in IT security. Risk Management, 51(7), 32–37.

Wang, D. W., Liau, C. J., & Hsu, T. S. (2006). Privacy protection in social network data disclosure based on granular computing. In Proceedings of the 2006 IEEE international conference on fuzzy systems (HICSS'05) (pp. 997–1003). IEEE CS.

Wang, P., & Petrison, L. A. (1993). Direct Marketing Activities and Personal Privacy: A Consumer Survey. Journal of Direct Marketing, 7, 7–19. .doi:10.1002/dir.4000070104

Wang, Z., & Lee, R. B. (2007). New Cache Designs for Thwarting Software Cache-based Side Channel Attacks. In Proceedings of the 34th International Symposium on Computer Architecture (ISCA) (pp. 494-505).

Warr, M. (1995). The polls – poll trends: public opinion on crime and punishment. Public Opinion Quarterly, 59(2), 296–310. .doi:10.1086/269474

Warren, P. (2005). Ten steps to secure networking. TechWorld. Retrieved November 30, 2007, from http://www.techworld.com/security/features/index.cfm?FeatureID=1862

Watts, D. J. (2003). Small worlds: The dynamics of networks between order and randomness. Princeton, NJ: Princeton University Press.

Webster, J., & Trevino, L. K. (1995). Rational and social theories as complementary explanations of communication media choices: Two policy capturing studies. Academy of Management Journal, 38, 1544–1572. .doi:10.2307/256843

Wei, J., & Li, Y. (2004). Forecasting on information systems attack sources. In Proceedings of the Information Resources Management Association, New Orleans.

Weible, R. J. (1993). Privacy and Data: An Empirical Study of the Influence and Types and Data and Situational Context upon Privacy Perceptions. Doctoral dissertation, Department of Business Administration, Mississippi State University.

Weise, E. (2005). Medical errors still claiming many lives. USA Today. Retrieved May 17, 2005 from http://www.usatoday.com/news/health/2005-05-17-medical-errors_x.htm.

Weitz, M., Drummond, N., Pringle, D., Ferris, L. E., Globerman, J., Hebert, P., et al. (2003). In whose interest? current issues in communicating personal health information: A Canadian perspective. The Journal of Law, Medicine & Ethics, 31(2), 292–301. PubMeddoi:10.1111/j.1748-720X.2003.tb00090.xdoi:10.1111/j.1748-720X.2003.tb00090.x

Weitzner, D. J., Hendler, J., Berners-Lee, T., & Connolly, D. (2006). Creating a policy-aware Web: Discretionary, rule-based access for the World Wide Web . In Ferrari, E., & Thuraisingham, B. M. (Eds.), Web and information security (pp. 1–31). Hershey, PA: IDEA Group Publishing.

Westin, A. F. (1976). Privacy and freedom. New York: Atheneum.

White, S. R. (1998, October 22). Open problems in computer virus research. In Proceedings of theVirus Bulletin Conference, Munich Germany.

Whitman, M. (2003). Enemy at the gate: Threats to information security. Communications of the ACM, 46(6), 91–95. doi:10.1145/859670.859675

Whitman, M. E., & Mattord, H. J. (2004). Making users mindful of IT security. Security Management, 48(11).

Whitman, M. E., & Mattord, H. J. (2005). Principles of information security (2nd ed.). Canada: Thomson Course Technology.

Whitmore, J. J. (2001). A method for designing secure solutions. IBM Systems Journal, 40(3), 747–768. doi:10.1147/sj.403.0747

Wilcoxon, F., & Wilcox, R. A. (1964). Some rapid approximate statistical procedures. Pearl River, NY: Lederle Laboratories.

William, C. W. (1996). Learning rules that classify e-mail. Paper presented at the AAAI Spring Symposium on Machine Learning in Information Access.

Willison, D. J. (1998). Health services research and personal health information: Privacy concerns, new legislation and beyond. [PubMed]. Canadian Medical Association Journal, 159(11), 1378–1380.

Wilson, E. V. (2003). Asynchronous health care communication. Communications of the ACM, 46(6), 79–84. .doi:10.1145/777313.777316

Winsborough, W. H., & Li, N. (2000). Safety in automated trust negotiation. In Proc. of 2004 IEEE symposium on security and privacy.

Winslett, M., Ching, N., Jones, V. E., & Slepchin, I. (1997). Using digital credentials on the World Wide Web. Journal of Computer Security, 5(3), 255–266.

Wisebrod, D. (1995). Visions of the Future: Governments Wrestle with Information Technology Directed Research Project, University of Toronto. Retrieved on March 12, 2007 from http://www.catalaw.com/dov/docs/dw-govts.htm.

Wong, M., & Schlitt, W. (2005, December 8). Sender policy framework for authorizing use of domains in e-mail, version 1. Retrieved December 18, 2007, from http://www.ietf.org/internet-drafts/draft-schlitt-spf-classic-02.txt

Wong, M., & Schlitt, W. (2006). Sender id: Authenticating e-mail. Retrieved December 18, 2007, from http://xml.coverpages.org/draft-ietf-marid-core-03.txt

Woodman, R., Ganster, D., Adams, J., McCuddy, M., Tolchinsky, P., & Fromkin, H. (1982). A survey of employee perceptions of information privacy in organizations. Academy of Management Journal, 25(3), 647–663. .doi:10.2307/256087

Woodman, R. W., Adams, J., Tolchinsky, P. D., & Fromkin, H. (1982). Asurvey of employee perceptions of information privacy in organizations. Academy of Management Journal, 25(3), 647–663. .doi:10.2307/256087

Wortley, R., & Smallbone, S. (2004). Child Pornography on the Internet. Retrieved June 21, 2007 from http://www.cops.usdoj.gov/mime/open.pdf?Item=1729

Wyer, R. S., Jr. (2004). The cognitive organization and use of general knowledge. In Jost, J. T., & Banaji, M. R. (Eds.), Perspectivism in social psychology: The yin and yang of scientific progress (pp. 97–112). Washington, D.C: American Psychological Association. doi:10.1037/10750-008

Xiao, X., & Tao, Y. (2006). Anatomy: Simple and Effective Privacy Preservation. Paper presented at the VLDB.

Xiao, X., & Tao, Y. (2006). Personalized privacy preservation. In Proc. Of the 25th ACM Intl. Conf. on Management of Data. (SIGMOD'06). Illinois, USA.

Xiao, X., & Tao, Y. (2007). m-Invariance: Towards privacy preserving re-publication of dynamic. In Proc. Of the 26th ACM Intl. Conf. on Management of Data. (SIGMOD'07). Beijing, China.

Xiong, L., & Liu, L. (2004). Peertrust: Supporting reputation-based trust for peer-to-peer electronic communities. IEEE Transactions on Knowledge and Data Engineering, 16(7), 843–857. doi:10.1109/TKDE.2004.1318566

Yadav, S. B. (2006). A six view perspective of system security—issues, risks, requirements, and mechanisms (Tech. Rep. #ISQSYadav2006-1). Lubbock, TX: Rawls College of Business, Texas Tech University.

Yaffee, R. (2000). Introduction to time series analysis and forecasting with applications of SAS and SPSS. Academic Press.

Yan, C., Rogers, B., Englender, D., Solihin, Y., & Prvulovic, M. (2006). Improving cost, performance, and security of memory encryption and authentication. In Proceedings of the 33rd International Symposium on Computer Architecture (ISCA) (pp. 179-190).

Yang, J., Zhang, Y., & Gao, L. (2003). Fast secure processor for inhibiting software piracy and tampering. Proceedings of the 36th International Symposium on Microarchitecture (MICRO), 351-360.

Yang, L., Kizza, J. M., Cemerlic, A., & Liu, F. (2007). Fine-grained reputation-based routing in wireless ad hoc networks. In Proceedings of IEEE Intelligence and Security Informatics Conference (pp. 75-78). IEEE Computer Society Press.

Yang, L., & Peng, L. (2006). SecCMP: a secure chip-multiprocessor architecture. In Proceedings of the first Workshop on Architectural and System Support for Improving Software Dependability (ASID) (pp. 72-76).

Yao, C., Wang, X. S., & Jajodia, S. (2005). Checking for k-anonymity violation by views. In Proc. Of the 31st Very Large Data Bases Conf. (VLDB'05). Trondheim, Norway.

Yao, W. T. M. (2003). Fidelis: A policy-driven trust management framework. In Proceedings of the First International Conference on Trust Management (LNCS 2692, p. 1071).

Yung, A., & Li, Y. (2009). A Literature Review of Developments in Health Information Banks, Retrieved on July 24, 2009 from ElectronicHealthcare, 7(4), e1-e5.

Zhan, &Li, Y (2007). Capture inference attacks for k-anonymity with privacy inference logic. In Proc. Of the 12th Intl Conf. on Database System for Advanced Applications (DASFAA'07). Bangkok, Thailand.

Zhang, J., & Tan, T. (2002). Brief review of invariant texture analysis methods. Pattern Recognition, 35, 735–747. doi:10.1016/S0031-3203(01)00074-7

Zhang, Y., Gao, L., Yang, J., Zhang, Z., & Gupta, R. (2005). SENSS: Security Enhancement to Symmetric Shared Memory Multiprocessors. Proceedings of the 11th Intl. Symposium on High-Performance Computer Architecture (HPCA), 352-362.

Zhu, Y., & Liu, L. (2004). Optimal randomization for privacy preserving data mining. Paper presented at the KDD.

Zhuang, X., Zhang, T., & Pande, S. (2004). HIDE: An Infrastructure for Efficiently Protecting Information Leakage on the Address Bus. Proceedings of the 11th International Conference on Architectural Support for Programming Languages and Operating Systems (ASPLOS), 72-84.

Zouridaki, C., Mark, B. L., Hejmo, M., & Thomas, R. K. (2005). A quantitative trust establishment framework for reliable data packet delivery in MANETS. In Proceedings of the Third ACM Workshop on Security of Ad hoc and Sensor Networks, New York, (pp. 1-10). ACM Press.

Zrebiec, J. F., & Jacobson, A. M. (2001). What attracts patients with diabetes to an Internet support group? A 21-month longitudinal website study. Diabetic Medicine, 18(2), 154–158..doi:10.1046/j.1464-5491.2001.00443.x

Zucker, L. G. (1986). Production of trust: Institutional sources of economic structure, 1840-1920. Research in Organizational Behavior, 8, 53–111.

Zunkel, D. (2005). The other side of privacy: protecting information with biometrics. Security Technology Design, 15(6), 46–53.

About the Contributors

Hamid Nemati is an associate professor of information systems in the Department of Information Systems and Operations Management at the University of North Carolina at Greensboro. He holds a doctorate from the University of Georgia and a Master of Business Administration from the University of Massachusetts. Before coming to UNCG, he was on the faculty of J. Mack Robinson College of Business Administration at Georgia State University. He has extensive professional experience in various consulting, business intelligence, and analyst positions and has consulted for a number of major organizations. His research specialization is in the areas of decision support systems, data warehousing, data mining, knowledge management, and information privacy and security. He has presented numerous research and scholarly papers nationally and internationally. His articles have appeared in a number of premier professional and scholarly journals.

* * *

Taner Pirim has a BS in Environmental Engineering from Marmara University in Istanbul, an MS in Civil Engineering emphasizing in river hydraulics and a Ph.D. in Production and Operations Management from the University of Mississippi. He has worked as an instructor and researcher in both the Engineering and Business Schools of the University of Mississippi and as a supercomputer consultant for the Mississippi Center for Supercomputing Research. He has also served as an adjunct professor for the department of MIS/POM at the University of Mississippi.

Tabitha James is an Associate Professor in the Department of Business Information Technology, Pamplin College of Business, at Virginia Tech. Her current research interests are in the areas of security and privacy in information technology, combinatorial optimization, heuristics, parallel computing, and social networks. She has published in journals such as *IEEE Intelligent Systems, IEEE Transactions on Evolutionary Computation, European Journal of Operational Research, Computers and Operations Research,* and *Engineering Applications of Artificial Intelligence.* She received a BBA in Management Information Systems and a Ph.D. in Business Administration with a major in Management Information Systems and a minor in Production and Operations Management from the University of Mississippi.

Katherine Boswell is an Assistant Professor at The University of Louisiana - Monroe. She received her Ph.D. in Business Administration in the area of Management Information Systems from the University of Mississippi. Her minor areas of study were Marketing and Production Operations Management. She received her Bachelor's in Accounting in 1990, and her Master in Business Administration in 1991 from Mississippi State University. Katherine has publications in the areas of system development/strategy and in appropriate use policies. Her current research interests lie in the following areas: e-commerce, security and privacy.

Brian J. Reithel, Ph.D., is a Professor of MIS at The University of Mississippi. Previously, he served as MIS/POM Department Chair, Dean of the School of Business, and as Associate Vice Chancellor for University Relations. He has received numerous awards for excellence in teaching, research, and service over the past 20 years. He is the author of more than 70 research articles and papers that have appeared in leading journals and conference proceedings. He has also appeared in major practitioner magazines, including ComputerWorld, PCWorld, and Inc. His research interests revolve around emerging technologies as well as factors that impact the success of systems development efforts, particularly in the area of strategic information systems.

Reza Barkhi is the Head of MIS department at American University of Sharjah (UAE). He is also a tenured Associate Professor and Research Fellow in the Accounting and Information Systems department at Virginia Tech (USA). While serving as the Head of MIS, he has been on leave of absence from Virginia Tech. His current research interests are in the areas of Electronic Commerce, Quantitative Business Modeling, and Behavioral implications of Technology. Reza has published in journals such as *Location Science, European Journal of Operational Research, Computers & OR, Group Decision and Negotiation, and Decision Support Systems, Communication Research, Communications of AIS, Information Technology and Management, Information & Management*, and *Journal of Management Information Systems.* He received a BS in Computer Information Systems, MBA, MA, and Ph.D. in Business focusing on Information Systems, Operations Management, and Decision Sciences all from The Ohio State University.

B. Dawn Medlin is the Chair and an Associate Professor in the Department of Computer Information Systems, John A. Walker College of Business, and Appalachian State University in Boone, NC. Her teaching and research activities have mainly been in the area of security, health care informatics, and e-commerce. She has published in journals such as The *Journal of Information Systems Security, Information Systems Security, International Journal of Electronic Marketing and Retailing*, and the *International Journal of Healthcare Information Systems and Informatics.*

Adriana Romaniello is currently serving as an Associate Professor in the Department of Management in Universidad Rey Juan Carlos in Madrid, Spain. She earned her PhD at Universidad Complutense de Madrid. Her teaching and research activities have been in the area of knowledge management, strategic management and information and communication technology. She has participated in many international conferences and in various projects supported with governmental funds.

John Cassini is a Professor in the Department of Management: Innovation Europe, Institue of Engineering Sciences (ISTIA) at the Université d'Angers, France. John has worked on several different international research projects and serves as the Board President at the English Library in Angers (EL-LIA). John has recently published in the *Journal of Information Privacy and Security.*

Jeff Barnett is an adjunct professor of health information science at the University of Victoria at Victoria (BC, Canada). He is also the Director of Clinical Informatics, Cancer Care at the British Columbia Cancer Agency, Victoria (BC, Canada). He holds a master's degree in health information science from the University of Victoria. He has extensive experience as an analyst, and project team member in health care systems deployment. Prior to his entry into the field of informatics he was a pharmacist with over 25 year's experience. His research specialization is in the area of privacy and confidentiality, data warehousing, computerized provider order entry and knowledge management. He has presented nationally and internationally on a wide range of topics relating to these research interests, and he has co-authored several papers in several health care journals.

Tamas S. Gal is a Ph.D. student at the Department of Information Systems at the University of Maryland at Baltimore County. He holds a Masters degree from the same department. He works for the Kentucky Cancer Registry as a Database Analyst. His research specialization is in the areas of privacy preserving data mining in the medical field, healthcare information systems and bioinformatics.

Zhiyuan Chen is an assistant professor at information systems department, UMBC. He has a PhD in Computer Science from Cornell University. His research interests include privacy preserving data mining, data navigation and visualization, XML, automatic database tuning, and database compression.

Aryya Gangopadhyay is a Professor of Information Systems at the University of Maryland Baltimore County (UMBC). He has a PhD in Computer Information Systems from Rutgers University. His research interests include privacy preserving data mining, data cube navigation, and core and applied research on data mining. He has co-authored and edited three books, many book chapters, and numerous papers in peer-reviewed journals. His research has been funded by the National Science Foundation, U.S. Department of Education, and various other Government and commercial entities.

Joseph A. Cazier is an Assistant Dean - Graduate Programs and Research at Appalachian State University. He has a keen interest in information ethics, trust and security and conducts research in this area. He has published in The *Journal of Information Systems Security, Information and Management, Information Systems Frontiers, International Journal of Electronic Marketing and Retailing, International Journal of Information Security and Privacy*, and the *International Journal of Healthcare Information Systems and Informatics.*

B. Dawn Medlin is the Chair of the Department of Computer Information Systems, John A. Walker College of Business, and Appalachian State University in Boone, NC. Her teaching and research activities have mainly been in the area of security, health care informatics, and e-commerce. She has published in journals such as The *Journal of Information Systems Security, Information Systems Security, International Journal of Electronic Marketing and Retailing*, and the *International Journal of Healthcare Information Systems and Informatics.*

Vance Wilson is a visiting faculty member of the Information, Operations, and Technology Management (IOTM) department in the University of Toledo College of Business Administration. His research focuses generally on organizational aspects of human-computer interaction with research streams in e-health, computer-mediated communication, online persuasion, and decisional guidance. Prof. Wilson is the supervising editor of the Information Systems and Healthcare Department at Communications of the AIS.

David Dobrzykowski enjoyed a 13 year career as a healthcare executive. Prior to returning for his Ph.D., he served in leadership positions in the provider and insurance verticals of the healthcare sector while also spending 8 years on the University of Toledo College of Business Administration's Adjunct Faculty. His research interests include supply chain management and information technology, specifically related to the application of supply chain concepts for value co-creation (service-dominant logic) in healthcare. David's research also explores comparisons of these phenomena in service and product-focused firms. His work has been published (accepted) in many conference proceedings as well as in refereed academic journals and books such as *Benchmarking: An International Journal, International Journal of Information Security and Privacy, International Journal of Services and Operations Management, Pervasive Information Security and Privacy Developments: Trends and Advancements, Service Science*, and *Strategic Outsourcing: An International Journal.* In Fall 2010, Dobrzykowski will join the business school faculty at Eastern Michigan University as an Assistant Professor in the Department of Computer Information Systems.

Joseph A. Cazier is an Assistant Professor in the Department of Computer Information Systems at Appalachian State University. He has a keen interest in information ethics, trust and security and conducts research in this area. He has published in journals such as The *Journal of Information Systems Security, Information and Management, Information Systems Frontiers, International Journal of Networking and Virtual Organisations, International Journal of Electronic Marketing and Retailing, International Journal of Information Security and Privacy*, and the *International Journal of Healthcare Information Systems and Informatics.*

Jingquan Li is an assistant professor of accounting and computer information systems at the Texas A&M University-San Antonio. His research focuses on data privacy and security, economics of information technology, supply chain management, agent technology, data mining and business intelligence. He has a PhD in business administration from the University of Illinois at Urbana-Champaign. He received the Seymour Sudmand Award and College of Commerce Fellowship from the University of Illinois at Urban-Champaign.

Michael J. Shaw is the Hoeft Endowed Chair in Information Technology Management and Director of the Center for Information Systems and Technology Management at the University of Illinois at Urbana-Champaign, where he has been on the faculty of the Department of Business Administration since 1984. He is also a member of the artificial intelligence group at the Beckman Institute for Advanced Science and Technology; He is the editor-in-chief of the journal *Information Systems and e-Business Management.* Among Professor Shaw's recent books are Handbook in Electronic Commerce, Information-Based Manufacturing and e-Business Management.

Zude Li is now a 4-year Ph.D. candidate at the University of Western Ontario, Canada. He received the M.S. degree (with honour of top-2 distinguished graduate) in 2004 from Tsinghua University, China. His research interests include database privacy (especially, microdata publication and privacy inference), software engineering, and information security. He has published over 10 papers in the privacy protection area in referred international journals and conference proceedings.

Syed Naqvi is a R&D Project Manager at the Centre of Excellence in Information and Communication Technologies (CETIC) Belgium. He holds a PhD in Distributed Systems Security and an MBA in Business Technology Management. He is a Senior Member of IEEE. His research activities circle around privacy, security and trust issues of large scale open distributed heterogeneous systems. He spent several years in industry before starting his research career in information and communication technologies. He has held a visiting scientist position at the University of Washington at Seattle; and was a research fellow at the Science and Technology Facilities Council of UK. He is an external reviewer of a number of international journals and has served a number of scientific symposia as technical program committee member. He has worked in various national, regional, and European funded collaborative research projects. He has also contributed for International Collaborations notably with NSF in USA and DEST in Australia. Syed is also active in the Future Internet (FI) initiative of the European Commission where he is contributing for the Security and Dependability issues of the Future Internet based Virtualisation Infrastructures. He is a member of the Scientific & Technical Programme Committee of the "Towards the Future Internet" Book series.

Lu Peng received his bachelor's and master's degrees in computer science & engineering from Shanghai Jiaotong University, China. He obtained his PhD degree in computer engineering from the University of Florida in Gainesville in April 2005. He joined the electrical and computer engineering department at Louisiana State University as an Assistant Professor in August, 2005. His research focus on memory hierarchy system, multi-core interconnection, power efficiency and other issues in CPU design. He also has interests in Network Processor. He received an ORAU Ralph E. Powe Junior Faculty Enhancement Awards in 2007 and a Best Paper Award from IEEE International Conference on Computer Design in 2001. Dr. Peng is a member of the ACM and the IEEE Computer Society.

Li Yang is an assistant professor in the Department of Computer Science and Electrical Engineering at University of Tennessee at Chattanooga. Her research interests include network and information security, databases, and engineering techniques for complex software system design. She authored papers on these areas in refereed journal, conferences and symposiums. She is a member of the ACM.

Balachandran Ramadass received his Bachelor's degree in electronics and communication engineering from Pondicherry University, India and his Master of Science degree in electrical and computer engineering in Louisiana State University. His research focuses are CMP architecture, on-chip power optimization, interconnection network, heterogeneous CMP architecture.

Hamid Jahankhani is the Associate Dean- Academic at the School of Copmputing, Information Technology and Engineering, University of East London. Hamid and his colleagues are pioneers in developing the context and frameworks for internationally recognised information security systems and computer forensics. With his own Masters and professional Doctorate in Information Security and Computer Forensics, Hamid stresses the importance of education in order to arm and deliver the next generation of future experts to tackle our global security issues. Dr Jahankhani is consulted internationally on global information security and cybercrime issues, and is also author of several influential articles published on the criminal investigation and forensics analysis of mobile devices. Hamid is the co-Founder of the VoICE, Victims of the Internet Crime Europe, a non-for-profit organisation. He directs the research of number of PhD students. Dr Jahankhani is the Editor-in-Chief of the *International*

Journal of Electronic Security and Digital Forensics published by Inderscience and general chair of the annual International Conference on Global e-Security (ICGeS), www.uel.ac.uk/icges. Hamid is, also member of the programme committee of number of International Conference on Security, Cryptography Digital Forensics and Incident Analysis. He has edited and contributed to several books and has over 80 conference and journal publications.

Elidon is a Senior IT Technician working at the University of East London with considerable experience in the industry. He holds a Bachelors degree in E-Commerce and a Masters in Information Security and Computer Forensics. He is a certified CompTIA professional and holds qualifications such as: CompTIA Linux +, CWNA (Certified Wireless Network Administrator), CWSP (Certified Wireless Security Professional) and CCNA (Cisco Certified Network Associate). Elidon holds also SSCP (Systems Secure Certified Professional) and CCAI (Cisco Certified Academy Instructor) qualifications. In addition to his work at UEL he is undergoing a PhD research degree in Computer Security looking at combining neural networks and Intrusion Detection Systems. His areas of interest area: computer security, wireless communications, Linux, computer forensics, ethical hacking and security tools development.

Xun Li earned her PhD. at the University of Kentucky and is currently an assistant professor of operations management at Nicholls State University. Her current research interests include work design in supply chains, supply chain agility, the effects of inter-organizational systems on supply chain performance, and privacy/security issues in information sharing. Her teaching activities include quantitative analysis in operation management, supply chain management, and IT applications in business.

Radhika Santhanam is a Gatton Endowed Research Professor in the Gatton College of Business & Economics at the University of Kentucky. Her research focuses on investigating human-computer interaction issues with particular interest in understanding how users learn new technologies. Her research findings are published in *MIS Quarterly, Information Systems Research, Journal of Management Information Systems, Decision Support Systems, Information and Organization, European Journal of information Systems* and other journals. She served as an associate editor of MIS Quarterly and is currently an Associate Editor at Information Systems Research and Decision Support Systems, and on the editorial board of Enterprise Information Systems.

S. E. Kruck is an associate professor in the computer information systems and management science department at James Madison University, Harrisonburg Virginia. Dr. Kruck earned a PhD from the Department of Accounting and Information Systems at Virginia Polytechnic and State University and is also a CPA in the state of Virginia. Dr. Kruck's research focuses on curriculum development and computer security. Dr. Kruck teaches management information systems and computer security courses at both the graduate and undergraduate levels and has been the recipient of the Madison Scholar Award and the Alumni Distinguished Faculty Award.

Faye Teer is professor of information systems in the computer information systems and management science department at James Madison University. Dr. Teer received the Doctor of Business Administration degree from Louisiana Tech University. She has taught statistics and computer information systems courses at James Madison University since 1986. Her research, primarily in the area of curriculum development and course design, has resulted in publications in a variety of journals. She has supported students' study abroad by serving as faculty member in residence during semesters in Paris, Geneva, London, and Antwerp.

Surya B Yadav is James & Elizabeth Sowell professor of telecom technology in Rawls College of Business at Texas Tech University, Lubbock, Texas. He received his Bachelor of Science degree in electrical engineering from Banaras University in 1972, the MTech degree in computer science from the Indian Institute of Technology, Kanpur, India in 1974 and the PhD degree in business information systems from Georgia State University, Atlanta, Georgia in 1981. He has published in several journals including *Communications of the ACM, IEEE Transactions on Software Engineering, IEEE Transactions on Systems, Man, and Cybernetics, Journal of Management Information Systems,* and *Decision Support Systems*. His research areas include information quality, intelligent information retrieval systems, and system security.

Ram Dantu (rdantu@unt.edu) has 20 years of experience in the networking industry where he worked for Cisco, Nortel, Alcatel, and Fujitsu, and was responsible for advanced technology products from concept to delivery. For the last five years, he has been researching prevention of DoS and spam attacks in VoIP networks. He has co-chaired three workshops in VoIP security. He is currently an assistant professor in the Department of Computer Science and Engineering at the University of North Texas (UNT). His research focus is on detecting spam, network security, and next-generation networks. He is the founding director of the Network Security Laboratory (NSL) at UNT. The objective of NSL is to study the problems and issues related to next-generation networks. Prior to UNT, he was technology director at Netrake, where he was the architect of the redundancy mechanism for VoIP firewalls. His additional experience includes as technical director in IpMobile (acquired by Cisco) where he was instrumental in the wireless/IP product concept, architecture, design, and delivery. In addition to more than 70 research papers, he has authored several RFCs related to MPLS, SS7 over IP, and routing. Due to his innovative work, Cisco and Alcatel were granted a total of twelve patents; another eight are pending.

João W. Cangussu (cangussu@utdallas.edu) received a BS degree in computer science from the Federal University of Mato Grosso do Sul-Brazil in 1990 and a MS in computer science from the University of São Paulo at São Carlos-Brazil in 1993. He received a PhD degree in computer science from Purdue University in 2002. He is currently an assistant professor in the Department of Computer Sciences at the University of Texas at Dallas. His research interests are software process modeling and control, software testing, adaptive systems, and network security. He is a member of the IEEE and the ACM.

Barbara Carminati is an assistant professor at the Department of Computer Science at the University of Insubria, Italy, where she is a member of the Database & Web Security group. Her main research interests are related to security and privacy for innovative applications, including XML security, Web service security, privacy and security for data stream management systems. On these topics, Barbara Carminati has published more than thirty scientific papers in international journals and conference proceedings. Barbara Carminati is a member of the Editorial Board of *Computer Standards & Interfaces Journal*, she has been Program Co-Chair of the First International ICDE Workshop on Secure Semantic Web, and she served as Program Committee member to several international conferences and workshops. She is a member of ACM and IEEE.

Elena Ferrari is a full professor of Computer Science at the University of Insubria, Italy, where she heads the Database & Web Security group. Her research activities are related to various aspects of data management systems, including Web security, access control and privacy, multimedia and temporal databases. On these topics she has published more than a hundred scientific papers in international journals and conference proceedings. Elena Ferrari is member of the editorial board of the VLDB Journal, the *Transactions on Data Privacy, and the International Journal of Information Technology* (IJIT). She has served as program chair of SACMAT'04, Software Demonstration Chair of EDBT'04, Co-Chair of the first and second ACM SIGKDD International Workshop on Privacy, Security, and Trust in KDD, and the first COMPSAC'02 Workshop on Web Security and Semantic Web. She has also served as program committee member of several international conferences and workshops. She is a member of ACM and senior member of IEEE.

Andrea Perego holds a Post-Doc position at the University of Insubria, Italy, where he is a member of the Database & Web Security group. His main research interests concern the Semantic Web, Web content rating and filtering, privacy, trust, and access control in social networks. Results of his research work have been published in proceedings of international conferences and workshops. He is an invited expert at the Protocol for Web Description Resources (POWDER) Working Group of the W3C.

Index

A

B

C

D

E

F

G

H

I

O

P

Q

R

S

Breinigsville, PA USA
20 August 2010
243711BV00002B/1/P

9 781616 9200